Pest control: A survey

Consulting Editor
Professor D. L. Lee
University of Leeds

Pest control:
A survey

Arthur Woods

Lecturer in Zoology
University of New South Wales

A HALSTED PRESS BOOK

John Wiley and Sons
New York

Published in the U.S.A.
by **Halsted Press**
A Division of **John Wiley and Sons, Inc., New York**

Library of Congress Cataloging in Publication Data
Woods, Arthur.
Pest control: a survey.
'A Halsted Press book.'
1. Pest control.   I. Title.
SB950.W56   1974   628.9'6   73–19629
ISBN 0–470–96001–9

FILMSET, PRINTED AND BOUND IN GREAT BRITAIN
BY COX & WYMAN LTD, LONDON, FAKENHAM AND READING

to my parents

Preface

We can no longer afford to lose a large proportion of our resources to other organisms, or to support the misery and costs of insect-borne diseases such as malaria and nagana. In this introductory text I have tried to outline the present methods of dealing with these problems and to suggest some of the techniques of pest control which may be of value in the future. The technology of pest control (as yet, it is hardly a science) is changing rapidly and, because of its impact upon our environment, there should be a general understanding of these changes. Thus, although written as a possible text for students in the biological sciences, agriculture, and related disciplines, the book should be understandable, in the main, to the general reader. The last part of the second chapter does presume an acquaintance with statistical ideas, but this is not essential for a general understanding of the remaining chapters. Here and there some knowledge of biological topics, such as genetics, is required; the lack of this may be remedied by reference to any of the many excellent texts of general biology now available.

No attempt is made to give directions for the control of specific pests. Such information can be found easily in many other places and, in any case, its value would be ephemeral. Various control methods are described, however, with sufficient examples for an estimate to be made of their worth. Most examples are taken from animal pest control, but other harmful organisms, such as weeds and plant pathogens, are also discussed. However neatly pest control can be divided in the classroom, there is no such schism in the field. The insect pests and diseases of fruit trees, for example, are controlled concurrently, and the two routines interact one with the other, and with the rest of orchard management.

I have tried not to give undue weight to one class of pest control, such as biological methods, at the expense of others. An attempt to survey the whole of pest control is, I believe, justified, as the practitioners are now trying to integrate the various methods available to them so that pest control may be carried out economically, with as little disruption to the environment as possible. The technology is thus changing from a collection of unrelated *ad hoc* procedures and is beginning to become an ecologically based science. This is the underlying theme of this book.

The text should be of value in introductory courses on pest control in universities and agricultural colleges; it will also provide background material for courses dealing with conservation, and with the impact of man on his environment.

I would like to acknowledge my debt to the writings of Dr B. P. Beirne and of Dr K. E. F. Watt. Dr Beirne has performed a great service to applied biologists by his analysis of the aims and methods of pest control, while Dr Watt has convinced me that the most valuable implement in pest control will not be the spray gun, but the digital computer.

Thanks are due to Mrs J. Westcott, Professor A. K. O'Gower, Drs E. Shipp and R. McLaughlin and Messrs J. Czyzewski and D. P. Sneads for reading and commenting upon various parts of the typescript; to the Queensland Department of Primary Industries for the provision of plates on insect diseases; to Messrs Chapman and Hall for permission to publish a summary of a table which appeared in *Biological Control of Insect Pests and Weeds*, and to the USDA (Agricultural Research Service); A. H. and A. W. Reed of Wellington and Butterworths & Co (Publishers) Ltd of London for permission to base certain drawings on illustrations published by them. I am particularly grateful to the Agricultural and Veterinary Chemicals Association of Australia for allowing me to use the schematic representation of a research and development programme which appeared in their Bulletin. The World Health Organization have also kindly allowed me to refer to certain of their mimeographed documents which have appeared recently. Miss D. Buxton and Miss J. A. Lawrence have helped with the more difficult sections of the typescript, and their help is gratefully acknowledged.

It is impossible to thank individually all the many people whose work has contributed to the preparation of a book dealing with such a wide field; I trust that they will accept this as an acknowledgement of my gratitude and debt to them. Needless to say, any errors are entirely my responsibility, and I would be grateful if my attention could be drawn to them. Finally, I must pay tribute to the staff of the McGraw-Hill Publishing Company in England for their help in bringing this work to a conclusion, despite the difficulties of our antipodean relationship.

Arthur Woods
Sydney, New South Wales, 1972

Contents

1

Pests and their importance

Some definitions

Pest control concerns everyone. It is the day to day business of a wide variety of professional men, including farmers and salesmen, chemists and biologists, physicians and physicists. Everyone, as a victim of pest attack, or as a consumer of treated produce, is subjected to its effects. Even those of us who pay high prices for food guaranteed unsprayed by chemicals can accumulate significant amounts of insecticides in the body.

Because so many different kinds of people are involved in the business of pest control, its terminology has become confused and any attempt to standardize it must surely fail. Nevertheless, we must, at the outset, define those terms which we are going to use frequently, although we recognize that, for the reason given, these definitions will not always agree with those found elsewhere.

Pest

A pest is an organism which harms man or his property, or is likely to do so. The harm must be significant; the damage of economic importance.

The last statement should be qualified by saying that the damage must at least be thought to be of economic importance. There are many organisms whose importance was not realized till recently and which were therefore not counted as pests. Conversely, a number of accepted pests rarely, if ever, cause damage which passes the economic threshold, but, because they are conspicuous in some way, they have earned the stigma of destroyers.

Most authors further qualify the definition by stating that before we can count an organism as a pest we must be able, at least in theory, to control it, for, otherwise, it is no more a pest, in the economic sense, than is an earthquake, a hailstorm, or unsuitable soil (Beirne, 1967).

In everyday usage, the word pest means some harmful insect or allied form, but here we shall include all those harmful organisms and viruses which are not left to the attention of the physician, the veterinary surgeon, or the food microbiologist. That is to say, our definition will not include those micro-organisms which cause disease in man and his domestic animals and those which damage processed products, but it will cover, besides the noxious arthropods, the nematodes, molluscs, and vertebrates which attack us and our property, the weeds which compete with our plants and those micro-organisms and viruses which cause disease in our crops.

It is difficult to fit into any tidy scheme the larger parasites, such as flukes, tapeworms, and roundworms, which attack man and his animals. Sometimes they are the concern of the physician or the veterinarian who treats the diseases they cause, but often their attacks may be avoided or lessened by techniques very similar to those used in the control of insect pests. When such techniques are available or possible the parasite will be regarded as a pest.

One important difficulty in the application of this definition is, as we hinted above, in deciding whether or not the damage is of economic importance. Two examples, the mangold fly and fruit pests, will illustrate this.

There is no doubt that in some seasons the mangold fly, *Pegomyia hyoscyami* var. *betae*, can reduce the yield of sugar-beet by two or three tons an acre by the mining of the maggots in the young leaves. The spectacular appearance of an attacked crop can be quite misleading, however. Field trials in which beet seedlings at the four- to eight-leaf stage were defoliated with scissors showed that the crop can undergo 50 per cent defoliation with no loss in sugar percentage and only a 5 per cent drop in expected yield. Even complete defoliation led to a loss of little more than a quarter of the normal yield (Jones *et al.*, 1955). It would seem likely that over the years sugar-beet crops have been sprayed many times quite needlessly.

Since the coming of the modern insecticides and fungicides, the public has come to expect, and demand, blemish-free fruit and vegetables. Thus, a number of organisms whose damage does not reduce the yield or nutritional value of fruit are now regarded as pests simply because of our newly won, but possibly impermanent, ability to control them, and as a result of our revision of the threshold of economic damage.

Control

This term rarely stands alone and is usually qualified in some way. Beirne has given much thought to the use of the term and its derivatives, and we shall adopt, in the main, his system (Beirne, 1967).

Beirne restricts the use of the term 'control' to any action which has, as its objective, the amelioration of the harm caused by pests, and in which man plays some deliberate role. Here he differs from many authors in that they also use the term for situations in which man plays no deliberate part, but in which population densities of both pests and non-pest organisms are regulated and kept within fairly definite bounds by environmental factors. Such a situation is called natural control by, among others, Varley (1947), Thompson (1956), Milne (1958), and DeBach (1964). For this kind of regulation of population density Beirne uses the term 'natural regulation'.

This restriction of the term 'control' to cases in which man participates, is useful and should be encouraged as it avoids lengthy qualifying statements when actual pest situations are discussed.

It should be noted that in the above definition of the term 'control' the object is not necessarily the destruction of the pest but rather the amelioration of its damage. The term control, however, implies to most people the actual killing of the pest population, or at least a high proportion of it. In fact, the destruction of the pest may often make nonsense of the real goal, the lessening of the damage. For this reason there is a tendency to substitute for 'pest control' other phrases which show more clearly the real aims of the measures taken. Beirne prefers the term 'pest management' and this, indeed, is the title of his book referred to earlier.

As we said before, the term 'control' rarely stands alone and is almost always associated with some other word or words. Examples are 'governmental control', 'chemical control', 'biological control', 'insect control' and so on. Beirne divided these phrases into nine major groups and for a full discussion the reader is referred to his book. Here we have space to discuss only a few of these categories.

Categories of control   The first refers to the way in which the objective, the lessening of the damage, is achieved. In preventive control the aim is to prevent the initiation of an attack, and the pest may or may not be killed. In corrective or curative control the aim is to destroy the organism either by killing it or preventing its reproduction after its attack has started.

The second category deals with the manner of application of the controls and it includes, for example, chemical control by spraying and biological control by the introduction of living material (other than resistant varieties) into the environment of the pest. Cultural control and ecological control are also listed here. The first refers to the adaptation of conventional practices so that the pest is destroyed or harmed. Ecological control has several meanings but here it will be used in the sense of using existing environmental factors so that they either harm the pest directly or benefit their natural enemies.

An obvious category refers to the agents used in control. Thus biological control involves living agents while chemical and physical control, collectively known as technical control, refer to the use of chemical and physical agents respectively.

In this book biological control is reserved for the employment by man of living material for the control of pests. Many authors, including particularly DeBach (1964a), use the term to include the action of organisms which harm the pest and occur naturally in the same environment.

All members of this category can be subdivided. Biological control, for example, may be effected by microbial pathogens, by predators, by parasites, or by the pest itself. The last technique is known as autocidal control.

It is often important to indicate the way in which the particular control measure works. The damage caused by the pest may be minimized by killing the pest outright, by hindering its reproduction, or by affecting its activity or behaviour in some suitable way.

Agents which kill the pest are known as pesticides, but there is an unfortunate tendency to restrict this term to chemical agents. There is a related tendency to call all chemical agents pesticides whether they kill the pests or not. Unless qualified in some way, for example 'biological pesticides' or 'physical pesticide', the term will mean a chemical agent which kills the pest.

Beirne's final category arises from the realization that pest control cannot be fragmented into several distinct techniques, each one of which should be applied for a particular problem, with the exclusion of the others. In the planning of a campaign against a particular pest all possible techniques should be considered, and a combination of those which will give the maximum control of the damage with the fewest undesirable side-effects, should be chosen.

Such approaches to the problem are known as pest management, integrated control, harmonious control or rational pest control. All these phrases mean essentially the same thing, 'to manage pests rationally in relation to their environments by applying controls that integrate with one another and with natural regulatory factors to produce a harmonious system. Or, to select and apply controls logically and intelligently, with commonsense and with foresight, to produce the optimum of good effects and the minimum of bad ones (Beirne, 1967).

Victim

We have defined the main terms to be used, and other necessary definitions will be given later as the need arises. We now have the opposite problem, the finding of a word to fit a definition.

Pests attack a wide range of objects, both living and dead, which are of value to us. There does not seem to be a suitable word in common usage which covers all these objects so, for want of a better term, we shall follow Beirne's suggestion and use the word 'victim' in this sense, although we realize that it will include inanimate objects as well as living ones.

The damage caused by pests

The ways in which pests damage their victims

In exploiting their victims for their own needs pests may damage them directly or indirectly. Unfortunately these terms are used in two different ways.

If the part of the victim which is damaged is the part in which man is interested then the damage is said to be direct, whereas if some other part of the victim is damaged, and not the useful part, except indirectly, the damage is indirect. Thus the larvae of the wheat bulb fly, *Leptohylemyia coarctata* burrow into the young shoots of wheat plants, and reduce the yield of the crop only indirectly. In this case they are said to be indirect pests.

The other use of the term is much commoner and will be employed here. If in exploiting the victim the pest damages it by virtue of this exploitation it is a direct pest. If the exploitation causes little or no damage, but the victim is harmed in some other way as a result of the pest's attentions or presence, then it is damaged indirectly.

Direct and indirect damage can take many forms, and since we have little space here to discuss them fully, the reader is referred to the many textbooks of economic entomology and plant pathology where the damage will be described under the specific causal organisms.

Direct damage usually results from the pest's use of the victim as a source of nourishment. Insects chew the tissue or suck the sap or blood of living organisms. Plant pathogens attack various parts of the plant for the same reason. But direct damage may be caused in other ways, for pests may exploit their victims as sites for egg-laying or shelter. Cicadas, when laying their eggs in young twigs, frequently damage them so severely that die-back follows. The apple twig cutter, *Rhynchites coeruleus*, having laid her eggs within the tissues of a twig, cuts it through just below the oviposition site so that it usually falls to the ground where the egg hatches, and the grub completes its development.

There are a number of beetles which will bore into any suitable hard material, such as wood, to provide themselves with pupation sites, and one species is even credited with the sinking of a wooden ship carrying a cargo of penguin carcases (*Hakluyt's Voyages*, ed. Goldsmid, Edinb. 1890, cited in Busvine, 1951).

The indirect damage caused by pests displays a much wider range of forms. Possibly the most important type is when the pest is a vector or alternative host of some economically important parasite or disease. Plant viruses of many kinds, and of great destructive power, are transmitted by sap-sucking aphids, leaf-hoppers and nematodes, and blood-sucking mosquitoes spread several important human diseases such as malaria, yellow fever, dengue and filariasis, to mention only four. The oriental rat flea, *Xenopsylla cheopis*, carries bubonic plague, essentially a disease of rodents, from the cooling corpses of rats to man. In this case both the rat and the flea are acting as indirect pests of their victim.

As is the case with the dead rat, the pests need not come into direct contact with their victims to harm them. Snails of the genus *Limnaea* do not attack sheep in any way, but they are the intermediate host of the liver-fluke, *Fasciola hepatica*, still an important parasite of sheep and cattle in waterlogged pastures.

Since very few weeds are directly parasitic on their victims, their main damage is indirect. Most importantly they compete for space, light, water, and nutrients, and some, such as couch grass, *Agropyron repens*, secrete substances which inhibit the growth of their rivals. Another form of indirect damage resembles that caused by the vectors and intermediate hosts mentioned before. Many weeds serve as alternative hosts for insect pests and pathogens, frequently carrying them over periods when the normally attacked crops are absent (Duffus, 1971). Thus several weeds support populations of aphids before potato crops are ready for attack. Some weeds, indeed, are obligate, or almost obligate, alternative hosts for pests. Barberry, *Berberis vulgaris*, is essential for the completion of the life cycle of black stem rust of wheat, *Puccinia graminis*, while, in British climatic conditions, few black bean aphids, *Aphis fabae*, overwinter except as eggs on spindle, *Euonymus europaeus*, and a few other shrubs. These aphids, incidentally, as well as being vectors of plant viruses, produce, like many other aphids, a copious sugary honey dew which coats the plant leaves, serving as a medium for sooty mould fungi. Although these do not attack the plant they do severely restrict its photosynthetic efficiency. The dried honey dew is also very difficult to remove from the paintwork of motor cars left standing under aphid-infested trees.

Returning to the indirect damage caused by weeds, a number of them are found to be troublesome stock-poisoners, the outstanding example in Europe being the common ragwort, *Senecio jacobaea*. In non-fatal poisoning a complicating feature is the craving which the victims, both stock and children, sometimes develop for the guilty plant which, on the first opportunity, they will actively seek out. This is useful in that in the cases where the offending plant has not been identified, the animal itself carries out its own diagnosis, but it does mean that the stock has to be excluded from the pasture in question. With children it is a little more difficult. The list of such plants includes, among others, laurel, rhododendron and woody nightshade, *Solanum dulcamara*.

Less serious, but still of economic importance to the farmer, is the tainting of milk and flesh by ramsons, *Allium ursinum*, and other members of the same genus growing in pastures.

Bartels and Cramer have published a valuable survey of poisoning by weeds, and of similar side-effects of plant diseases and animal pests on the health of animals and man (Bartels and Cramer, 1966). In this survey they also discussed the effects of pests on the quality of harvested products.

Many plant pathogens produce toxic principles in the crops they attack and they effect the health of the consumers of the product. Outstanding among these is the fungus known as ergot, *Claviceps purpurea*. The active principles of this pathogen are used in medicine for the relief of migraine and for the excitation and contraction of the muscles of the uterus. The alkaloids stimulate the involuntary muscles and paralyse the sympathetic nervous system. The ergots are most often found among rye grains but fortunately, with modern screening methods, their presence in human food is now quite rare. The most recent serious outbreak of human poisoning took place in the small French village of Pont St Esprit, about twenty miles north of Avignon, in 1948. In this case the entire population of the village suffered from ergotism, the source of the infection being traced to a black-market supply of rye used by a village baker. The hallucinations and visual abnormalities arising in this case and others may explain the stories of witchcraft and other manifestations when, in the Middle Ages, whole districts obtained their bread from one source.

Ergot also infests certain pasture grasses, especially rye grass in Europe and paspalum in Australia. People that have walked through badly infected pastures sometime show mild ergotism, and stock feeding on the grasses display typical gangrenous symptoms in the tail, the hooves, the lips, and the tip of the tongue, all due to the constriction of the smaller blood vessels.

The various rusts of cereals have also been implicated in outbreaks of disease but few plant pathogens, if any, have such serious effects as ergot. In recent years, however, there have been widespread deaths among turkeys in England resulting from their being fed with peanuts and peanut meal infected with certain strains of *Aspergillus flavus*. Subsequent experiments showed the fungus can be toxic to other animals. The symptoms of poisoning included hepatitis, tissue necrosis and kidney damage. It is also probable that the active principles, which are referred to as aflotoxins, are carcinogens. The toxic nature of this fungus appears to have been long recognized by a native tribe in Guyana who used ground-nuts for the execution of offenders against tribal laws (Moody and Moody, 1963; Bartels and Cramer, 1966). Other fungi, such as *Aspergillus parasiticus* and *Penicillium puberulum*, produce aflotoxins identical with, or similar to, those of *A. flavus*.

Sweet clover is occasionally infected by certain moulds which convert the coumarin present in the hay into oxycoumarin or dicoumarin. Oxycoumarin inhibits the formation of prothrombin in the blood, and since this is necessary for the coagulation of the blood, animals eating the spoiled hay may die from internal haemorrhage.

A number of animal pests infesting crops and stored products are troublesome because of their allergic effects on sensitive people. House mites have been implicated, for example, in many cases of asthma. The dust arising from the remains of insects in stored products is also responsible for many allergic conditions. It is also possible that, because of their content of quinones, these remains, if consumed, would be carcinogenic.

Weeds, and other kinds of pests, can make farming and forestry operations more difficult and costly. Cleavers, *Galium aparine*, and stem borers such as the wheat stem sawflies *Cephus pygmaeus* and *C. cinctus*, as well as various plant pathogens, cause the lodging of cereal crops, with consequent difficulties in harvesting. The cabbage aphid, *Brevicoryne brassicae*, by virtue of the sticky waxy nature of its colonies makes the harvesting of brassicas unpleasant and their preparation for market more difficult.

A similar form of damage is the contamination of stored products and packaged goods by the dead bodies and exuvia of insect pests and the faeces of insects, birds, and rodents.

Rodents, especially rats, cause considerable trouble by their gnawing of various materials in their search for food. This can cause serious structural damage to buildings, and may even be the cause of a quarter of the fires of unknown origin in the United States of America (Dykstra, 1967).

The cost of pest damage

While nobody denies that pests are costly, both in terms of money and in human suffering, no one has calculated exactly how expensive they are on a global basis. Good estimates have been made in some of the more advanced countries, notably the USA and the United Kingdom, and extrapolations have been made from these to cover other parts of the world. These are, however, countries with advanced agricultural and public health techniques who suffer a smaller proportionate loss than, for example, some of the Asian, South American, and African states. On the other hand, because of their generally higher yields per acre and greater total production, the larger advanced countries probably suffer more in absolute terms through pest damage to crops and stored products.

It is also difficult to find a suitable unit in which to express crop losses. A monetary unit is unsatisfactory because of the elasticity of prices. In times of shortage resulting from pest attack, a product will command a high price on the market, much greater than it would fetch in times of glut when pests are few. To measure pest damage in terms of the higher value would give an inflated estimate. There is also the difficulty of converting values from units in one country's currency to those of another.

Expressing the damage in terms of lost yield, either absolutely or as a percentage, does not allow for losses in quality of the product, a disadvantage not shared by the monetary measure. Nor does it take into account the costs of pest control operations and research, whether these be successful or not.

In the literature, the percentage loss can be expressed in two ways, either as a fraction of the yield actually obtained or as a percentage of the yield which would have been obtained in the absence of the pests. Unfortunately, some authors do not state which method they are using. Another difficulty which is usually glossed over is how one determines what the yield would be in the absence of pests – a crop without its troubles is probably even rarer than a man in perfect health.

Ordish has tried to circumvent some of these difficulties by the introduction of a new unit, the untaken acre (Ordish, 1952). This is, simply, an extra acre of the crop which must be grown to, as it were, satisfy the pests; an acre, incidentally, which, in the absence of pest damage, could be devoted to something else.

This method does avoid one pitfall into which agricultural economists sometimes fall, that of double counting. In adding up the losses in a crop it is easy to forget that an individual plant or fruit can only be destroyed once, even if it is suffering from the attentions of several pests. Similarly a man can only die once even if he is suffering from both malaria and sleeping sickness. It was this sort of error that led an official of the US Department of Agriculture to conclude, facetiously, that the USA harvested no wheat or maize at all in 1944 (Ordish, 1952).

The trouble with Ordish's measure is that it is only meaningful with respect to the particular area and time under consideration. It is difficult to 'translate' India's untaken acres of wheat into English ones because of the greater productivity of the English fields. In Europe and other advanced areas there has been a steady increase in yields per acre of most crops over the last century, thus a straight comparison of the untaken acres of the 1860s with those of the 1960s would be meaningless.

This general increase in productivity causes another difficulty. It is obviously more meaningful to express pest losses as an average over a period of say, ten years, than to express the losses in terms of what happened in a single season. If the yield has increased through the use of better fertilizers, increased irrigation, new varieties, more efficient crop protection techniques and so on, the interpretation of the average becomes difficult.

One of the most ambitious attempts at counting the costs of pests in agriculture has been made by Cramer (1967). He has surveyed much of the world's literature on the subject and arrived at a total cost of between seventy and ninety thousand million American dollars a year. This estimate is based on the prices paid to the growers in the various countries, the data being taken from the Production Yearbook of the FAO (1965).

To this total must be added a large part of the annual value of world production of pesticides, about one thousand million dollars. Cramer estimates that without these pesticides world food prices would be between fifty and one hundred per cent higher than they are now.

His estimate for the percentage losses are, as a fraction of the actual production, 55 per cent and as a fraction of the potential production, 35 per cent. The latter figure can be broken down as follows: Insect pests, 13·8 per cent; Diseases, 11·6 per cent; Weeds, 9·5 per cent.

This survey only follows the crop as far as the harvest. Further heavy losses follow during the storage and transport of the products, from the depredations of such organisms as rodents, insects, and fungi. On a world basis these losses are probably, in absolute terms, just as great as those suffered before harvest. In the Congo, for example, insect attack during the course of one year led to a loss of weight of 50 per cent in stored sorghum, 20 per cent in beans, and 15 per cent in ground-nuts.

In Sierra Leone stored paddy rice lost a quarter of its weight in one year from insect attack. What remained was not all rice. A large part of what was left in this stack, and in those of the Congo, consisted of the living and dead bodies of insects, and frass (Herford, 1961).

Fruit, of course, suffers particularly badly from rotting when stored and during transport. A 25 per cent loss of stone fruit and imported pineapples was, till the late 1950s, regarded as normal by fruiterers in the USA, and the familiar green mould of oranges, *Penicillium digitatum*, used to destroy 12 per cent of the Australian domestic supplies, though new chemical treatments have reduced this loss considerably (Turner, 1959).

It is impossible to overestimate the importance of the disease-carrying pests. Mosquitoes, as vectors of malaria, have influenced the course of history. Much of Rome's strength lay in the surrounding malarious marshes where besieging armies were weakened by the disease. Plague, always endemic among the rodents of eastern Asia, has frequently erupted out of China and may yet do so again. Aided by the Oriental rat flea it brought the Black Death and possibly the end of the Feudal system in Europe, though this is probably an over-simplification.

The tsetse flies, vectors of sleeping sickness and nagana – a Zulu word meaning the sickness of cattle – halted the northward trek of the Boers with their oxen-drawn waggons. The flies still hold sway over some 6·3 million square kilometres of Central Africa, and the economic atlases show little overlap in the distribution of tsetse and cows.

Tsutsugamushi disease, or scrub typhus, a disease long said by the Japanese to be passed on to man by the 'akamushi' or 'dangerous bug' and now known to be transmitted by larval trombiculid mites, probably incapacitated more Allied soldiers in the Pacific area than did the Japanese. The same war nearly finished with a typhus pandemic and would have done so if the Allies had not deloused the entire population of Naples with the then new insecticide DDT. Typhus is the companion of war but for an account of its history the reader cannot do better than consult Hans Zinsser's classic, *Rats, Lice and History*, fortunately recently reprinted.

To measure the human suffering caused by these and other arthropod-borne diseases in terms of money is both impertinent and well-nigh impossible, but something must be said about the economic effects of the diseases. It is sometimes argued, ethical considerations being left aside, that these diseases help to slow down the population explosion and, untreated, would allow us a little more time to catch up with food production. In reality most of these diseases take a considerable time to kill their victims, who may be made incapable of working for years but who, while awaiting their deaths, still need food, clothing, and fuel. All too frequently the victims are children who die before they can begin their productive life.

Malaria is such a disease. Although it kills two and a half million people a year (Metcalf, 1965) this is a small number compared with the total number of people suffering from the disease. Malaria and many other diseases also deny the use of vast areas of otherwise productive land. Nagana is an outstanding example of a disease of this kind.

How pest problems arise

Of the two or three million organisms in existence probably not more than a few thousand are pests. If we are to avoid creating new pest problems and if we are to control efficiently those pests which we have inherited, it is obviously important to know how these exceptional species became harmful.

As Edwards and Heath have pointed out, there are both long- and short-term factors which control the transition of populations of organisms from the innocuous state to the harmful condition. The long-term factors are those evolutionary pressures which alter the pest population so that it exploits its victims more efficiently. These arise from the interplay of mutations and natural selection. The short-term factors are changes in environmental resistance to population growth. This environmental resistance is composed of abiotic and biotic factors, but they cannot be clearly separated since they interact with each other. Climatic conditions, for example, have a direct effect on both the pest species and on the various parasites, predators, and pathogens which attack it (Edwards and Heath, 1964).

Without a doubt, one of the most important causes of an innocuous population changing into a damaging one is the provision of an almost unlimited food supply by man. Since Neolithic times man has grown his food on comparatively large areas in more or less pure stands, and has stored what he does not immediately use. This has provided organisms, which previously depended on scattered specimens of their victims, with food in such large quantities that their population densities could increase rapidly. There could be little danger of the food supply running out and the population starving, nor could there be much chance that most of the offspring would perish in a vain attempt to find a host. Such pest organisms have subsequently adapted themselves to the microclimates and other special conditions afforded by the crop or stored product.

Man has also selected his crop plants for high yields and nutritional value and this has often made them even more suitable as a source of food for the pests. In this selection he has also unwittingly eliminated some of the defences against other organisms which the plants had evolved when growing in the wild.

The tillage operations carried out in farming alters the soil conditions so that they become more suitable for the growth of various plants that cannot compete well in crowded environments. Many of these plants have become troublesome weeds.

As farming has progressed man has become more and more enamoured with clean farming. He has not only tried to grow nothing but the crop within the field, he has also tried to eliminate every 'weed' on the headland and in the hedgerows. This has, in many cases, destroyed various requisites of many of the parasites and predators of the pests within the field. Many adult parasitic wasps, for example, obtain much of their food from umbelliferous plants and other flowers in the hedgerows. It is impossible at the moment to say what is the correct balance of clean and 'dirty' farming for we know all too little about the biology of these natural enemies, and, in any case, the right combination will vary with the season, the district, the crop concerned, and the pests attacking it. What is well established is that the more complex an ecosystem is, the more stable it is. Outbreaks of insect pests are much less common in tropical rain-forest with its multitude of species than they are in a fauna and flora-impoverished conifer plantation. Thus the 'old fashioned' English landscape, with its patchwork of smallish fields, hedgerows, and copses, is less likely to suffer from serious pest outbreaks than would the same landscape with larger fields and with most of the hedgerows grubbed out for the sake of more efficient mechanical farming (Elton, 1958).

The discovery and exploitation of new continents, such as the Americas and Australia, led to the introduction of exotic plants into new environments. All too often native organisms found the new plants very suitable for their development. The cottony cushion scale, *Icerya purchasi*, was, and still is, an insect which uses native Australian shrubs as its hosts. Since the introduction of agriculture by the Europeans it has extended its host range to include a number of cultivated European and Asiatic fruits, in particular, *Citrus*. It can even be regarded as a pest on some of the native shrubs since Australians found it desirable to cultivate these in their gardens and parks.

The Colorado beetle, *Leptinotarsa decemlineata*, was, till the 1850s, an unimportant insect living on wild *Solanum* species in a restricted area of the Rocky Mountain region. In 1839 settlers first grew potatoes in this area and ever since the beetle has been spreading along its man-made lines of communication. By 1874 it had reached the Atlantic coast. In 1876 it turned up in Germany, but was destroyed. In 1920 it began its successful occupation of Europe (Elton, 1958).

An analogous situation is found where vectors of a disease occur in its absence. Malaria is not endemic in Britain but it was once fairly common, certainly frequent enough to kill Sir John Falstaff. A number of anopheline mosquitoes which are capable of transmitting malaria are still widespread in Britain where, fortunately, they are now merely a nuisance. They could presumably spread the disease and thus become pests once again if sufficient malaria sufferers were to enter the country. In most countries, introduced plant viruses would also find sufficient aphid vectors to increase the latters' importance as pests.

Probably the most important pest problems are created nowadays when organisms are introduced into a new environment. This is more dangerous than the introduction of new hosts into the organism's environment, for the latter usually travels without its normal complement of predators, parasites, and pathogens. If the new climate is favourable, or if the organisms can adapt quickly enough to the climate, there is little to stop an explosive population increase on suitable hosts. Elton's work, cited above, gives numerous examples and case-histories. The Colorado beetles' invasion of Europe has already been mentioned and it is only strict quarantine measures allied with immediate blanket spraying with insecticides when these fail, that has prevented it from devastating the British potato crop. Quite recently fire blight of pears, *Bacterium amylovorus*, has arrived in Britain. While this was not an unexpected visitor (Wormald, 1955) and eradication measures were taken immediately, the disease has still established itself.

Possibly most of the important pests in the USA, and certainly approximately half of the six hundred or so insect pests in Australia, are introduced species (Waterhouse, 1971). Among these we can list the cosmopolitan codling moth and most of the aphids. Australia has a mere handful of native species of aphids, yet members of this family are among the most important pests on this continent.

This interchange of fauna and flora between the different regions of the world has been going on since prehistoric times, but the pace quickened with the voyages of discovery of the Portuguese and Spaniards in the fifteenth and sixteenth centuries. Modern high-speed travel, both surface and air, has greatly increased the danger, especially of vectors of human and animal diseases which are harder to detect than pests depending upon plant materials.

Some organisms, particularly those which live in close association with man, such as the house-fly, *Musca domestica* and the various species of *Rattus*, have spread to most parts of the inhabited world. Some pests, including *M. domestica*, are such seasoned travellers that it is now impossible to be sure of their country of origin.

Fortunately, the great advantage that many of the introduced pests have enjoyed, namely their lack of natural enemies, can be lessened by the introduction of some of these from the pest's country of origin. If care is taken, these, in their turn, can be introduced without their own complex of natural checks and thus prove more efficient as regulatory factors than they did in the original area of distribution. This is, of course, the classical technique of biological control which will be discussed in a later chapter.

It is not only man's agricultural operations at home and in new countries that has led to the building up of pest populations. Many other of his activities have led to new pest problems. Large-scale ecological changes have been brought about by such activities as deforestation and large-scale engineering so that the environment has been altered in such a way that certain pest populations can increase rapidly. The changes may also destroy many of the natural checks or their requisites, as when the felling of a deciduous woodland reduces the numbers of nesting sites for small insectivorous birds. Large-scale irrigation works and dam building increases the breeding sites for molluscs and, in countries like Egypt, this could be followed by an increase in the numbers of the blood flukes causing Bilharzia. Mosquitoes are also likely to increase in these conditions unless care is taken to avoid shallow water and weeds which provide shelter for the larvae. This technique has been used successfully on Lake Burley Griffin, an artificial amenity lake in the centre of Canberra.

Man's buildings provide shelter for an assortment of organisms, some of which are rarely or never found far from them. His rubbish dumps also provide cover and food for various organisms and may even serve as reservoirs of diseases among rats and other rodents.

During the last thirty years man's success in the control of some pests has, para-doxically, provided him with others for, unfortunately, many of his pesticides kill organisms other than the target species. They have destroyed many of the natural enemies of potential pests, allowing these to develop into ones of economic import-ance. Thus the tar oil winter washes and the chlorinated hydrocarbons have killed many of the predators of red spider mites while having little effect on the mites them-selves. A century ago, spider mites were almost unknown to the fruit grower, now they are his most important orchard pests in many parts of the world. This will be discussed at greater length in later chapters.

Although we have gained many pests over the last century we have also lost a few. The disappearance of such pests as bedbugs can be correlated with increased sales of soap and paint and the introduction of new building methods, but it is difficult to explain the decline in importance of a number of horticultural and agricultural pests. Old texts such as John Curtis's *Farm Insects* (1860) and Eleanor Ormerod's *A Manual of Injurious Insects* give prominence to pests such as the turnip sawfly, *Athalia rosae*, which today are of little economic importance. Obviously, the pest complex is in a state of flux, with some species becoming important and others becoming rare. It would be just as advantageous to know why a pest disappears as it is to know why another becomes more important. Unfortunately, if a species is no longer there, it is impossible to study. Only rarely can we give a firm reason, as does Massee when writing of the apple blossom weevil that it was considered by many fruit growers, until the advent of DDT in 1946, to be one of the most important apple pests. It is no longer regarded as important in commercial orchards (Massee, 1954). Yet the loss of this useful knowledge stresses one feature of the practice of pest control, namely the importance of knowing as much as possible about the biology of the pest and about the factors which influence it, if we are to control efficiently the damage it causes.

The economics of pest control

In agriculture and forestry, pest control measures are used in order to bring an increased profit, and, theoretically at least, this profit can be expressed in monetary terms. In public health pest control ethical considerations also play a part; it is difficult, and often irrelevant, to try to measure the profit in terms of cash. Some pest control is carried out without any accounting at all: the private gardener rarely pays any attention to the costs of the pesticides he uses, and the profit which they bring.

The simplest situation concerns a farmer who applies a chemical pesticide to control an infestation in his crop. We will assume that the effects of the pesticide are confined to that crop in that particular year. In other words we are ignoring any possible long-term environmental effects, or long-term beneficial effects that might arise from the protection of perennial crops. We will also presume that the market price which the farmer receives for each unit of his product will not be affected by the control measures that he and other farmers take. These are, of course, unrealistic assumptions, as will be seen later, but most farmers, concerned with an annual crop, would start with these assumptions.

Pests can be present in a crop at such a low level that they cause no economic damage. At some population density – the economic injury level – economically important damage will occur, and pest control measures are needed if the farmer is to avoid a loss in his crop. Infestations can, however, develop quickly, and the farmer may have to take measures before the economic injury level is reached, because, otherwise, the population present (at the economic threshold level) would probably reach economically important levels. In some cases a farmer may apply chemicals even when the pest is absent, as an insurance measure. This would be done, for example, in areas where potato blight, *Phytophthora infestans*, damages the crop significantly in most years. In situations of this kind, the costs of pest control should be assessed over a number of years; years in which spraying proved, in retrospect, unnecessary, and years in which it proved to be essential.

There is rarely, if ever, a rectilinear relationship between the number of pest individuals attacking a crop and the damage they cause (measured, for example, as loss in yield). At very low densities the crop is usually able to recover from the damage. The yield may even be increased if a low pest infestation or infection reduces the competition between individual plants. At high pest densities competition between the pest individuals increases, reducing the average impact on the crop of the individuals. The situation is, of course, analogous to the economist's law of diminishing marginal returns, if we regard the pests as the input which varies, and the damage they cause as their return (Fig. 1.1). Justesen and Tammes (1960) present four mathematical models to describe the self-limiting effect of injurious or competitive organisms on crop yield. The first concerns multi-infections of rootlets by a soil pathogen, the second the invasion of seeds by weevils, and the third the infection of leaves by a fungal disease. Interestingly, the third model was derived from one used for bombing patterns during war-time. The fourth model is concerned with the spacing of the crop plant.

These relationships have been illustrated by experiments in which plants were grown in soils containing different population densities of nematodes. In some cases the relationship could be made approximately rectilinear by converting the population densities to their logarithms, an indication that the relationship between damage and initial population density is exponential.

It is clear, therefore, that it is extremely difficult to forecast how much loss of yield a given infestation is likely to cause, especially when it is realized that many factors will influence the population density of the pest later. Yet this is what the farmer must do if he is to make a rational decision about the use of pesticides for control.

If the pest population rises above the economic injury level control measures will bring some benefit which, theoretically at least, can be measured in terms of increased yield. The greater the infestation, the greater are the benefits which control measures will bring. We can, in fact, make use of cost–potential benefit ratios. The costs, of course, include the costs of pesticide, labour, and machinery depreciation. If this ratio is less than one then, clearly, the control measures are worth taking. In practice, however, a cost–potential benefit ratio should be considerably less than one because of the uncertainties always involved in farming. There may be, for example, a

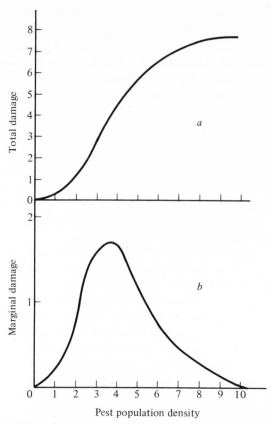

Fig. 1.1. The relationship between pest density and the damage caused to a crop is rarely rectilinear. The usual curve resembles that of the Law of Diminishing Returns. As the pest population density increases the total damage increases (a) but eventually levels off. As pest population density increases the marginal damage (the additional damage caused by each new unit increase in pest population density) at first rises, then falls (b)

reduction in yield, unconnected with pest attack, as a result of inclement weather. This could increase the actual cost–benefit ratio to more than unity.

The use of pesticides is, of course, itself subject to the law of diminishing marginal returns. The first units of input, pest control measures, will probably result in little extra yield. Further units of input will give substantially greater increase in yield. Eventually, however, still further units will bring little or no increase, as the pest's population density falls to, and below, the economic injury level. With most pests there is no point in trying to achieve complete control, although many growers attempt to do so. Each unit increase in pest control costs will thus bring an additional increase in benefit, though this will diminish. Ideally, the farmer should stop increasing his pest control input at the point where the ratio of the additional costs to the additional yield value is unity, although, of course, it is exceedingly difficult to judge where this point is.

Hillebrant (1960a, b) has pointed out that the curve displaying the relationship between pesticide dosage and percentage increase in yield has a distorted sigmoid shape, and thus resembles the dosage response curve in insecticide bioassays, and that transforming the dosage to log dosage produces a conventional sigmoid curve (cf. chapter 3).

At this point it should be remembered that pests can be roughly divided into two groups, one containing pests which attack directly those parts of the victim which man uses or consumes, and that containing the pests which attack other parts of the victim. In the first case the increased revenue resulting from the pest control measures may be quite closely related to the degree of control achieved. The greater the proportion of the codling moths that are killed, the greater is the proportion of saleable apples. If, however, the pest is a root-feeder which does not attack directly those parts of the plant used by man, the relationship between the degree of control achieved and the revenue is not necessarily so simple.

Wheatley and Coaker (1970) describe a trial on the control of the cabbage root fly attacking the roots of summer cauliflower plants. Different doses of diazinon were applied around the roots of the plants as drenches. About ten weeks after the application the plants were harvested and scored for the degree of root damage. The percentage increased yield (compared with untreated plots) increased only slightly as the diazinon dosage rose from 3 mg per plant to 50 mg per plant, whereas the reduction of root damage (compared with control plots) rose from about five per cent to about ninety per cent. In other words increasing the dosage from 3 mg per plant (one tenth of the recommended dose) to 50 mg per plant hardly increased the yield, although it decreased the damage to the roots by a factor of eighteen. Supplementary trials showed that the lowest dosage was sufficient to protect the plants during the critical period just after transplanting. After establishment the plants were able to tolerate considerable root damage.

The same paper presents the idea that pest damage can be expressed as a function of two components, namely the proportion of victim units which are attacked, and the degree of damage which a unit suffers. The severity of a pest attack can thus be graphed in two dimensions, with the percentage of units damaged being one axis (y), and the degree of damage to each unit (x) along the other. Thus the point $(0, 0)$ would represent zero damage, $(\infty, 100)$ complete loss and $(\infty, 5)$ the complete loss of a small proportion of the units. The last would represent, for example, a situation which often occurs with codling moth: only a small proportion of the apples is attacked, but those apples which are attacked are unsaleable. The space of the graph can be divided into appropriate zones which may be used as a basis for determining rational goals for pest control operations.

The greater the initial infestation of the crop, the greater will be the potential benefit–cost ratio of the pest control measures. There is thus a series of input–benefit curves, one corresponding to each level of infestation. Ordish (1952) tabulates the cost–potential benefit ratios for a variety of crops in a variety of situations. He considers, for example, three different kinds of apple orchard, each with its own potential value in terms of yield of crop. For each of these he tabulates the cost–potential benefit ratios for three types of control programme (full, partial, low) for severe, medium, and slight attacks by fungal and arthropod pests. There are thus twenty-seven ratios which range from $1:0{\cdot}5$ (a loss) to $1:13{\cdot}6$. He also tabulates ratios for potato, wheat, turnip, sugar cane, and tea crops in various situations. The actual costs quoted are, of course, out of date, but the table illustrates the principles involved.

Headley (1972) uses microeconometric models to give a rather more sophisticated treatment than that described above. If the amount of land and labour devoted by a grower to a particular crop is fixed, but the costs can be varied, then the value of the output will vary with the cost input. The output will first rise but will eventually level off, and, perhaps, even fall, in accordance with the law of diminishing marginal returns. If the farmer is trying to achieve a maximum profit his total cost should be such that the cost of the final unit produced should equal its market price (farmers being in 'perfect competition'). In other words at the point of profit maximization the slope of the cost–output line is unity, if both axes are scaled in monetary units. Let the relevant total cost and total output (in terms of market value) be C_A and O_A.

In the above situation it is presumed that there is no attack by pests. If pest damage now occurs, and no control measures are carried out, the costs remain the same (C_A) but the output drops to O'_B. The grower has thus lost, as a result of uncontrolled damage, $(O_A - O'_B)$, (Fig. 1.2a).

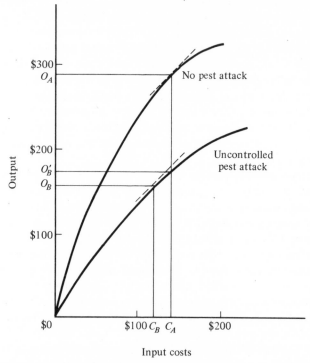

Fig. 1.2a. Costs–output curves for a crop grown in the absence of pests, and in their presence, but without pest control. For full explanation, see text. (Adapted from Headley, 1972)

The point O'_B lies on another cost–output line, that applying to the situation where pests are present, but where pest control is not carried out. This point lies below the first, with its point of profit maximization below and to the left of that of the first curve. The total costs and total output at this point are C_B and O_B. If, therefore, the grower had been aware that damage would occur from pests and that he would thus be working on a lower cost–output line, he would have obtained a larger profit if he had reduced his input costs from C_A to C_B. He would, in fact, have made a profit of $(O_B - C_B)$ which is larger than $(O'_B - C_A)$, but smaller, of course, than $(O_A - C_A)$. His lost profits are therefore greater than $(O_A - O'_B)$ if we measure from points of profit maximization.

There is, of course, a third curve possible, namely that which is obtained when the

16 Pest Control

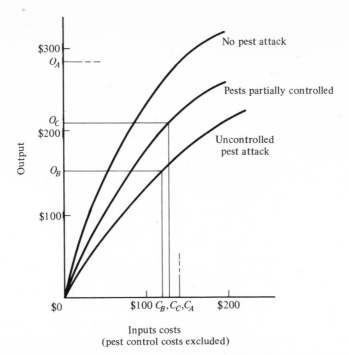

Fig. 1.2b. Costs–output curves for a crop grown in different conditions: in the absence of pests; in the presence of uncontrolled pests; in the presence of partially controlled pests. For full explanation, see text. (Adapted from Headley, 1972)

pests are present and pest control is carried out (Fig. 1.2b). If the pest control costs are not included in the input costs, and if complete control is not achieved, then this third curve will lie between the first two. If the total costs (excluding pest control costs) and the total output value at the point of profit maximization are C_C and O_C, then

$$O_A > O_C > O_B \quad \text{and} \quad C_A > C_C > C_B$$

With pest control the grower's maximum profits will be $(O_C - C_C) - (\text{pest control cost})$, and the extra profit he will obtain above the maximum obtainable without pest control will be

$$(O_C - C_C) - (O_B - C_B) - (\text{pest control cost})$$

In other words the cost benefit ratio is

$$\frac{(O_C - C_C) - (O_B - C_B)}{(\text{pest control cost})}$$

Since there is still some damage by pests there is still some loss in profits (even when pest control costs are disregarded). The minimum loss of profits measured from the maximum profit obtainable in the absence of pests is

$$(O_A - C_A) - (O_C - C_C) + (\text{pest control cost})$$

This model, as Headley points out, is a 'simple, timeless, perfect certainty model' but it does show that there are various methods of measuring losses from pests. The

Pests and Their Importance **17**

model given by Headley is the relevant one with respect to economic efficiency, whereas other models tend to overestimate pest losses and 'perhaps encourage overinvestment in pest control'.

Few growers make use of this model or, indeed, of any similar model. We cannot be sure, for example, that a particular grower is actually profit maximizing; he is more likely to be output maximizing – increasing his yield to the point where the cost–output curve begins to fall – provided, of course, that in doing this he makes a satisfactory profit. In fact, as the psychologist Simon (1956) observes, psychological studies of learning reveal that people have a much smaller capacity for obtaining information and performing computations than is demanded by the models of rational behaviour postulated by the economists. In general, 'Organisms adapt well enough to "satisfice"; they do not, in general, "optimize"'.

In short, few growers would rationalize in the way postulated by the model. Furthermore, there would not be just three curves, but a very large number. The pest control input can be varied (and the pest control itself is subject to the law of diminishing returns) and the level of attack by the pests can also vary. Furthermore, this level cannot be predicted with any certainty. Clearly, the computation of the costs of pest damage and the prediction of likely benefits from pest control measures is beyond the capacity of even the most economically rational grower. Each crop would, in fact, need an intensive study by economists and biologists armed with ample computational facilities and with adequate biological data about the damage caused by various levels of pest infestation in varied growing and weather conditions. Needless to say, such biological data are almost completely lacking. In the meantime, growers will have to be satisfied with the somewhat simpler models postulated by Ordish and others. Eventually, perhaps, computers will be established at regional advisory headquarters with relevant information stored in them about the conditions on, at least, the largest farms in the area. It would then be possible for a grower to get rapid advice on the optimal control measures that he must take if an infestation appears. This will, however, need far more accurate long-range weather forecasting than is available now.

The above models apply to situations in which the pest control measures are restricted to the one crop in one particular year. It would be more difficult to extend the computations to other kinds of pest control, especially those which are self-sustaining. The initial costs of biological control are likely to be too high for the individual grower to take the necessary steps, even if it were physically possible for him to do so. As biological control is, in general, self-sustaining, it is not of interest to conventional firms which would receive little revenue after the first stages of the project. The potential benefits, which accrue over a wide area, which increase the output of a whole industry, and which extend over a long period, are also high. Because of these special characteristics of biological control methods, and because there is also a high risk of failure, the costs are likely to be borne by government bodies.

One feature of successful biological control is that, once it is established, nothing more need be done to control the pest. In fact it is soon forgotten that the pest control is in operation. Some estimates have been made of the cost–benefit ratios of biological control and related methods. Simmonds (1967) costed seven biological control projects which gave permanent control. The costs were estimated to be £25 000 but these gave an annual return of over 1000 per cent per annum. DeBach (1964b) examined the records of the Department of Biological Control at the University of California. Between 1929 and 1959 about one and a half million pounds were spent on five successful projects which brought a saving of forty million pounds up to 1959, with further savings later. These costs compare very favourably with the current research and development costs of a single chemical pesticide. Hussey (1970) discusses the economic considerations in the future development of biological and related pest control. This paper, together with that of Wheatley and Coaker cited above, forms part of an important monograph devoted to the technological economics of crop protection and pest control (Society of Chemical Industry, 1970).

It is now realized that real pest control situations are far more complicated than that considered at the beginning of this section. The effects of the chemical pesticide application are not confined to the place and year of application. High rates of use, for example, are likely to produce resistant populations with increased pest control costs in subsequent years. Headley (1972) suggests, therefore, that the pest control costs in any year should include a term for the depreciation of the method, just as terms are included for the depreciation of farm machinery in the costing of cultivations. Similarly, the destruction of beneficial insects makes pest outbreaks more probable in subsequent years. This destruction may also lead to the appearance of new pests (such as red spider mites). As this leads to extra expense in developing new control methods to deal with both the original pests and the new pests in subsequent years, part of this expense must be included in the costs of using the first control method. Headley develops a mathematical treatment of this depreciation and of other costs in his review.

We have, so far, considered mainly only private costs and benefits of pest control. There are also social and community costs which must be considered when assessing the benefits of a particular control method. The chlorinated hydrocarbon insecticides, for example, are so stable that they act as pollutants at places far from where they were used (chapter 4). The cost–potential benefit ratio of the grower is thus becoming unsatisfactory, at least from the community point of view. It will have to be replaced, or supplemented, by a potential risk–benefit ratio.

Finally, it must not be forgotten that pest control is not simply an additional technology superimposed upon an existing agricultural or similar technology. Its use has, in fact, made possible the introduction of other innovations (the reverse, of course, also applies). Many modern crop varieties with high yields, including those which are bringing the 'Green Revolution' in the underdeveloped countries, could not be grown without modern pest control because they are especially susceptible to pest damage. They also demand large applications of fertilizer and adequate irrigation. Chemical weed control has released many men from time-consuming hand cultivation and this, when combined with other innovations, has allowed the farmer to use a smaller labour force. The men thus set free are available for other industries, especially light industry situated in rural areas and various mining concerns. Headley cites cotton-growing as an industry which has changed greatly over the last few decades as a result of new technology, including advances in pest control. Formerly, cotton was grown only in the humid, sub-tropical areas of America, but the introduction of mechanical picking, new varieties, fertilizers, irrigation, and pesticides has allowed the crop to spread to arid regions. All these factors are interdependent. Strong (1970) estimates that it took about twenty man-hours to grow an acre of cotton in 1970, compared with about one hundred and twenty in 1950.

Ironically, the failure to control a cotton pest (the boll weevil, *Anthonomus grandis*) in Coffee County, Alabama, brought the farmers there higher profits. Forced to abandon cotton they turned to other crops, such as ground-nuts, potatoes, maize, and sugar-beet, which proved more profitable. They accordingly erected, in 1919, a monument to the weevil in the county town, Enterprise (Madel, 1971).

Without doubt pest control has made the growing of many crops easier, though possibly many growers, faced with the bewildering variety of proprietary chemical pesticides, would find it difficult to agree. Yields have been increased, and year to year fluctuations in the produce sold in the markets have been smoothed out. Each grower will attempt to produce as high a yield as possible, but if this leads to over-production of a crop in the country as a whole his real income may fall. If the overproduction becomes extreme either many growers will abandon the crop voluntarily, or, if the crop is one which is suitable for such a control, a quota system may be imposed. Martin (1970) gives an excellent example of this process. At the beginning of the century English hop growers were troubled by two major pest problems, hop damson aphid, *Phorodon humuli* and hop powdery mildew, *Sphaerotheca macularis*. Quassia and, later, derris and nicotine were introduced for the control of the aphid and 'Made hop growing too easy'. The resulting overproduction of hops, coupled with certain changes in brewing methods, eventually led to the establishment of the Hop Marketing Board and to the introduction of quotas on the planting of new gardens. These are, however, troubles of highly developed countries. Overproduction, resulting from the use of pesticides, is not likely to be a problem in the underdeveloped countries for many years.

We have only touched upon another important aspect of the economics of pest control, namely the research and development costs involved in the marketing of chemical and microbial pesticides. This topic is discussed in chapter 3.

Most of the remainder of this book deals with the techniques of pest control, both actual and possible, and, for convenience, the introduction closes with a summary of these methods.

Summary of pest control methods

The following is a simplified summary of pest control methods. The techniques are not listed in the order of present-day practical importance, although there has been an attempt to do this in the body of the text. The classification is not clear cut as some of the methods could fall under more than one main heading. Several of the techniques are still in the experimental or developmental stage.

VICTIM PROTECTION (Preventive Control)
1. The reduction of the number of contacts between the pest and the victim.
   (i) By the erection of barriers.
       (a) Inspection and quarantine measures.
       (b) Physical barriers (fences; fly-screens; insect-proof packaging).
       (c) Sensory barriers (repellents, physical and chemical; attractants, physical and chemical – in conjunction with destruction of the pest; antifeedants).
       (d) Temporal barriers:
           Crop rotation and adjustment of time of sowing, etc.: cultural control, in part.

2. The reduction of the effects of contacts between the pest and the victim.
   (i) By the use of resistant varieties of the victim.
   (ii) By the use of protective (as opposed to curative) chemicals.
   (iii) By the use of measures to promote the general vigour of the victim.

PEST DESTRUCTION (including curative control when the pest is attacking the victim, and the destruction of the pest in the general environment).
1. Cultural control (in part).
   (i) Cultivations and similar operations to destroy the organisms.
   (ii) General hygienic measures such as the destruction of refuse and infected material.
2. Ecological control.
   (i) The manipulation of existing environmental factors either to harm the pest directly, or benefit natural enemies of the pest.
3. Biological control – the introduction of living material (other than resistant varieties) into the pest's environment in order to destroy it.
   (i) The introduction of parasites, predators, and pathogens of the pest either by the 'inoculation' of the environment with small quantities of living material (persistent), or by the 'inundation' of the environment with very large quantities of living material (non-persistent).
   (ii) Autocidal methods – the use of the pest species itself, or some very closely related form which will mate with it, to destroy the pest population: or the use of some characteristic of the pest species to destroy the population.
      (a) Sterile insect release methods (sterility obtained by irradiation, chemosterilants or genetic characteristics of the introduced individuals).
      (b) Introduction of sterilizing agents into the environment to induce sterility in the wild population.
      (c) The use of pheromones as attractants (overlapping with preventive control, and trapping).
      (d) Genetic control.
   (iii) The replacement of the pest population by an innocuous or less harmful competitor.
4. Destruction by pesticides, introduced into the environment.
   (i) Chemical pesticides, including insect hormones and hormone mimics. These may be also used in autocidal control.
   (ii) Physical pesticides consisting of either electromagnetic energy, or mechanical energy.
5. Trapping (with and without attractants and repellents), shooting, collection by hand, and similar methods.

INTEGRATED CONTROL, ETC.
The simultaneous and integrated application of all suitable methods to reduce pest damage, with as few deleterious effects on the environment as possible.

2

The numbers of animals

Whenever we try to control a pest we interfere in some way with its environment. If we are to get satisfactory results by design rather than by good luck we must understand how this environment 'works' and particularly how its component factors, biotic and abiotic, determine the pest population's density. A pest's environment is taken to include the pest population itself.

There have been a number of theories which undertook to explain the dynamics of insect populations. Starting with certain assumptions, some of which were founded on what appears to be commonsense, and some on laboratory experiments or field studies, deductive mathematical models were produced which were then used to make predictions about events in the field. Unfortunately, such a large number of factors could play a role that it was extremely expensive and time consuming to collect all the necessary data, and, in any case, the computations themselves became too unwieldy using traditional pencil and paper methods.

Digital computers have helped to solve this difficulty. It is now possible to use computers to develop sub-models which attempt to explain a part of the situation, and which are then tested in the field. After modifying as necessary they are again tested, and the process repeated till each sub-model is satisfactory. Later the sub-models can be integrated to form a model for the whole system, or those parts of it which are important in determining population density. This is the approach which is being used successfully by Holling in his studies of predation.

Accurate and copious field data are essential for the success of this approach, and the provision of these has been facilitated by modern developments in sampling theory and techniques, and in methods of analysing these data. Many of these techniques, applicable to insects in general, are given by Southwood (1966), and methods for soil animals by several authors in Murphy (1962).

An essential step in the analysis of the population dynamics of an insect is the construction of life-tables, in which the fate of a cohort of insects is followed from oviposition to oviposition. They resemble in some ways the mortality tables used by actuaries for insurance purposes, but the time scale is usually developmental stage rather than age in days or months. The life-tables of a pest, and those of other species which influence its mortality, serve as a basis for the development of the mathematical model or models. They are discussed in more detail in a later section of this chapter.

The approach outlined above will, eventually, bring the objectivity needed in the study of the population dynamics of insects and other organisms. As Harcourt points out in his review on the use of life-tables in insect population studies (Harcourt,

1969) the ultimate comparison of observed population densities with those predicted by a model will allow us to calculate the proportion of variance in the system that is explained by the model; the greater this proportion, the more satisfactory the model and 'this is the most quantitative, objective, and scientific method of determining how much we understand about the population dynamics of a species'.

The growth of populations

Since Malthus and Darwin it has been realized that the numbers of an organism are capable of increasing exponentionally. The change in population size with time, given unlimited resources, can be expressed by the equation

$$\frac{dN}{dt} = (b-d)N \quad \text{or} \quad \frac{dN}{dt} = rN \tag{2.1}$$

where N represents the size of the population at any given instant, t is time, b the instantaneous birth rate and d the instantaneous death rate. r, or $(b-d)$, is the intrinsic rate of natural increase.

Integration of this equation gives the size of the population after a time t,

$$N_t = N_0 e^{rt} \tag{2.2}$$

where N_0 is the size of the population at the beginning of the period and e is the base of Napierian logarithms, $2 \cdot 718. \ldots$

Almost all populations gain and lose members by immigration, emigration, and dispersal, and terms for these can be included in the above equation.

No population can grow indefinitely at this rate. Sooner or later food, or some other requisite, would be exhausted, or the environment would furnish some other kind of resistance to further growth. By making certain simplifying assumptions, such as the rate of increase at any time depending only upon the numbers present, and that all the animals are effectively identical, the following model can be put forward: that the rate of increase at any time is linearly proportional to the difference between the numbers present at the time, and the numbers which the habitat can contain at saturation. If K is taken to be the maximum number which the habitat can contain then

$$\frac{dN}{dt} = rN \frac{(K-N)}{K} \tag{2.3}$$

where r and N have the same meanings as in eq. (2.1).

It will be seen from eq. (2.3) that, in this model, the maximum rate of growth occurs when the population size is half the maximum possible (this may be confirmed by differentiation). As N approaches K, after passing the half-way point, the slower the rate of growth becomes.

Integration of (2.3) yields the equation

$$N = \frac{K}{1+e^{(a-rt)}} \tag{2.4}$$

where a is a constant which defines the position of the curve relative to the abscissa. The derivation of this curve may be found in, *inter alia*, Andrewartha and Birch (1954).

Equation (2.4) is known as the logistic equation, and the curve derived from it, the logistic curve (Fig. 2.1). The growth curves of many populations have been shown to fit it rather well. These include fruit fly (*Drosophila*) populations in laboratory cultures, and the population of sheep in Tasmania between 1814 and 1934. For several examples see Allee *et al.* (1949).

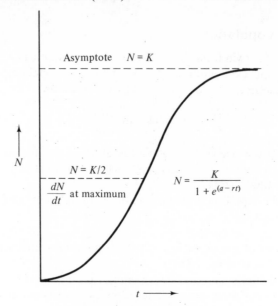

Fig. 2.1. The logistic curve

The logistic curve is, however, almost divorced from reality. Too many simplifying assumptions are made in its derivation and, in any case, it is deterministic, rather than stochastic. That is to say, it does not allow for the many chance happenings that affect a population during its growth; catastrophic happenings such as epizootics of disease and acute shortages of food; favourable occurrences such as prolonged periods of favourable weather. Modern models are stochastic. It is here that computers are important for they can be used to explore, in a comparatively short time, the probable results of a wide variety of chance happenings. Yet, despite its short-comings, the logistic relation does provide a valuable starting place for thinking about population dynamics.

To summarize, as a population grows in size the environment tends to resist this growth more and more so that the rate of growth lessens. It would seem obvious at first that the resistance increases because the supply of food or some other requisite is approaching exhaustion, or at least reaching a level which is insufficient for the satisfactory development and reproduction of all the members of the population. This does not seem to happen very often however, except among certain pest populations which are, of course, unusual in other ways as well. Most non-pest populations level off or decline long before such a stage is reached. Furthermore, rare species tend to remain rare for long periods, and common species common, unless there is some gross ecological change caused by man's activities.

Ignoring for the moment changes resulting from immigration, emigration, and dispersal, the only factors which can influence the size and growth rate of a population are those which affect the birth rate or death rate. These comprise characteristics of the population as a whole, characteristics of individual members of the population, and factors outside the population. The difficulty lies in determining which of these factors regulate the population by keeping it within fairly well defined upper and lower limits of density, and which of them are disturbing factors, causing more violent changes in density. It is also important to learn which first and multiple order interactions between these, and between these and movement into and out of the area, are significant.

Factors influencing the birth rate

For our purposes the birth rate may be taken as the ratio of births or ovipositions during a short period to the total number of individuals in the population at the beginning of the period.

It hardly needs pointing out that all necessary requisites, food, shelter, suitable microclimatic conditions and so forth, must be present for there to be any reproduction at all. The maximum number of offspring a female can produce in a given time, when all these requisites are available in optimal quantity and quality, is known as the fecundity of the female. Her fertility is the number she actually succeeds in producing in a given set of conditions. Thus, fecundity is the result of the heredity or genotype of the female, modified possibly by the history of her parents, and the fertility is the result of the modification of fecundity by various environmental factors, and by the success of fertilization when reproduction is bisexual.

We can consider the average fertility of the individual females, resulting from various factors acting upon those females, and the fertility of the population as a whole, which is a function of the composition of the population.

Enough food of the right kind must be available to the female if her fertility is to be high. The climatic conditions, especially the temperature range and relative humidity, must be suitable as many organisms can survive easily in conditions that make reproduction impossible.

The population density (itself partly the result of the birth rate) can influence the fertility of the individual females. As Allee observed several years ago, if the population density is low then the mean fertility of the females is also low because of the smaller probability of the female finding a mate. 'Undercrowding' may also reduce the protection afforded to the reproductive females by the group in colonial or gregarious forms, and thus lessen their chances of a successful mating. Conversely, birth rate will fall off when the density becomes so high that interference with mating becomes important, or the animals alter the environment in some way (often by excretory products, for example, copious honey dew), which makes it less favourable. Crowding may also produce various stresses that produce a lowered fertility by hormonal or some other physiological means. Finally, there may be a shortage of oviposition sites, though the importance of this may be modified by the prevailing microclimatic conditions.

These factors, and others, influence the average fertility of the females in a population, but the fertility of the population as a whole can be affected in other ways. Most important are the effects of the sex ratio, and the proportion of the females in the population which are reproductive. The sex ratio is the proportion of females in the population, and its influence will depend largely on the mating behaviour of the species, and on the possibility of facultative parthenogenesis.

The proportion of the females that are reproductive depends upon the length of the reproductive period compared with the periods before and after, and on the birth rate itself. The greater this rate, the larger will be the proportion of immature females in the population. Furthermore, if favourable conditions shorten the pre-reproductive and reproductive periods of the female, without reducing her total fertility, the birth rate will rise.

Factors influencing the death rate

The death rate may be regarded as the counterpart of the birth rate, that is to say, as the ratio of the number of deaths during the given period to the number of animals at the beginning.

Some causes of death are particularly associated with restricted age groups, while others are more general. Thus the impact of an agent upon a population depends upon the age structure of the population. Certain factors affect one sex more than the other and, in these cases, their effects on the population as a whole will depend upon the sex ratio. Both age structure and sex ratio will vary from place to place, from time to time, and from species to species.

Death from disease, predation or parasitism will come to mind at once. Predation, in the narrow sense, is difficult to study quantitatively because once the prey is killed it is usually rapidly consumed. Various indirect methods are of use, such as the study of faecal pellets of owls, or the study of stomach contents of mammals, or, for invertebrate predators, the employment of prey tagged with radioactive tracers, and the precipitin technique. For details of these and other methods the reader should consult Southwood (1966).

The most important parasites of insects are other insects, and nematodes. Parasitic insects are unusual in that they are generally relatively large, and almost invariably destroy their hosts. Their relationship with the hosts may almost be regarded as extended predation although, unlike predators, few of them kill more than one host, being associated with only one individual during their whole pre-adult life. The resemblance is close enough for both phenomena to be regarded as one in many mathematical models.

Some parasitic nematodes of insects cause substantial mortality. At least one important species, *Neoaplectana carpocapsae*, cooperates with a bacterium which it introduces into the insect's body cavity. The bacterium multiplies and serves as food for the nematode.

Predation and parasitism are often age-specific; this is probably more marked with parasitism. Some pathogens also are age-specific, attacking mainly the larval stages. The most important pathogens which cause death are fungi, viruses, bacteria, and protozoa. Many of these diseases are most severe when the host's population density is high. Furthermore, many diseases, especially those caused by fungi, are largely weather dependent, being most damaging during humid periods.

Weather itself kills many organisms, especially when shelter is in short supply. Certain stages, such as the eggs or the pupae, are often more resistant to adverse conditions than are the other stages, so that even climatic extremes such as drought or flood kill different proportions of the various stages. The weather can also affect the supply and quality of the food.

There remains death by accidents of various kinds (such as the malfunction of moulting), and the interaction between the various agents mentioned above. An animal weakened by disease, or made more conspicuous by some resultant change in behaviour, falls more easily to a predator. An organism suffering from starvation succumbs more easily to cold or drought. A virus disease, latent in the population for several generations, springs to life when the insects are subjected to various stresses.

The regulation of population density

Populations are usually in balance or equilibrium with their environment. Only rarely do they increase for many years in succession, or disappear completely. The fluctuations may have great amplitudes but, in general, the rare species stay rare and the common, common. Writing in 1961, Richards cited the case of the sixty-eight species of British butterflies, a group of insects whose fortunes have been chronicled better than most because of their general appeal. Seven may be classed as rare migrants, and three as common ones. Twenty-eight resident species are common, seventeen local and ten very local. One, the black-veined white, *Aporia crataegi*, is almost certainly extinct, although it still rates as a pest in continental Europe. Two are extinct, but one of these, the large copper, *Lycaena dispar*, probably disappeared as a result of the draining of the Fens. Most of the others have kept their status for at least 150 years, though it is difficult to be confident about their future with modern farming techniques (Richards, 1961).

We have few, if any, records for other invertebrates as detailed as this, but there is no reason to think that they would be very different. Unfortunately, there is no doubt that many larger animals, such as certain birds and mammals, are much less common than they were a century or so ago: man is a far more efficient predator of the buffalo and the passenger pigeon than of the insects, but our interests, at the moment, are restricted to these and other invertebrates.

Over long periods of the order of several generations, and sometimes over many decades or hundreds of years, insect populations seem to keep at roughly the same level of abundance, although with some peaks and troughs. This applies to populations living in habitats with suitable average climatic conditions, and sufficient resources. Populations living in areas in which the climate does not permit survival every year will sooner or later disappear, but the areas may be restocked by immigration when conditions improve. Very small populations may become extinct because they are very small; some catastrophe destroys too large a number for the population to survive a catastrophe that would be merely a temporary set-back to a larger population.

It seems well established, therefore, that most populations are regulated in some way, but, at the same time, there are disturbing mechanisms which cause (or allow) large, but temporary, increases in numbers.

The nature of the regulatory factors has long been disputed by population ecologists, and the arguments have, at times, been heated ('vituperative' was the description by Varley and Gradwell in their recent review). Richards considers that, possibly, 'some authors are more interested in proving their opponents wrong than in providing evidence for alternative theories' but he concedes that 'disagreement probably arises from the absence of sufficient evidence of the right kind'. Fortunately, such evidence is now being accumulated and possibly, within twenty or thirty years, many of the differences will be resolved (Richards, 1961; Varley and Gradwell, 1970).

It is tempting to go deeply into the various theories concerned, but this would take far too long and a sketch must suffice. The reader who is interested in following the arguments in more detail should read the articles which enliven the *Annual Review of Entomology* almost every year, and the following papers: Nicholson (1954, 1958); Thompson (1956); Solomon (1957); Andrewartha and Birch (1960); Richards (1961); Klomp (1964); Huffaker and Messenger (1964a, b). In addition the following texts should be consulted: Andrewartha and Birch (1954); Wynne-Edwards (1962); Slobodkin (1962); MacArthur and Connel (1966); Clark, Geier, Hughes, and Morris (1967); Andrewartha (1970). Solomon's (1969) short text is a valuable introduction, and it includes a section of plant population dynamics. A useful history of ecology and ecological ideas will be found in the classic text by Allee and his colleagues (Allee *et al.*, 1949). These texts and papers give an ample bibliography of the important publications in the field, up to about the beginning of 1970.

There are also three symposia of interest: the first on the abundance of insects, edited by Southwood (1968), the second on the numbers of men and animals (Cragg and Pirie, 1955) and, finally, the Cold Spring Harbour Symposium of 1957 (Anon, 1957). Hazen (1970) has also gathered together a number of interesting papers by various authors in the field.

Most ecologists believe that if a population is to be regulated there must be at least one mortality factor which kills a higher proportion of the population at high densities than at low. Thus as the population density increases the rate of increase is checked, but as the population density falls the severity of the mechanism is relaxed, so that the population is not eradicated. Such a factor of mortality is called a density-dependent factor, but it should be pointed out that it is the mechanism which is density-dependent rather than the agent itself. This may be unaffected by changes in the population density. Furthermore, it is possible that some agents have a density-dependent effect only in certain circumstances, or over restricted ranges of population density. A predator, for example, might slow down the population increase of a pest when the pest's density is low, but fail to do so when this passes a certain level. Pests are pests partly because they often reach such an escape point.

Certain factors may have a negatively density-dependent action, so that as the population density increases, their effects become less proportionately. Such factors are disturbing rather than regulatory. Yet again some agents are density-independent in their action, either killing roughly the same proportion whatever the population density, or a proportion that is unrelated to it.

Concepts such as these were first put forward by Howard and Fiske in 1911, greatly developed by Nicholson from about 1935 onwards, and further modified since by several workers including Richards, Solomon, Varley, and Gradwell.

These density-dependent actions are not, of course, restricted to causing mortality: some act through their effects on the fertility of the members of the population.

Those ecologists who stress the importance of these concepts point out that mathematical models which do not include density-dependent mortality or fertility reduction factors are unstable. They consider that a negative feedback mechanism is just as important in population regulation as it is in engineering control systems, or biochemical and physiological processes.

Examples of factors with proposed density-dependent actions are parasitism, predation and disease, and inter- and intraspecific competition for food, shelter, and other requisites. Some workers (for example, Milne (1958)), regard intraspecific competition as automatically density-dependent in its action, for it must have a depressing effect sooner or later as the population density increases. The other factors are organisms of other kinds, and their effect is thus probabilistic, as their density depends upon yet other factors, as well as upon the density of the subject species.

Some factors will show a density-dependent effect immediately the population density begins to rise. Cannibalism and the impact of the various agents which reduce fertility are likely to have such an effect, and in many cases the functional response of predators will increase, i.e., each individual predator will consume more of the prey. The functional response depends, however, on the availability of alternative prey, and upon the capacity of the predator. Diseases also can respond quickly to increases in population density, but there is a time lag which depends upon the incubation period of the disease.

Numerical responses of predators and parasites, resulting from increased reproduction as opposed to immigration, show a time lag which corresponds to generation time of the species concerned. Varley and Gradwell have stressed the difference between those factors which have an immediate or almost immediate effect, and those with a delayed density-dependent response. They point out that a delayed density-dependent response is often very difficult to detect, but they have developed a graphical method which is of value, and will be discussed later.

Nicholson was responsible for the greatest development of the theory of population regulation by density-dependent factors. Starting from certain basic assumptions, which he regarded as axiomatic, he employed deductive reasoning to develop his ideas which he examined and tested with various laboratory experiments and the field data available at the time. He considers that populations are self-governing systems which persist in environments containing various elements which have relations with the population density of the species. These functional relations, and the relations between the biological characteristics of the species and its density, are described as density factors.

In Nicholson's development there are two main kinds of density factor: legislative density factors (non-reactive and independent of density) and governing factors (reactive and related to density). Legislative factors do have an effect upon the density of the population. They are held to be related to such things as the general climatic conditions, shelter, and the quantity of food, when these are not influenced by the numbers of the animals. Such factors cannot hold a population in balance, as they do not react to changes in the population density.

Nicholson believes that 'The mechanism of density governance is almost always intraspecific competition, either amongst the animals for a critically important requisite, or amongst natural enemies for which the animals concerned are requisites.' Such governing reactions hold the population in a state of balance in the environment, and provide a compensatory mechanism which ensures the persistence of the population even when there are violent changes in the environment. The level about which the population fluctuates, being held in balance by the governing reactions, is largely determined by factors independent of density which modify either the properties of the animals or of their environment.

Nicholson does not consider that governing density factors are equally effective in all the places in which the animals live. In favourable places the mechanisms described allow the populations to persist indefinitely, but at or near the edge of the distribution reactive processes must be slight or absent. In some areas the animals would not persist at all but for immigration from more favourable regions. Part of this diffusion from these areas must be the result of crowding, so that even the mortality caused by the harsh conditions of the fringing zones is caused, ultimately, by density-governed reactions.

The views of Nicholson outlined briefly above have been attacked by several ecologists, notably Thompson, and Andrewartha and Birch.

Andrewartha and Birch reject 'the traditional subdivision of the environment into physical and biotic factors and "density-dependent" and "density-independent" factors', for they do not consider that they constitute a precise or useful basis for the discussion of problems of population ecology. They acknowledge the probable existence of density-dependent mechanisms but believe that they rarely, if ever, play a part in the regulation of populations.

Their main objection to the Nicholsonian theory appears to be that Nicholson has developed it from certain axioms (which they are not prepared to accept without further investigation) by a deductive process whereas the chief technique of scientific investigation should be chiefly inductive. Such a conceptual model, they argue, should be tested by observation and experimentation in the field. Nicholson tested many of his hypotheses by a brilliant series of laboratory experiments with blowflies, but he was limited by the paucity of field data which were available to him. The main objection to laboratory trials is that so many of the factors which are highly variable in nature are held fairly constant. Andrewartha and Birch are also distrustful of mathematical models of population dynamics since they consider that these are far too simplified to be of real value as explanatory models.

A conceptual model is not necessarily incorrect because it is based on deductive reasoning. The development of most scientific theories depends upon the application of deductive and inductive reasoning in cycles. Deductive reasoning provides an hypothesis to test. This test provides data for the modification of the original hypothesis. The process continues in this circular fashion until the theory provides a satisfactory model of the system. A conceptual model is only wrong if it is retained after extensive experimentation shows it to be inconsistent with reality.

Models take many forms – physical, verbal, mathematical and so on. Mathematical models for population dynamics are discussed later in this chapter. Andrewartha and Birch present verbal and graphical models as alternatives to those of Nicholson and his co-workers. From their examination of published work, and from their own field research, they conceived of the environment of the individual as consisting of four components: food; weather; other animals and pathogens (including animals of the same kind, and of other kinds, but excluding those which are the food of the subject species); a place in which to live. They propose that animals in a natural population may be limited in three ways. The most important is the lack of time in which the rate of increase, r, is positive. Briefly, during favourable periods the population rises but these favourable periods are limited, so that sooner or later the rate of increase becomes negative. Conversely the unfavourable periods rarely last so long that the population becomes extinct throughout its range. A population is split into a number of sub-populations, each inhabiting a separate habitat which differs from those of the others. The possibility of all the animals in all these sub-populations being destroyed during an unfavourable period is considered to be very slight. With the return of favourable conditions some or all of the areas where the animals were destroyed will be recolonized. Thus a second way in which animals may be limited is by the inaccessibility of material resources (such as food, nesting sites, places in which to live), in relation to the animal's powers of dispersal and ability to search for these. The third limitation, considered to be the least important by these authors, is the absolute shortage of material resources such as food and living space.

Andrewartha and Birch have been attacked on several grounds. Many ecologists consider that the extinction of populations would be much more common than it appears to be if regulation depends upon suitable sequences of favourable and unfavourable periods, or, alternatively populations would exhaust one or other of their requisites much more frequently than they do.

Nicholson (who refers to this theory, somewhat unkindly, as the 'chaos hypothesis' of population regulation) considers that Andrewartha and Birch, and Thompson, who expresses similar opinions, have not appreciated the great changes in population size that could take place in a very few generations if density-dependent mechanisms do not operate. He gave, as an example, a population which could multiply at the rate of 100 times per generation when unrestricted. If 99·5 per cent of one generation is destroyed, its numbers in the next generation would be one half of those of the original population; if the destruction is 98 per cent, the population would double. There is little difference between the favourabilities of the two periods concerned, yet a short run of such favourable periods, or of unfavourable periods, would lead to massive changes in population numbers.

Andrewartha and Birch base many of their arguments on two species which they have studied extensively in the field, namely the thrip, *Thrips imaginis*, on roses, and the grasshopper, *Austroicetes cruciata*, in South Australia. They concluded from their grasshopper studies, which extended over several years, 'that the distribution and abundance of *A. cruciata* are determined largely by weather; there is no evidence for "density-dependent factors"'. It is true that the dry weather of the summer destroys most of the green feed that is available to the insects so that Birch (1957) writes: 'The small amount of green food that is left at the end of spring is related not to the number of grasshoppers but to the onset of the hot dry summer,' and that there is so little food available that, 'the chance of a grasshopper finding food is independent of the number searching for it.'

Several ecologists disagree with the interpretation of these biologists. Clark (1947), for example, points out that swarm formation does not occur with this species until the population density becomes high, and that this swarming and the following mass migration are density-related processes in which density and numbers in relation to food play a part.

Davidson and Andrewartha (1948) used multiple regression techniques to relate the maximum population density reached by the thrips each year to certain features of the weather, and their analysis led them to conclude that the fluctuations in maximum density from year to year are controlled almost entirely by density-independent mechanisms. Andrewartha and Birch (1954) accounted for 78 per cent of the variance by four quantities which were calculated entirely from meteorological records, and they considered that the residual variance of 22 per cent did not give enough room for any other systematic cause of variation.

Other ecologists have quarrelled with the choice by Davidson and Andrewartha (1948) of a multiple regression model for the analysis of the observations. Smith (1961) re-analysed the results using correlation methods and the comparison of variances and found a strong inverse correlation between population size and population change, i.e., evidence for the operation of density-dependent mechanisms.

Nicholson (1958) points out that these calculations referred to certain periods each year during which the populations grew rapidly to reach their peak numbers, and that the changes took place each year from some approximately constant number, which was the mean of all 'peak' numbers observed. Thus some additional factor or factors must be operating to bring the numbers to an approximately constant value each year, and Nicholson believes that this factor, or factors, must have a density-dependent component.

Milne (1957a, b, 1958, 1962) has attempted to synthesize elements of Nicholsonian theory and that of Andrewartha and Birch, which he considers to be almost identical with Thompson's concepts. Milne believes that Andrewartha, Birch, and Thompson do not give density-dependent mechanisms the importance due to them, but he disagrees with Nicholson about their nature. In Milne's opinion the only perfect density-dependent mechanism is that which results from intraspecific competition, that is, competition between individuals for requisites such as food, living space, nesting sites, and so forth. When the population rises above a certain density this density-dependence mechanism automatically comes into play. In other words, the rate of increase begins to slow down when one or more requisites are in short supply. Usually, however, these mechanisms are not operating, because the population is held, fluctuating, by density-independent and imperfectly density-dependent mechanisms, well below the critical level. In short, 'the environment rules' with the combined factors 'each supplying the lack of the other' keeping the population level below that at which intra-specific competition is evoked.

As the population falls during an unfavourable period density-dependent mechanisms lessen their depressive effects, but they always remain depressive and never become elevating, so that extinction results unless the conditions imposed by density-independent factors become less harsh. This extinction resulting from prolonged unfavourable conditions would not occur, however, unless the area is very small and immigration and emigration is precluded.

Milne's main disagreement with Nicholson seems to be about the role of natural enemies, and of competing animals of other kinds. Most of these will be imperfectly density-dependent factors since their impact will always depend upon components of the environment in addition to the density of the subject population. Thus their numbers will depend, to some extent at least, upon their own natural enemies, upon density-independent mechanisms in their environment, and upon the availability of alternative food or other requisites. In general, however, they will tend to damp down the upward fluctuations of the subject species, and to increase its downward fluctuations. They will thus make the 'outbreaks' of the subject species, with their heightened intra-specific competition, less common. This is consistent with the frequently observed fact that agricultural operations, such as spraying, which destroy natural enemies, lead to pest outbreaks. This observation is also, of course, consistent with the Nicholsonian theory which postulates a closer relationship between the impact of natural enemies and the population density of the subject species.

Many other ecologists have made valuable contributions to population dynamics theory. Chitty (1960, 1965) elaborates the idea that populations are self-regulating through the induction of changes in the 'quality' of the individuals as the population density rises, and that these changes in quality or average vitality are genetically induced. Thus crowding in one generation can influence the viability of the individuals of the next. He bases his theory mainly on studies of voles and other rodents, but he suggests that the mechanism may be common in other kinds of organisms as well. Vole populations often build up to very high densities, then suddenly collapse, and, 'According to field evidence the individuals in a declining vole population are intrinsically less viable than their predecessors.'

He suggests that possibly all species can regulate their numbers in this way before they destroy all the renewable resources of their environment, and that they can do this without the help of enemies and adverse weather. The theory also makes meaningless the idea that physical forces have a density-independent effect since when the quality of the individuals declines, these factors will have a greater impact. Their effect is thus not independent of population density.

The weather may also place stress on organisms carrying a heavy burden of parasites which, in favourable conditions, have little effect. Lack (1954) reports that the red grouse, *Lagopus lagopus scoticus*, are often heavily infested with internal parasites, but without injury, except when their food, *Calluna vulgaris*, is damaged by winter weather or extensive attack by the heather beetle, *Lochmaea suturalis*.

Klomp (1966) carried out a study on the dynamics of a field population of the pine looper moth, *Bupalus piniarius* in the Netherlands for a period of about fifteen years. He discovered that in years following extremely high population densities larvae and eggs were less viable than in years following comparatively low population densities. Although density-dependent predation effects were observed they were found to be of less importance than this self-regulating mechanism. It is not apparently necessary for the insects to reach levels at which severe intra-specific competition for food occurs; Kennedy and Way (1964) have given several instances of interference between individual insects at quite low densities.

Klomp's findings are similar to those of Chitty. Wynne-Edwards (1962) gives further examples. During a decline of the snowshoe hare, *Lepus americanus*, in Minnesota, captured animals died after a very short time in captivity although, at other times, they survived for long periods in similar conditions. At capture the animals appeared to be quite healthy, moving about normally, and eating well. Such an animal would, however, suddenly sink into a coma or develop convulsions and die. On dissection the hares were found to have shrunken livers and were apparently unable to store glycogen. The immediate cause of death was apparently hyperglycaemia. It seems, therefore, that overcrowding often produces, through stress, massive changes in the physiology of the animals of a population, especially in the endocrine system, and this complex of changes reduces their chances of survival and reproduction.

Wynne-Edwards also describes a number of other self-regulating mechanisms which can affect population density, and which are based on intrinsic characteristics of the organism. Many birds, for example, lay more eggs than are needed for population replacement. Some of these may be destroyed by the adults, or the fledgelings may be killed and eaten by their parents or siblings, and this is more likely to happen when food resources are in short supply. Calhoun (1962) describes the pathological behaviour of *Rattus norvegicus* when these were kept in an overcrowded colony. The behaviour of the mothers greatly reduced the chances of survival of many of the offspring.

In some overcrowded populations the individuals release some factor which slows down the growth of their fellows, thus postponing maturity and reproduction, or, in some cases, reducing the final biomass of the population. Large tadpoles of

Rana pipiens release a factor which inhibits or arrests the growth of small tadpoles; the agent was eventually found to be cells which passed from the guts of the larger tadpoles into the water. When ingested with the food by the other tadpoles the cells checked growth. It is worth noting that this negative feedback mechanism can operate independently of the quantities of food available; it is more closely related to crowding.

Such mechanisms are apparently common in aquatic environments where the water is a suitable medium for the factors. It can also occur in terrestrial environments. The conditioning of the food by flour beetles, *Tribolium* spp. is such an homoeostatic adaptation.

Wynne-Edwards postulates an evolutionary mechanism for the development of some population self-regulatory mechanisms. If the short-term advantage of an individual endangers the safety of the population, then group selection will win. Otherwise the population will eventually disappear and be replaced by another. Klomp (1966), however, 'meets difficulties in his understanding of the concept' and believes that many cases in which group advantages are said to be involved have been misinterpreted. A low density is often an advantage to the group, as it avoids over-exploitation, but the existence of a group advantage does not rule out the existence of an individual advantage in low density. Klomp suggests one possible reason, based on this concept, for *Bupalus* larvae spreading out on the trees.

Pimentel (1961) has also proposed a genetic mechanism for the regulation of population numbers, a mechanism which interacts with several of the others mentioned before. By a feedback process the relationship between a predator and its prey, or between a herbivore and its food plant, tends to evolve to a homoeostatic condition.

Many of the difficulties which have bedevilled the study of population dynamics have been semantic in origin. Clark and his colleagues discuss some of the various definitions given by authors of population and environment. They note that only one author they considered emphasizes the inseparable existence of the population and environment, and to emphasize the unity even more they have introduced the concept of the 'life system'. The life system of the population is 'composed of a subject population and its effective environment which includes the totality of exernal agencies influencing the population, including man (in practice, those biotic and abiotic agencies whose influence can be observed and, preferably, measured)'. This concept is less restrictive than many of the definitions of population and environment. Its spatial delimination, for example, depends upon the spatial delimination of the population which, in turn, is a matter of convenience.

They point out, in their discussion of the functioning of life systems, that insect species differ greatly in their characteristics, and in the places in which they live, and that, accordingly, generalizations about the determinations of insect numbers are only useful as guides when studying a particular population – a truism which seems to have been forgotten at times by some of the earlier authors. Their approach, therefore, is to achieve a synthesis 'by describing in terms of the general characteristics of life systems, the complex webs of interactions that determine the persistence and abundance of populations'.

The inherent properties of the members of the population, and the intrinsic attributes of its environment which determine the existence and population density are the co-determinants of abundance. Their interactions can only be assessed as ecological events (such as births and deaths – the primary events, and secondary events which modify the primary events, either by altering the supplies of requisites, or by direct action on the existence of individuals).

They recognize that processes may function independently of population density, or in some way that is related to it. Density-independent processes have either an almost constant effect, or one that fluctuates independently of the density. Density-related processes are either immediate or delayed in their effect, and they may decrease or increase proportionately with population density. They agree with Milne that some processes are automatically density-dependent in their action while others behave in a probabilistic fashion. A larger part of the text by Clark and his colleagues is taken up by descriptions of the life systems of a number of insect species, and these illustrate their approach excellently (Clark, Geier, Hughes, and Morris, 1967).

Huffaker and Messenger (1964b) also stress the importance of studying each population-habitat entity separately. In some habitats, especially those near the limits of the ecological range of the species, the physical environment is favourable for population growth only for relatively short periods. If this habitat is also impoverished in the sense that there are relatively few species of animals and plants present, and if it does not offer many micro-habitats, then density-dependent mechanisms will not be of great importance. During favourable physical conditions, if requisites are ample, the population density of the subject species is likely to become very high. It will be remembered that this description fits a typical crop – a habitat that is relatively poor in species and in diversity, and which is relatively transitory. Pests are opportunistic species which can thrive in such conditions; they are also often species whose changes in population density are not greatly influenced by density-dependent mechanisms.

The subject species can also occur in another kind of habitat: one in which the climatic factors and availability of food and other requisites rarely check population growth. If this habitat is also rich in plant and animal species, offering ample cover and micro-habitats, then density-dependent mechanisms, arising from predation, parasitism and so on, will regulate the population density.

In Huffaker and Messenger's concept, weather and other density-independent functioning factors, are of supreme importance for they dictate whether or not a population can exist and grow in a given habitat. Conversely, no habitat is without density-dependent mechanisms which play a part, large or small, in the regulation of the population densities of the species present. Their ideas are presented, verbally and graphically, in a more extended form in their 1964b paper. As this forms a section of a text on biological control their outlook is largely that of the economic entomologist.

In this last contribution we have returned to the idea that a complex community, complex, that is, in the sense of there being many trophic interconnections between the constituent species, is a stable one. It also stresses the importance of considering the community when studying the population dynamics of the subject species. This interest in the community seems to be a trend in the most recent work on population dynamics. The study of communities and their stability is considered later in this chapter.

It will be realized that the summaries of the various theories that have been given have grossly condensed the thoughts of the various authors and, in fairness, the reader should consult their original works.

An understanding of the mechanism of the regulation of population density is important to the economic biologist because it will allow him to cooperate with the existing regulating agents when controlling a pest. It will also help to avoid some of the pest outbreaks that have resulted from the inconsiderate use of pesticides and other agents. He is also interested, however, in the causes of the fluctuations in population numbers. Among other things such knowledge helps him to forecast outbreaks and to know when best to apply control measures.

At first examination the pinpointing of the factors which are responsible for the fluctuations would seem to be extremely difficult. It is important to realize that these factors may not have a great mortality effect. An agent which kills a large number of the species population each year, but always an approximately constant proportion, will not induce fluctuations. On the other hand an agent which kills a relatively small proportion, but a proportion which fluctuates markedly from generation to generation, could induce marked fluctuations. It is often surprisingly simple to isolate these factors, or at least the most important factors, in the way Watt (1963a) suggests. The first step is to determine the stage in the life cycle that is most responsible for increases or decreases in pest population densities, and the second is to find the causative factor. Statistical techniques related to multiple regression analysis, and the analysis of variance may be used but 'in fact the result is so clear cut that these techniques will in practice rarely be required'.

Watt illustrated his procedures with studies on five agricultural crop pests in Canada. In three of these (the fruit tree leaf roller, *Archips* (*Cacoecia*) *argyrospila*; the diamondback moth, *Plutella maculipennis*; the European corn borer, *Ostrinia nubilalis*) about three-quarters of the variance in the trend indexes was due to the proportion of the potential egg supply actually laid. The important factor in the case of the diamondback moth was the weather, and in the cases of the other two moths, probably adult migration. The survival of winter larva was the most important ecological event in the life cycles of the other two pests considered, namely the eye spotted bud moth, *Spilonota ocellana*, and the pistol casebearer, *Coleophora malivorella*, and this accounted for 68·4 per cent and 90·8 per cent of the variance respectively. Frost was responsible for the mortality of the bud moth larvae, and parasitism by *Epilampsis laricinellae* and bird predation for that of the casebearer.

Morris has also suggested that often a single factor, or, at the most, two or three, will account for most of the variation from generation to generation, and he has developed an analytical method for identifying these factors. Before these methods can be applied, however, it is necessary to construct life-tables for the species, and to develop the sampling techniques which enable us to do this.

Life-tables and sampling

A life-table is a systematic tabulation of survival and mortality in a population. It is not an end in itself, but forms the basis for further study and analysis. Life-tables should be replicated both spatially and in time, so that they may be used for the study of population dynamics, and in the case of pest species, as an aid to pest management (Harcourt, 1969).

Life-tables are derived from the mortality tables of demographers and actuaries, but differ from them in various respects. Human life-tables, for example, take no note of the initial population density and, as is well known, this information is very important to the population ecologist.

Many insect populations offer a great advantage in the construction of life-tables in that the various generations do not overlap. In such a case an *age-specific* life-table can be constructed in which the fate of a real cohort is followed. In age-specific studies the population may be stationary or fluctuating. In *time-specific* life-tables, applicable to organisms with overlapping generations, a sample of individuals, constituting an imaginary cohort, is taken from a population which is assumed to be stationary. It is, of course, essential to be able to determine the ages of the individuals of the sample.

Richards has suggested the alternative name of 'budget' to a study which tabulates the absolute population densities at different stages of the life cycle, together with the actions of the mortality factors.

The construction of a budget has been well described by Southwood in his text on ecological methods, cited above. The following is a brief summary of the relevant chapters in which the reader will find a full description of the techniques mentioned below.

Ideally, the budget will contain absolute estimates of population density in as many stages of the life cycle as possible. The total numbers entering a stage can sometimes be estimated (for example, by the use of emergence cages to trap winged insects that pupate in the soil) but often we can only make estimates of the population density on successive sampling days during the stage, and from this data calculate an estimate of the recruitment to the stage. Southwood gives several techniques that can be used for this purpose but they all rest on the assumption that mortality is constant throughout the stage. The choice of method also depends upon the synchronization of the individuals in the population. It is most convenient when all, or nearly all, individuals are passing through the same stage simultaneously, but often there is considerable overlapping of two or more stages.

A graphical method used by Southwood and Jepson (1962) estimates the total population of a stage. Successive estimates of population (usually made daily during the stage) are plotted against time, and the area under the plotted line measured. This is then divided by the mean development time for that stage in appropriate units (usually, of course, days). This gives an estimate of the total population. If the mortality is constant and heavy this will be the total population at the median age of the stage, and thus well below the recruitment numbers. When, however, most of the mortality occurs towards the end of the stage, the total population estimated in this way approaches the recruitment population.

Waloff and Bakker (1963) developed a partly graphical technique to determine the total population change due to migration in a mirid population in which there was a stage with a well-marked peak, and in which the migration was restricted to the middle of this stage. Southwood (1966) suggests this technique could also be used for other populations in which migration or a particular mortality factor occurred in similar circumstances.

Richards and Waloff (1961) have described two techniques for the determination of numbers entering a stage. The first method applies to stages with a well-marked peak, and an approximately steady mortality rate; the second to cases where there is no well-marked peak but a wide overlap of recruitment and mortality. In this second technique the numbers passing through the egg stage and entering the first larval instar can be calculated if a number of samples is taken during the egg stage, and if the number of eggs laid, and the duration of the egg stage, are known. This process can then be repeated in a stepwise fashion for the successive stages so that the life-table may be completed.

Southwood also summarizes Dempster's method for the estimation of natality, mortality, and migration for populations in which the corresponding stages of successive generations do not overlap. It is thus widely applicable, but requires a large number of accurate population estimates. Harcourt's review on life-tables lists other relevant techniques (Richards and Waloff, 1954, 1961; Dempster, 1961; Southwood and Jepson, 1962; Waloff and Bakker, 1963: these techniques are summarized in Southwood, 1966; Harcourt, 1969).

During the sampling for, and compilation of, life-tables, the ecologist usually tries to collect other data, especially relating to mortality factors. It is simple, for example, to examine samples brought into the laboratory to assess the incidence of parasites or predators. These results will, however, be biased as some of the hosts thus removed from the environment would probably have been destroyed by other mortality factors before they could be killed by the parasites, had they remained there. Such counts, therefore, give an assessment of the potential mortality from parasites, rather than actual mortality, unless a suggestion made by Miller is followed, namely that the assessment is made only during a period when death from parasitism is imminent.

Another possibility is to estimate the population densities of other pests which occur in the same habitat. It is probable that this will entail some additional sampling days if the two life cycles do not coincide exactly but the information is well worth gathering for the comparatively small extra cost involved. The second pest may, for example, compete with the one of primary interest for food or other requisites. Harcourt (1969) sampled for two cabbage caterpillars simultaneously at little extra cost, and there are other examples among forest pests.

There are two important aspects of sampling which, although interdependent, may be discussed separately.

The first concerns the physical methods used, and the second, statistical considerations. Both are covered in a number of texts including Morris (1960), Strickland (1961), Murphy (1962), and Southwood (1966). Statistical aspects are also discussed briefly in many standard statistical texts, but in detail in Cochran (1963) and in Mendenhall et al. (1971).

The sampling plan and methods depend upon the objects of the study. In age-table studies densities per unit area are essential, and these are obtained directly, wherever possible, by sampling a unit of the habitat (a single host animal, a core of soil or litter, a plant or part of a plant, and so on), or by capture–mark–release–recapture methods. Relative methods (for example, catch per unit effort) are used if they can be converted to estimates of absolute density. As they stand relative methods are useful for wide surveys or for following fluctuations of populations from year to year. When densities in different stages must be compared, as in life-table studies, they may not be satisfactory.

Capture–recapture methods are extensions of the Lincoln index technique, and are useful for the study of active and mobile organisms. They have been used, for example, in the estimation of tsetse fly densities. In the simplest application a number of organisms are captured, marked in some way which will not reduce their mobility or chances of survival, then released at the point of capture. After a period of time which is judged to be sufficient for them to mingle with the remainder of the population a second sample is taken, and the numbers of marked and unmarked individuals noted. A simple calculation gives an estimate of the population, if it is assumed that no births, deaths, immigration, or emigration have taken place in the intervening period. Furthermore, empirical confidence limits for the estimate can be calculated. The four assumptions are usually unjustified, but estimates of such changes in population density can be made by capturing, marking, and releasing on a number of days so that some of the individuals carry more than one mark, each corresponding to a particular capture date.

Such a method can be used, in some cases, for the estimation of adult population density, but this is often calculated indirectly from the estimated recruitment (obtained from a study of the pupal stage), the expected egg production, and the egg production actually achieved.

Whenever it is technically and economically feasible the population density of a given stage will be estimated in more than one way. Most of the techniques not only give an estimate of the absolute population density, but also allow the calculation of confidence limits for the estimate.

When units of the habitat are sampled it is possible to determine whether the organisms are randomly distributed, clumped or spread more uniformly. This information is important for it dictates what, if any, transformations of the raw data must be made before statistical analysis. Some sampling schemes are especially designed for the detection of non-randomness in distribution. In the nearest-neighbour method an organism is detected, the search usually starting from a randomly selected point. Systematic searching is then continued, usually along a spiral path, until the distance of the nearest neighbour is found. This, and similar methods work well when the organism is conspicuous, but is unreliable for cryptic species such as soil animals. Hughes (1962) has developed a method for such circumstances in which paired samples are taken, the first randomly, and the second at a fixed distance from the first, in a random direction. Although non-random distributions can be detected readily by the comparison of the observed sampling distribution with the theoretical random distribution, calculated from the Poisson distribution, it is difficult to find a mathematical distribution that describes aggregated populations in a way that is satisfying both biologically and mathematically. The negative binomial distribution is often employed.

Finally, sequential sampling may be used. In this comparatively new development samples are taken and examined only until a definite answer has been obtained to a question about the population density. As soon as an estimate has been obtained with satisfactory confidence limits, sampling is stopped. The method depends upon detailed knowledge of the pattern of distribution of the organisms.

When the ecologist is mainly interested in the dispersion of the organism, and is sampling units of the habitat, he will not be particularly concerned about minimizing variance and he will probably sample over all the area of interest. If his aim is to collect reliable results for the construction of life-tables then he will be anxious to minimize the variance and will therefore use some stratified random sampling procedure, taking few, if any, samples from areas in which the organism is known to be rare. In sampling sawfly cocoon populations in a forest, for example, he would sample intensively beneath the trees, and not at all between them.

An initial sampling survey is made to study the composition of the variance. To take a simple example from forest entomology, several trees are selected randomly, and the canopy of each divided into three levels: top, middle, and bottom. A few branches are selected from each level and the insects present on, say, five leaf clusters on each branch are counted. After a suitable transformation of the data (if needed) to fulfil the assumptions of the analysis of variance, the counts are analysed. The following variances are then compared: leaf clusters within branches; branches within levels; levels within trees. If the variance of the leaf clusters within branches is found to be relatively small, then only one leaf cluster per branch will be sampled. The higher levels are examined in the same way. Cost (usually in terms of man-hours) must, however, be considered: if the trees are long distances apart it may be more economical to take more samples per tree on fewer trees than fewer samples per tree on many trees. For life-table studies there is a further qualification, as Varley and Gradwell (1970) point out. When results are to be compared from year to year in the study of temporal fluctuations it is better to concentrate on a small number of neighbouring trees, if resources are limited, because otherwise it would be difficult to separate temporal and locality effects.

Generally, it is better to use a large number of small sampling units rather than a smaller number of larger ones. Morris (1960) has listed a number of requirements for the ideal sampling unit, and these may modify the choice. In orchard sampling, for example, the leaf cluster has proved to be a more stable unit from year to year (and thus more comparable from year to year) than the single leaf.

Most of the sampling techniques outlined above (and in particular that of Varley and Gradwell) are concerned with plots. Price (1971) has criticized such a concentration on well-defined plots, and argued convincingly the case for an holistic approach to population studies. He bases his argument for a broad approach on the grounds that the community is more complex than all the subunits of which it is composed. Plot techniques cannot be replicated sufficiently to cover the great variations in density, quality, and selection pressures present in the population. He suggests, therefore, that in addition to such conventional plot methods, transect sampling should be carried out to reveal strong density gradients and foci of population increase. Furthermore, studies should be made of the individual organisms, of the population, and of the community in order to understand the adaptation and evolution of the population.

If the ultimate aim of the study is to produce a mathematical model which will describe the dynamics of the population, or which can be used for prediction, the type of model envisaged will also influence the kind of sampling carried out. It would be obviously uneconomical to collect more data than are needed if the model is a simple one, and even more disastrous if it is one that requires more information than is collected.

These sampling methods will allow us to calculate confidence limits for our population estimates for the various stages of the organism's life but, as Southwood (1966) notes, life-tables provide another check on our estimates. A population cannot increase unless birth, hatching, or immigration takes place, and if our knowledge of the biology of the organism allows us to rule this out at certain times, an apparent increase must be the result of faulty estimation. An absence of such spurious increases in the life-table gives us additional confidence in our estimates, although this extra confidence cannot be expressed in statistical terms.

The physical aspects of sampling cannot be discussed here as there are too many techniques in use. Often a new set of methods has to be devised when studies on an organism begin. Sometimes, however, an existing technique can be pressed into service with little modification. The apparatus that was used to sample wire-worms during the war-years (Salt and Hollick, 1944) was later found to work well for estimations of the density of the eggs of the wheat bulb fly, *Leptohylemyia coarctata*. Most methods based on the examination of a unit of the habitat suffer from one obvious disadvantage; some individuals are bound to be overlooked, and the estimate, as a result, will be too low. Recovery experiments are only a partial solution as we cannot be sure that the human hand will insert, say, a wheat bulb fly egg into a crumb of soil in the same way as a female fly's ovipositor. A high recovery of 'planted' individuals does not guarantee a correspondingly high yield of naturally occurring individuals from a field sample. This trouble is particularly serious when sampling soil animals and other cryptic organisms.

Time-specific life-table studies require both estimates of numbers and estimates of the distribution of age classes within the samples, for, it will be remembered, they concern populations with overlapping generations. Obviously they are of prime importance in the study of vertebrate populations for only rarely do adult vertebrates die as soon as they reproduce. Possibly some of the most important applications have been in resource management in commercial fisheries, and in conservation.

When insect populations have overlapping generations the ages of the larval stages are easily determined, by such features as the structure of the anal spiracles in Diptera. In some cases it is even possible to age the nymphs of Heteroptera within an instar. The greatest difficulties arise in determining the age of pupae and adults. In pupae the degree of pigmentation is often useful; in adults the daily growth layers in the cuticle of some species can be used as a measure of the age in days. These layers can be observed microscopically with crossed polaroids. Other methods depend upon the changes in the male genitalia (Nematocera), in the ovaries and other internal organs such as the fat body. The amount of coloured excretory material deposited in the malpighian tubules has proved a useful measure of age in some species.

The description of life-tables

The results of the sampling of a population in the way outlined in the last section may be summarized either graphically, or in a tabular form, the life-table proper.

The graphs simply show the number of survivors from the initial population at different ages. Slobodkin (1962) has described the following four basic types:

Type 1: the population remains fairly steady until shortly before the maximum age is reached, when it falls rapidly.

Type 2: a constant number die per unit of time so that the graph is a straight line when arithmetical scales are used.

Type 3: the mortality rate remains constant so that the graph is a straight line only when the numbers of survivors are plotted on a logarithmic scale.

Type 4: high mortality is restricted to the early stages after which the remaining population dies off slowly.

The most commonly used headings in a life-table are as follows (Harcourt, 1969):

    x: The age interval of the class, in units of time, or as developmental stage

    lx: The number alive at the beginning of the age interval.

    dx: The number dying within the age interval.

dxF: The mortality factor responsible for dx.

$100qx$: Percentage mortality within the age interval.

S_x: Survival rate within the age interval.

There are variations from author to author. Some include additional columns for special purposes, some correct the data so that the initial population is regarded as one thousand, while others use the absolute population numbers. Table 2.1 shows an artificial life-table, based on Harcourt's scheme.

Table 2.1   Hypothetical life-table

x	lx	dxF	dx	100qx	S_x
Eggs (N_1)	6000	Parasitism	3000	50	0·50
First larvae	3000	Weather	2000	67	0·33
Second larvae	1000	Parasites	200	20	
		Predation	300	30	
			500	50	0·50
Pupae	500	Parasites and predation	100	20	0·80
Adults	400	Sex (40% ♀♀)	80	20	0·80
Females×2	320	Fecundity reduction	160	50	0·50
Normal females×2	160	Adult dispersal and death	140	88	0·13

Expected eggs = 160×200/2 = 16 000.

Actual eggs (N_2) = 20×200/2 = 2000.

$I = N_2/N_1 = 2000/6000 = 0·33$.

In this hypothetical case the sex ratio of the adults deviated from 1:1. As this deviation was in favour of the males, this was regarded as mortality. The number of females was doubled to restore balance in the table. Previous trials had shown that the maximum number of eggs laid by females of the species was two hundred. A sample of the current adult females gave a mean egg production of one hundred. There was thus a 'fecundity reduction mortality' 160, and the number of 'normal females' became 160. The expected number of eggs was, therefore, 16 000. At the beginning of the next generation, however, the actual number of eggs was 2000, the discrepancy arising from dispersal and adult mortality before egg laying.

As it stands the life-table can be used to calculate r, the rate of increase, or one of its derivatives. A series of life-tables for a population will provide information on the dynamics of the population. This topic is discussed in the next section (Southwood, 1966, Harcourt, 1969).

The analysis of life-table data

The life-table itself is little more than a systematic description of the fate of a group of organisms during one generation. To be of further use it must be combined with other life-tables for the population, constructed from data collected over a number of years or generations.

The trend index, I, is the ratio of the number of eggs of one generation (lx for eggs) to that of the preceding generation, and it is used as a measure of population growth or decrease. In our hypothetical example it is, of course, 2000/6000 = 0·33. There is, however, no obligation to base the trend index on egg numbers and, indeed, the use of some other stage will often give quite a different I value. Watt (1968) has suggested that the adult stage should be used for the basal age group, but Morris (1960) points out that the egg stage is probably the most uniform in quality and should therefore be chosen. This will also reveal changes in fertility and their effects. In key factor analysis, the use of two different trend indices will often result in the selection of different key factors. Thus, the economic entomologist, interested in a pest problem, would usually measure the trend index from the damage-causing stage in one generation, to the same stage in the next (Varley and Gradwell, 1965).

The trend index will probably be influenced, partially at least, by the population density of the base population. The influence can be detected quite simply by a graphical method suggested by Watt (1968) in which the population density of one generation is plotted against that of the following generation (Fig. 2.2). It would be misleading to

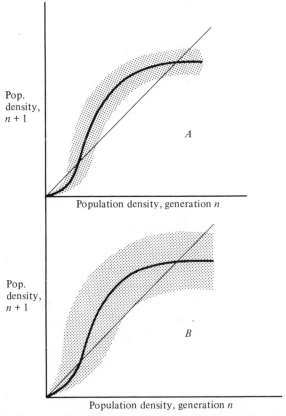

Fig. 2.2. Watt's graphical method for the demonstration of density-dependent mechanisms. A: Population regulated by density-dependent factors; B: Population regulated by density-independent factors. For explanation, see text. (After Watt, 1968)

plot the trend index against the basal population density for this would give a spurious appearance of density-dependence which would occur even if the consecutive population densities were merely random numbers.

The underlying curve of the suggested graph has a more or less sigmoid course, but the actual points are scattered in a band around it. The degree of scatter is informative; if the band is narrow density-dependent mechanisms are relatively important; if wide, the population is not tightly regulated by such mechanisms. In the first case the I values will not vary greatly from generation to generation; in the second case they will oscillate greatly. Unfortunately, such an approach requires a large number of points and involves the collection of data over a long period (Watt, 1968).

An alternative method is to use simple linear regression. The population densities of the chosen stage (see above) are converted to logarithms to stabilize the variance and to provide linearity. The regression of the log populations for each generation on the log populations of the preceding is then computed, giving the equation

$$\log P_{n+1} = \log F + b \log P_n \qquad (2.5)$$

This is the first step in Morris's key factor analysis. The assumption is made that the relationship between $\log P_{n+1}$ and $\log P_n$ is rectilinear, and this may not be valid, especially when extreme population densities are involved. Watt's graphical method does not make such an assumption, but it is difficult to express the results quantitatively.

The correlation coefficient, r (not to be confused with the rate of increase), is also calculated for by squaring it the coefficient of determination, r^2, is obtained, and this may be used as an estimate of the proportion of the variance of the dependent variable, $\log P_{n+1}$, which is accounted for by the variations of the independent variable, $\log P_n$. The regression coefficient, b, is a measure of the rate at which the population increase declined with increasing density. If b attains a value of one it is an indication that there is no density-dependent mechanism operating upon the population. Lower values of b furnish evidence of density-dependent mechanisms and the next step is to try to identify these from biological knowledge and from the effects of introducing them into the regression calculation. The P_n values in each generation are multiplied by the proportions surviving the effect of a particular factor, and the adjusted $\log P_n$ values are used in the regression computation in place of the original values. This modification produces a new b value, usually closer to one than the original. Candidate density-dependent factors are incorporated in the equation, one after the other, until the b value does reach one. It is then presumed that all the density-dependent factors have been accounted for. Sometimes, however, the final value of b remains significantly different from one and then a b value must appear in the final equation. After each repetition of the individual steps the coefficient of determination is determined to find how much of the variance in $\log P_{n+1}$ has been explained.

It is not likely, however, that any of the factors identified so far by this procedure will be responsible for the larger fluctuations. This will be some disturbing factor or factors which can be identified by the next part of the process. This consists of relating the deviation of an observed point from the regression line to the magnitude of a candidate factor. The value of this factor which causes no deviation from the line can be determined graphically, or by simple regression if the relationship between deviation and magnitude is rectilinear. The factor can now be incorporated into the equation which now takes the following form:

$$\log P_{n+1} = \log F + b \log P_n \, p_1 p_2 \ldots p_n + b_1 (q_1 - \bar{q}_1) + $$
$$+ \ldots + b_m (q_m - \bar{q}_m) \qquad (2.6)$$

in which $p_1, p_2 \ldots p_n$ are the proportions surviving the effects of each of the n density-dependent factors included, q_1 to q_m the magnitudes of the m disturbing or density-independent factors, and $\bar{q}_1$ to $\bar{q}_m$ the respective magnitudes corresponding to zero deviations from the line. The regression coefficients for the deviations from the line are represented by b_1, etc.

As each of the disturbing factors or density-independent factors is built into the equation the coefficient of determination is recalculated. The change in this with each successive step indicates how much of the variance of the log P_{n+1} is accounted for by the factor just added: it is thus an easy task to determine which of these factors are key factors. Once these have been identified the equation may be used for predictive purposes, and as a guide for possible control methods. It would be dangerous, however, to use it, unmodified, for other populations, and it may be desirable, as Solomon (1964) points out, to make separate studies for the periods when the population is, in general, increasing, and when it is declining.

It will have been noticed that the key factor analysis requires an accurate population density estimate for only one stage, but the life-table is, nevertheless, valuable in that it provides the necessary data on mortality for the computation of the predictive equation.

Varley and Gradwell (1970) have criticized this approach on several grounds, although they recognize that the resulting equation can have predictive value. The main objection is that correlation and regression methods are only valid when there is no intercorrelation between the variables. Furthermore, the incorporation of the indexes for the mortality factors of the various age stages indicates only their effects on the population density of the next generation, and thus provides no information on their parts in the regulation of the population. They also feel that the multivariate analysis used in the Canadian work on the spruce budworm was bound to work 'in the sense that the statistical procedure would produce some kind of formula'. Their further criticisms may be found in their 1970 review, together with a bibliography on their own approach, which is outlined below.

Their method is to carry out an analysis of the life-table data to identify the key factors and to determine their contributions to population change. Sub-models are then constructed for the mortalities caused by these factors. Finally, the sub-models are combined into a single equation which is tested against observed values in the field. If resources are limited they prefer to restrict the observations to a small area. In their work on winter moth, *Operophtera brumata* (Plate 2), for example, most of the studies were limited to five oak trees in Wytham Great Wood, near Oxford. Thus, they were able to separate temporal from spatial fluctuations which was impossible in the spruce bud-worm studies in Canada.

The life-tables constructed from several years of observations are then converted into a form more suitable for analysis. The usual conversion is to express generation survival for each generation as the product of the survival rates of the successive stages:

$$S_G = S_1 . S_2 . S_3 \ldots S_n \tag{2.7}$$

A logarithmic transformation is used to produce linearity and reduce variance:

$$\log S_G = \log S_1 + \log S_2 + \log S_3 + \ldots + \log S_n \tag{2.8}$$

An alternative method, more useful for the analysis under discussion, is to use k

values which are the differences between the successive values of log l_x. As these are logarithmic measures of the killing power of mortality factors, eq. (2.8) may be expressed in the alternative and equivalent form:

$$\text{Total mortality} = K = k_1 + k_2 + k_3 + \ldots + k_n \qquad (2.9)$$

Sometimes, however, a k value cannot be based on a direct measurement of mortality, but is derived from the difference between successive estimates. In such a case it is referred to as a loss – for example, as 'overwintering loss'.

When the data have been cast in the form of a number of equations of the form of (2.9), one for each year, they can be tested graphically to reveal which mortality factors are most closely correlated with changes in the generation population density, from the chosen basal age of one generation, to the same age of the next. This is achieved simply by plotting K, k_1, k_2, etc. against generation and noting which of the k values vary markedly, in phase with the K values.

In their winter moth experiments Varley and Gradwell (1968) listed six k values, namely:

k_1: 'Winter disappearance' – including mortality of adult moths before they have completed egg laying, together with egg mortality, and early larval mortality.

k_2: Mortality from the parasitic tachinid fly, *Cyzenis albicans*.

k_3: The summed mortality from attacks by various non-specific parasitic Hymenoptera and Diptera.

k_4: Mortality from infection by a microsporidian, *Plistiphora operophterae*.

k_5: Mortality from pupal predation.

k_6: Mortality from parasitism by the ichneumonid *Cratichneumon culex*.

Their graphs show clearly that when population fluctuations from egg stage to egg stage are considered, most of the changes are due to fluctuations in k_1, that is from variations in the amount of winter disappearance from year to year, and that pupal predation compensates for this to some extent. If, however, the generation mortality is measured from larval stage to larval stage, k_5 and k_1 are more or less in phase. They also applied a key-factor analysis similar to that of Morris (1963) to the data for the years up to 1963. This indicated that measurements of pupal predation were more valuable for prediction than those of winter disappearance. When, however, the data for the years 1964 to 1966 were incorporated later k_1 was once again seen to be the most important.

The k values are next tested for direct density-dependence by plotting each k value against the numbers entering the age interval in which it acts. If the regression of k on the numbers entering the stage is significant, it may indicate density-dependence but further calculations are needed to confirm this as the two variables are not independent and the conditions do not therefore fulfil the assumptions of the regression model. The two variables are, in fact, $\log P_i$ and $\log P_i - \log P_{i+1}$. The regression of $\log P_{i+1}$ on $\log P_i$ *and* the regression of $\log P_i$ on $\log P_{i+1}$ must be calculated: if the regression coefficients are both significantly different from $b = 1$, and if both are greater than 1, or less than 1, then density-dependence is demonstrated. Returning to the original regression calculation, that of the k values against the numbers entering the stage, we find that the regression coefficient indicates the nature of the density-dependence. If the regression coefficient is one, the factor compensates completely for any change in population density, if it is less than one the factor cannot compensate completely for such changes, and if it is greater than one, the factor overcompensates. The closeness of the coefficient to one is a measure of the factor's stabilizing efficiency.

Varley and Gradwell's method gives even more information about the role of population factors when the k value is plotted against log initial density, and the resulting points joined in their proper time sequence. Direct density-dependent factors give a pattern which is a fairly straight narrow band; delayed density-dependent factors circles or spirals; density-independent factors irregular or zig-zag patterns. Herein lies one of the great advantages of the Varley and Gradwell method: there is no other convenient way of demonstrating delayed density-dependent effects.

Luck (1971) compared Varley and Gradwell's method with that of Morris on a simulated life-table which was constructed with mortalities with known modes of action. His analysis showed that while Morris's method detects mortality variations between generations, it was not able to distinguish the density relationships. Varley and Gradwell's method, on the other hand, was able to detect the density relationships of the model.

Watt (1968) also criticizes the use of multiple linear regression analysis for the description of highly complex ecological systems. He points out that what is really needed is a means of describing the ways in which a group of factors can interact and he supports Mott's (1967) suggestion that all possible pairs of interactions between independent variables should be treated as if they are, themselves, additional independent variables. He is also concerned that the ease of calculation furnished by modern computers will encourage research workers to go on what he calls 'fishing trips', searching for possible independent factors which might be significant. Such significance would not necessarily mean that the factor is important. It may simply be highly correlated with a factor that is; relative humidity and rainfall, or rainfall and cloud cover are two such possible pairs. Furthermore, such analyses give no understanding of the mechanics of the relationship, although they may be of value for prediction.

Watt's most telling criticism, however, is that there are many relationships in ecological and other biological systems which cannot be forced to fit a straight line, whatever the transformations that are used. Many relationships which are asymptotic cannot be expressed rectilinearly, and these are all too common in ecological systems.

In such cases the worker is forced to use curve-fitting methods if his equations are to be realistic (Bliss, 1970). Unfortunately, such methods involve extended reiteration, computing and recomputing until estimates of population parameters are obtained which fit the data sufficiently well. This is extremely tedious but it is now feasible as computers may be used for all the routine calculations. Watt (1968) gives a brief description of curve fitting and of methods for maximizing the rate of convergence of the best parameter values.

Watt (1966) approaches the problem from the viewpoint of the systems analyst. He gives, as a suitable operational definition of a system, 'An interlocking complex of processes characterized by many reciprocal cause–effect pathways.' Thus the system should be studied as a whole, with each step of the research so planned that its findings can be fitted together with all similar results at the end of the study. An underlying belief in the approach is that extremely complex processes can be broken down most easily into a large number of small simple components, rather than into a small number of complex units. Consequently, when describing population dynamics systems Watt attempts to break down the model into sub-models and sub-sub-models which are amenable to study, and finally to construct the total model from its constituent parts and sub-parts.

A further concept is that complex historical processes, such as those of population ecology, in which all variables change with time, are best expressed in terms of recurrence equations in which the state of the system at time $t+1$ can be given as a function of its state at time t. An extension of the approach enables us to deal with situations in which the state at time $t+1$ is not only a function of that at time t, but also of those at time $t-1$, $t-2$ and so on. Matrix notations and the ability of the computer to perform the necessary computations are of great value here.

A further underlying concept of systems analysis is feedback, and the relevance of this to ecological systems will be appreciated when density-dependence is considered. Other important features of systems analysis are discussed in the paper just cited, and examples of the applications of the techniques are given in the book to which it forms an introduction (Watt, 1966). Further examples will be found in the text edited by Waterman and Morowitz (1965). Finally, it should be remembered that systems analysis has been developed in many fields ranging from business management, military organization, and economics to physiology, and the findings and techniques of these apparently unrelated fields may be of great value to the ecologist.

In Watt's approach the research worker again starts with life-table data, cast in the form of eq. (2.7), but with additional terms for the proportion of females which oviposit, and for mean fecundity. His goal is to produce an equation in which the trend index is expressed as a function of the independent variables which influence it. The equations describing the survival rates for the various stages form the sub-models; these, in their turn, are composed of sub-sub-models each of which describes some component, that is, the influence of one or more of the independent variables. It must be remembered that an independent variable may appear in some or all of the sub-models, and, indeed, since several of them may affect the survival of the subject species in several ways, they may appear more than once in the same sub-model, or even sub-sub-model. Temperature, for example, affects the rate of metabolism and growth of the subject species, and of various natural enemies, but at extreme values it may have a deleterious effect on some or all of these.

The raw data for each sub-model are best entered on to punched cards to facilitate sorting, an essential because of the large number of observations needed for a reliable model.

The first step is the examination of the roles of the independent variables in the sub-model to see how this is best split into sub-sub-models. If, for example, two synchronous factors are being considered it must be decided if the probabilities of their having an effect are independent. If so the probability of neither event occurring is the product of the two probabilities of non-occurrence. If the two factors are mutually exclusive, the probability of neither having an effect is one minus the sum of the individual probabilities. In other cases the two factors under consideration may operate one after the other, with or without an overlap in time. Other applications of probability theory will then have to be used to calculate the combined probabilities.

Such reasoning may lead the ecologist to split a sub-model into a form such as:

$$S_i = A.B.C+D \qquad (2.10)$$

where S_i is the survival rate of stage i, and the independent variables A, B, C, and D

signify the constituent sub-sub-models. These, in their turn, may be written as functions of one or more of the ultimate independent variables, for example

$$A = f(X_1, X_2, X_3) \qquad (2.11)$$

If the ecologist has no hypotheses about the causal relationships within his sub-model he must embark on the construction of an inductive model. The techniques to be employed are described in Watt's important 1961 monograph, and in his recent text (1968), and, in summary, in Southwood's book (1966). Only a brief outline of this, and of his methods of constructing inductive–deductive models, will be attempted here.

The cards are sorted for each independent variable in turn, ranked, and divided into batches of fifty (it is presumed in the following example that there are 500 cards in all). The mean S value and the mean magnitude of the independent variable are determined for each batch, and the resulting values plotted against each other. This grouping gives a convenient number of points to plot, and helps to smooth out the curve. If the factor is not correlated with any of the other factors each of these will have approximately the same mean value in each group, and will thus have no influence on the mean values of the dependent variable, the survival rate.

Conventional statistical analysis of the data will then indicate which of the independent factors is the most important.

The next step is to consider the combined effect of the important factors on the survival rate. The cards are ranked in the order of magnitude of the most important independent variable, and again divided into batches. Each of these batches is then ranked with respect to the second factor, and then divided into sub-sub-batches. The process is repeated for the final factor so that, at the end of the sorting, the 500 cards are divided into 100 ranked sub-sub-sub-batches, each of five cards. The means of the dependent variable, and of the independent variables, are then calculated for the sub-sub-sub-batches, and a family of graphs constructed, showing how the dependent variable varies with the three factors. Each of the family of graphs is constructed with the survival plotted along the Y-axis, and the level of the most important factor along the X-axis. A separate line is plotted for each mean value of the second most important factor. The whole graph is repeated for each level of the third factor. Sometimes the mean value of, say, the third factor in a given sub-sub-sub batch in one sub-sub-batch, differs markedly from the corresponding sub-sub-sub-batch means in other sub-sub-batches. In such cases ranks would be used in place of means.

The set of curves is now examined to determine in what way, if any, the variables should be transformed for further analysis, and, if possible, to find a mathematical equation to describe their relationships. One possible method is to find transformations which will give a rectilinear relationship, and then to use normal multivariate regression analysis to find the best parameter estimates for the straight line. Watt's objections to the ecologist, aided and almost abetted by the computer, using these techniques uncritically, have already been mentioned and, consequently, his 1961 monograph outlines other mathematical devices for curve fitting.

The fault with the purely inductive approach is that it produces an arbitrarily chosen equation which accounts for the variation in the experimental data. A purely deductive model in which certain assumptions are made, and the consequences deduced, usually in the form of a set of differential equations, is just as unsatisfactory as the model is only as good as its foundations, namely the initial assumptions. The best approach is a mixture of induction and deduction in which a mechanism is suggested, and the deductions derived from it tested against the observations. The model is refined, on the basis of the observations, and the cycle repeated. Much of the value of Watt's work is that he provides a workable operational scheme.

Additional data is needed to construct a model in this way, namely the apparent mortality caused by the significant factors. The apparent mortality due to a factor is the number dying as a proportion of the numbers entering the stage, as compared with real mortality which is calculated on the basis of the population numbers at the beginning of the generation. The apparent mortality is graphed against the magnitude of the factor causing it, for each factor (population density of parasite or predator, maximum rain drop velocity, and so forth). It is now necessary to determine, from the course of the curves, the relationship between the factor and the mortality it causes. To the mathematician this is largely intuitive, but with an intuition based on much experience. The ecologist is usually at a loss at this stage but Watt has published a key, similar to those used for the identification of organisms, which leads to the choice of the most appropriate equation, in a differential or partial differential form. The number of possible equations is, of course, extremely large, but Watt lists the thirty-two which have proved to be the most commonly useful in biological applications. The logical tree is used simply, by answering in sequence five questions about the way in which the two variables change together (Watt, 1961, 1968; Southwood, 1966).

Once the equation has been chosen it is integrated and transformed into a form suitable for testing graphically. Whenever possible a transformation which would give linearity is used. If the data fit the equation the ecologist proceeds to the next stage: if not, the equation is modified and the cycle repeated. Once a satisfactory fit has been obtained the other factors are incorporated into the model, sequentially, again using the above method for choosing the appropriate relationship.

Once the final equation has been chosen for the sub-sub-model the parameters are estimated, using appropriate computer programs. The observed data and the equation are then tested statistically for goodness of fit. Unfortunately, the sub-sub-model is now usually so complex, with several parameters contained within it, that unless there are many degrees of freedom available for the test of significance, it becomes very difficult to demonstrate a significant difference between the observed data, and those predicted from the equation. In other words, the observed data may well fit even when the equation is not completely satisfactory. This is a consequence of the loss of one degree of freedom for each parameter estimated from the data. There are various ways around this difficulty. The data may be obtained from a number of observation sites so that there are several places-within-year replications available for model-building and testing. Alternatively, one set of observations may be used to develop the model, and another to test it.

Although this outlines the main steps in the production of the sub-sub-model, using Watt's techniques, further refinements are possible, details of which may be found in the publications cited.

Finally, when all the sub-sub-models have been constructed, they are assembled into sub-models, and these, in turn, into the full model describing and explaining the variation in population trends from year to year, and from site to site within a year.

Experimental component analysis

The model building described above concerns systems for which a large body of information has already been collected. Despite the complexity of the resulting mathematical models, they may not be detailed enough to allow us to manipulate the system in the way we desire, and for this we obviously need more detailed models for ecological processes such as predation and epidemics of disease. The necessary information cannot be obtained from field observations alone so the analyses depend heavily upon experimentation. This is closely interlocked with hypotheses building and testing, using computer simulation, so that finally mathematical models are obtained which are biologically realistic, and which give satisfactory statistical goodness of fit to observations and experimental data. The models must also be general: in predation component analysis, for example, a model is sought which applies to all cases of predation.

Holling has been a leader in this approach which he has called experimental component analysis. It will be appreciated that this method differs fundamentally from life-table analysis in that an attempt is being made to construct an holistic model by building it up from its component parts. Ideally, as Holling points out, such studies should precede the gathering of data for life-table studies as they indicate the features of the system that should be concentrated upon.

Holling has concentrated most of his energies so far upon predation. This was a happy choice because of the vast amount of data published in the literature, and because a successful component analysis of predation indicates the most desirable characteristics for a biological control agent.

The act of predation is composed of a number of component processes, some of which are basic in that they are a part of all cases of predation, and some of which are subsidiary, being peculiar to certain examples of predation. The components of predation (or of some other ecological process) are determined by studying the literature, from general observation, and by experimentation. An experimental situation is selected which contains only the basic components, and, using the technique outlined above, a model is constructed. This is then modified experimentally to include the various subsidiary components, and their interactions with other components, in turn, until, finally, the complete analysis has been achieved.

Holling concluded that there are five groups of variables which affect predation, namely the population density of the prey, and of the predator, characteristics of both the species, and of the environment.

Prey density and predator density are both basic in that they affect all kinds of predation. The total number of prey individuals killed is the product of the mean number killed by a predator and the number of predators. The response of a predator species to a change in the number of prey can be either a change in the number of prey individuals killed by a predator or a change in the number of predators. Actually, both responses will occur in most cases, but the change in numbers (numerical response) is largely the outcome of the effects of food consumption on reproduction, movement, and mortality. Thus the functional response of a predator to changes in prey density is of basic importance, and has therefore been studied first by Holling.

The effects of prey density on functional response of the predator have the following components which must be included in the final model: the rate of successful search by the predator; the time the prey is exposed to the predator; the handling time; hunger; learning by the predator; inhibition by the prey. The first three are clearly basic; the last three are found only in some situations and are thus subsidiary. Similarly, the effect of predator density has the following components: exploitation (basic: functional response declines as more predators exploit the prey); interference between predators (basic); social facilitation (subsidiary: individuals stimulate each other's response as density increases; hunting in packs, etc.); avoidance learning by prey (subsidiary: increases with increased contacts between prey individuals and predators).

The components listed above may be subdivided to sub-components, some of which are basic, and some subsidiary. Thus the rate of successful search depends upon: (a) the maximum distance from a prey individual at which the predator will attack; (b) the speed of the predator relative to that of the prey; (c) the capture success, that is the proportion of the prey 'at risk' that is successfully captured. Similarly, the time the prey is exposed to predators is composed of the time spent in feeding activities of various kinds, and the time not so spent. The time spent handling the prey is composed of the time spent in pursuing and killing the prey, the time spent in eating it, and the digestive period during which the predator is not hungry enough to attack (Holling, 1966).

Holling's procedure, as stated before, is to find an experimental situation in which only the basic components are present, and then by experiment and hypothesis building and testing, to produce an equation which precisely describes the situation. After this has been completed other experimental situations are studied so that the subsidiary components can be built into the model. It is clearly difficult to find a natural situation in which the basic components only are found, but an artificial situation was created which served to provide the necessary data. A blindfolded subject took the part of the predator, and the predation consisted of searching for, and removing, sandpaper discs tacked to a table. The only components are therefore the rate of search, the time of exposure of 'predator' and 'prey', and handling time after discovery.

This experiment led to the following mathematical model:

$$N_A = a(T_T - T_H N_A)N_0 \qquad (2.12)$$

which may be rewritten as:

$$N_A = \frac{aT_T N_0}{(1 + aT_H N_0)} \qquad (2.13)$$

in which N_A is the number of prey attacked, a the rate of successful search, T_T the time the prey are exposed to the predator, T_H the time spent in handling each prey (pursuit, capture, killing, and eating), and N_0 the prey density.

Holling next extended his analysis to include the subsidiary component, hunger, calling the resulting model the 'invertebrate' functional model. He showed that it describes the responses of all of the invertebrate responses (eleven different species) he used in his experiments. His 1966 paper describes, in the main, the incorporation of the hunger component, using, as experimental animals, the mantid *Hierodula crassa*, *Drosophila* spp. and the house-fly, *Musca domestica*. This is clearly a case of 'ambush predation' so that the subsidiary components of speed of the predator and speed of the prey are unimportant.

Hunger is most conveniently defined as the weight of food, in grammes, needed to satiate a mantid. Holling measured how hunger varies with the time that a mantid is deprived of food after complete satiation. Hunger rose rapidly at first, then levelled to a plateau, the descriptive equation being found to be:

$$H = 1 \cdot 00 \{1 - \exp[-(0 \cdot 0464 \ TF + 0 \cdot 0670)]\} \qquad (2.14)$$

where H signifies hunger, as defined above, and TF the time of food deprivation, in hours. Experiments with *Mantis religiosa*, and the analysis of data supplied on *Phormia regina*, feeding on sucrose solutions, produced similar equations but with, of course, different parameters. Examination of published curves on locusts, a salamander, and rats indicated that the equation form may have even wider applicability.

Having expressed hunger as a function of time of food deprivation, Holling investigated how it interacted with other components, and expressed these interactions mathematically. With increasing hunger the maximum distance at which a mantid would stalk or strike at flies increased, but no reaction at all was obtained for the first eight hours after satiation. Thereafter, the distance increased with time, in a curvilinear fashion. When the distance was plotted against hunger, however, a straight line fitted the data very well, after the initial threshold. Thus, the following equation could be written:

$$r_m = GM(H - HT) \qquad (2.15)$$

where r_m is the maximum distance at which *H. crassa* stalked or struck at a fly, in centimetres, H the hunger, HT the attack threshold (in this case, $0 \cdot 369$), and GM the slope of the line ($11 \cdot 73$).

This example must serve to illustrate how the interaction of hunger with the various components is studied. Holling's original papers should be consulted for details of the procedures used with the other components. At the beginning of the work it was hoped that the effects could be finally incorporated into a differential equation which could be integrated to a form which would give an estimate of the number of prey attacked at different prey densities, but this soon proved impossible. The expression became far too complex and, furthermore, expressing the results in such a form would have made it difficult to incorporate expressions for further components later. Calculus is not, in any case, the ideal language for the description of the events which occur in predation, as Holling points out. The events are basically discontinuous, with relatively few contacts between the predator and prey, and each time a mantid consumes a fly there is an abrupt change in the condition of the mantid in that its hunger level falls. Calculus is not suited to dealing with thresholds (for example, hunger threshold) and, finally, the historical element is an important feature of predation, current events depending on both present conditions and past conditions. It was for these reasons that Holling and his colleagues turned to an alternative mathematical language more suitable for the description of such events, namely Fortran and this, of course, led to the use of computers. Fortran, and similar computer languages, can handle cyclical events easily and succinctly, so that the repeated sequences of prey detection, capture, handling, digestion, and so forth present no difficulty. The IF statement deals effectively with thresholds, and can also be used to differentiate between daylight hours when mantids are active, and the night. For a brief discussion of the applicability of Fortran to ecological problems the reader should consult section 8.2 of Watt's textbook (Watt, 1968). Computer simulation is also discussed by Garfinkel (1965) in a book devoted to theoretical and mathematical biology. Applications of a more recent language than Fortran, namely S/360 CSMP, are given in a paper by Brennan and his colleagues (1970).

The key reference to Holling's experimental component analysis is his 1966 monograph already cited, but the following are also important: Holling, 1961, 1965, 1968.

Community stability

Many ecologists have remarked that the more complex an ecological community is, the more stable it is. There are fewer wide amplitude fluctuations of species in tropical forests, for example, than in the fauna and flora-impoverished arctic regions. The concept is at least as old as the Victorian philosopher, Herbert Spencer (the concept of stability of the heterogeneous state), and it has been eloquently expanded more recently by Elton (1958).

Complexity only exists, of course, if there is an intricate web of interrelationships between the member species. Margalef (1957) has expressed this by stating that the greater the number of avenues by which energy may flow through a complex system, the more stable that system is. MacArthur (1955), using the modern technique of information theory, reached similar conclusions.

Community stability is, of course, of great interest to applied biologists, for one aim of pest control is to change a population of a pest that displays wide fluctuations in density, to a population with a lower mean density, and with dampened fluctuations. It is true that some insecticidal treatments have had the opposite effect, presumably by simplifying the community but, nevertheless, the principles are of importance in biological control, as will be explained below.

It would seem that an animal that has a large number of natural enemies, many of which are present at one time, would be stable in numbers. Conversely, we would expect that an unstable pest would have few natural enemies. Field observations show that these assumptions are often incorrect. Locusts, for example, have a large number of natural enemies – so many in fact that some ecologists have expressed surprise that they can persist at all! Similarly, many of the forest defoliating caterpillars are attacked by a wide range of other organisms. As Watt (1965) has put it, this 'raises the question as to whether such pest species have a wide variety of attackers because they are so productive and unstable, or whether they are so unstable because they have such a wide variety of attackers'.

Watt analysed a large mass of data collected by the Canadian Forest Insect Survey, and from this work and from the study of results of other workers, he put forward the following hypotheses; stability at any trophic level increases with the number of competitor species at that level, but decreases with the number of competitor species that feed upon it. It also decreases as the proportion of the environment that contains useful food increases. A pest of an agricultural crop, or a deciduous forest (i.e., in a more or less pure stand of plants) is thus likely to fluctuate, particularly if there are few competing species. If there are many natural enemies of the pest in competition for it as a source of food, then this entomophagous level is likely to be stable, and unable to expand in numbers if the pest population increases rapidly. Thus a pest population with a large number of natural enemies attacking it is likely to escape from their regulation, if Watt's hypotheses are valid. Watt's findings are in accord with those of certain other ecologists, such as Zwölfer (1963), and Turnbull and Chant (1961). Zwölfer studied six species of Lepidoptera, each of which is attacked by one to three specialized parasites. Two of them are also attacked by a number of other parasites, so that there is considerable competition. The populations of the four species attacked by three or fewer well-synchronized parasites are much more stable than those of the two species attacked by a variety of parasites.

This is clearly relevant to the practice of biological control. The question often arises whether it is better to release a single effective predator or parasite, or a number of parasites and predators. Turnbull and Chant have suggested that before a release is made there should be a careful prescreening of the candidate control agent. It is here that the work of Holling will be of great value. Of equal importance is the development of methods for assessing the effects of the agent after release.

A further problem is raised by these studies of stability, namely whether or not it would be worthwhile to release a natural enemy of a pest which is already being subjected to the attentions of a number of parasites and predators. It could, in fact, add to the instability of the pest. Nevertheless it would be of value if the introduced species did not come into direct competition with the species already present. In other words, it should fill some functional niche not already occupied, by attacking, for example, some stage of the pest not attacked by other parasites or predators. Alternatively, it may do this more effectively than the species already occupying the niche (Turnbull and Chant, 1961; Zwölfer, 1963; Watt, 1965, 1968).

Conclusions

This chapter has become much longer than was originally planned, but with good reason. Pest control has, so far, been an unscientific procedure, and this has largely been due to a lack of understanding of population dynamics. It seems to the author that the only way in which such an understanding can be acquired is by the development of mathematical models of population and community dynamics. If this is done ecologists will be in a similar position to the physicists in their relation to their discipline. Unfortunately, ecological systems are far more complex than the systems with which the physicists have had to deal and for this reason the ecologist long had to make do with verbal models. In recent years, however, the development of high-speed computers with large memory stores, and of suitable computer languages, has eliminated this disadvantage, and now models with adequate reality and generality are being developed. This chapter has attempted to outline some of these developments. There is a further advantage in the construction of adequate models, and the use of computers. An ecological study can take many years. In fact it must take many years if a population is followed through a sufficient number of periods of increase and decrease. By the use of computer simulation methods, and Monte Carlo techniques, experiments can now be carried out on a computer, if an adequate mathematical model is available, in a matter of a few hours. Furthermore, the effects of various strategies, such as the introduction of a predator or parasite, or the application of an insecticide, can be studied in a short time. An additional advantage of computer experimentation of this type is that strategies which might destroy the natural system, or lead to a pest outbreak, can be tested without ill effects. We shall return briefly to this topic in the final chapter, but first it is necessary to discuss the methods of pest control now available.

It must be stressed that mathematical modelling is in its infancy, but there is already ample evidence to indicate that it will be paying substantial dividends within the next few decades.

3

Chemical pesticides and their uses

Toxic chemicals are still the main defence against pest attack and they are likely to remain so for many years to come. This is not to say that alternative methods with less danger to the public do not exist, and which are also less disruptive in their ecological effects. There are many such alternatives to chemical control, and the greater part of this book is devoted to them. Unfortunately, many of these methods are fairly specific, each particular technique controlling only one or two pest species. Often, too, their research and implementation is costly, although once established they cost little, unlike control by chemical pesticides, which usually have to be applied regularly, season after season. But because they are expensive at the beginning, and need the talents of many kinds of specialists, the work is usually carried out by government agencies and similar bodies. Naturally, the most pressing problems are tackled first and, meanwhile, we have to contend with the less important pests (less important, that is, from the national point of view, but extremely important to the individual growers) and with those major pests which have so far resisted alternative methods. The wide spectrum of use of most pesticides is, in many ways, an advantage, but as will be seen later, it is also their undoing.

Finally, even when alternative methods are available, they are often not sufficient to give a satisfactory control, and have to be combined with some applications of chemicals.

A chemical pesticide may be defined as a substance which kills harmful organisms which are free-living at some stage in their life cycle, and which is usually applied during the free-living period of the pest's life. They are thus distinguished from those drugs which are used to kill organisms which are endoparasitic throughout their life cycle. The definition follows naturally from the definition of a pest given in the first chapter. Sometimes, however, a pesticide may be directed against a stage of the pest which lives within the victim or its tissues.

Chemical pesticides are primarily classified by referring to the type of pest they control, and thus we speak of insecticides, acaricides, nematicides, rodenticides, fungicides, and weed-killers or herbicides. Some chemicals fall into more than one of these categories, parathion, for example, being an insecticide, an acaricide, and a nematicide.

It is not difficult to list all the characteristics we would wish for in a pesticide, but it is rare to find one that comes near to the ideal. It should kill pest organisms efficiently without being markedly toxic to other species. At the same time it must kill a worthwhile number of different kinds of pests if a manufacturer is going to find it profitable to market.

Damage to the victim is unacceptable. Some pesticides are phytotoxic, some taint foodstuffs, others stain materials. Benzene hexachloride when first introduced as an insecticide frequently gave to such crops as potatoes a most unpleasant earthy taste. This was caused by isomers other than the insecticidally active γ isomer.

Speed of action is not usually important, unless damage must be checked quickly. This would occur, for example, when an aphid vector of a plant virus has to be controlled before the disease is widely disseminated, or if a crop such as apples has to be protected from blemishes a day or two before harvest. It does play a part in the acceptance of a product by the consumers. A housewife expects an aerosol spray to knock down flies quickly, and a farmer may suspect the efficiency of any material which does not produce a good kill within a few days.

Whether or not a product should be persistent depends upon its use at the time. It should not be so stable that it leaves undesirable residues in food or soil, but it must remain long enough to control the pest. In general, there is a swing towards pesticides with a short half-life, especially among the insecticides, but in some cases, such as when house spraying is used in malaria control, great persistence is essential.

Stability in storage is, however, necessary. This problem is made more complicated by the necessity of formulating the material in some way. The active ingredient is very rarely used in the pure form, or merely mixed with water, but is almost always mixed with other materials that facilitate its dilution, application, and retention. The resulting product must be stable both chemically and physically over a wide range of storage conditions, which vary from those found in a rough hut in the tropics, to those of a modern air-conditioned storehouse.

Finally, a pesticide must be reasonably cheap to manufacture. Fortunately, few products are made in isolation. Pesticide manufacture is usually a small section of a large chemical company with by-products of one division being used as raw materials in others. Such integration allows the costs to be spread more widely.

Toxicity

It is common knowledge that people differ from one another in their susceptibility to a toxic substance, whether it be alcohol, nicotine, strychnine, or any other poison. Populations of other organisms vary in a similar way, some individuals being very susceptible while others are tolerant to a particular poison. If the numbers of organisms just killed by each of the various dosages are plotted against the logarithms of the dosages, a curve of the kind shown in Fig. 3.1 usually appears. This curve approximates to the well-known normal distribution curve of the statisticians, since susceptibility to a poison is randomly distributed in most populations. The dosage is expressed as its logarithm along the abscissa, since a biological response changes in magnitude in proportion to the logarithm of the stimulus.

It is obviously extremely difficult (if not impossible) to find the exact dosage which kills an individual, as the poison must be applied in increasing amounts by discrete steps. All that can be said of a value obtained in this way is that it is certainly not less than the exact lethal dosage, and is almost certainly somewhat greater. Nevertheless, a close approximation to the lethal dosage can be obtained in this way,

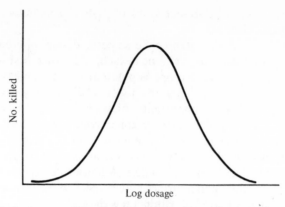

Fig. 3.1. The distribution in susceptibility to a toxic substance in a population of organisms. The 'No. killed' signifies the number of organisms killed by exactly the given dosage: it does not include organisms killed by any smaller dosage

but the method is time-consuming and is rarely used unless the test organisms are scarce.

An alternative method is to apply several fixed dosages (usually increasing geometrically to facilitate subsequent computations) to groups of randomly selected individuals. The responses to the various dosages may be assessed by taking some quantitative measurement such as the time taken by each individual to die, or some quantal measurement – the numbers dead in each group, and the numbers alive, or the numbers displaying some other 'all or none' response. Such a quantal measurement is usually made when the experiment is believed to have reached a steady state, that is, when the poison will have no further effect, but before natural mortality becomes significant. In insecticide bioassays this is often taken to be 24 or 48 hours after application.

If the percentage kill for each dosage is plotted against the logarithm of the dosage a curve of the kind seen in Fig. 3.2 results. If, as was suggested above, the dosages are increased geometrically, they will be spaced at equal intervals along the abscissa. The sigmoid curve which is obtained is, of course, the cumulative frequency curve corresponding to the normal curve shown in Fig. 3.1. An examination of the curve shows that the dosages corresponding to 100 per cent kill and to no kill cannot be read from the graph, and those corresponding to 95 per cent and to 5 per cent cannot be read with any accuracy. The most reliable dosage will be that corresponding to 50 per cent kill, hence the most commonly met measure in toxicology, the Lethal Dose 50 or LD 50. This is the dosage which, when administered to each of the individuals in a population, would kill half of them.

In an actual bioassay the experimental points do not fall neatly on to a curve of the kind shown in Fig. 3.2. It is the biologist's task to find the curve which best fits his imperfect data. It is difficult to fit data to a curve of the kind shown in Fig. 3.2, so a transformation which places a set of perfect data on to a straight line is used. The appropriate method is to replace each percentage by its corresponding probit value. Tables of probit values will be found in standard collections of statistical table, or they may be calculated from tables of the areas of the normal curve. The most convenient table is one which tabulates deviations from the mean against areas between $-\infty$ and the deviation. The proportional area which corresponds to the

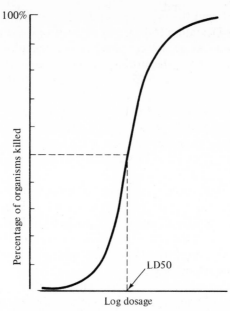

Fig. 3.2 The cumulative distribution of susceptibility to a toxic substance in a population of organisms. The percentage killed includes the individuals killed by smaller dosages than that indicated

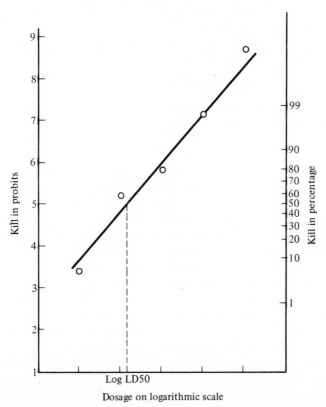

Fig. 3.3. Log dosage response line, fitted to data after the transformation from percentage kills to probits

percentage kill is found, and the deviation read. This deviation, which is known as the Normal Equivalent Deviation (NED) will be positive if the percentage is greater than fifty, and negative if less, so, for convenience in calculation, five is added to the NED to give the probit. Thus the probit for 40 per cent is $5+(-0\cdot25) = 4\cdot75$, and for 60 per cent, $5+(+0\cdot25) = 5\cdot25$. The probits for 50, 0 and 100 per cent are $5\cdot00$, $-\infty$ and $+\infty$.

The line which gives the best fit for the experimental data, $y = a+bx$, is now computed from the transformed data, using a modified regression technique. In the equation y represents the probit kill, and x the long dosage (Fig. 3.3). The calculations form a part of the section of statistics known as probit analysis (Finney, 1952; Busvine, 1957).

In addition to the computation of the line the calculations also give confidence limits for the estimated dosages corresponding to various kills, and a means of testing the homogeneity of the samples of organisms used in the bioassay. The narrowest confidence limits are those of the LD 50, as would be expected from the shape of the sigmoid curve. The probit transformation merely facilitates computations; it cannot make the raw data any more accurate.

A useful device is to make those experimental groups which are expected to receive dosages near the LD 50 larger than the other experimental groups for in the full probit analysis an adjustment is made for the size of the sample. This modification gives a more accurate estimate of the LD 50 of the whole population.

LD 50s of various pesticides for both pests and vertebrates are widely quoted in the literature as an indication of their effectiveness and their potential danger to man and wild life. The route of entry (through the mouth or skin, by injection, etc.) is usually stated, and since there are often differences between males and females the sex is generally specified. The dosage may be expressed as milligrammes of toxicant per individual or per kilogramme of body weight, or it may be expressed, where appropriate, as a concentration.

By itself the LD 50 has only a limited use as a guide to the safety of a product. Much depends on the age (physiological or absolute) of the test organisms, their state of nutrition, the way in which they have been reared, and the environmental conditions at the time of application and during the holding period before the mortality is assessed. Furthermore, bioassays are usually carried out with highly purified active ingredients which are often suspended or dissolved in some simple solvent such as acetone or water, whereas practical applications involve highly formulated products. The additives may increase the toxicity markedly. Oily formulations of chlorinated hydrocarbons, for example, have greater dermal toxicity to mammals than do the active ingredients.

The slope of the probit line is an important statistic which is not indicated by the simple LD 50. It is, of course, a measure of the variance of the lethal dosages of the toxicant to the population. The smaller the value of b in the calculated log probit line, the shallower the slope, and the greater is the variation in the susceptibility of the individuals of the population to the poison. When considering the killing of a pest population a pesticide with a shallow log probit line is at a disadvantage because a dosage considerably higher than the LD 50 would have to be used to achieve a reasonable kill. If the slope of the line for mammalian toxicity is also shallow then a dosage much smaller than the LD 50 could kill susceptible individuals. It is to be regretted that pesticide manufacturers rarely quote more than the LD 50 in their technical literature. An estimate of b or of the LD 90 and the LD 10 would increase the value of these documents immensely at very little cost to the manufacturer.

Experimentally obtained LD 50s with man as the test organism are not, for obvious reasons, available. Toxicity to man can only be assessed from cases of accidental fatality, and from data obtained for other mammals. Unfortunately, there are important differences between the various species of mammals in their susceptibilities to various poisons. If, however, a poison has similar LD 50s for a well-chosen range of mammalian species it can usually be assumed that the LD 50 for man will be of the same order.

These disadvantages of the simple LD 50 are quite unimportant compared with its main drawback. The LD 50 is an estimate of the acute toxicity only, that is the toxicity experienced from a single application. Many poisons are cumulative in that their poisoning effect builds up with a long series of small doses each of which, by itself, would be harmless. Possibly the best-known example is lead poisoning, once an important disease of plumbers. An allied problem is the possibility of a pesticide being a carcinogen. The virtually compulsory testing for both these dangers entails many months, or even years, of trials with consequent increased development costs.

A phenomenon frequently met with is potentiation. If two chemicals which are administered together, or within a short period, produce a toxic effect greater than would be expected from their toxicities when applied independently from one another, they are said to potentiate each other. This can be of immense value when the poisoned organism is a pest, as when pyrethrum is potentiated by a synergist such as piperonyl butoxide, but it can be dangerous when the animal concerned is man or one of his domestic animals. The LD 50s of the compounds give no indication of the danger.

In this section the word 'dosage' has been used advisedly. The amount of the substance which is administered to the organism, the dosage, is not the same as the amount which finds its way to the site of the toxic action in the organism, and which thus takes part in the actual poisoning. There is probably not even a linear relationship between the two quantities. It seems worthwhile, therefore, to distinguish between the dosage, as defined above, and the somewhat hypothetical dose which bears a linear relationship to the quantity actually involved in the poisoning. It is, however, conventional to speak of the Lethal Dose 50, rather than of the Lethal Dosage 50.

Mechanisms of toxicity

Apart from the purely academic interest the investigation of the ways in which pesticides actually poison organisms is an active field of research for two practical reasons. An ability to relate chemical structure to mode of action will help in the design of new toxicants, and in the development of rational treatment in cases of accidental poisoning.

Among the numerous pesticides now in use there are many different classes of poisons, each class with its own particular mode of action. Few of these are understood, despite intensive research, and the various theories can only be dealt with very briefly here. Up-to-date summaries will be found in O'Brien (1967) and O'Brien and Yamamoto (1970), and the reader may follow subsequent developments in review publications such as the *Annual Review of Entomology*. A further valuable reference is Martin's *Scientific Principles of Crop Protection* (1964).

O'Brien, Martin, and many other authors divide poisons into two groups, the physical toxicants and the chemical toxicants. The distinction given is that it is only in chemical toxic actions that chemical bonds are broken, or new ones formed. Modern concepts of the nature of the chemical bond blur this distinction, but it may be retained for convenience.

An indication that a group of poisons which arouse similar symptoms are physical toxicants rather than chemical is a closer relationship between toxicity and some physical property such as fat solubility than between toxicity and chemical structure. An excellent example is furnished by a number of narcotics which are used as anaesthetics and fumigants. They include many of the simple paraffin hydrocarbon derivatives and closely related chemicals. The symptoms they produce and their physical properties suggest that they accumulate in some fat-rich part of the nervous system which has been called the biophase. When they reach a certain activity threshold at this site narcosis occurs. For a further discussion of this topic, and for the relevant bibliography, the reader should consult O'Brien (1967).

Most of the poisons which are fatal to organisms in relatively small dosages are chemical toxicants which interfere in some important biochemical process. The biochemical lesion involves some chemical change in which bonds are broken and formed.

It is believed that many chemical toxicants exert their effects by inhibiting important enzyme-substrate reactions. The toxicant occupies the reactive site or sites of the enzyme and thus prevents the formation of the normal enzyme-substrate complex. There appears to be two possible kinds of inhibition: competitive inhibition in which the toxicant occupies the sites normally used by the substrate, and non-competitive inhibition in which the toxicant fits on to other sites of the enzyme. In both cases the normal functioning of the enzyme is interrupted, and since a very small quantity only of an enzyme is normally present, as very little is needed for a rapid turn-over in a biochemical process, a correspondingly small quantity of the inhibitor will inactivate all the enzyme and thus interrupt the reaction.

The theory that the toxicant attaches itself to some complementary site on the enzyme can be extended to attachment to membranes. Electron microscopy has shown the great wealth and importance of membranes within the cell. Besides those which surround the cell and nucleus, controlling the ingress and egress of materials, there is a folded complex within the cytoplasm, the endoplasmic reticulum, and the membranes of sub-cellular bodies such as mitochondria and lysosomes. An attachment of the toxicant to such membranes, or to the surface of neurones, could disrupt some vital process in the body but, unfortunately, such mechanisms are more difficult to demonstrate experimentally than is the inhibition of enzymes.

Selective toxicity

Selective toxicity has been defined as the injury of one kind of living tissue without harm to some other kind with which the first is in intimate contact (Albert, 1968). Selective toxicity is thus not unique to pesticides for it can be found in drugs used to destroy disease organisms, and in those drugs, such as pain killers, which affect cells in the nervous system, but not those in other parts of the body. The definition should be extended for our purposes since in pest control, the two kinds of living matter, for example man and insects, need not be in intimate contact. It is sufficient that there should be a possibility of both coming into contact with the pesticide but that one should be susceptible, and the other not.

If a pest organism and a man are both exposed to a spray of some pesticide the great difference in the sizes of the two makes it likely that the pest will be injured more, for its small size brings with it a much greater surface-area to volume ratio, and thus a much higher dosage in terms of milligramme per kilogramme of body weight.

It has often been suggested that the lipophilic property of many of the modern contact insecticides would facilitate the entry through the insect's cuticle, but that it would not aid it markedly to penetrate mammalian skin. LD 50 estimates seem to confirm this hypothesis. The toxicity of DDT to the rat, by the dermal route, is 3000 mg/kg, and by the intraperitoneal route, 150 mg/kg. The corresponding values for *Periplaneta americana* are 10 and 7 mg/kg. Direct measurements of the passage of DDT through mammalian skin and through insect cuticle show that the insecticide penetrates the two kinds of integument at essentially the same rate. It is possible that the mammal can detoxify the material almost as quickly as it penetrates, but that insects cannot do this, while even the mammal succumbs to a sudden and massive influx by injection. This would account for the different toxicities cited above.

Gerolt (1969) has used autoradiographic techniques to study the mode of entry of labelled contact insecticides and found that little penetrated the cuticle to enter the haemolymph. There was strong evidence that topically applied material spread laterally and entered the insect through the spiracles. The central nervous system is well supplied with tracheoles and thus topically applied material could reach it readily.

If the insecticide can reach the nervous system in such a direct way there will be few barriers on its passage to the sites of action. If it must take some other route (as indeed it must in animals that do not have a tracheal system), then it will pass through a number of membranes whose lipoid content would store a considerable amount of the material.

These and similar mechanisms of selective toxicity are important in practice, but they are probably of less intrinsic interest than those depending upon differences in biochemistry. Organisms differ in the ways in which they metabolize a toxicant that has entered the body. Sometimes the metabolites are less toxic, sometimes more toxic, than the original compound. Many organophosphate insecticides, such as malathion (see below), contain a $P=S$ group which, in insects, is rapidly converted to a $P=O$ group. This change takes place slowly or not at all in mammals. The resulting compound (in the case of malathion, malaoxon) is the actual toxicant which inhibits cholinesterase, a reaction which is believed to be the basis of the toxicity of these compounds. The $P=S$ compound has a low toxicity to mammals, but the $P=O$ compound is quite dangerous, thus the former is preferred as an insecticide. There is a further mechanism which contributes to the selective toxicity of these $P=S$ compounds. Mammals can rapidly de-esterify them to an acid which is quickly eliminated, but insects can carry out this process only slowly.

Selective toxicity can also arise from differences between the targets in two different kinds of animals. The cholinesterases concerned in insects, for example, react more readily with many organophosphates than do those in vertebrates. It is also possible that some organisms can withstand greater damage to the target than can others, but this is more likely to favour insects.

The whole field of selective toxicity, both in pest control and medicine, is well covered in the text by Albert and in insecticides by O'Brien (Albert, 1968; O'Brien, 1967).

The formulation and application of pesticides

Most pesticides are sprayed or dusted, in a diluted form, on to the subjects they are to protect or kill. Others are used as vapours, gases, smokes, or fogs. Apart from the gases, which are usually used in their pure form unless they are inflammable, almost all pesticides are formulated in some way before they are used as control agents. Good formulation is important: it is no exaggeration to say that a pesticide is only as good as its formulation.

Most active ingredients are more or less insoluble or immiscible with water, but are reasonably soluble in organic solvents such as oils and xylene. These, in their turn, are immiscible with water and would soon separate if mixed with water in the spray tank. To get round this difficulty various agents are added to the dissolved pesticide which enable it to form a stable emulsion when mixed with water. These emulsifying agents are surface-active agents (surfactants) which form a 'surface film' on the dispersed oil globules in the water. Most insecticidal emulsifiable concentrates, as the products are called, contain mixtures of non-ionic surfactants such as the polyglycolethers, and anionic surfactants which include dodecylbenzene. These additives can be used in much smaller quantities than soaps which formerly were commonly employed, and which, in any case, form an insoluble scum in hard water.

Cationic surfactants are sometimes used in pesticide formulation, and, furthermore, certain fungicides are themselves cationic surfactants. If these are mixed with anionic agents an insoluble grease is formed and the surfactant properties are destroyed. For this reason pesticides should be mixed with one another only after it has been established that they are compatible.

If the instructions are followed carefully the emulsifiers and other agents added to a particular pesticide should not cause any damage to the object sprayed, but if two products are mixed together in order to control two different kinds of pest at once, there is a danger of damage.

Wettable powders are less likely to be phytotoxic than are the emulsifiable concentrates. In these products the finely ground pesticide and inert diluent are treated with wetting and dispersing agents which produce a fine suspension when the mixture is stirred in water.

Both kinds of spray should give a good coverage when they hit the target surface. The emulsion is designed so that it separates into two phases on impact, the water running away, and the pesticide solution remaining to cover the surface. This is a property at variance with the quality desired in the tank, namely stability, a conflict of interests which makes the formulator's task even more difficult. Coverage is improved by the addition of wetters and spreaders, and persistence of the deposit by the use of stickers. Wettable powders also contain deflocculating agents to prevent the aggregation of particles. Even so the presence of solid particles in the suspension erodes the nozzles and prevents their use in very low volume spraying where nozzle blocking would be frequent (Anon, 1960).

A few pesticides are soluble enough in water, or in a solvent which is miscible with water, such as butyl cellosolve or diethylene glycol, to be made up as solution concentrates. Another limiting factor, however, is the hydrolytic stability of the active ingredient and, in practice, this limits the method to two scarcely used organophosphate insecticides, to the salts of the so-called hormone herbicides, and to the basic diquat and paraquat.

Often it is extremely inconvenient to use water as the ultimate diluent, either because the terrain is too rugged (a spray tank of water being very heavy) or because it is too arid. Powder formulations are of use here but, in general, they do not give as good a coverage as sprays, and they are more difficult to handle in bad weather. It also seems to be much easier to design an efficient sprayer than to produce a good dusting machine. One of the main difficulties is to make a metering device which feeds out the powder evenly.

Pesticide powders are usually made by treating some finely ground carrier with the pesticide dissolved in a suitable solvent. The resulting dust concentrate may be further diluted, or it may be left in this convenient form for shipment, to be followed by dilution with some local inert carrier. The final product should not be too fine or the particles will clump together in the air blast, giving a poor distribution. Conversely, coarse dusts do not penetrate into the interior of plants, and there is also a danger that the particles carrying the pesticide will separate from the remainder. This does not occur if the particle sizes are all less than 20 μ in diameter.

Other requirements are that the product should not cake in humid conditions, or be so hard that it abrades the machinery. Furthermore, it should have the right electrostatic charge. In dry conditions the dust acquires a charge by friction as it passes through the machinery. For example, if it is to be attracted to a leaf surface, which is usually electronegative, this charge should be positive.

Small quantities of pesticides can be applied from aerosols. These produce a very fine droplet size as a liquid of very low viscosity propels the active ingredient through a very fine orifice. The droplets are further broken up as the propellant liquid boils when the jet enters the air. Aerosol cans are expensive, and for safety reasons their size is limited. They are thus used for small-scale applications only – in the house or garden. The foam dispenser, a familiar example of which delivers shaving cream, has been used for herbicides for the spot treatment of weeds in lawns and garden beds.

Some pesticides are sufficiently stable to be used in smoke generators. One potent insecticide, γ BHC, was actually used as a smoke for cover in the front line during the First World War: it is surprising that its effects on lice were not noticed! A smoke generator contains the active ingredient, an oxidant and a combustible material which gives rise to a hot non-inflammable gas when it emerges from the orifice of the container. Sodium chlorate and a solid hydrocarbon are often used in such proportions that the issuing gas consists of the active ingredient, water vapour, carbon dioxide, and a little carbon monoxide. Smoke generators are used in glasshouses, dwelling houses, and stores.

A more recent development is the granular formulation. In these the pesticide is impregnated into particles of an inert substance such as Fuller's earth, vermiculite, attapulgus clay, corn-cob chips, or walnut-shell particles. The particles range in diameter from about 0·5 mm to 1·5 mm. Being comparatively heavy they are less subject to drifting than sprays or dusts, and when applied to a crop many of them can force their way through quite dense foliage to the ground, while others lodge in leaf axils and other crevices. They can also penetrate foliage overhanging waterways when applied from the air. A further advantage is that the active ingredient is released from the granules slowly (the rate depending upon the chemical and physical properties of the granule) so that one application can often replace two or more spraying or dusting operations. Special applicators have been developed to scatter the granules or to drill them into the soil, and they also lend themselves to aerial application.

In agriculture and horticulture pesticides may also be used as seed dressings. This restricts the area which is affected by the pesticide, with a reduction of contamination problems, and a saving in the quantities of active ingredients.

There are a number of other possible means of dispensing pesticides – for example, herbicides may be impregnated in a long cylinder of wax which is dragged across a lawn, insecticides may be mixed in baits or paints, or impregnated in strips of paper for the control of domestic pests. Many of these special techniques provide a product which can be used without any spraying machinery or special applicators, and they are usually intended for small-scale use.

The main task of the formulator is to provide a product which can be easily applied, and in this section we have concentrated on some of the ways in which this is done. He also has many other problems to solve, such as producing a material which can be packaged attractively and cheaply, which will store and travel well, and which poses no fire or explosion risk (Telle, 1970).

There is little space to discuss machinery for pesticide application as this may range from a perforated tin can on a stick, as used in some peasant communities, to the large air-blast machines used in orchards and plantations. Various machines are discussed in the MAFF Bulletin *Farm Sprayers and their Use* (Bulletin No. 182, HMSO, London, 1961), in the World Health Organization publication *Equipment for Vector Control* (WHO, Geneva, 1964) and similar publications. The Imperial College Field Station near Ascot in Berkshire, England, also maintains a collection of equipment suitable for use in developing countries. The MAFF Bulletin cited above also deals with the legal aspects of spraying, the regulations concerning the approval and use of the various pesticides, and the economics of spraying and dusting. Since these vary from year to year, and from country to country, the reader should consult the official publications of the countries in which he has an interest.

In crop and plantation spraying water is used to ensure an even coverage, and also as a visible indication that the chemical is reaching the target. Clearly the larger the volume applied to the acre, the easier it is to ensure a uniform distribution. Large volumes of liquid soak the crop so that there is a considerable loss from 'run-off'. A common practice is to gauge the quantities of water and active ingredient so that run-off is just achieved at the required dosage of active ingredient per acre.

Generally speaking the higher the volume of water used, and the lower the pressure, the coarser is the spray. A coarse spray is easier to control and direct than a fine one – it is possible, for example, to spray seedling weeds between the rows of a crop without any drift to the near-by plants. A very fine spray, on the other hand, can drift many hundreds of yards in the slightest wind, and it is thus clearly unsuitable for a herbicide which could have disastrous effects on a growing crop in an adjacent field.

There is, nevertheless, a continuous move towards low-volume spraying, for obvious reasons of economy. This has been facilitated by improvements in machinery which ensure a constant output, but there are great demands on the skill of the operator. Possibly the extreme has been reached in the United States where undiluted malathion has been used as a spray, and in the Sudan where DDT formulations have been used on cotton at less than one gallon per acre. Both these applications are made from the air using special apparatus to produce the very fine droplets.

A typical low-volume, low-pressure farm sprayer would hold 20 to 50 gallons (91–227 litres) in the tank, and have an output of up to 1363 litres/hr. Such a machine, operated from a tractor power take-off, would apply rates of up to 20 gallons per acre (91 litres per 0·4 ha) when moving at 4 to 6 km/h. A low/medium volume machine would hold 181 litres or above, and apply spray at up to 60 gallons per acre (272 litres per 0·4 ha). A low/high volume machine would be used up to 100 gallons per acre (454 litres per 0·4 ha). Such machines would be used for the application of 'hormone' weed-killers, and occasionally other products. Larger sprayers holding up to 1363 litres and using pressures of 10 to 42 kgf/cm² are used for other kinds of spray materials both on field crops, or in orchards.

All these machines use water as the propellant and it requires high pressures to project the droplets to the tops of standard trees in an orchard. For such purposes air blast machines have been developed in which the product is mixed into a comparatively small quantity of water. This liquid is atomized by a nozzle or a spinning disc, and the spray blown on to the trees by a powerful blast of air. A small motorized knapsack sprayer which works in the same way has been developed, and can be used to apply liquids at a few gallons per acre. Such a machine is, of course, much pleasanter to use than the older types of knapsack sprayer which, when filled, were very heavy, and usually uncomfortable. Unfortunately, they are noisy, but they are now being widely used by small farmers in tropical and sub-tropical regions.

The control of pests indoors and in stored products raises different problems and calls for different formulations and methods of application. Fumigants are widely used for insect control in stores, silos, ships, and rolling stock, and the fumigant is usually released from pressure cylinders. Since most buildings cannot be made completely air-tight best results are usually obtained during still weather. Small amounts of material can be treated by vacuum fumigation in which the pressure is reduced before the fumigant is released. This ensures a rapid and thorough penetration of the fumigant. Some materials can withstand fairly high temperatures so that they can be used as a smoke. A recent technique is the impregnation of the material in a plastic such as PVC so that the vapour is released slowly over a period of months.

Insecticides and acaricides

Insecticides and acaricides may be classified in the way that the chemist would adopt but it is useful also to categorize them by the way they enter the target species. A stomach poison, for example, is one which is eaten by the pest together with its food, and which enters the tissues through the walls of the alimentary canal.

Stomach poisons are usually sprayed on to the surface of the potential victim and are thus usually directed against pests with biting and chewing mouthparts. They are only effective when the surface can be reached by a spray or a dust and, when used on growing crops, they must be applied frequently to protect newly emerged growth. They give little protection against boring and tunnelling species, unless these ingest enough surface material as they enter, or against pests with sucking mouthparts which get their food from within the victim's tissues. The latter can be attacked by a special group of stomach poisons; the systemic insecticides and acaricides, and those with a local penetrating action. A systemic insecticide is one which enters the plant and either it or one of its metabolites is carried to all parts of the plant after application to the leaves, stems, or roots. Some are so effective that they can be applied to seeds or tubers and give protection for several weeks after the emergence of the plants. This is particularly useful when it is necessary to protect the plants from aphid-borne virus diseases just after emergence when it is difficult to apply sprays effectively. Most of the systemic insecticides belong to the organophosphate group though the phenomenon was first observed in some selenium compounds.

Contact insecticides and acaricides need not be ingested by the pest for poisoning to occur. Such toxicants can penetrate the cuticle (or enter the spiracles) of the pests when they come into contact with the material. The pests may absorb the poison when walking on treated surfaces, when flying through a mist of fine droplets, or when hit directly during spraying or dusting. Many of them can kill pests which are exposed on a surface but which have their mouthparts probing deeply within the host plant or animal. They do not come into contact with an organism living within a plant's tissues (though some, such as aldrin and dieldrin can be absorbed by the plant when used as seed dressings), but such pests are often destroyed between the emergence from the egg and entry into the tissues. Alternatively, the free-living, egg-laying adults may be destroyed. A few contact insecticides and acaricides have marked ovicidal action.

Fumigants enter the pest's body through the spiracles. As fumigants are most effective in closed spaces they are widely used for the control of stored-product pests and in public health work. They find some use in horticulture in glasshouses or on orchard trees enclosed in temporary tents.

The above classification is not clear cut. Most systemic insecticides, for example, are also contact insecticides, but the divisions are useful in practice.

The chlorinated hydrocarbons

DDT and related compounds DDT is an abbreviation of dichlorodiphenyl trichloroethane which is more accurately described, chemically, as 1,1,1-trichloro-2,2-di-(4-chlorophenyl) ethane (1). First synthesized in 1874 its insecticidal activity

(1) DDT

was not discovered until 1939 but since then it has rapidly made up for lost time. Although it was used by the Allied troops after 1942, and doubtless saved many from insect-borne diseases, it first forced itself into the news during the threatened typhus outbreak in Naples in 1944. In less than one month one million people were dusted with the insecticide and, as a result disaster averted. Since then its use has increased so that now it is the most used insecticide. This may not continue for much longer; many countries are considering banning or severely restricting its use, for reasons which will be explained in the next chapter. Some countries such as Sweden, Great Britain, Canada, and the USA have already done so. The World Health Organization, however, is opposed to a world-wide ban on the chemical as there is, as yet, no satisfactory substitute cheap enough to use for vector control in warmer countries. Rudd estimated, in 1964, that the annual world production approached 114 000 000 kg, and Moore suggests that the total amount on the surface of the earth may reach one million tonnes (Moore, 1967; Rudd, 1964).

DDT has a wide spectrum of use among the insects, that is, it kills a wide variety of species. It is much less effective against arachnids, especially mites. Its main uses have been against biting and chewing pests, domestic insects, and mosquitoes. Many species have developed marked resistance, some quite rapidly, but besides this there is a wide variation in susceptibility among the various groups of insects, even where resistance has not been acquired. The bed bug, *Cimex lectularius*, non-resistant strain, has an LD 50 of about 10 mg/cm^2 and the yellow fever mosquito, *Aedes aegypti*, 0·0001 mg/cm^2.

Although such a large amount of DDT has been used since the war little is known about its mode of action. This is a very disquieting feature, for we do not fully understand the weapon we have been using. There is a general agreement that it is mainly a poison of the nervous system. It produces symptoms in both insects and vertebrates which indicate this: DDT 'jitters' (tremors) and a loss of the powers of movement in the insects, and convulsions and tremors in the vertebrates. Furthermore, it produces toxic effects in nervous tissue at much lower concentrations than are needed to produce toxic effects in other tissues and enzyme systems.

In neurophysiological experiments it has been shown that DDT has a multiplicative effect on a nervous impulse. When an impulse reaches a treated section of a nerve it initiates a whole train of further impulses. The sensory axons have proved to be tens of thousands of times more sensitive than motor or central nervous axons. The trains of impulses travel to the central nervous system where they are indirectly responsible for the uncoordinated stimulation of the motor nerves.

This is all well established but so far no one theory has satisfactorily explained why DDT is more toxic at lower temperatures. A summary of recent and not so recent theories will be found in O'Brien's text (O'Brien, 1967). Readers lacking a knowledge of the transmission of impulses along nerves and across their junctions (synapses) should consult Katz's *Nerve, Muscle and Synapse* (Katz, 1966).

A number of workers subscribe to a theory, originally put forward by Mullins (1954, 1955) who suggested that a number of molecules of DDT or related compounds fit in a certain way into a nerve membrane pore channel. Their presence distorts the adjacent pores and induces the leakage of sodium ions through the nerve membrane. Recently Holan (1969) has modified this theory to take into account the confirmation that DDT forms a complex with a protein matrix, and he has successfully used his hypothesis to predict the toxicities of several new halocyclopropane insecticides on the basis of their molecular form. These compounds are particularly interesting because the cyclopropane groups are chemically more stable than the trichloroethane group of the DDT molecule, whose place they take. Because of this stability the new compounds are not degraded by DDT dehydrochlorinase, the enzyme which confers resistance to DDT. Holan suggests that some of the applied insecticide (DDT or its cyclopropane relatives) distributes itself at the lipid-protein interface of the nerve membrane. The base with the phenyl ring complexes with the protein while the smaller apex (the cyclopropane ring or the trichloroethane group) fits into a pore in the lipid part of the membrane. Because of the complexing the molecule is locked in position, and the apical group keeps the pore or receptor open to sodium ions. The formation of such a complex which would tend to dissociate at higher temperatures may explain why DDT is less toxic to insects at higher temperatures.

The cyclopropane relatives of DDT are promising insecticides, but at the time of writing they are not generally available. Two established compounds, closely related to DDT are TDE (also known as DDD) and methoxychlor (2 and 3). TDE has a

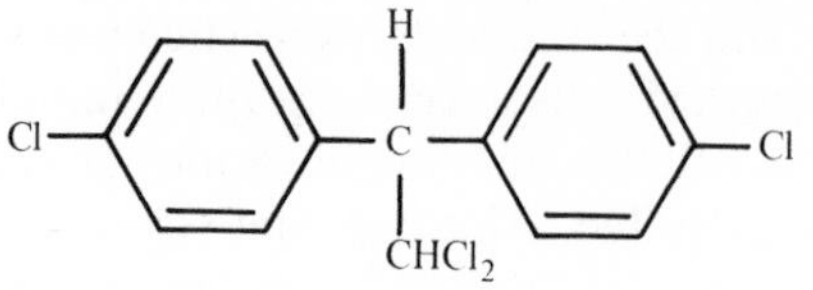

(2) TDE

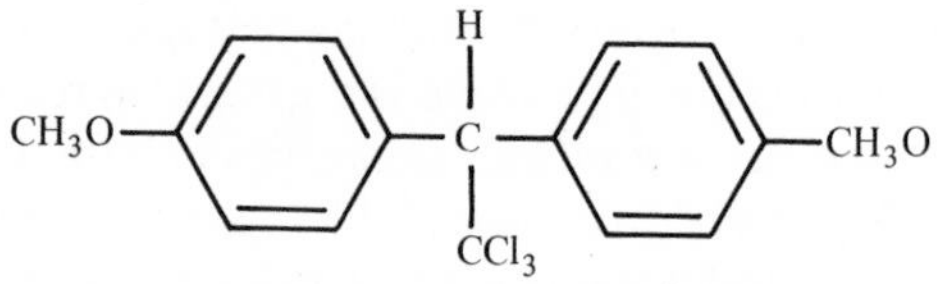

(3) Methoxychlor

narrower spectrum of use than DDT but it is more toxic to certain difficult-to-control pests such as the light brown apple moth, *Epiphyas (Tortrix) postvittana*. Methoxychlor too has certain specialized uses, but its main virtue is its relatively low level of accumulation in the fatty tissues and milk of mammals. These compounds are, like DDT, contact insecticides.

Gamma BHC (lindane) This chemical is the gamma isomer of 1,2,3,4,5,6-hexachlorocyclohexane. Lindane is a preparation containing not less than 99 per cent of this isomer (4). BHC is an abbreviation of benzene hexachloride, a misleading

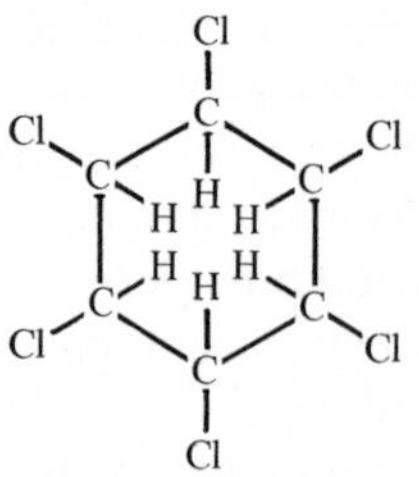

(4) γ BHC

name for the carbon ring is not the flat benzene ring, but the non-planar cyclohexane one. This non-planarity makes possible the existence of eight isomers although only five are known. Of these the γ isomer is the most active as an insecticide.

Gamma BHC has a wide range of action but it is expensive when it is used as lindane while the crude product has a tendency to taint foodstuffs. These faults limit its use in horticulture and agriculture to high-value crops although crude BHC has been widely used in India. It is, on the other hand, very stable when heated, a property which explains its use in smoke 'bombs'. Its main value lies in its control of domestic pests, of ectoparasites and of soil insects such as wire-worms. It is used as a spring spray to control apple aphids, and it is probably responsible for the decline of the apple sawfly, *Hoplocampa testudinea*, as an orchard pest.

As with DDT the mode of action is unknown but like DDT it is probably a poison of the nervous system and produces similar symptoms. There are additional symptoms such as wing fanning in house-flies, and this suggests that the mechanisms of action are not identical.

The cyclodiene insecticides This series of chlorinated hydrocarbons is prepared by the Diels–Alder reaction, a well-known process in organic chemistry. The starting compound is hexachloro*cyclo*pentadiene (5). The isomerism of the resulting com-

(5) Hexachlorocyclopentadiene

pounds is complicated, and the American nomenclature differs from that used in Europe. The British usage is employed here.

Aldrin is the common name of the product containing at least 95 per cent of HHDN, 1,2,3,4,10,10-hexachloro-1,4,4a,5,8,8a-hexahydro-*exo*-1,4-*endo*-5,8-dimethano-naphthalene (6). Dieldrin is the common name of a product containing not less than 85 per cent of HEOD, 1,2,3,4,10,10-hexachloro-6,7-epoxy-1,4,4a,5,6,7,8,8a-octahydro-*exo*-1,4-*endo*-5,8-dimethanonaphthalene (7). Dieldrin is the epox-

(6) Aldrin   (7) Dieldrin

ide of aldrin and in some plant and animal tissues, and in certain soils, the second is partially oxidized to dieldrin. It is perhaps worth noting that dieldrin is pronounced as two syllables, not three, as it is named after the chemist Diels.

Both insecticides have been widely used against a wide range of insects, especially cotton pests, soil insects, grasshoppers, and public health pests. Dieldrin is widely used by the World Health Organization in its efforts to control mosquitoes. Resistance has, however, built up in many strains and the compound is being superseded by organophosphates and carbamates. Aldrin is less persistent than the more toxic dieldrin as it is a more volatile compound. Part of the disappearance is due, however, to epoxidation.

Following the widespread death of birds in Britain during the early 1960s, part of which could be atributed to aldrin and dieldrin poisoning (see the next chapter), the uses of these chemicals have been limited in the United Kingdom by the agreement under the Pesticides Safety Precautions Scheme. The main uses are now against wire-worms in potatoes, cabbage root flies, and as a seed dressing on wheat sown up to the end of December, but only if there is a real danger of an attack by the wheat bulb fly, *Leptohylemyia coarctata*. It is worthwhile pointing out that to use a seed dressing for this purpose on wheat sown much later than this is, in fact, of little value (Gough *et al.*, 1961).

Related compounds are chlordane, heptachlor, isodrin, endrin, and toxaphene. Of these chlordane has the lowest mammalian toxicity while endrin and isodrin are hazardous. Chlordane has been used for the control of household pests such as cockroaches and ants, but resistance in the former has become serious.

The cyclodienes are probably nerve poisons, though once again the mechanisms are unknown, and may well differ greatly from compound to compound. O'Brien considers that they form a complex with the neural membrane, and do not interfere directly with enzyme systems. The symptoms resemble those of gamma BHC poisoning in some ways, and it should be noted that both kinds of poisons possess, in their molecular models, a pentagon of chlorine atoms. This arrangement is found only in the gamma isomer of BHC.

The organophosphate insecticides

These compounds are closely related to the nerve war gases such as sarin and tabun which were being developed during the Second World War. Their insecticidal activity was first discovered in Germany where they were used as a substitute for the almost unobtainable nicotine.

All active and economic organophosphate insecticides fit the general formula:

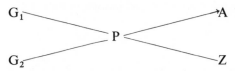

where G_1 and G_2 are usually the same and almost always CH_3O- or C_2H_5O-, A is O or S, and Z is one of a wide range of possible groups. Z is acidic and is easily split off from the rest of the molecule by alkaline hydrolysis. For this reason it is often known as the 'leaving group'. In some compounds Z is the remainder of the molecule, so that the compound is a symmetrical anhydride or derivative of pyrophosphoric acid, but in most organophosphates it is quite a different group. Hundreds, possibly thousands, of these compounds have been synthesized and examined for biological activity. Of these perhaps one hundred are, or have been, available as commercial products.

Ethyl and methyl parathion were among the first organophosphate insecticides to be marketed and they are still the most widely used examples of the group. It is unfortunate, therefore, that their mammalian acute toxicity is rather high for this, together with their military ancestry, has brought some disrepute to the group. The more recently introduced compounds are, in general, much safer to use and, further-more, they are much less persistent than the organochlorines. For the second reason alone they are rapidly replacing the organochlorines for many purposes. Some, such as demeton, are systemic and are much used for the control of aphids and spider mites.

We have space to consider only the better known products. Details of organo-phosphate insecticides and of other pesticides appeared regularly in the supplements of the *World Review of Pest Control* which also published, from time to time, a *Dictionary of pesticide common names*. This journal has now ceased publication.

Contact organophosphate insecticides Parathion (8), diethyl-4-nitrophenyl phos-

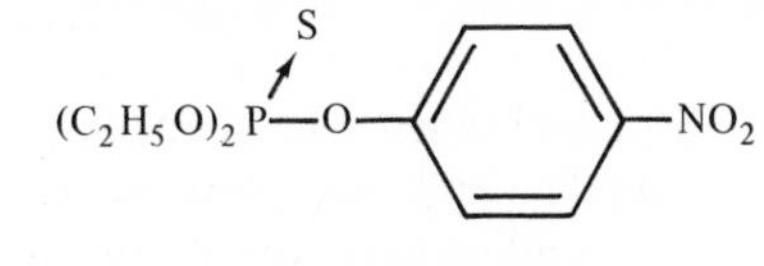

$$(C_2H_5O)_2 \overset{\overset{S}{\nearrow}}{P} - O - \underset{\underset{}{}}{\bigcirc} - NO_2$$

(8) Parathion

phorothionate, despite its high mammalian toxicity (LD 50 rats, acute oral, 6·4 mg/kg) is widely used because of its wide spectrum of activity, its moderate persistence of two to three weeks, and its cheapness. In the United Kingdom it is mainly used on sugar-beet and glasshouse crops. It is stable for short periods at high temperatures so, unlike most organophosphates, it can be used as a smoke as well as a spray or a dust. Its performance against mites is rather poor but it does have some nematicidal activity. It can be used, for example, for the control of the leaf and bud eelworms, *Aphelenchoides fragariae* and *A. ritzema-bosi*, in strawberries.

It is more widely used abroad than in the United Kingdom, being sprayed on many fruit, vegetable, and field crops. It has been responsible for several deaths in warmer climates, largely because of the natural reluctance to wear, and probable inability to afford, the necessary protective clothing and gas masks. It is classed under the British Regulations as a Part II substance so that full protective clothing and a face mask or respirator must be worn for most applications.

In contrast, malathion is such a comparatively safe compound that it is not listed as a scheduled product. It is, in fact, widely used by amateur gardeners, and in the protection of stored grains. Malathion (9) S-[1,2-di(ethoxycarbonyl)ethyl] dimethyl

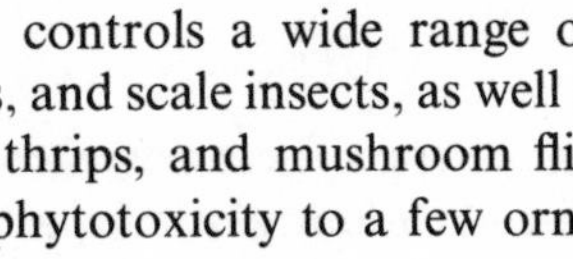

$$(CH_3O)_2 \overset{\overset{S}{\nearrow}}{P} - S\,CHCOOC_2H_5$$
$$\underset{CH_2COOC_2H_5}{|}$$

(9) Malathion

phosphorothiolothionate, has the low mammalian toxicity of between 480 and 5800 mg/kg (rats, acute oral). It controls a wide range of sucking insects and mites, including aphids, mealy bugs, and scale insects, as well as such pests as codling moth, fruit flies, sawflies, beetles, thrips, and mushroom flies. It can be used on a wide variety of crops but shows phytotoxicity to a few ornamentals ranging from antir-rhinums to zinnias.

Diazinon (10), diethyl 2-isopropyl-6-methyl-4-pyrimidinyl phosphorothionate, is of great value for the control of sheep blowflies in Australia, since the unacceptable build up of organochlorine residues in the fat of the animals. Unfortunately, some species of blowfly are beginning to show some resistance to this insecticide, though not enough as yet to make it necessary to change to some other compound.

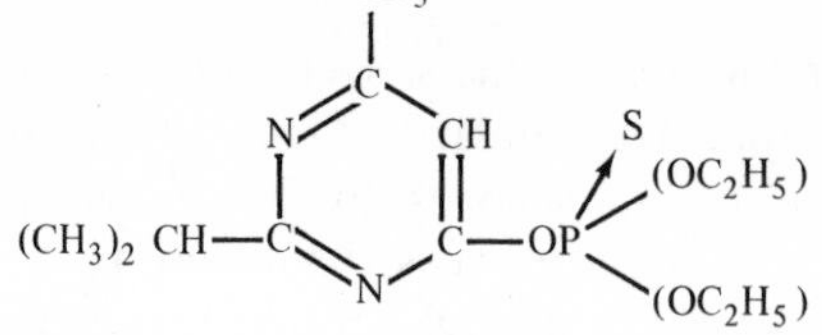

(10) Diazinon

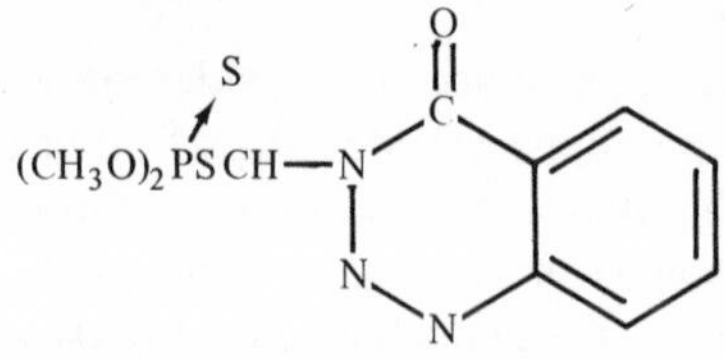

(11) Azinphos methyl

Azinphos methyl (11), S-(3,4-dihydro-4-oxobenzo[d]-[1,2,3]-triazin-3-ylmethyl) dimethyl phosphorothiolothionate, and azinphos ethyl, though fairly toxic to mammals, give good control of codling and tortrix moths, and of red spider mites in orchards. They are also used on cotton crops.

Another useful compound is dichlorvos (12), 2,2-dichlorovinyl dimethyl phosphate, which is particularly valuable as a contact and stomach poison against house-flies and other domestic pests. It has a marked fumigant and penetrant action, and it is becoming a powerful competitor of pyrethrum as an aerosol can filling. It has even been formulated as a dog-collar and cat-band for protection against fleas, lice, and ticks.

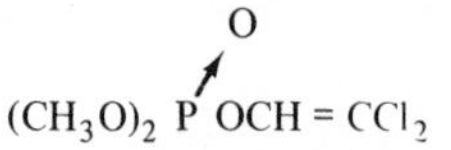

(12) Dichlorvos

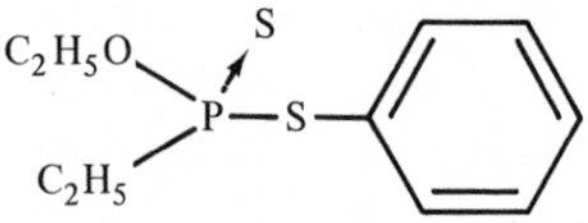

(13) Fonofos

A number of interesting organophosphates have been introduced within the last four or five years. The highly persistent soil insecticide fonofos (13), ethyl phenyl ethylphosphonodithioate, shows promise against such pests as root maggots, wireworms and symphylids, but possibly of even greater interest is its control of the

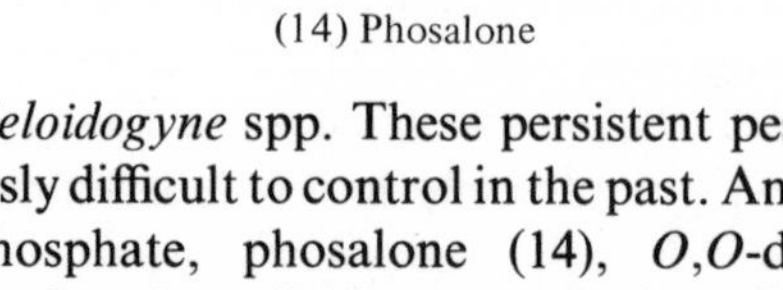

(14) Phosalone

root knot eelworms, *Meloidogyne* spp. These persistent pest nematodes of warmer soils have been notoriously difficult to control in the past. Another recently developed non-systemic organophosphate, phosalone (14), *O,O*-diethyl S-(6-chlorobenzoxazolone-3-yl-methyl) phosphorodithioate, may be of value against aphids, larval flies, beetles and Lepidoptera, and against active stages of mites. Chloro-

74 Pest Control

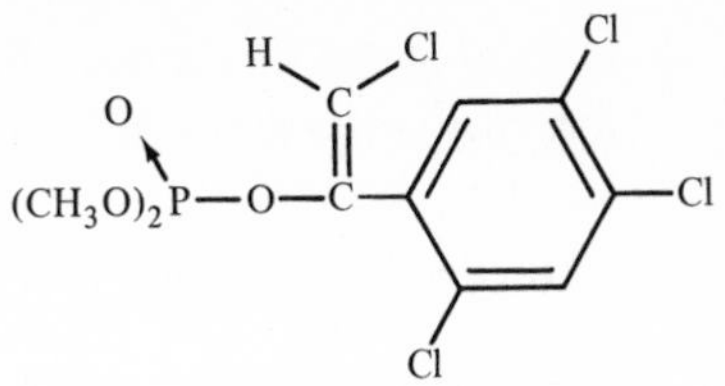

(15) Tetrachlorvinphos

trichlorophenylvinyl dimethyl phosphate (tetrachlorvinphos, 15) is a compound with a low mammalian toxicity (4000 to 5000 mg/kg acute oral, male rats) but with high toxicity to house-flies and related species. It is thus potentially useful for dairy and livestock farmers.

Systemic organophosphate insecticides   There are obvious advantages in a pesticide which, when applied to restricted areas of a plant, can find its way to the remainder of the plant, either in its original form, or as a toxic metabolite. The topic has been mentioned briefly at the beginning of this chapter. It is not so difficult, of course, to apply such a material as it is to spray a contact insecticide which must cover all the susceptible parts. If a systemic insecticide is moderately persistent it will also protect parts of the plant which have not emerged at the time of spraying. Careful spraying is, however, essential; failures have been caused by too-hurried spraying when too much reliance was put on the efficiency of the systemic action. It hardly needs pointing out, perhaps, that systemic organophosphate insecticides are also usually contact insecticides. Most of them are rapidly absorbed by the plant and a few hours after spraying the surfaces are relatively non-toxic to beneficial species. Those individuals which are hit directly by the spray, or which land on the plant soon after, are usually killed, and bees and other insects have been killed some time after spraying when feeding on poisoned nectar.

Most of the compounds below are used for the control of spider mites and aphids, especially when the latter are vectors of virus diseases. The outstanding example is against *Myzus persicae* and other vectors of sugar-beet yellow viruses. There has also been some success in the production of seed potatoes with very low virus infection rates in ware potato districts.

Demeton-S-methyl is one of the current representatives of the demeton methyl group. Demeton methyl is a 70:30 mixture of demeton-O-methyl (16), 2-(ethylthio)-ethyl dimethyl phosphorothiolate and demeton-S-methyl (17), S-2-(ethylthio) ethyl

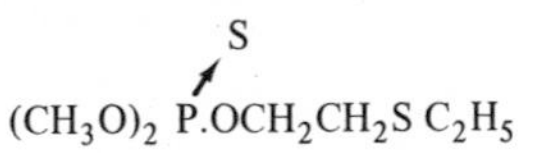

$(CH_3O)_2 \overset{S}{P}.OCH_2CH_2S\ C_2H_5$

(16) Demeton-O-methyl

$(CH_3O)_2 \overset{O}{P}.SCH_2CH_2SC_2H_5$

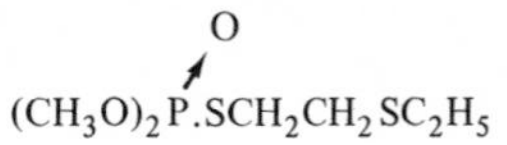

(17) Demeton-S-methyl

dimethyl phosphorothiolate. The former, the thiono form, spontaneously isomerizes to the thiol form, demeton-S-methyl (17), so this is now marketed separately.

Within the tissues of the plant these compounds are partially oxidized to other active insecticides, and partially hydrolysed to inactive compounds. One way of avoiding the loss by hydrolysis is to use the oxidized substances, hence the introduction of oxydemeton methyl, the sulphoxide of demeton-S-methyl and demeton-S-methyl sulphone, the sulphone.

$$\text{(CH}_3\text{O)}_2\overset{\displaystyle O}{\overset{\displaystyle \nearrow}{P}}\text{SCH}_2\,\text{CH}_2\text{SOC}_2\text{H}_5$$

(18) Oxydemeton methyl

Oxydemeton methyl (18) is S-[2-(ethylsulphinyl) ethyl] dimethyl phosphorothiolate. Demeton-S-methyl sulphone differs from this in the replacement of the SO group by an SO_2 group. These compounds of the demeton methyl group are used for the control of aphids and red spider mites on most agricultural and horticultural crops, but oxydemeton methyl is not recommended for brassicas. Oxydemeton methyl is formulated as a spray and as 5 per cent granules for the systemic treatment of trees and shrubs. The demeton methyl insecticides have proved particularly useful for the protection of sugar-beet from yellow virus diseases. Some protective clothing is needed when diluting the concentrates, but not during spraying out-of-doors.

Disulfoton is another insecticide of this group, being the dithioate (or, in some conventions, the thiolothionate) of the ethyl homologue of demeton methyl. Being a dithioate it cannot isomerize in the same way as demeton methyl and demeton. Technically it is diethyl S-[2-(ethylthio) ethyl] phosphorothiolothionate (19). It is

$$\text{(C}_2\text{H}_5\text{O)}_2\,\overset{\displaystyle S}{\overset{\displaystyle \nearrow}{P}}.\text{SCH}_2\text{CH}_2\text{SC}_2\text{H}_5$$

(19) Disulfoton

considerably more toxic to mammals than are its above relatives but it is applied to foliage and soil in a granular form which lessens the hazards. It gives a fairly long protection against aphids and the carrot fly, *Psila rosae*.

Phorate is the thiomethyl derivative of disulfoton, and is even more toxic to mammals. Both are oxidized in the plant to soluble sulphones and sulphoxides, in a way similar to demeton-S-methyl. Oxydisulfoton, for example, has a solubility of about 10 mg/100 ml in water at room temperature, and has been released as an experimental insecticide in a granular formulation.

Dimethoate (20), dimethyl S-(N-methylcarbamoylmethyl) phosphorothiolothion-

$$\text{(CH}_3\text{O)}_2\,\overset{\displaystyle S}{\overset{\displaystyle \nearrow}{P}}.\text{SCH}_2\overset{\displaystyle O}{\overset{\displaystyle \|}{C}}\text{NHCH}_3$$

(20) Dimethoate

ate, is less toxic to mammals than any of the above systemic insecticides. As well as controlling aphids and red spider mites it is effective against mangold fly, *Pegomyia hyoscyami* var. *betae* and various fruit flies such as the olive fly, *Dacus oleae*, and the Queensland fruit fly, *D. tryoni*.

Under British conditions a period of at least two or three weeks must pass between spraying with these compounds, and the harvesting of sprayed edible crops. The exact preharvest interval depends upon the particular compound and the time of the year. In contrast, the systemic compound mevinphos, 2-methoxycarbonyl-1-methylvinyl dimethyl phosphate (21) can be applied a mere three days before harvest,

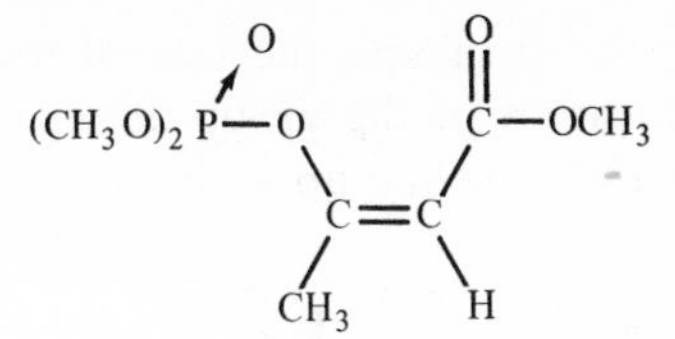

(21) Mevinphos

for it is soluble in, and decomposed by water. Although of high mammalian toxicity it presents no danger to livestock which are let into the treated areas twenty-four hours after spraying. It is one of the few synthetic insecticides which needs no complicated formulation. Mevinphos was introduced originally as an acaricide but it is now probably used more as a close-to-harvest, wide-spectrum pesticide.

Fenchlorphos (22), dimethyl 2,4,5-trichlorophenyl phosphorothionate, is also used for plant sucking insects and for flies, but its remarkably low mammalian toxicity has suggested another use. It can be employed as a systemic insecticide in livestock for the control of such parasites as the warble flies, *Hypoderma* spp.

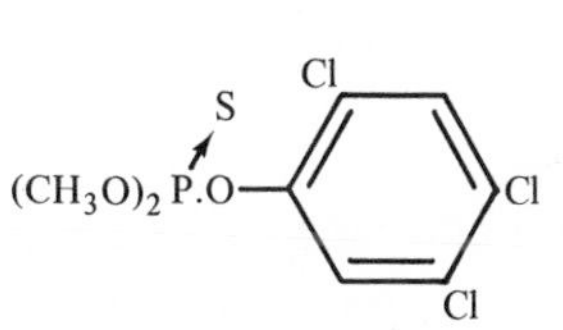

(22) Fenchlorphos

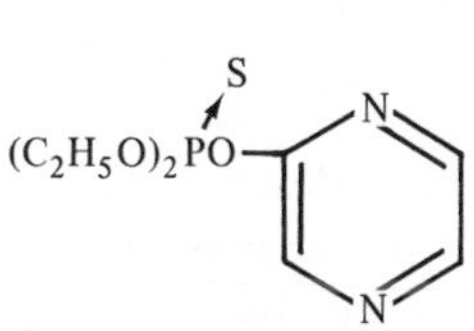

(23) Thionazin

Thionazin (23), diethyl *O*-2-pyrazinyl phosphorothionate, has systemic insecticidal properties against sucking insects and some lepidopterous larvae, but, in addition, it will kill soil nematodes when applied at a rate of 1 to 2 kg active ingredient/acre, though with some risk of plant damage to some crops.

Finally, to show the versatility of the organophosphates, phenkapton (24),

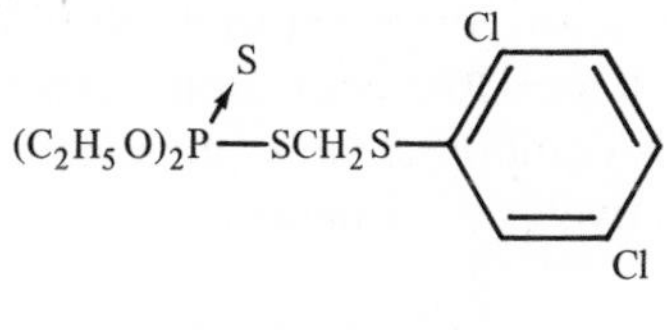

(24) Phenkapton

O,O-diethyl *S*-(2,5-dichlorophenylthiomethyl) phosphorothiolothionate, should be mentioned, although it is not a systemic. In fact, it is not particularly insecticidal, but it does kill plant feeding mites and their summer eggs. It is rather toxic to mammals, a fault which limits its use on edible crops.

The mechanism of the action of the organophosphates is much better understood than that of the organochlorines. Most workers agree that they are neurotoxins which inhibit cholinesterases. This leads to an accumulation of acetylcholine at the neural synapses with a consequent disruption of the nervous system. These compounds inhibit several other ester splitting enzymes, but the reaction with cholinesterase is believed to be the most important.

Organophosphates can be hydrolysed, the rate of the reaction being dependent upon the electronic configuration of the group attached to the P atom, upon the steric characteristics of the molecule and upon the hydroxyl ion concentration in the medium:

$$
\begin{array}{c}
\text{G} \\ \\ \text{G}
\end{array}
\!\!>\!\!P\!\!<\!\!
\begin{array}{c}
\text{A} \\ \\ \text{Z}
\end{array}
\quad + \text{OH}^- \rightarrow \quad
\left[
\begin{array}{c}
\text{G} \\ \\ \text{G}
\end{array}
\!\!>\!\!P\!\!<\!\!
\begin{array}{c}
\text{A} \\ \\ \text{O}
\end{array}
\right]^{-}
\quad + \text{H}\cdot\text{Z}
$$

The meaning of the term 'leaving group' now becomes apparent.

The reaction of the organophosphate with the enzyme is similar to hydrolysis but, since the enzyme is a very large molecule, the steric configuration of the insecticide molecule is relatively more important. Strictly speaking it is the enzyme which attacks the organophosphate, the overall course of the reaction being:

$$
\text{EnzOH} + \quad
\begin{array}{c}
\text{G} \\ \\ \text{G}
\end{array}
\!\!>\!\!P\!\!<\!\!
\begin{array}{c}
\text{A} \\ \\ \text{Z}
\end{array}
\quad \rightarrow \quad
\begin{array}{c}
\text{G} \\ \\ \text{G}
\end{array}
\!\!>\!\!P\!\!<\!\!
\begin{array}{c}
\text{A} \\ \\ \text{OEnz}
\end{array}
\quad + \text{HZ}
$$

where EnzOH represents the uninhibited enzyme.

The nature of the leaving group is clearly of importance in determining the rate of the reaction but has nothing to do with the final nature of the inhibition. The phosphorylation of the enzyme at the active site involves the formation of covalent bonds, an unusual occurrence in the inhibition of enzymes. The inhibition of the cholinesterase is thus not normally reversible over short periods. After some time there is, however, a dephosphorylation of the inhibited enzyme, and this can be accelerated by the use of organophosphate antidotes. The most important of these are oximes, compounds containing the $CH\!=\!NOH$ group, an excellent example being 2PAM (24a). Atropine may also be used to relieve the symptoms of poisoning.

It is surprising that the majority of organophosphates are poor inhibitors of cholinesterase *in vitro*, but are found to act rapidly at very low concentrations *in vivo*. These latent inhibitors are converted to direct inhibitors in the tissues of the organisms. There are several possible activation reactions but the most common is the conversion of a $P\!=\!S$ compound to a phosphate or to a $P\!=\!O$ compound. Typical examples are the activations of parathion and malathion to paraoxon and malaoxon.

The carbamate insecticides

These insecticides are among the derivatives of carbamic acid, $HOC(O)NH_2$. The pharmacological properties of some of these compounds have been recognized for many years, the first to be discovered being physostigmine, the active principle of calabar beans which were used for trial by ordeal in West Africa. The discovery of the anticholinesterase activity of the organophosphates stimulated the examination of this group for possible insecticides, an investigation which led to the discovery of carbaryl and similar compounds.

In general, the carbamates are rather better insecticides than acaricides, though some are quite effective and may replace organophosphates for the control of resistant strains.

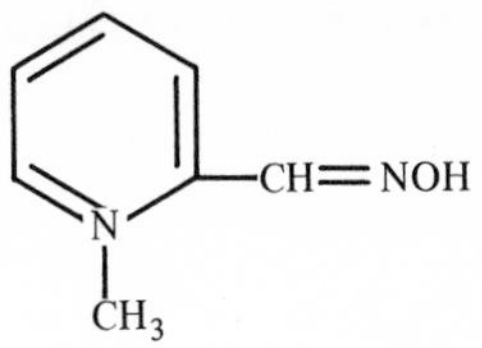

(24a) 2-PAM-antidote for Organophosphates

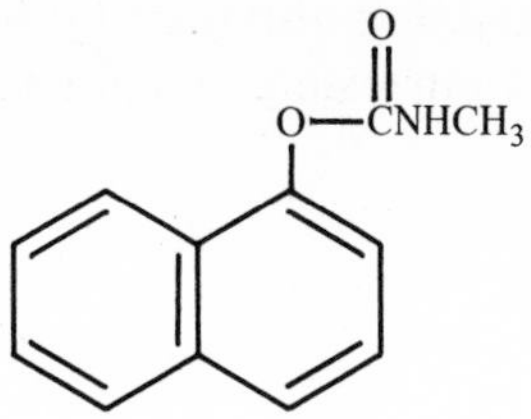

(25) Carbaryl

Carbaryl (25), 1-naphthyl *N*-methylcarbamate, is probably the most widely used of the group so far. It is a product with a low mammalian toxicity and is used in the United Kingdom for the control of apple pests, especially codling moth and other Lepidoptera. Unfortunately, it does not control mites, which puts it at some disadvantage to organophosphates such as azinphos, and which also leads to increases in mite populations as a result of the killing of insect predators. In other countries it is also used for the control of the cotton boll weevil and caterpillar pests of cotton. In general, it shows toxicity to leaf-feeding caterpillars, beetle larvae and some aphids, but little to adult house-flies. Like many carbamates it has some systemic activity, but not enough to be of much practical value. The insecticide is of moderate persistence.

In contrast, two of the earliest carbamates to be introduced, Pyrolan® (26), 3-methyl-1-phenyl-pyrazolyl-(5)-dimethylcarbamate and Isolan® (27), 1-isopropyl-

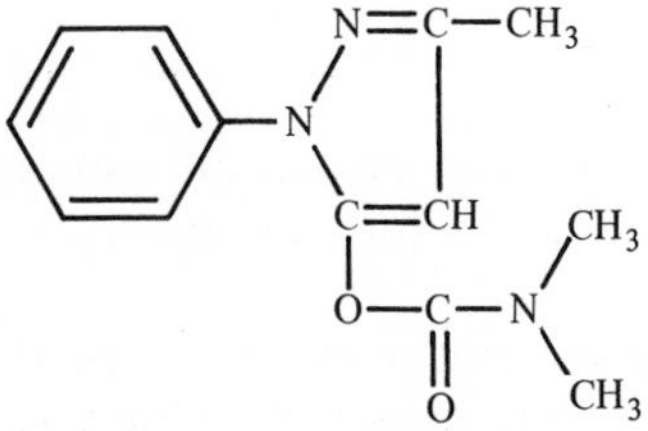

(26) Pyrolan ®

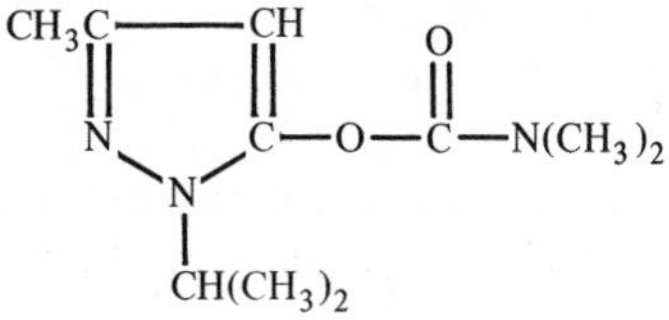

(27) Isolan ®

3-methyl-5-pyrazolyl-dimethylcarbamate, were found to be highly toxic to house-flies and aphids. Pyrolan® has also been used in methyl euganol baits for the control of the fruit fly, *Dacus dorsalis*.

Zectran®, 4-dimethylamino-3,5-xylyl *N*-methylcarbamate (28) is a systemic carbamate with a broader spectrum of use, having acaricidal activity as well as

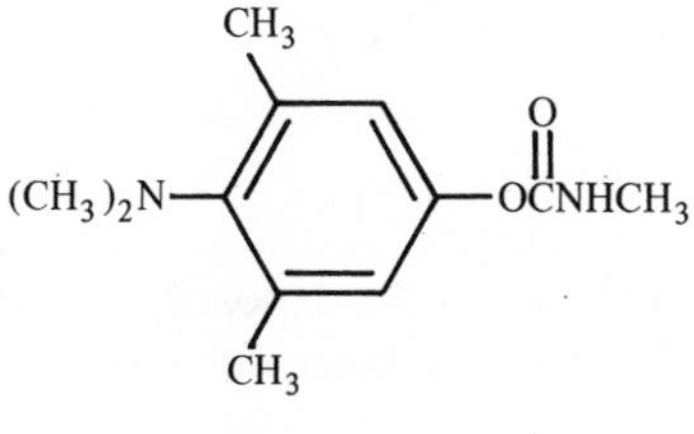

(28) Zectran ®

insecticidal. It is also toxic to slugs and snails, but is hazardous to mammals. One of the most interesting compounds, however, is arprocarb (29), 2-isopropoxyphenyl

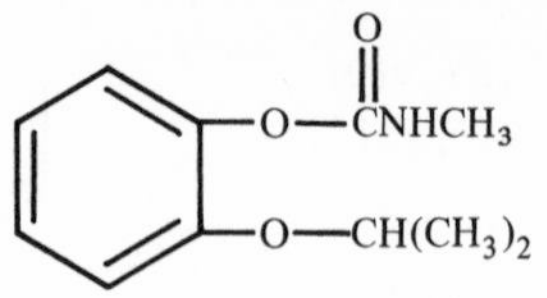

(29) Arprocarb ®

N-methyl-carbamate, a compound which is being extensively tested by the WHO for the control of mosquitoes. It is unusual in that it combines a rapid knock-down, close to that of pyrethrum and dichlorvos, with a long residual effect, a useful combination of properties in a pesticide for the control of vectors of diseases. It is also effective against muscid flies and cockroaches. In addition, it has systemic properties. Its mammalian toxicity is acceptable (rat, oral, 150 mg/kg).

The carbamates appear to be cholinesterase inhibitors, but it has proved difficult to correlate anticholinesterase activity *in vitro* with insecticidal activity. There is some dispute about the biochemistry of the inhibition of the enzyme, but it may be analogous to the reactions of the organophosphate insecticides with cholinesterase (O'Brien, 1967; Casida, 1963).

A further interesting feature of the carbamates is that several of them are potentiated by compounds which are also synergists for pyrethrum. Thus sexoxane increases the activity of carbaryl against house-flies fifty-fold.

The hydrocarbon oils

Refined tar and petroleum oils have long been used in crop protection, especially on orchard and glasshouse crops. They are particularly useful for scale insects which are difficult to control by other chemical pesticides.

Tar oils are generally applied as winter washes when fruit trees are dormant, as they are phytotoxic, and they control the overwintering stages of aphids, some moths such as the small ermines, *Yponomeuta* spp. and the winter moth, *Operophtera brumata*, and also mealy bugs and scales. Petroleum oils are effective against capsid bugs and fruit tree red spider mites, *Panonychus ulmi*.

The toxicity of dormant sprays is sometimes enhanced by the addition of dinitro-o-cresol (DNOC, 30). DNOC is also used as a herbicide (it is a general biocide) and

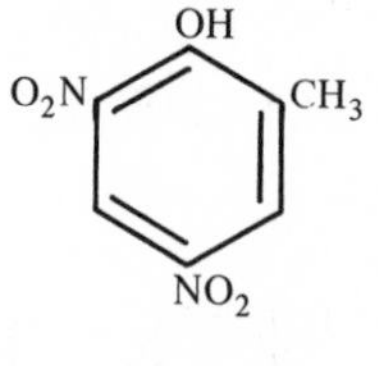

(30) DNOC

thus it cannot be used when foliage is fully exposed, though it may be applied at bud-break. It also has a fungicidal effect, against, for instance, peach leaf curl, *Taphrina deformans*.

Summer washes are available for use when fruit trees and shrubs are in leaf, and in some formulations they may also be used on certain glasshouse crops. They can be applied, for example, against red spider mite, *Tetranychus urticae*, especially when resistance to the insecticides discussed earlier is a problem.

Oils are widely used for the control of scale insects on citrus trees, and suitable formulations must be used to avoid damaging the ever-present foliage. In Australia, for example, two sprays are usually applied to coincide with the crawler stages of armoured scales (California red scale, *Aonidiella aurantii*, purple scale, *Lepidosaphes beckii*, etc.), and unarmoured scales (white wax scale, *Ceroplastes destructor* and *C. rubens*, cottony cushion scale, *Icerya purchasi*, etc.). Once again, other insecticides, such as malathion, azinphos methyl and dimethoate, may be added to the spray to improve the kill.

The hydrocarbon content of the oils is of importance in determining which insects are killed, as well as in influencing the degree of phytotoxicity. Some pests are more susceptible to oils with high contents of paraffin hydrocarbons, while others succumb more easily to the aromatic constituents. Despite this degree of selectivity it is generally thought that the mode of action of oils is mainly physical – a smothering of both the active stages and the eggs.

Organotin compounds and anti-feedants

In 1949 the International Tin Research Council established a research team to seek out practical uses for organotin compounds. This resulted in the introduction of various compounds into industrial processes. The team also found that a number of them, such as triphenyltin acetate and hydroxide, could be used as agricultural fungicides, and possibly as molluscicides. Ascher and Nissim (1964) noted that triphenyl acetate (fentin acetate, 31) which had been applied as a fungicide, inhibited

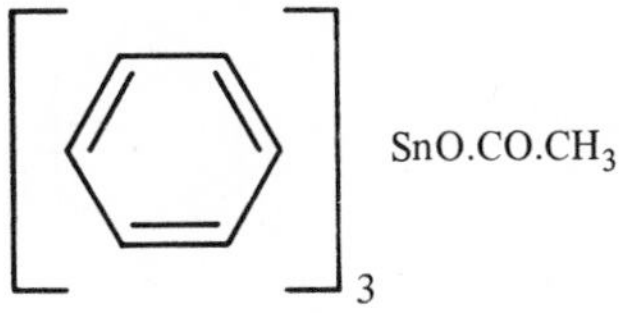

$$\left[\bigcirc \right]_3 SnO.CO.CH_3$$

(31) Fentin acetate

the feeding of insects on treated foliage. The effect was not that of a repellent, for that would drive the insect away completely, and, furthermore, it was effective only against surface-feeding species; insects which pierced the plant tissues to suck out liquids were not affected.

Strictly speaking, such anti-feedant compounds are not pesticides in the sense used in this text, for the insects are not killed directly by the material, but only indirectly as the result of the denial of food. Some conventional pesticides do have an anti-feedant effect at sublethal dosages, arprocarb being a notable example. This carbamate has been shown to have such an action against the boll weevil, *Anthonomus grandis*, at rates below those used for insecticidal control.

The only compound which has been tested on a large scale for anti-feedant properties is 4′(dimethyltriazino) acetanilide (DTA), and this performed well against pests which feed on the surface of foliage, and against some pests of fabric. It was essential, however, to cover the surface completely, and the material was not very persistent in the field. A number of naturally occurring substances deter feeding: they constitute defence mechanisms in plants. The larval European corn borer, *Ostrinia nubilalis* is prevented from feeding by 6-methoxybenzoxazolinone extracted from maize: various alkoglycosides from *Solanum* spp. inhibit the feeding of the Colorado potato beetle, *Leptinotarsa decemlineata*. Clearly there are opportunities for finding useful new materials of this kind in the plant kingdom.

Anti-feedant compounds or insecticides with such an effect at sublethal dosages are potentially valuable in integrated control for they will have little effect on entomophagous insects. If they work in such a way that the pests cannot feed on the treated crops and move to some alternative, non-economic host, they will support a population of valuable parasites and predators. Unfortunately little is known of the mode of action of these compounds. In some cases at least it seems that a small quantity of the material is swallowed and this inhibits further feeding, even on untreated surfaces, with consequent starvation.

It is possible that the phenomenon is commoner among conventional pesticides than is sometimes thought, for in many field trials it is often assumed that damage control (conveniently measured inversely as yield increase) is the result of killing the pest, whereas it may, in fact, be wholly or partially due to the inhibition of feeding.

Other organotin compounds have been investigated in several countries for their direct insecticidal properties, and they show some promise. Often, however, they are too phytotoxic for use in agriculture (e.g., the trialkyl derivatives).

Organotin compounds are also used as molluscicides, and in antifouling compounds to prevent the settling of marine organisms on the bottoms of ships. A further marine use is the protection of timbers from attack by boring molluscs.

Moth-proofing agents, such as Eulan CN® (32) and Mitin FF® (33), are applied

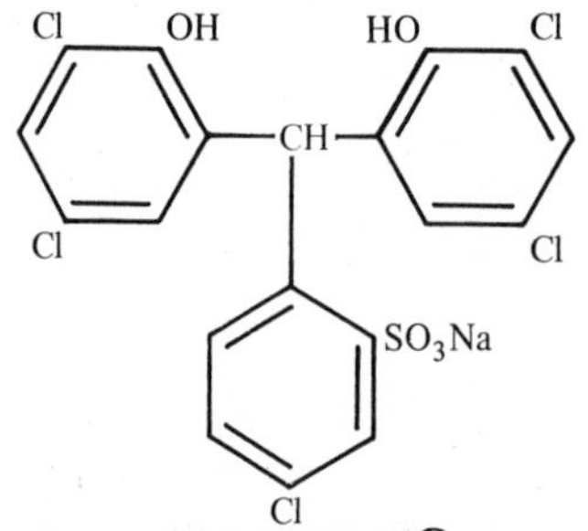

(32) Eulan CN 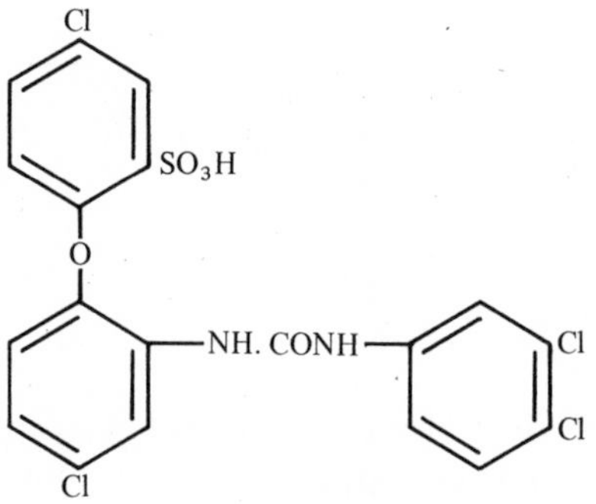 ®   (33) Mitin FF ®

to the fabric in the dye bath, and are both repellents and stomach poisons. A quite different approach is to render the wool molecules indigestible to the insects. Normally the insects split the disulphide cross-linkages in the keratin molecules, but these may be broken by reduction and then replaced by short hydrocarbon chains. This manufacturing process has little effect on the physical properties of the wool (Busvine, 1966; Ascher and Nissim, 1964; Wright 1963, 1967).

Botanical insecticides
Several effective insecticides are derived from plants, the most important being pyrethrum, nicotine, rotenone, and ryania. Pyrethrum is holding its place against

competition from the synthetic insecticides, while the use of nicotine and rotenone is dropping steadily. Ryania is a comparatively new product which fits well into integrated control programmes on apples. A few synthetic compounds which are closely related to naturally occurring insecticides are included here for convenience.

Pyrethrum Pyrethrum is a contact insecticide prepared from the flower heads of *Chrysanthemum cinerariaefolium*. The varieties currently grown in the Highlands of Kenya yield the highest contents of active ingredients. The crop is also grown in several other countries including Iran and Japan. It has been used as a source of insecticide in the Caucasus for at least a century and a half, and possibly much longer.

The main active ingredients are complex compounds known as pyrethrin I and II, and cinerin I and II. Their structural formulae, and that of some important related compounds are given in (34). Since these substances are never separated in commercial products they are known collectively as 'total pyrethrins'.

These insecticides are unstable, being rapidly oxidized, especially when exposed

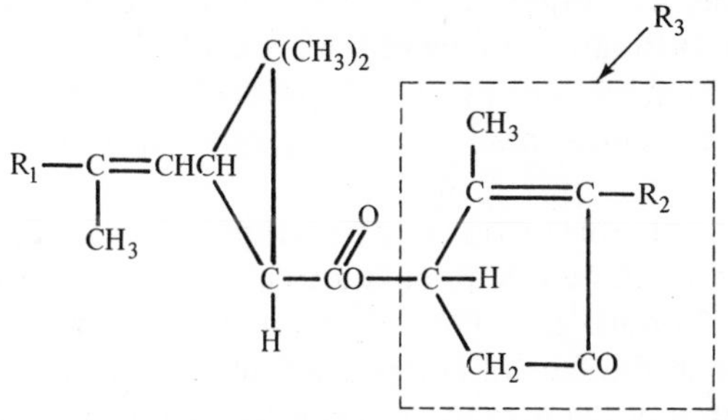

$-CH_3$ = A $\quad -CH_2:CH:CH:CH:CH_2$ = B$_|$$-COOCH_3$ = C

$-CH_2 \cdot CH:CH \cdot CH_3$ = D

$-CH_2CH:CH_2$ = E

	R_1	R_2	R_2
Pyrethrin I	A	B	
Pyrethrin II	C	B	
Cinerin I	A	D	as above
Cinerin II	C	D	
Allethrin	A	E	
Barthrin	A	~	F
Dimethrin	A	~	G

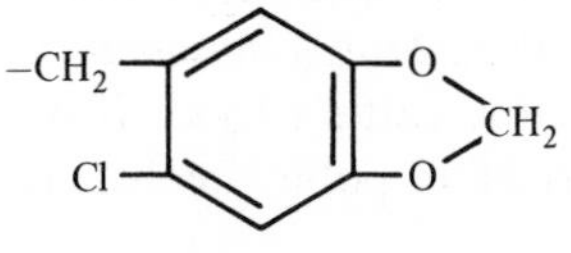

F

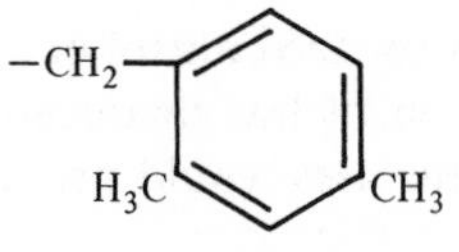

G

(34)

to light with a high ultra-violet content. For this reason they are rarely used out-of-doors, and are of little or no value in crop husbandry. Before the discovery of the insecticidal activity of the chlorinated hydrocarbons pyrethrum was often employed in anti-malarial campaigns.

Unfortunately, because of the method of production which involves much manual labour, the commercial product is comparatively expensive. In 1938, however, the first effective synergist was introduced, and since then a whole range has been developed. The addition of these to pyrethrum preparations has allowed the content of the active pyrethrins to be reduced drastically without loss of insecticidal activity. Probably one of the most commonly used synergists is piperonyl butoxide (35).

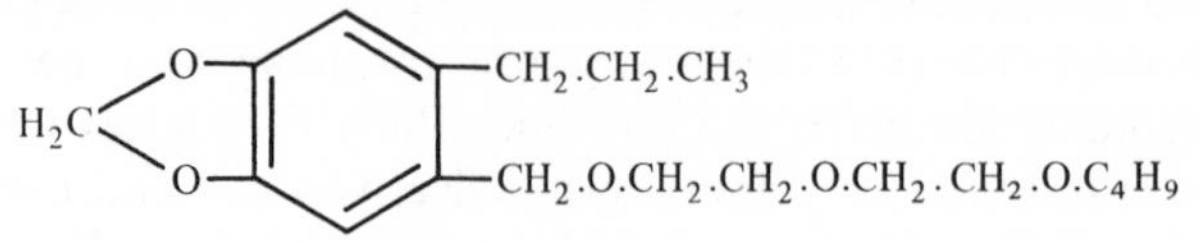

(35) Piperonyl butoxide

Pyrethrum is a nerve poison with a very rapid 'knock-down' effect. This, and its low mammalian toxicity and rapid break-down, has led to its popularity as a pest killer in the house. Unfortunately, many of the insects knocked down recover some time later, usually without the knowledge of the housewife. Because of this, most domestic formulations contain a small quantity of some other insecticide to complete the work.

One outstanding and valuable feature of pyrethrum is that no significant resistance has developed towards it among wild populations of insects, with the possible exception of some *Tribolium* populations. This, together with its low mammalian toxicity, explains its persistence in the face of competition of cheaper synthetic insecticides.

Beyond the recognition of the nervous action little is known of its mode of action. There is an interaction with the neural axon in which there is an increase of the negative after-potential, inducing abnormal excitation and convulsions; and a blockage of sodium and potassium movement, causing paralysis. It is not established, however, whether the primary site of action is in the central nervous system, or the peripheral, nor is it known how knock-down differs from lethal action. Some of these problems have been discussed recently by Yamamoto (1970).

A number of synthetic relatives of the pyrethrins have been prepared and tested. Despite the high costs of their manufacture a number of countries are interested in their production, mainly as a precaution against being isolated from pyrethrum sources in time of war. The first important one of these compounds was allethrin (34), an allyl homologue which is more stable to heat and light than is the natural compound. It is about as toxic to house-flies as is pyrethrum, but less toxic to several other pests. Like pyrethrum, it has a low mammalian toxicity (920 mg/kg of the rat, oral).

At present most interest is in two compounds with extremely low mammalian toxicities, namely dimethrin and barthrin (34). The LD 50, acute oral to rats, of barthrin is, for example, 15 000 to 40 000 mg/kg. Both compounds have been used successfully in trials for the control of face flies, *Musca autumnalis*, breeding in cow-dung, by the addition of the chemicals to the cattle's food. If manufactured on a large scale their main uses would probably be in public health entomology and on livestock.

Nicotine Insecticidal washes prepared by steeping tobacco leaves in water were in use in the eighteenth century. Tobacco smoke has long been employed to kill

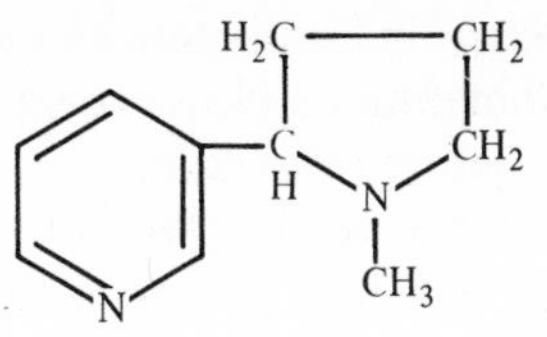

(36) Nicotine

insects, especially aphids. Commercial extracts were on the market by the late nineteenth century. The insecticidal alkaloid itself, nicotine (36), was first isolated in 1828, its structure determined in 1893 and its synthesis achieved in 1904. Although it occurs in a variety of plants the main source is the genus *Nicotiana*, especially the species *tabacum* and *rustica*.

Despite its long history as an insecticide its use is falling off rapidly. This follows from its high mammalian toxicity and its comparative inefficiency in cold weather. The main danger is skin absorption, and splashes must be washed off immediately.

Nicotine acts mainly as a fumigant although it can also pass through the insect cuticle and gut wall. The volatility of free nicotine can be a disadvantage in sprays and it is thus usually formulated as the sulphate, or in some other 'fixed' form. If the rapid fumigant action is to be utilized alkaline materials are added just before application.

Nicotine is one of the least phytotoxic of the insecticides and is effective against most soft-bodied insects. It is especially useful against the codling moth as it kills the eggs. Edible crops may be harvested two days after spraying, a much shorter pre-harvest interval than those required by most other insecticides.

Nicotine kills insects by blocking the motor nerves in the ventral nerve cords. It has been established that in vertebrates it blocks the receptor for acetylcholine in cholinergic synapses, and it is probable that it acts in a similar way in insects. Little is known about this protein receptor and interest in it has been eclipsed by research on acetylcholinesterase. It has not been described chemically, and it has only recently been detected in broken cell preparations (O'Brien, 1970).

Other important molecules which combine with acetylcholine receptors in vertebrates are atropine, curare, and muscarine: it will be remembered that atropine is used for the treatment of the symptoms of organophosphate poisoning.

The study of other compounds which block the acetylcholine receptors in insects should prove fruitful, for, as O'Brien observes, they probably differ sufficiently from those of vertebrates for selective toxicity to occur.

Rotenone (37)   This material is contained in those leguminous plant roots which

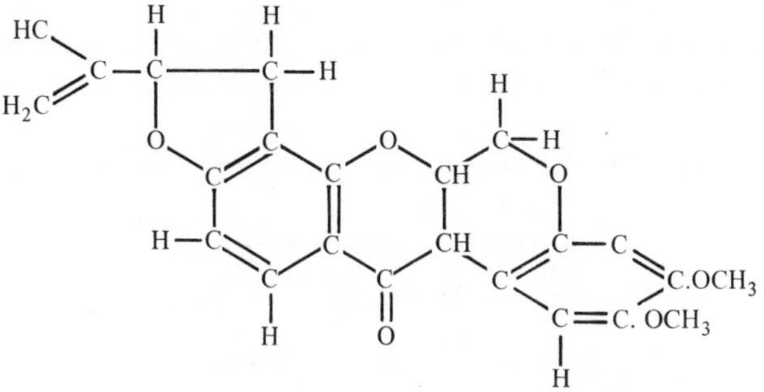

(37) Rotenone

have also been used as fish poisons in South America and South East Asia for many years. Fishery organizations sometimes still use them as fish killers for sampling or for clearing land-locked water before restocking.

Rotenone is usually formulated as finely ground roots of *Derris* or *Lonchocarpus* mixed with some inert diluent. It is used on crops to a limited extent, and for the control of animal parasites. It was once thought to be of extremely low mammalian toxicity but this has been questioned. Pigs are reported to be susceptible. A recent report on research on the selective toxicity of rotenone and on its mode of action has been published by Fukami and his colleagues (Fukami *et al.*, 1970).

Ryania Ryania is a product derived from the stems and roots of the South American shrub *Ryania speciosa*. The main active ingredient is the alkaloid ryanodine. The product acts as a contact insecticide and as a stomach poison for a number of lepidopterous pests, including stem borers and the codling moth. It is thus a fairly selective insecticide, and since it has a low mammalian toxicity (rat, acute oral, 750 to 1000 mg/kg) it is being included in integrated control programmes on apples.

Acaricidal compounds

Mites and ticks differ sufficiently from the insects in their biochemistry for certain compounds to be highly toxic towards them, yet relatively harmless to insects. Oddly many of these acaricidal compounds are also efficient fungicides, a combination of properties which is particularly valuable in fruit pest control. The lack of insecticidal activity can also be an advantage for many of the natural enemies of mites are insects.

The bridged diphenyl group A casual examination of the structural formulae of the members of this group would suggest that they are possible insecticides for they closely resemble DDT. Dicofol, for example, differs only in the substitution of an hydroxyl group for the hydrogen on the bridge between the two phenyl groups. Surprisingly they lack insecticidal properties yet they are much more toxic to Acarines than is DDT. Dicofol (38) is effective against all stages, except winter eggs, of plant-feeding mites.

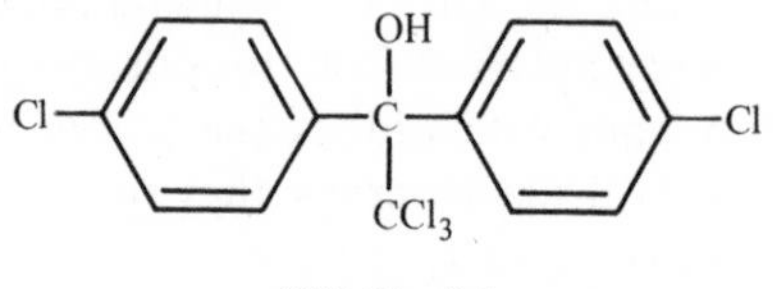

(38) Dicofol

The related chlorobenzilate (39) is perhaps the most extreme member of this group for it has been used for the control of *Acarapis woodi*, a mite which lodges in the tracheae of honey bees.

Other acaricides in the group are azobenzene, fenson, chlorfenson, and tetradifon.

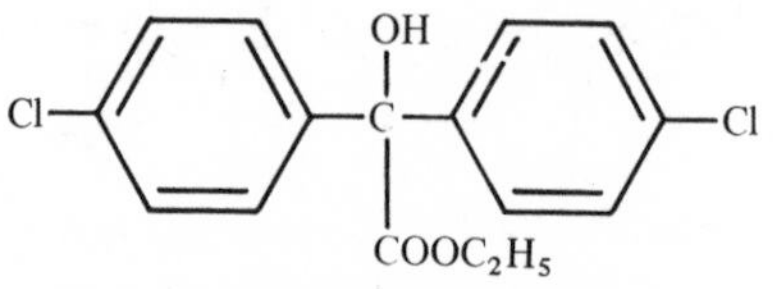

(39) Chlorobenzilate

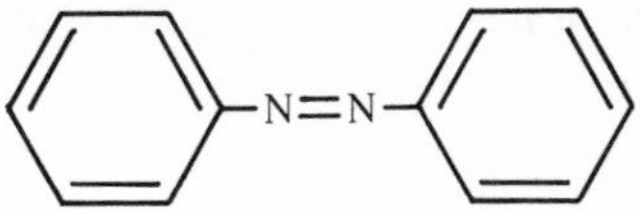

(40) Azobenzene

Azobenzene (40), one of the first to be used, is sufficiently volatile to be used either as an aerosol or in a paint which is applied to the hot water pipes of glasshouses. Tetradifon is a particularly useful product as it kills all stages of plant-feeding mites, including winter eggs (41).

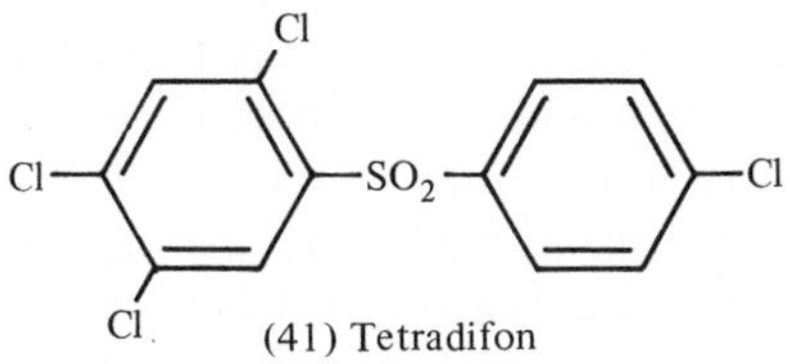

(41) Tetradifon

Little or nothing is known of the mode of action of these compounds, though this is far from surprising as the study of the physiology of mites lags far behind that of insect physiology.

Thioquinox and chinomethionat These closely related compounds (chinomethionat is also known as oxythioquinox) are examples of acaricides with fungicidal properties against apple mildew, *Podosphaera leucotricha*, and other powdery mildews. This is especially convenient as the fungicides which had replaced sulphur for fruit spraying (for example, captan against apple scab, *Venturia inaequalis*) did not give a good enough control of mildew. It was, in fact, the search for fungicides effective against apple mildew which led to the discovery of some of the dual-purpose pesticides.

Thioquinox (42) is an acaricide with a low mammalian toxicity which is effective

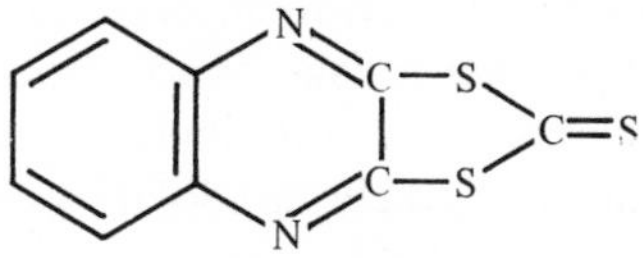

(42) Thioquinox

against the summer eggs of spider mites. It does irritate the skin of some subjects and this lessens its value in orchard spraying. Chinomethionat (43) is also of low mammalian toxicity and has applications similar to those of thioquinox.

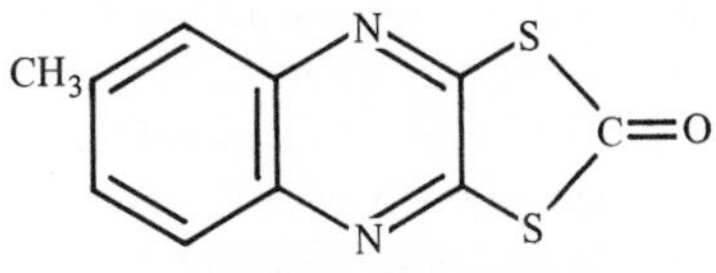

(43) Chinomethionat

Dinocap This product, a mixture of two isomers (44a and b) of 2-(1-methyl-n-heptyl)-4,6-dinitrophenyl crotonate, was introduced just after the war as an acaricide and fungicide. It is toxic to all active stages of spider mites, and acts as a contact

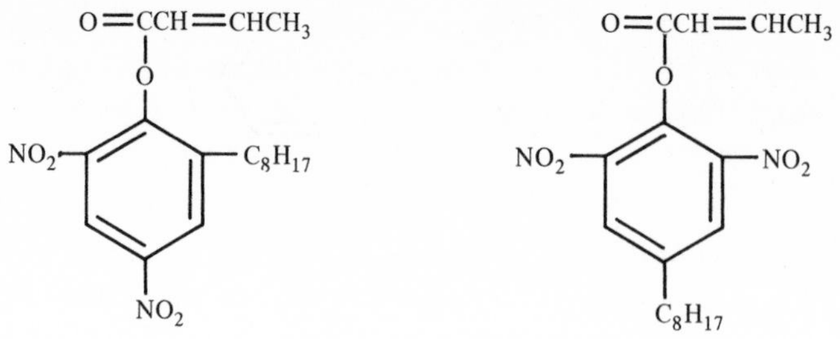

$$O=CCH=CHCH_3$$

(44a & b) Dinocap

fungicide against the powdery mildews. It is, however, a dinitrophenyl compound and, as such, it can be phytotoxic, especially in hot weather. Dinocton-*o* is a related mixture of homologues with similar uses; it may cause russeting on some apple varieties if applied too early in the season.

Sulphur   This element was mentioned by Homer as 'all-curing', and it still finds use as an acaricide and fungicide. For orchard use it may be formulated as a dust, a wettable powder, or a paste (colloidal sulphurs). It is not toxic to mammals or to most plants, though some varieties of fruit, such as Lane's Prince Albert apple, are sulphur 'shy'.

Sulphur will control eriophyid mites – difficult to kill by any other means – as well as tenuipalpids and tetranychids. In public health work it is still used against scabies (*Sarcoptes scabiei*) and follicle mites (*Demodex folliculorum*) as a soap or ointment. It is used similarly for the mange mites of livestock, and as a dip for the depluming mite of poultry (*Chemidocoptes laevis* var. *gallinae*).

'Third generation pesticides'

This is the name given by Carroll M. Williams to insecticides that are either insect hormones or compounds closely related to them (Williams, 1967). The two drawbacks of most of the synthetic insecticides which have been discussed above are the lack of selectivity, and the rapidity with which insects can develop resistance towards them. It would be rash to claim that resistance to a hormonal preparation would be impossible, though its development is not likely to be rapid, but selectivity is a reasonable expectation. The most likely candidate at the moment is the juvenile hormone.

This material, released from the corpora allata, plays an important part in the orderly development of the insect. A second hormone, circulating in the blood, the moulting hormone, initiates the laying down of new cuticle by the epidermal cells, and the amount of juvenile hormone present determines the nature of this cuticle. In the larval stages, when a considerable amount of the juvenile hormone is present, further juvenile cuticle is laid down. In the last larval instar little or no hormone is present and adult cuticle is laid down. In those insects which pass through a pupal stage an intermediate amount of the hormone controls the laying down of the pupal cuticle. The juvenile hormone also plays a role in the adult insect, controlling the formation of yolk material in the fat body cells of females, and the formation of spermatophore material in the accessory glands of the males.

The amount and timing of the juvenile hormone is extremely important. If it is present at the wrong time, or in too large quantities, gross abnormalities of growth may occur. Thus if corpora allata from a young nymph, or material with juvenile hormone activity, are placed within the last larval instar of the bug *Rhodnius*, an extra instar, giant in form, results. The growth abnormalities usually result in the death of the insect. Juvenile hormone must be absent from the eggs of insects if these are to develop normally and hatch.

Juvenile hormone is known to act at the cellular level. Wigglesworth, a pioneer worker on insect hormones, suggested that the juvenile hormone may be a coenzyme for those enzymes which control larval development, or it may affect permeability in some way so that these enzymes can be brought into action. Alternatively, it may act directly on the nuclei of the epidermal cells, affecting the action of the moulting hormone, ecdyson, through effects on the gene system. Another view will be found in a collection of papers dedicated to Sir Vincent Wigglesworth (Novák, 1967).

Interest in the application of these substances to pest control was quickened by what is likely to become one of the classics of serendipity in the biological sciences. Williams attempted to rear specimens of the European bug *Pyrrhocoris apterus* in his American laboratory, but the culture repeatedly failed. The fifth larval instar, instead of moulting to the adult stage, produced an extra larval instar (and sometimes eventually a seventh) or changed into adultoids with many larval characteristics. The trouble was eventually traced to the paper towelling used to line the cages. This was manufactured from balsam fir, *Abies balsamea*, a pulp wood not used in Europe. The 'paper factor' was isolated and analysed by Bowers who found it to have a structure similar to that of the juvenile hormone (juvenile hormone or neotenin, 45; paper factor or juvabione, 46).

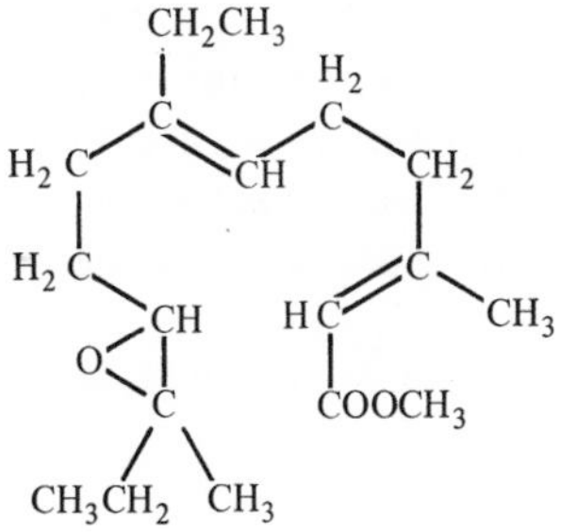

(45) Neotenin 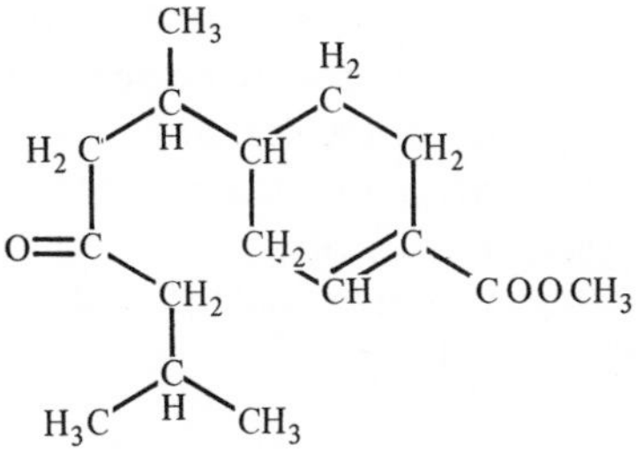   (46) Juvabione

Juvabione was found to prevent the emergence of fifth instars of the red cotton bug, *Dysdercus fasciatus*, from their cuticles. This important pest of cotton is a member of the same family as *Pyrrhocoris*, the Pyrrhocoridae. The hormone mimic also prevented the hatching of the eggs of *Pyrrhocoris*, but it was not toxic to eggs of *Rhodnius* and *Oncopeltus*. Juvabione therefore shows considerable specificity but Bowers doubts if its toxicity is restricted to the one family. Nevertheless, it seems likely that some analogues of the juvenile hormone that may be found in some plants in the future will be fairly specific, killing some insects feeding on plants, but not their parasites and predators.

One intriguing question is the function of juvabione in the balsam fir, especially as most of the economic species of the Pyrrhocoridae are pests of cotton, and none are found on balsam fir. Williams suggests that it is possible that such a bug did once attack this tree but that it is now extinct, or has changed to some other food.

The discovery of the juvenile hormone mimic in balsam fir has initiated the search for other mimics in other plants, although some positive results has been obtained even before the paper factor episode. In 1960, for example, Schneiderman and his colleagues reported definite hormone-like activity in ether extracts of soy-beans. The following year Schmialek (1961) isolated a material with similar activity from yeast and from *Tenebrio* faeces. This substance was identified as the terpenic alcohol farnesol (47). The related aldehyde farnesal was also found in the same materials,

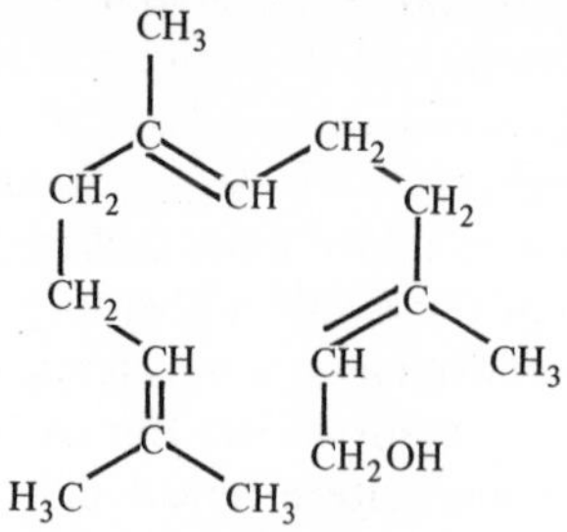

(47) Farnesol

and this too was active. The potency of these compounds is so high that it was thought at first that one of them must be the actual hormone.

Sláma (1969) summarizes the results of the search for juvenile hormone mimics in plants. The results have been a little disappointing (Sláma and his colleagues examined sixty plants chosen at random and found significant activity towards *Pyrrhocoris* in only two of them). He concludes that substances with juvenile hormone activity for a given species are not so widespread as had been hoped, but he pointed out that the probable specific nature of the compounds makes the search more difficult. It will probably be necessary to test candidate plants on a variety of insect species. Much of the search has, so far, been confined to coniferous trees since these are rich sources of terpenes. Mimics are not present, apparently, in the volatile fractions, i.e., turpentine.

In 1968 Bowers reported that compounds containing a methylenedioxy-phenyl group,

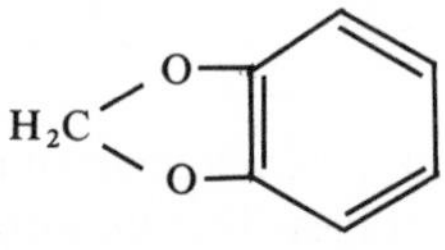

common in plant material, show juvenile hormone activity, often to a marked degree. This applies to even large molecules such as sesamin which shows slight activity. It will be remembered that this radical is common in synergists for pyrethrum.

The second hormone concerned in moulting has, like neotenin, been synthesized, but it seems less promising as a practical pest control agent. Its precursor is cholesterol, to which it is structurally very close.

Two moulting hormones (possibly three) have been extracted from *Bombyx* pupae which, of course, can be obtained in large quantities. The original extraction on which the analysis was based started with about four tons of fresh pupae. The two hormones are known as α and β ecdysone, and the second is probably 20-hydroxy-ecdysone.

Closely related steroids have been extracted from various plants, such as the conifer *Podocarpus* (Ponasterone and 20-hydroxyecdysone) and the fern *Polypodium* (20-hydroxyecdysone). The ecdysone analogues comprise over 1 per cent by dry weight of the fern's rhizomes.

The ecdysone analogues in plants do not seem to have any contact toxicity to the insects which feed upon the plants, but they may act as growth inhibitors. Certainly very few species feed upon *Polypodium*. The structures of ecdysone and some of the analogues are shown in (48).

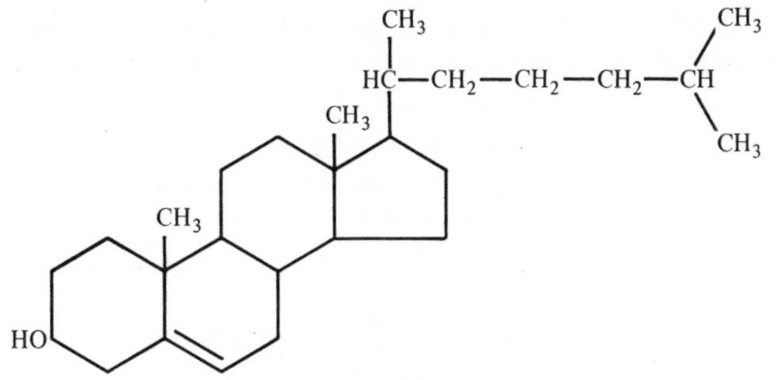

Cholesterol

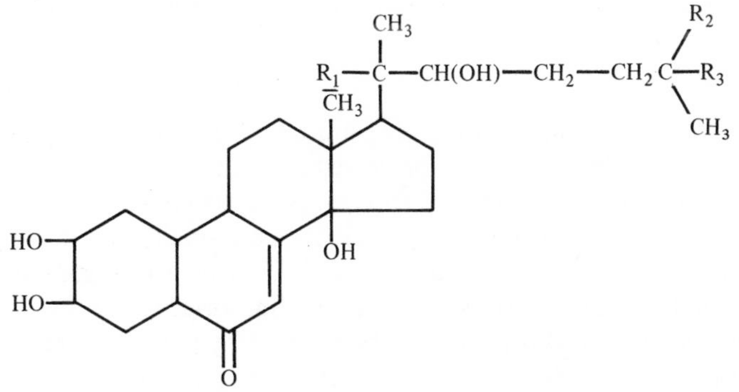

	R_1	R_2	R_3
α-ecdysone	H	CH_3	OH
β-ecdysone	OH	CH_3	OH
Ponasterone	OH	CH_3	H
Inokosterone	OH	CH_2OH	H

(48) Ecdysones and related compounds

Juvenile hormone and its mimics are more promising as pest control agents. When they are toxic to an insect (there is often a high degree of specificity) the toxicity is much greater than that of conventional insecticides such as DDT. At the same time they seem to present no risk to other organisms. They may be of value as direct insecticides, but they are more likely to be used as chemosterilants. A general discussion of the control of pest populations by the release of sterile insects will be found in chapter 8. A dose of less than 1 μg of the mimic DMF, the dihydrochloride of methyl farnesoate, causes a female *Pyrrhocoris apterus* to lay inviable eggs for the rest of her life, and males cultured in the laboratory can be dosed with a load of the material which does not damage them but which is sufficient to 'booby trap' the females with which they mate. Other insects which possibly could be controlled in a similar way, using juvenile hormone analogues as chemosterilants, are the human louse, *Pediculus humanus humanus*, and the yellow fever mosquito, *Aedes aegypti*.

An excellent account of insect endocrinology will be found in the text by Highham and Hill (1969), which also gives the most important references to work by Wigglesworth and others.

The defensive secretions of spiders, insects, and snakes, and their toxic salivas and venoms, are often efficient insecticides, for this is one of their main functions. The crude venom of the Australian bulldog ant, *Myrmecia gulosa*, when injected into house-flies shows approximately the same toxicity as DDT applied topically (Woods, unpub.). Man has not yet, to my knowledge, used any of these products as insecticides, but a study of their mechanisms of action could lead to new methods of poisoning insects. Unfortunately, most of them are complex mixtures of proteins and other substances, so it is unlikely that the same materials would ever be synthesized on a commercial basis.

As Frontali and Grasso (1964) have shown, venoms can display a high degree of selective toxicity. They isolated three toxic components from the venom of *Latrodectus tredecemguttatus*, an Italian relative of the notorious black widow spider. Two of these were toxic to house-flies, but not to mice, and the other toxic to mice, but not to house-flies.

Fumigants

The requirements of a fumigant are that it should be toxic to insects and mites, that it should penetrate the material to be protected easily, and that it should not damage this or leave harmful residues. As most fumigants are also toxic to mammals (and are often used for rodent control) a strong characteristic smell is an advantage. If it is to be used on seed corn, propagation material, or fruit a fumigant should not be phytotoxic, but for many purposes this is not important. Ideally, it should not present a fire risk, though this may often be reduced by the admixture of substances such as carbon tetrachloride.

Most fumigants are considered to be physical toxicants, but a few, such as methyl bromide and hydrogen cyanide, are chemical toxicants. It must either be a gas at normal working temperatures, or vaporize readily.

The following are some of the chief fumigants, together with their most important uses.

Carbon disulphide, CS_2
  Grain: usually in non-flammable mixtures.

Ethylene chlorobromide,
$$CH_2Br$$
$$|$$
$$CH_2Cl$$
  Fruit.

Ethylene dibromide,
$$CH_2Br$$
$$|$$
$$CH_2Br$$
  General fumigant.

Ethylene oxide,

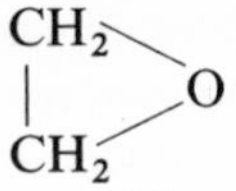

$$\begin{array}{c} CH_2 \\ | \quad\diagdown \\ \quad\quad O \\ | \quad\diagup \\ CH_2 \end{array}$$

Grain, cereals and some plant products. Also toxic to some bacteria, viruses, and fungi.

Hydrogen cyanide,   HCN

General fumigant.

Methyl bromide, CH_3Br

A widely used, non-flammable general fumigant, but with high mammalian toxicity.

Phosphine, PH_3

A grain fumigant generated from aluminium phosphide tablets.

Propylene oxide, $CH_2 \cdot CH \cdot CH_2 \cdot O$

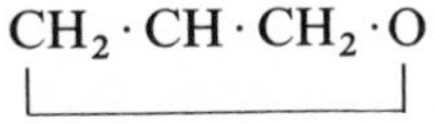

Used for the fumigation of dried fruit in packets.

A valuable booklet on insect fumigation has been published by the Food and Agriculture Organization of the United Nations (Monro, 1961).

This survey of insecticides has been brief and several pesticides which are still in use have been omitted. Lead arsenate, for example, is still employed in pest control, though on a diminishing scale. The treatment of fungicides and weed-killers must be even more condensed and readers seeking further information should consult the references given below.

Fungicides

Most agricultural and horticultural fungicides are used in one of two ways. They may either be applied to kill the mycelium of the fungus when this is exposed on the surface, or they may be applied as a protective coating to prevent the germination of spores, or penetration by the organism. Most fungi, of course, grow within the tissues of the host, and some are never exposed on the aerial parts of the plant, so eradicant fungicides are of little use for their control. There has been an intensive search for systemic fungicides, especially for the control of vascular diseases, and some promising materials have been found in the last few years.

Fungicides may be applied as foliage sprays, or to the soil, or as seed dressings. Many of the chemicals which are applied to the soil are used to kill nematodes, and fungal control is considered to be of secondary importance.

Sulphur and its derivatives

Sulphur has already been discussed briefly as an insecticide and acaricide, but it was probably first used as a fungicide and disinfectant. McCallan believes that Achilles used sulphur to cleanse a libation cup that had been contaminated by mildew from the clothing with which it had been stored (*The Iliad*, XVI, see McCallan, 1967). McCallan suggests that its fungicidal properties were forgotten or ignored from Classical times till the beginning of the nineteenth century when Forsyth, the Royal gardener, included it in a recipe for mildew on fruit trees.

During the nineteenth century it was increasingly used for powdery mildew control on fruit, but it was given a great impetus when the powdery mildew of grapes, *Oidium tuckeri* (*Uncinula necator*), was introduced into England in 1847. The fungus was named after Edward Tucker, a gardener in Margate, who first used sulphur as a remedy for the disease. The pathogen spread to the European vineyards where, after 1850, control was achieved by sulphur dusting (Large, 1940).

Sulphur in its various forms and formulations is still widely used for the control of powdery mildews and other fungi on fruit trees and bushes (e.g., *Podosphaera leucotricha* on apple; *Sphaerotheca mors-uvae*, American gooseberry mildew; apple and pear scab, *Venturia inaequalis* and *V. pirina*).

The powdery mildews are fungi which grow superficially on the surface of fruit and leaves with only non-reproductive haustoria penetrating below the surface. The 'powder' is the growth of mycelium and fructifications. Such fungi are, of course, susceptible to suitable curative fungicides applied to the surface, and among these are the sulphur preparations. *Erysiphe graminis* on cereals and *Phyllactinia corylea* on ash and hazel are exceptional in that their mycelia can be endophytic.

Sulphur can be applied in various forms as dusts or sprays. The more finely divided the sulphur, the greater are its fungicidal properties. Thus colloidal sulphurs are formulated so that there is at least a 40 per cent content of sulphur, 90 per cent of which must consist of particles of 6 μ or less in diameter, and 55 per cent of particles of 2 μ or less.

Lime sulphur has long been prepared on the farm or at the orchard by boiling together sulphur and lime water, but it is now available as a standard product. It consists mainly of calcium polysulphides, $CaS.S_x$, which break down to release elemental sulphur.

A number of varieties of fruit are 'sulphur shy' and may be damaged by sulphur sprays. Varieties of apples, pears, gooseberries, and blackcurrants are all affected and the manufacturer's instructions should be consulted before application.

The copper fungicides

Although copper sulphate had been used since the eighteenth century as a seed treatment for bunt of cereals, the copper compounds, like sulphur, were really launched as fungicides by a disease which threatened the vineyards of France. The introduction of American rootstocks to confer resistance against *Phylloxera* merely traded one pest for another for some of the imported material carried the downy mildew fungus, *Plasmopara viticola*. Millardet, the botanist responsible for the introduction of the *Phylloxera* resistant rootstocks also found a remedy for the new malady. He noticed that along the roadsides of the Gironde the vines were daubed with a mixture of lime and copper sulphate to deter pilfering, and that only those leaves which had been treated in this way escaped the ravages of the disease. Three years later, in 1885, after experimenting with various mixtures of the two chemicals he introduced the fungicide which came to be known as Bordeaux mixture.

For high-volume spraying the standard mixture is $10:12\frac{1}{2}:100$ ($4.5:5.5:454$) (10 lb (4.5 kg) copper sulphate, $12\frac{1}{2}$ lb (5.5 kg) hydrated lime, 100 gallons (454 litres) water). The mixture is difficult to prepare and must be used on the same day. It is also difficult to spray since it causes wear and blockage of spraying nozzles. It is, however, extremely fungitoxic and tenacious, and is thus still widely used despite the fact that some of the later copper formulations are more convenient to use.

The lime is the main source of difficulty in spraying the standard Bordeaux mixture, so for low-volume spraying the lime may be reduced to give an $8:4:20$ ($3\cdot6:1\cdot8:90$) mixture. An alternative way of avoiding the troubles caused by lime is to substitute washing soda or soda ash, the resulting preparation being known as Burgundy mixture.

The chemistry of Bordeaux mixture and its mode of action is still, after almost a century, far from being understood. The fungicidal activity is thought to arise from water-soluble copper compounds which are slowly released from the precipitate by the chemical action of the pathogen.

Other copper compounds are used as fungicides, but the most important is copper oxychloride, a complex substance which approaches, in composition, $3Cu(OH)_2\text{-}CuCl_2$. This can be applied in a dust formulation, or as a spray. In colloidal copper formulations the oxychloride is so finely ground that the particles, with the aid of dispersing agents, remain permanently in suspension in the paste and in the spray tank. Copper fungicides are also formulated mixed with mercury compounds or with dithiocarbamates, two groups of fungicides mentioned later.

In Europe the main use for copper fungicides is the protection of potatoes from attack by late blight, *Phytophthora infestans*. The object of spraying is to keep the foliage healthy as long as possible. If a blight attack does develop the foliage has to be destroyed some time before lifting, usually by a chemical treatment, to prevent spores from infecting the tubers and causing a rot in store. The grower has to decide, in a blight year, when is the best time to stop spraying with fungicides and to destroy the haulms. If left too late the disease spreads enough in the crop to cause tuber infection; if done too soon there will be a shortened growing time and a loss in yield.

Formerly blight sprays were regularly applied as an 'insurance' but the passage of machinery through the crop, and the copper spray itself, damage the crop. It is now possible to issue blight warnings through the radio and television whenever the weather conditions are suitable for an attack to start. Also infra-red aerial photography can be used to pick out foci of infection, and it has been suggested that orbiting satellites may eventually be used for this work and for similar disease scouting. With these aids the farmer is able to restrict his spraying to those times when they are absolutely necessary. Unfortunately, blight warnings come in bad weather and it may be impossible to get on to the land to spray. Alternatively, if the grower depends upon contract spraying, there may be such a large demand from other growers that the disease spreads considerably before the crop is sprayed. Ordish and Mitchell estimate that twenty years ago about 20 per cent of the potato acreage (presumably, of the world, though this is not stated explicitly) was sprayed against blight, and that now spraying is restricted to about 15 per cent each year. Presumably the number of annual applications on sprayed fields has also decreased (Ordish and Mitchell, 1967).

Other major diseases which are controlled by the copper fungicides are grape downy mildew (see above), black pod of cacao *Phytophthora palmivora* (which also attacks a wide range of other crops, including rubber, cotton, and various palms), blister blight of tea, *Exobasidium vexans*, and various seed-borne diseases of cereals (Ordish and Mitchell, 1967).

Copper fungicides tend to be alkaline in reaction so that they are incompatible with most organophosphate insecticides. If they are used together spraying must be carried out immediately after mixing.

Organic fungicides

Sulphur has long been used in the vulcanization of rubber, and a theory that this process might resemble its action on fungi instigated an examination of various organic compounds that are used as rubber accelerators. This led to the discovery of the fungicidal properties of the derivatives of dithiocarbamic acid,

$$\text{H}_2\text{N}-\overset{\overset{\displaystyle S}{\|}}{\text{C}}-\text{S}-\text{H}$$

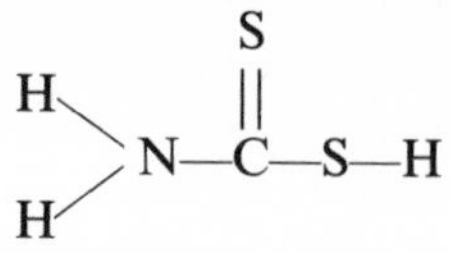

the first patent in America being granted in 1934.

The first of the compounds to be used commercially was thiram (tetramethyl-thiuram disulphide, 49). This compound had been known earlier as an insect

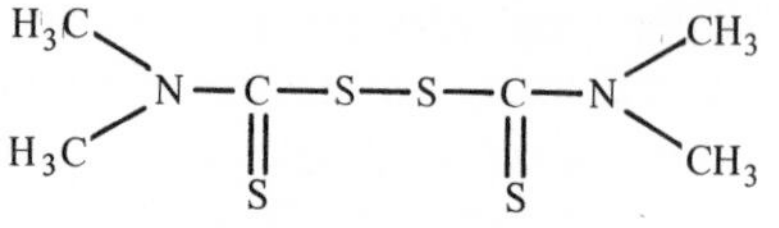

(49) Thiram

repellent (and is now being investigated as a mammal repellent) but this property was soon to be overshadowed by its uses as a fungicide. It was probably first used as a seed dressing to prevent the seedling blight of flax, *Colletotrichum linicola*, and then as a seed dressing for many other fungal diseases. By about ten years later thiram had also established itself as a protectant spray.

Thiram is still widely used against *Botrytis cinerea* on lettuce and strawberries, *Uromyces dianthi*, carnation rust, apple scab (as a post-blossom spray) and various other fungal parasites. As a seed dressing it is used as a protectant against damping-off, foot rot, and seed-borne *Verticillium* in such crops as flax, linseed, lucerne, and peas. It has now been partially replaced by metallic salts of dithiocarbamic acid and its analogues. Among these are the zinc salts ziram and zineb, and maneb and nabam, which contain, respectively, manganese and sodium.

Maneb and zineb are now widely used for the control of potato blight, *P. infestans*. They are less phytotoxic than the copper preparations, are equally or more fungi-toxic, at least just after application, but they are not as persistent. A copper–maneb mixture is also marketed, a preparation which combines the better qualities of the two fungicides.

Captan (50) is an organic fungicide which has come to rival the dithiocarbamates

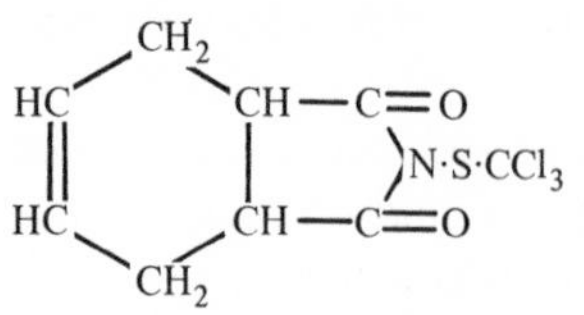

(50) Captan

in usefulness. Chemically it is *N*-(trichloromethylthio)cyclohex-4-ene-1,2-dicarb-oximide. A very persistent fungicide, it is somewhat specific in its action. Thus it can be used for the control of *Venturia inaequalis* on apple, but it has little effect on powdery mildew, *P. leucotricha*. It does, however, have the very useful property of reducing the incidence of bitter rot of apples, *Gloeosporium* spp., a troublesome disease that affects fruit during late storage. It may also be used to protect tomatoes from *Didymella* stem rot, soft fruit from *Botrytis* spp. and roses from black spot, *Diplocarpon rosae*.

The ineffectiveness of the earlier organic fungicides against powdery mildews was a serious disadvantage which prevented them from replacing sulphur on the apple crop, even though many varieties are sulphur shy. Some more recently introduced organic compounds have given excellent control of this troublesome disease, with, often, the added advantage of red spider mite control. Two examples are binapacryl (51). 2-(1-methyl-n-propyl)-4,6-dinitrophenyl 2-methyl-cronate, and chinomethionat (43), and others have been mentioned earlier.

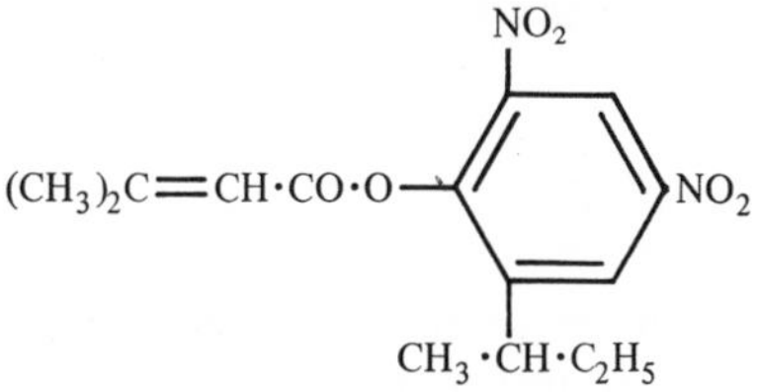

(51) Binapacryl

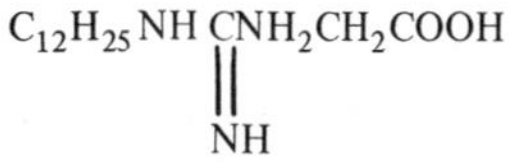

(52) Dodine

A further interesting fungicide for use on apples is dodine, dodecylguanidine acetate (52), a compound particularly effective against apple scab. Its great value lies in its ability to eradicate a recent infection, but, because it is a cationic surfactant, care must be taken when mixing it with other pesticides.

Organophosphates as fungicides

There has been considerable interest recently in the use of certain organophosphate compounds which can be used to control the rice disease blast, *Piricularia oryzae*. Organomercurials, at fairly high rates, have been used to control this widespread and destructive disease, but as some of the organophosphates being investigated also control the leaf-hopper vectors of the dwarf virus, it is possible that these compounds will replace the mercurial compounds. Mercury compounds are, of course, also toxic to mammals. Much of the work has been carried out in Japan with the compounds bearing the trade names Hinosan® (53), Cerezin® (54) and Kitazin® (55).

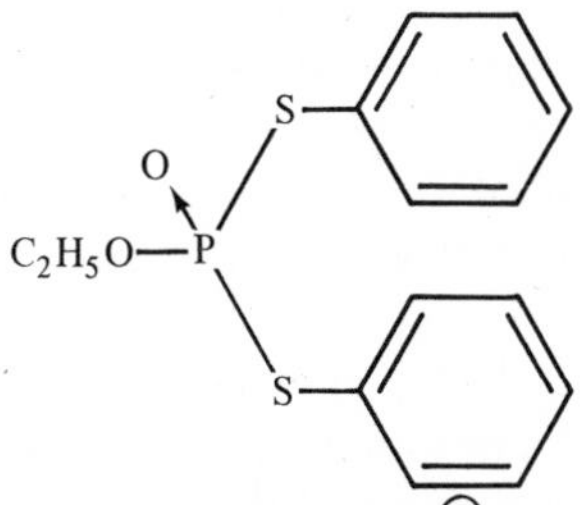

(53) Hinosan ®

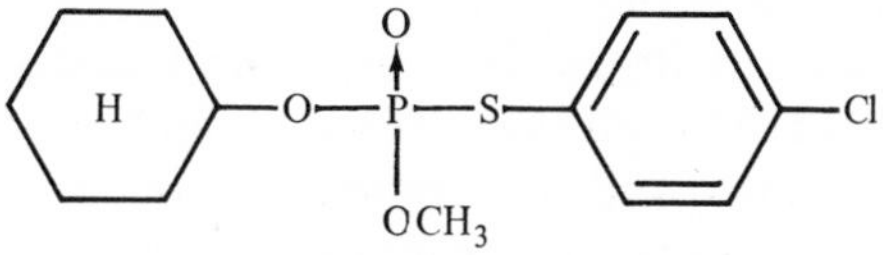

(54) Cerezin ®

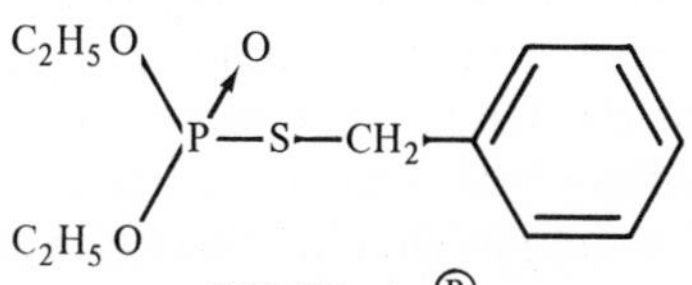

(55) Kitazin ®

A number of organophosphates show promise against powdery mildews and at least one of them, Wepsyn ®, has been marketed for some time. Wepsyn ® shows systemic activity against apple mildew and cereal mildew, *Erisyphe graminis*, in laboratory and field trials and, of course, it also has insecticidal properties.

A further organophosphate reported to have excellent activity against a number of powdery mildews is Dowco 199 ® (56), and it is also fungitoxic to apple scab, rose black spot, and the brown rot of peaches, *Monilia fructicola*.

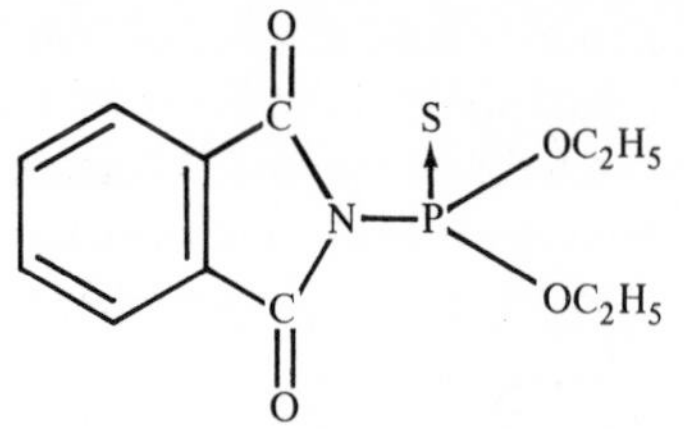

(56) Dowco 199 ®

Several of the organophosphates mentioned in the section on insecticides are also reported to inhibit the growth of mycelium of various fungi such as *Pythium debaryanum*, *Rhizopus nigricans*, *Rhizoctonia solani* and *Verticillium albo-atrum*.

Various reports have suggested that systemic organophosphates also suppress potato blight but after a series of glasshouse trials McIntosh and Eveling (1965) concluded that the effect was indirect, possibly resulting from the reduction of the amount of honey dew produced by aphids.

The use of organophosphates as fungicides is reviewed by Scheinpflug and Jung (1968) in a collection of papers presented to the pioneer worker, Gerhard Schrader.

Organomercurial fungicides

Organomercurials find two main uses in British agriculture and horticulture: spray materials for the control of apple scab in certain areas where other fungicides give poor results, and as seed dressings. The sprays can cause some damage to certain tree varieties, and especially to Cox's Orange Pippin.

In seed dressings they are used alone or mixed with insecticides such as lindane or dieldrin. The fungicide controls certain of the seed-borne diseases of sugar-beet and cereals, while the insecticide protects the seedlings from the attacks of wheat bulb fly and wire-worms.

Outside the British Isles organomercurials are commonly used to protect cotton from seed- and soil-borne diseases, and rice from blast, *Piricularia oryzae*. The rice seed is dressed with a mercurial compound at the rate of 200 g/100 kg, and during growth the plants are treated with mercury sprays, or with mercury–lime dusts, in the evening before dew-fall.

Phenyl mercuric acetate, $C_6H_5HgOCOCH_3$, is probably the most commonly used organomercurial in Britain, but various other compounds have also been employed in the United Kingdom and elsewhere. In Sweden, for example, alkyl compounds have been widely used, and they are probably responsible for the serious mercury poisoning of wild life there (see chapter 4).

Organotin fungicides

In the section on organotin compounds as insecticides it was mentioned that they also have fungicidal properties. Triphenyl tin salts are used to some extent on ornamentals and root crops, but the trialkyl derivatives are too toxic for use on growing crops.

Tributyl tin was first used for the stabilization of some transparent plastics against photochemical change, but was soon found to have biocidal properties. It is now also used for rot-proofing of canvas and other fabrics, in anti-fouling preparations and for the protection of marine timbers.

Antibiotics in the control of plant diseases

Many of the several hundred antibiotics now known are fungicidal, but most are too unstable for use in crop protection (Martin, 1964). Others, such as streptomycin, are employed in agriculture for their bactericidal properties. Most of the Gram positive plant pathogens are seed-borne, and some of them may be controlled by streptomycin. It can also be used against the Gram negative bacterium which causes fire-blight of pears, *Bacterium amylovorus* and against the bacterial canker of cherry, *Pseudomonas mors-prunorum*. Formerly, Bordeaux mixture was used for bacterial canker, but it was found to be unsatisfactory because of the difficulty in spraying the mixture, and the necessity for a complete cover. Streptomycin can be applied much more easily and it is also translocated to some extent through the leaves. Despite this Crosse (1962) found that it was not as effective as even low-strength Bordeaux mixture because the latter is more destructive to the leaf surface inoculum.

Cycloheximide is obtained from the same micro-organism as streptomycin, namely *Streptomyces griseus*. The discovery of its fungicidal properties in 1946 raised hopes that antibiotics would be widely used in pest control, but although it controls wheat rust it is possibly too phytotoxic and too poisonous to mammals for use on food crops (Martin, 1964).

Griseofulvin, an antibiotic obtained from the fungus *Penicillium griseofulvum*, is translocated in plants and has been used to protect tomatoes from *Alternaria solani* and lettuce from *Botrytis* spp. It seems to be more effective against powdery mildews when taken up by the roots than when used as a spray. In Japan it is said to be (sometimes) used against apple blossom blight, *Sclerotinia mali* and *Fusarium* wilt of melons (Dekker, 1971). Formerly, it was thought to be toxic only to fungi containing chitin but it is now believed to interfere with nucleic acid metabolism.

In 1963 there were few practical applications for antibiotics in agricultural pest control but since then there has been a great development in their use in Japan where, in 1968, antibiotics made up almost seven per cent of the entire pesticide production. This arose from the need to find replacements for mercurial compounds for the protection of rice from attacks of rice blast, *Piricularia oryzae* (see the two

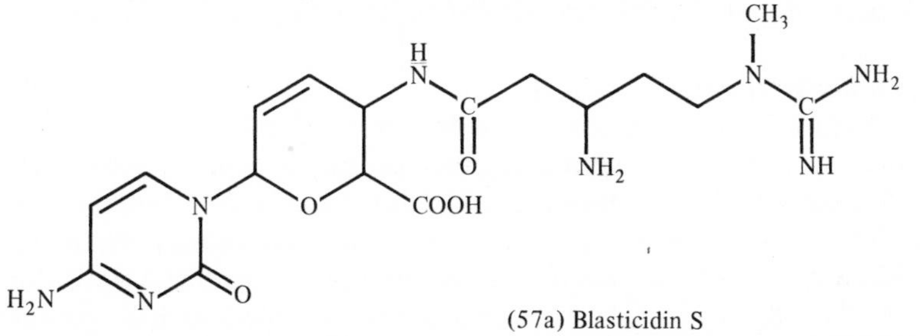

(57a) Blasticidin S

preceding sections). The most important antibiotics to be used for the control of this disease have been, so far, blasticidin S (57a) and kasugamycin (57b). Blasticidin S

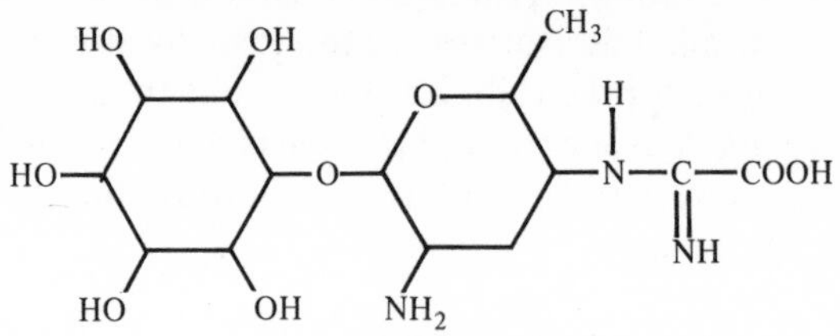

(57b) Kasugamycin

contains a nucleoside, cytosinine and an acid – blastidic acid – which were previously unknown. This antibiotic is produced by the soil actinomycete *Streptomyces griseochromogenes* which also synthesizes the structurally unrelated blasticidins, A, B, and C. In passing it should be noted that the term antibiotic does not refer to a chemically homogenous group of compounds; antibiotics are simply the products of micro-organisms which can inhibit the growth or even kill other micro-organisms at very low dilutions.

Although blasticidin S checks the growth of many species of bacteria it is very selective towards fungi. Fortunately, rice blast is very sensitive. It is, however, phyto-toxic but its effects on the fungus disease more than compensate for the slight plant damage which it causes. It is also toxic to mammals (LD 50 for mice, 39 mg/kg) and fish.

Blasticidin S, like some other antibiotics, has anti-viral activity. It reduces the incidence of the rice stripe virus but it is possible this is due to some action against the insect vector of the disease, *Laodelphax striatellus*.

The more recently discovered antibiotic kasugamycin (from *Streptomyces kasugaensis*) is less damaging to plants than blasticidin S, and its acute and chronic toxicity to mammals is very low. It has little bactericidal activity and, like blasticidin S, it is very selective towards fungi. It appears to be inactive against rice blast cultures *in vitro*; the juices of the rice plant apparently activate it in some way.

Blasticidin S and kasugamycin have been the antibiotics most used for the control of rice blast, but there are several other promising substances, including the polyoxins and aureofungin. Some antibiotics show promise against viral diseases of plants. They include blasticidin S, cycloheximide, and bihoromycin. The latter appears to have a translocatory effect against tobacco mosaic virus. Unfortunately, no antibiotic as yet appears to be suitable for the practical control of plant virus diseases.

Dekker (1971) has recently reviewed the use of antibiotics in agriculture, and much of this section is derived from this review. Jung and Scheinpflug (1970) also discuss various aspects of pest control in rice in Japan, including the use of antibiotics.

Miscellaneous systemic fungicides

There has been a long search for fungicides with systemic properties, particularly for the control of soil-borne pathogens, such as the *Verticillium* wilts, which cause vascular diseases of plants. During the last few years a few promising synthetic compounds have been discovered. Among these are two closely related oxathiins, namely Vitavax® (58), 2,3-dihydro-5-carboxanilido-6-methyl-1,4-oxathiin, and Plantvax®, its dioxide, with two oxygen atoms attached to the sulphur of the

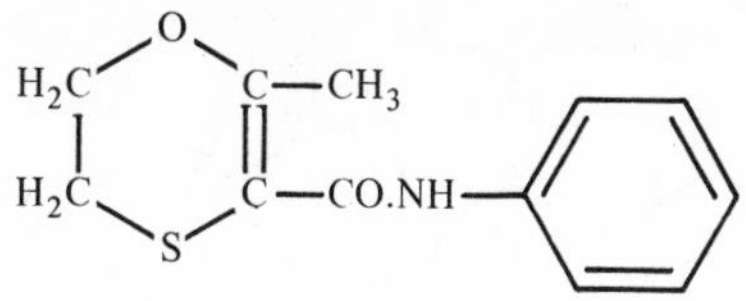

(58) Vitavax ®

oxathiin ring. Vitavax® shows promise as a seed dressing or soil drench for the treatment of cereals against loose smuts (*Ustilago avenae, Ustilago nuda* etc.), seed-borne pathogens which could not be destroyed by chemical surface treatment of the seed. It may also be used in the same way to control *Rhizoctonia* diseases of cotton, sugar-beet, and vegetables. Plantvax® can be used as a seed dressing, and there are indications that it may be of value in the control of rust diseases of cereals and vegetables. The oxathiins are currently being tested against a number of basidio-mycetes which cause crop diseases.

The pyrimidine fungicide, methyrimol (59), is a further systemic compound.

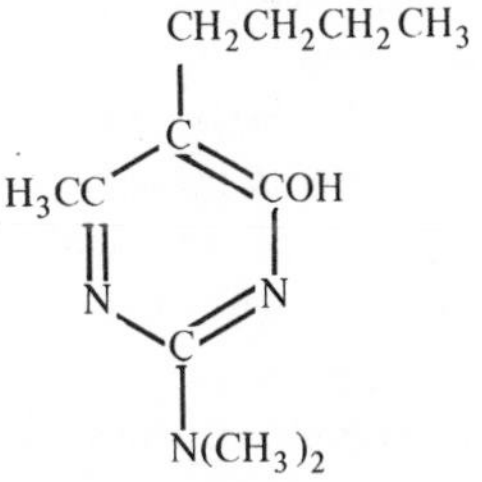

(59) Methyrimol

Chemically it is 5-butyl-dimethylamino-4-hydroxy-6-methylpyrimidine. It has been shown to be systemically active against powdery mildews of curcubits, chrysanthe-mums, cineraria, and sugar-beet. It is less active against the powdery mildews of apple, pears, cereals, and roses and, unfortunately, it is without effect on vine powdery mildew. Although it breaks down fairly rapidly within the plant after being taken up from the soil, it is fairly persistent in the ground.

Benlate® (Du Pont Fungicide 1991) (60), is methyl 1-(butylcarbamoyl)-2-

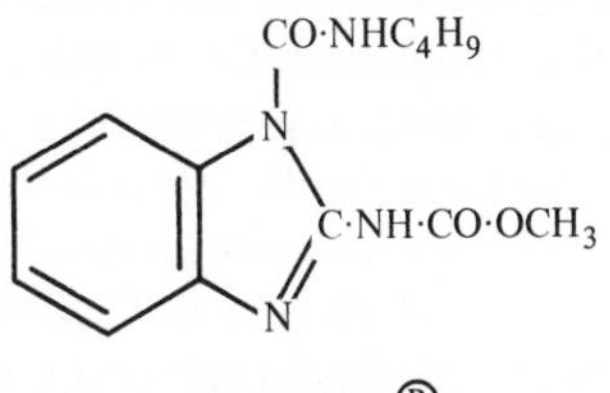

(60) Benlate ®

benzimidazolecarbamate, and is thus a systemic fungicide from yet another chemical group. It has controlled foliar diseases when applied to the plant or to the soil in which the plant was growing. As a spray it was effective against apple scab, and it also appears to act systemically against this disease, and against *Cercospora* leaf spot in sugar-beet, rice blast, various powdery mildews, and *Verticillium* and

(61) TBZ ®

Rhizoctonia within stems. TBZ® (61) is a second benzimidazole compound with systemic properties against *Verticillium*. Chemically it is 2-(4-thiazolyl) benzimidazole.

Nematicides

It has already been pointed out that certain of the organophosphate insecticides have nematicidal properties, but they will not be discussed further in this section. Here we shall consider those compounds which are used to treat the soil, as fumigants, before planting. Most of them have wide spectra of activity so that they may be considered as general sterilants. The treatments are thus effective against fungal pathogens, many bacteria, and weed seeds.

Treatment is carried out sufficiently early for the material to break down or disperse before the crop is planted. This is largely to avoid plant damage, but it is also to kill the nematodes while they are free within the soil. Certain species of nematodes spend all or part of their life within the tissues of the host, where they are difficult to reach with nematicidal compounds.

The effectiveness of the treatment depends upon several factors, the most important of which are soil type and conditions, and the temperature. As the materials work as fumigants a warm soil is necessary, so that the chemical can disperse rapidly. It disperses evenly if the soil tilth is as good as that used for drilling the seed, and if there is not too much free water which would block some of the channels. Soils vary in their ability to absorb the fumigant: organic soils can often inactivate so much of the material that the control of the nematodes fails.

Too rapid a dispersion is, of course, undesirable, as it leads to too great a loss from the top soil to a depth of about 7 cm, where nematicidal concentrations are difficult to maintain. Much can be done by covering the soil with polythene sheeting, or by using a sealant such as water or xylenol, but even then there is a rapid loss laterally. Conversely, the lower layers of the soil are not reached by the chemical, so these serve as sources for the eventual recolonization of the upper layers by both harmful and beneficial organisms. It seems that, in general, the beneficial saprophytic species are the most resistant among the fungi, and these recolonize the treated parts first.

Soil treatments are often carried out more or less routinely without a detailed knowledge of the species concerned. This is to be regretted as there is a well-marked seasonal migration of some species between the various soil levels, and this should be taken into account when timing the treatments.

Certain of the compounds used are either soil nutrients, or break down to nutrients (for example, calcium cyanamide to urea), while others, such as D-D® increase the availability of nutrient nitrogen through their action on the soil micro-organism composition. Such actions lead to an increase in plant growth, which is difficult to distinguish from that resulting from nematode control.

Some of the fumigants used are also employed for the control of insects. Methyl bromide is an example, but its volatility makes it difficult to contain, even when the soil surface is sealed off in some way, and it is also very toxic to mammals. It is more useful as a nematicide when it is used to treat infested seed or plant material in the conventional way, that is, in an airtight chamber. Chloropicrin (nitrochloroform: $NO_2 . CCl_3$), the tear gas of the First World War, is a substitute for methyl bromide, because it is both nematicidal and fungitoxic, and, unlike methyl bromide, gives ample warning of its presence.

Other simple substituted hydrocarbons are ethylene dibromide, and the constituents of D-D ®. Strictly, ethylene dibromide is not an ethylene compound: chemically it is 1,2-dibromoethane, CH_2BrCH_2Br. It is usually applied to the soil in an organic solvent. D-D®, the most widely used nematicide so far, is a 1:1 mixture of 1,2-dichloropropane, $CH_2Cl . CHCl . CH_3$ and 1,3-dichloropropene, $CHCl:CH . CH_2Cl$. It is usually injected into the soil at rates of between 220 and 670 kg per ha. It is said to be somewhat more effective than ethylene dibromide for the control of *Heterodera* spp. and lesion nematodes, and just as effective against *Meloidogyne* spp.

A more recent related compound, 1,2-dibromo-3-chloropropane (Nemagon®) is less phytotoxic than D-D®, and has been used, under controlled conditions, on established crops. It has been applied, for example, to citrus trees in California. One side of the row was treated in one year, and the other side the next. This allowed the plant roots to recover from any damage on the first application before the second was made.

Soil fumigation is expensive, as the material must be injected in a grid pattern at about 40-cm spacing, and an expensive cover is often needed. It is not commonly practised with field crops, but is a routine procedure with many glasshouse crops. Obviously a non-volatile compound which could be worked into the soil, or applied as a soil drench, would be more useful, but cheap compounds of this type are elusive. An inorganic mercury salt has been used on high-value potato land, but the most successful drench so far has been an organophosphate (here we must make an exception) with the trade name Nellite®, *O*-phenyl-*N*,*N'*-dimethylphosphorodiami-

(62) Nellite ®

date (62). One reason for its effectiveness is that it is not markedly adsorbed by soil organic matter as it is an extremely hydrophilic chemical.

Molluscicides

There are, from the pest control point-of-view, two kinds of snails: vectors of diseases such as bilharzia (schistosomiasis) and pests of crops. In general, different kinds of chemicals are used for the control of the two groups. Manufacturers appear to be concentrating at the moment on the search for molluscicides for the control of aquatic snails which are vectors of pathogens, but there is still interest in chemicals for agricultural pests.

Snails and slugs are particularly harmful to seedling crops, and to winter-sown wheat before emergence. The best-known chemical for their control is metaldehyde, a polymer of acetaldehyde. Metaldehyde (63) is also used as a convenient solid fuel

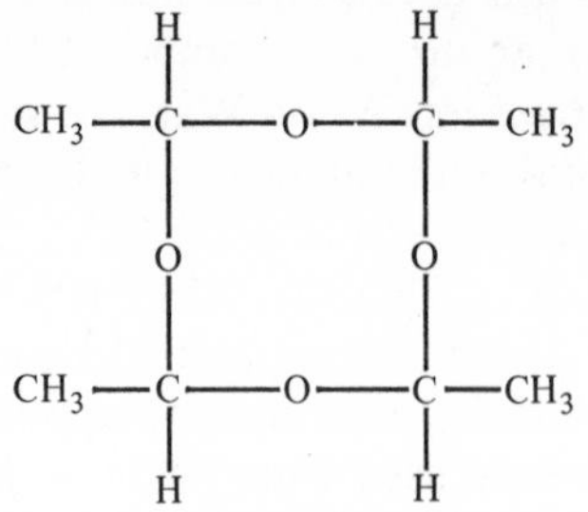

(63) Metaldehyde

for picnic stoves, and its biological activity is said to have been accidentally discovered on just such an expedition (Hartley and West, 1969). It is toxic to snails by contact and by ingestion, but it appears to work by immobilizing them and inducing them to produce a copious slime. The snails eventually die through desiccation. Its mammalian toxicity, about 500 mg/kg oral, is quite low. Normally, it is supplied mixed with some bait material such as bran.

Metaldehyde may be used to control damage by the field slug, *Agriolimax reticulatus*, on winter-sown wheat. The chemical is applied at a rate of $\frac{1}{2}$ lb (900 g) in 28 lb (31 kg) of bran per acre (ha), and it is usually most effective when broadcast on a warm, damp evening two or three days after the end of a dry spell (Edwards and Heath, 1964).

DNOC and dinex (dinitro-*o*-cyclohexylphenol) also kill some snails when they are applied as herbicides, but it is doubtful whether the control is usually sufficient or not. Copper sulphate solutions are sometimes used on grass in liver-fluke meadows to kill small *Limnaea* snails. A more effective compound is Frescon®, *N*-trityl morpholine (64), a product which was released in 1966.

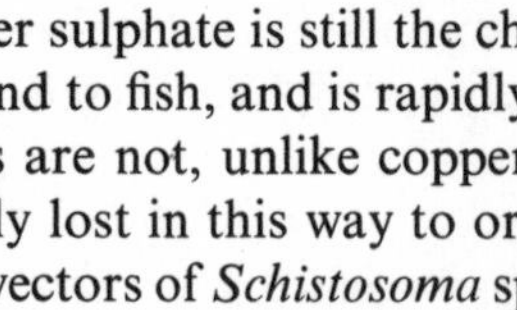

(64) Frescon®

Both copper sulphate and Frescon® are unusual in that they can also be used against aquatic snails. Copper sulphate is still the cheapest of aquatic molluscicides, but it is very toxic to algae and to fish, and is rapidly precipitated in alkaline waters.

The organic molluscicides are not, unlike copper compounds, adsorbed by clay particles, but they are rapidly lost in this way to organic muds. Unfortunately, the aquatic snails which are the vectors of *Schistosoma* spp. occur in stagnant water or in slow-flowing rivers with large amounts of such muds. It is also difficult to disperse molluscicides in these slowly moving waters. The organic compounds such as Frescon® and niclosamide are, however, less phytotoxic than copper sulphate and another commonly used chemical, pentachlorophenol.

Niclosamide, 5,2'-dichloro-4'-nitrosalicylanilide (65) was introduced in 1960 and has since been much used in Egypt for the control of vectors of bilharzia pathogens

(65) Niclosamide

(*Schistosoma* spp.). It is toxic to most of these at concentrations of less than 1 p.p.m., but its mammalian toxicity is low – rats, oral, 5000 mg/kg. An unusual application was the killing of snails in a swimming pool where the bathers often suffered from swimmer's itch. This annoying complaint is caused by the fruitless efforts of cercariae which are normally parasites of birds, to penetrate the human skin.

The application of molluscicides for the vectors of bilharzia is only one approach to the problem. If the waterways and ponds in the areas in which the fluke is epidemic were not used for both drinking and excretion, and if irrigation canals could be kept free-flowing, with little or no vegetation, bilharzia would decline rapidly.

Pesticides for the control of vertebrate pests

Many vertebrates, including even the elephant at times, can be pests, but among the mammals by far the most important are the rodents and the rabbits. Rats are, of course, direct pests, consuming and spoiling vast amounts of food, but they are also concerned with bubonic plague.

Almost all poisons that can be used against vertebrate pests are also highly toxic to man and his domestic animals, but there are one or two remarkable examples of specificity. Even with generally poisonous materials the danger can be lessened by careful use and, in any case, many are refused by man and domestic animals.

In the early 1950s the anticoagulant compounds were introduced for rodent control, but the older acute poisons were not completely displaced. The anticoagulants are chronic poisons; the baits must be laid at intervals so that the rats and mice will feed on them several times in order to accumulate lethal amounts. This repeated application is reasonable on a small scale, but it might prove too expensive for an eradication campaign in a city.

As they are cumulative poisons the anticoagulants are slow in action so that when an outbreak of plague threatens it may be necessary to use acute, rapidly acting poisons in their place, after the destruction of the fleas by insecticide application to the habitat. A further disadvantage of the anticoagulants is that resistance to them has appeared in a number of rat populations.

A desirable feature in a rodenticide is that it should induce the rats to emerge and die in the open, so that rotting carcases are not left in inaccessible places. If the rodenticide does have this property it should also break down in the poisoned animal so quickly that it does not affect cats or dogs that may devour the corpse.

Red squill, the powdered inner flesh of the bulbs of *Urginea* (*Squilla*) *maritima*, has been known to be poisonous to rats since the Middle Ages, and it is still used in many parts of the world. Although it is poisonous to other mammals it usually makes them vomit (which the rodents apparently do not do) so that the material is not particularly hazardous. It also decomposes fairly rapidly in the body of the rat, although it has a reasonable shelf life.

The rodenticide antu, 1-naphthylthiourea (66), was introduced in 1946 and proved

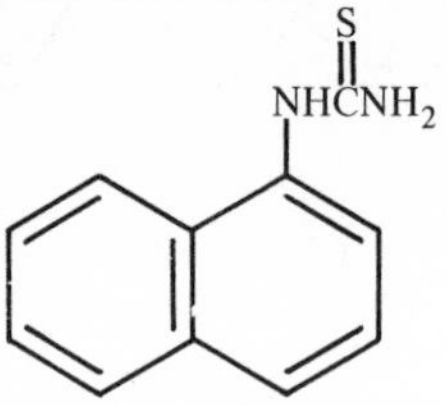

(66) Antu

effective against *Rattus norvegicus*: *R. rattus* and mice are much more tolerant. Furthermore, it induces extreme bait shyness in those individuals which do not consume a fatal dose, so it is rarely worthwhile baiting with it more than once a year.

Various fluorine compounds such as sodium fluoroacetate (1080 ®) and fluoroacetamide are used in several countries but in the United Kingdom their sale to farmers and growers is banned. There is, unfortunately, no specific antidote for these chemicals. Their high mammalian toxicity also resulted in their being abandoned as aphicides in Britain.

These fluorine compounds are best applied by trained workers, and then only in places such as sewers, ships holds, and closed warehouses where entry can be prevented. The liquid baits are preferred to the solid as the rats cannot transport them from the baiting station.

Thallium sulphate is not only used as a rodenticide, but also for the control of coyotes, jackals, and some pest birds. It is a dangerous compound: it is readily absorbed through the skin and often symptoms do not appear until some time after a dangerous dose has been taken in.

Other acute rodenticides that are, or have been used, are yellow phosphorus, zinc phosphide and arsenic compounds, but the most remarkable compound is norbormide, 5-(α-hydroxy-α-2-pyridylbenzyl)-7-(α-2-pyridylbenzylidene-norbon-5-ene, 2,3-dicarboximide (67), for this appears to be specific to the genus *Rattus*. The LD 50s for the Norwegian rat and for the black rat are 12 and 60 mg/kg oral, but

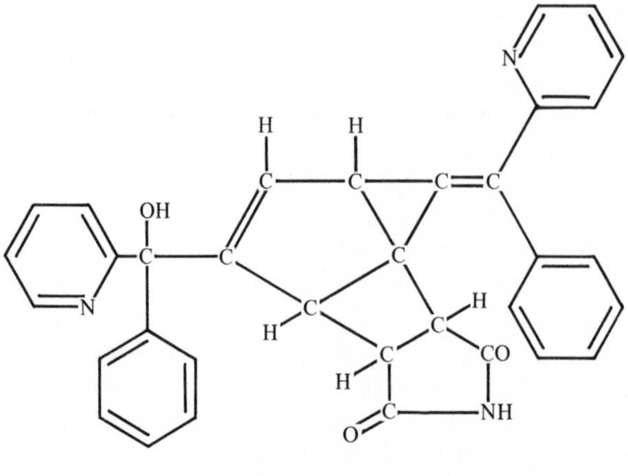

(67) Norbormide

dosages of 1000 mg/kg produced no effects on dogs, cats, and monkeys. The results in the field have ranged from poor to excellent, largely because of its poor acceptability to rats and because it induces bait shyness. As Drummond (1966) has observed what is needed is less specific toxicity than specific acceptability.

The advantages of the anticoagulants, which appeared in the early 1950s, are the non-development of bait shyness in the rats when used at the proper rates, the lack of danger to birds and other mammals, and the existence of a good antidote, vitamin K, if needed.

The anticoagulants interfere with the action of vitamin K, and they reduce the coagulating powers of the blood so that the rats eventually die from internal haemorrhages. This kind of poisoning had long been known from cattle which had fed on spoiled sweet clover, *Melilotus alba*, and in 1941 the active principle, dicoumarin, $3:3'$ methylene *bis*(4-hydroxycoumarin), was identified. Dicoumarin (68) has since been used as a rodenticide, and for the treatment of coronary thrombosis.

Most of the anticoagulant rodenticides are hydroxycoumarin derivatives, though some are indandiones. The most widely used is warfarin, $3(\alpha\text{-acetonylbenzyl})$-4-hydroxycoumarin (68), which takes its common name from the Wisconsin Alumni Research Foundation. Later competitors include coumachlor, coumafuryl, and coumatetralyl (68). Diphacinone, 2-diphenylacetyl indane-1,3-dione (69), is an

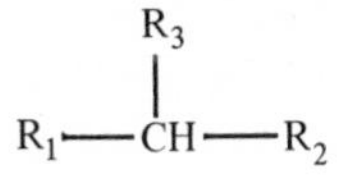

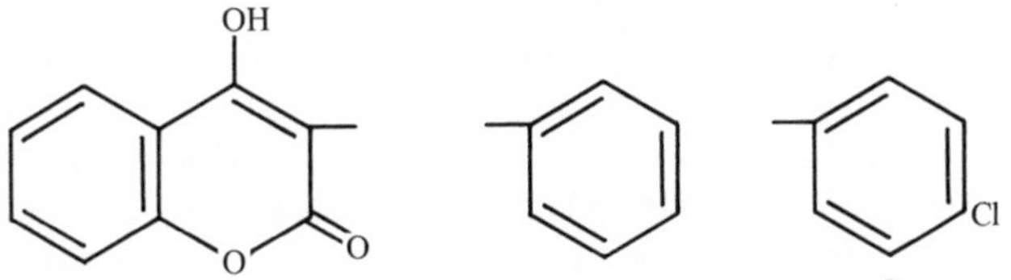

$$R_1 - \underset{\underset{R_3}{|}}{C}H - R_2$$

A

$-CH_2CO.CH_3$
E

B

C

H
F

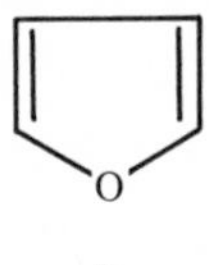

D

	R$_1$	R$_2$	R$_3$
Dicoumarin	A	A	F
Warfarin	A	B	E
Coumachlor	A	C	E
Coumafuryl	A	D	E

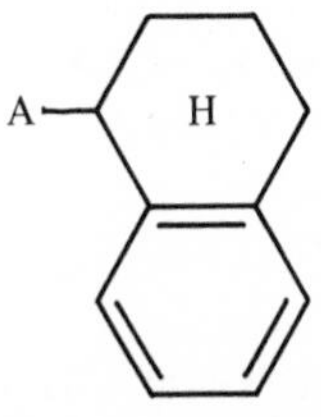

(68)

A — H

(68) Coumatetralyl

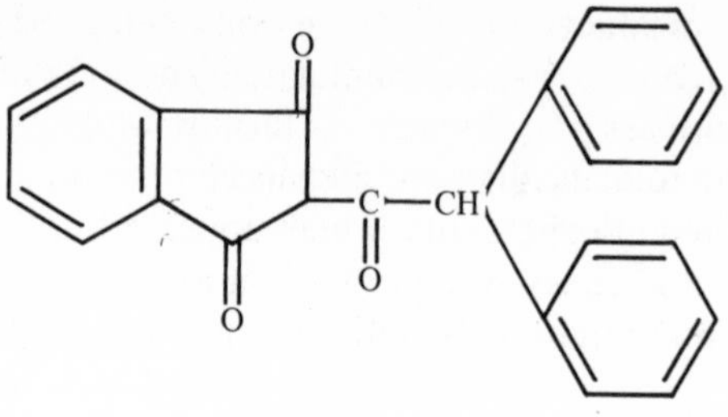

(69) Diphacinone

example of the other group. There are differences in the performance of the various anticoagulants – diphacinone is reported to be one of the most effective of these poisons against *R. rattus*, whereas warfarin is more toxic to *R. norvegicus*.

It is unfortunate that resistance to anticoagulants is appearing in scattered populations of rats, for the alternative poisons are unsatisfactory in many ways. It appears that rodents are, in general, increasing in numbers, unlike most other mammals, and the danger of plague outbreaks is always present. There is, without doubt, a great need for new, selectively acting rodenticides which kill rapidly.

Rabbits are most easily killed by fumigating the burrows, hydrogen cyanide being often used. It is usually formulated with magnesium carbonate and anhydrous magnesium sulphate to give a powder with good storage properties in dry conditions (Cymag®). An alternative is phosphine, released from aluminium phosphide tablets. This has also been used for the control of paddy field moles. Moles can also be controlled by baiting with earthworms into which strychnine has been injected, but there are cases of secondary poisoning of dogs.

In most parts of Europe there is reluctance to use poisons against birds, although dieldrin-dressed seed corn has been used illegally against woodpigeons. In many parts of Africa, however, there is a desperate need to control the weaver finch known as the quelea (*Quelea quelea*), a bird which is probably the most important single pest in that continent. Most methods used at present seem to do little more than remove the breeding surplus from the vast flocks.

In Britain the public will accept the use of narcotics such as chloralose (70), the

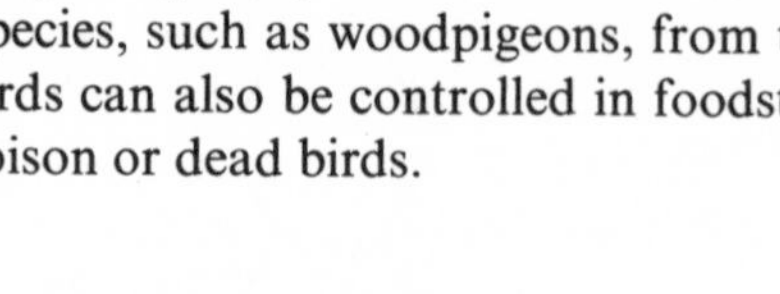

(70) Chloralose

effects of which are only temporary. Such a substance enables the controller to separate the harmful species, such as woodpigeons, from the useful ones, and then to release the latter. Birds can also be controlled in foodstores without any danger of contamination by poison or dead birds.

Herbicides

Weed control appears, at first sight, to be intrinsically more difficult than most kinds of pest control as the pest is usually closely related to the victim. This relationship calls for a high degree of selective toxicity in the techniques used for weed control, but, fortunately, this selectivity need not be biochemically based. This is because most weeds, unlike fungi and insect pests, are not directly dependent upon their victims, but merely share the same habitat. The difficulties arise when they are sharing the habitat at exactly the same time. It is then difficult to apply chemicals so that they affect only the weed plants. Even this can be achieved in some cases by using suitable spraying machinery.

A common situation is the occurrence of broad-leaved weeds growing in a monocotyledonous crop such as cereals. Such a crop has leaves which are difficult to wet, largely because of the upright habit and the fine, longitudinal corrugations. The spreading leaves of the dicotyledonous weeds, on the other hand, readily retain small droplets. Furthermore, the sensitive growing points of cereals and most grasses are protected by enfolding leaves, while those of broad-leaved weeds are fully exposed at the tips of shoots and in the leaf axils. Biochemical selectivity is, however, still important and it will be mentioned, where appropriate, below.

Weed-killing has gone far beyond the mere control of weeds in a growing crop. It has led to a quiet revolution in the farming year. Many of the cultivations which have been carried out annually for hundreds of years, and such practices as fallowing, evolved largely as a means of suppressing weeds. Apart from the labour involved there are certain disadvantages in such cultivations. They can, for example, aggravate soil erosion in wet seasons or districts. Chemical weed control can make many of these cultivations superfluous, and there is a strong movement towards omitting them.

Weed-killing may have gone past the main purpose of pest control in some farming communities, to a stage which has been aptly called 'cosmetic weed-control', though many of us who can remember the pre-war English corn fields will doubt that killing the poppy has (to misquote) gilded the lily. A weed-free farm has come to be taken as the mark of a good farmer and, as Pirie (1969) has pointed out, cleanliness can bring you closer to your bank manager. The possible advantages of 'dirty' farming are discussed elsewhere, but we cannot yet expect bankers to have an ecological outlook.

Herbicides are applied to the foliage of the weed or to the soil at one of three times: before the planting of the crop (pre-planting); after planting, but before emergence (pre-emergence); after emergence (post-emergence). They may damage the weed either by contact with the roots or with the foliage, or after translocation through the roots or through the foliage. There are thus twelve possible categories $(3 \times 2 \times 2)$: for example, pre-sowing, contact foliage treatment of weeds or post-emergence, translocated soil treatment of weeds.

Separate categories are complete, or total weed control, where there is no crop to be considered, and water treatment for the control of aquatic weeds.

A full explanation of these categories, with appropriate listings of commercial herbicides, may be found in current editions of the *Weed Control Handbook* which is published at regular intervals by the British Weed Control Council. This deals only with weed control in temperate, mainly European, countries.

A pre-planting, contact foliage treatment is usually carried out with a non-persistent herbicide to kill annual weeds shortly before planting or sowing, whereas a translocated foliage herbicide would be chosen for the pre-planting control of a wide range of perennial weeds. This could be carried out several months before planting, often just after harvesting the previous crop.

Some perennial weeds are difficult to control in this way, an outstanding example being couch grass, *Agropyron repens*. Pre-planting soil treatments with a persistent translocated herbicide will control some of these. This technique may also be used for annuals which will emerge during the early stages of crop growth before post-emergence sprays can be applied safely. The precise time of treatment naturally depends upon the rate of breakdown of the herbicide, and upon the susceptibility of the crop.

Pre-emergence contact foliage sprays, applied in order to kill weeds that have emerged, are fairly straightforward, provided that the crop is slow to germinate. Alternatively, sowing may be delayed till most of the weeds have emerged, and the herbicide is then applied immediately after sowing.

The soil application of a residual herbicide after sowing, but before emergence is, naturally, hazardous, but it may be used when the seed is sown deeply, and the herbicide is one which will not move from the top layers of the soil. Alternatively, the chemical may be biochemically selective.

In post-emergent treatment, both weed and crop are likely to come into contact with the herbicide, so selective killing depends either upon differential retention of the chemical, or upon differences in susceptibility to the material, or both. Contact foliage herbicides fall into the first category, and translocated foliage materials into the second. Post-emergence residual soil treatments are uncommon, but they may be used during the dormant season of such crops as blackcurrants.

Inorganic herbicides

The most important of these is sulphuric acid, generally obtained as BOV – brown oil of vitriol. Although an effective selective herbicide (by differential retention) for the control of annual weeds in cereals and onions, it is extremely corrosive to spraying machinery and clothing at the concentrations normally used. Since the withdrawal of sodium arsenite in the United Kingdom, sulphuric acid has become important as a potato haulm killer for the prevention of blight infection of the tubers.

Sodium chlorate, borax, and crude sodium borate are the other important inorganic herbicides. All are non-selective weed-killers, but while sodium chlorate is very rapid in action the boron compounds work only slowly. Sodium chlorate is, however, hazardous in use because it makes any organic matter extremely inflammable. Particular care must be taken with clothing that has been soaked with the spray.

Organic herbicides I: growth regulators

Many of these are misleadingly called hormone weed-killers, mainly because of their effects on the growth of plants. They are better called auxin-type growth regulators, but the first name is well established and likely to persist.

MCPA (71), 2-methyl-4-chlorophenoxyacetic acid and 2,4-D (72), 2,4-dichloro-

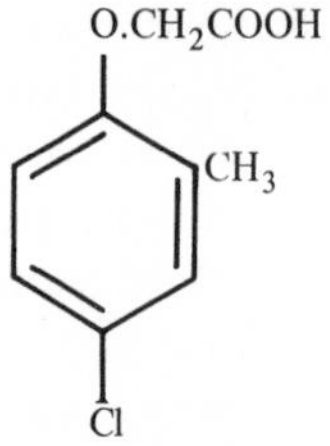

(71) MCPA 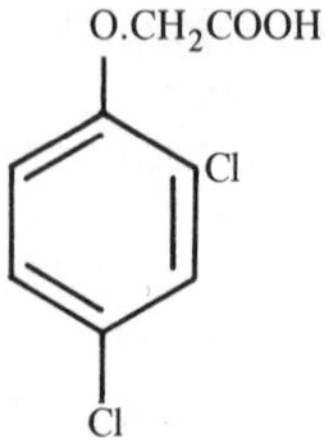   (72) 2,4–D

phenoxyacetic acid, were the first auxin-type weed-killers to be introduced, and their advent in 1942 ushered in a new era in weed control. There were several reasons for their success, the chief being their ability to kill many important annual and perennial weeds in cereals without damaging the crop, the ease of application, and their cheapness. Furthermore, they are not hazardous in use apart from the danger of the drift of vapour or spray on to near-by susceptible crops.

They were soon to be called hormone weed-killers, for in susceptible tissues they behaved in a similar way to the natural plant auxin, 3-indoleacetic acid (73) which

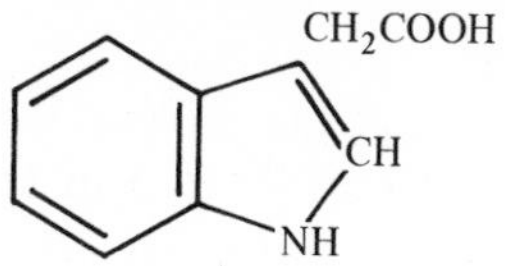

(73) 3 indole acetic acid

causes the elongation of cells. In the natural situation this elongation is controlled by various biochemical processes such as the oxidation of the indoleacetic acid. These processes do not, however, affect the action of the weed-killers, so the familiar distorted growth of weeds poisoned by these materials results.

Both materials can be used in various forms, such as potassium or sodium salts, amines, or esters. The differences between the various types are reflected in their uses. Thus, cereals are less likely to be damaged by the 2,4-D amines than by the esters, but the two forms are effective against somewhat different spectra of weeds.

2,4-D and MCPA are used as post-emergence sprays in cereal crops (though 2,4-D should not be used on spring oats), grassland, and turf. MCPA is also used on asparagus, linseed, and flax. These weed-killers can make certain weeds such as rag-wort, *Senecio jacobaea*, more palatable to stock. Cattle should therefore be kept out of fields containing such weeds for at least two weeks after the fields have been sprayed.

There have been a number of developments from the MCPA and 2,4-D starting points. MCPB and 2,4-DB are phenoxybutyric acid derivatives which some plants can readily convert into the corresponding phenoxyacetic acids, i.e., MCPA and 2,4-D. In this way, further selectivity is achieved, and put to use as, for example, when MCPB is applied as a post-emergence spray to peas. Both chemicals can, of course, be applied to the same crops as the phenoxyacetic acids with even less likeli-hood of crop damage. They may be sprayed, for example, on cereals. but at an earlier stage of growth. Naturally MCPB and 2,4-DB kill a smaller range of weeds than do MCPA and 2,4-D, so they are sometimes formulated as compromise mixtures with the corresponding phenoxyacetic acid.

Other related compounds are: 2,4,5-T (74), used against woody plants; MCPP

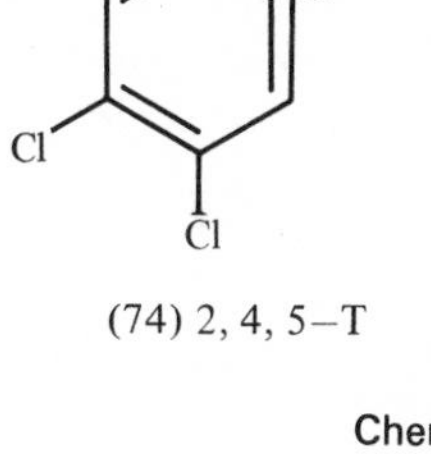

(74) 2, 4, 5—T

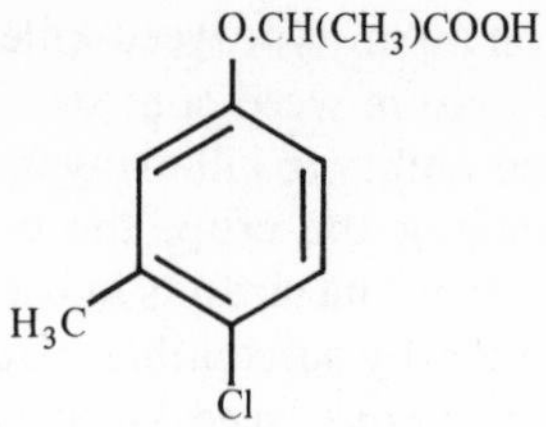

O.CH(CH₃)COOH

(75) Mecoprop

or mecaprop (75), for the control of the difficult weeds *Galium aparine* (cleavers or sweethearts) and *Stellaria media* (chickweed) in cereals; dichloroprop or 2,4-DP (76). The last controls most of the weeds that MCPA can deal with and also *Polygonum*

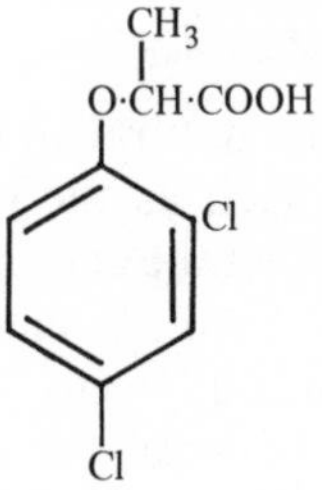

(76) Dichloroprop

aviculare (knotgrass), *P. convolvulus* (black bindweed), *P. lapathifolium* (pale persicaria), and *P. persicaria* (redshank).

2,4-DES or disul (77) recalls some of the organophosphate insecticides in that it

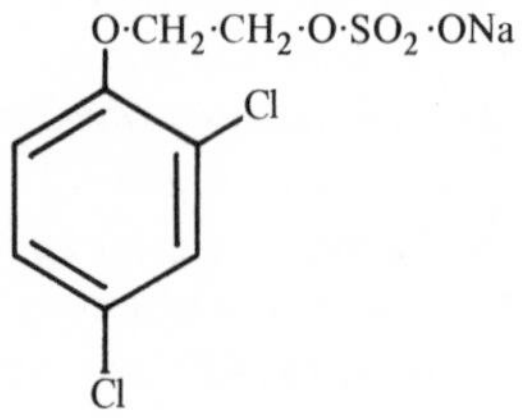

(77) Disul

is not a herbicide, but is converted into one. The conversion takes place in the soil, the product being 2,4-D, which prevents the growth of weeds which would arise from seeds present in the soil. It is mainly used in horticultural crops.

There are several growth-regulating weed-killers which are not of the auxin type. Dalapon,2,2-dichloropropionic acid (78), is particularly interesting because it is a translocated herbicide which is selectively toxic towards monocotyledons. It is most

$$CH_3 \cdot CCl_2 \cdot COONa$$

(78) Dalapon

effective when applied to actively growing grasses in mild or warm seasons. Typical applications are the control of couch and other perennial weed grasses around and between apple trees or fruit bushes, the killing of grass swards so that they can be reseeded without ploughing, and the clearing of irrigation canals and other waterways. A valuable characteristic, especially when it is used for aquatic weed control, is its low toxicity to fish.

Maleic hydrazide (79) is a growth-regulating herbicide but is usually put to a rather

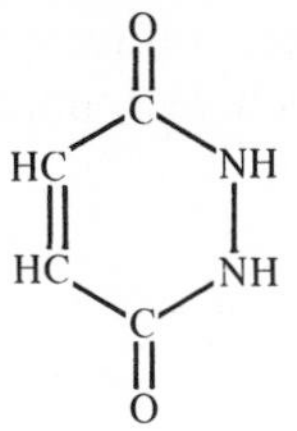

(79) Maleic hydrazide

different use, namely the inhibition of the growth of grasses without killing them. It obviates the need for the frequent mowing of roadside verges or graveyards, without producing the unsightly browning of other weed-killers.

Two closely related herbicides, diquat (the 9,10-dihydro-8a,10a-diazoniaphen-anthrene cation, 80), and paraquat (the 1,1'-dimethyl-4,4'-bipyridylium ion, 81),

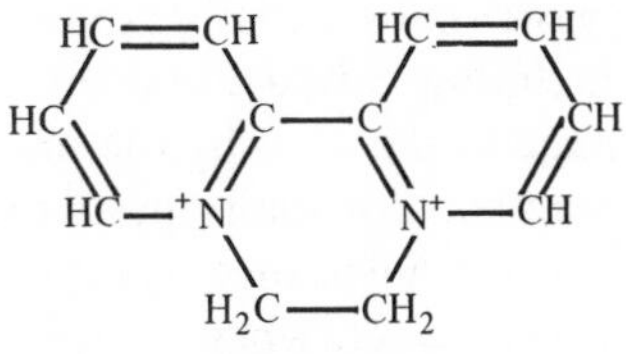

(80) Diquat

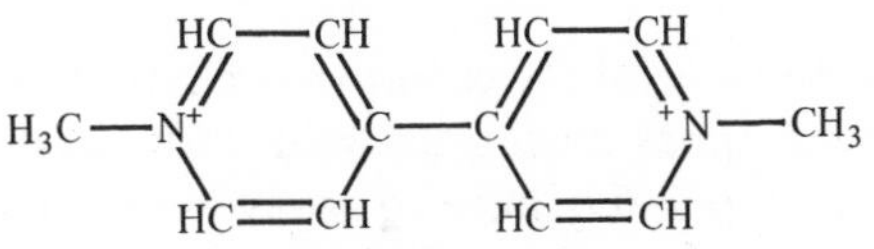

(81) Paraquat

Bipyridylium herbicides

were introduced in 1955 and have become increasingly important ever since. Their value arises from their inactivation immediately they come into contact with the soil, although they kill the aerial parts of the plant quickly. This property allows the planting of a crop immediately after spraying. The herbicides deal effectively with annual weeds but, of course, they kill only the tops of perennials, and regrowth often takes place.

Paraquat is rather more toxic to grasses than is diquat, and it is often used in a similar way to dalapon, namely to kill swards before reseeding, and to control grass weeds in orchards.

Organic herbicides II: contact herbicides

The dinitrophenols are the most important representatives of this group. DNOC (30) and dinoseb are still widely used despite their general biocidal effect. They are, for example, extremely toxic to man and mammals by ingestion, inhalation, and skin absorption. DNOC came into prominence during the agricultural campaigns of the war, and was somewhat misused, as evidenced by the number of yellow-stained farmworkers and farm dogs in those days. It has been responsible for a number of deaths among spray-men. Despite this it is a valuable post-emergence cereal herbicide as it kills a number of weeds which are not controlled by MCPA or 2,4-D. It is, however, likely to be displaced by the much less unpleasant herbicides mecaprop and 2,3,6-TBA.

Pesticides in practice

The brief descriptions of pesticides given above mentioned the commonest uses of the various substances, but they gave little idea of the ways they are actually used in agriculture and horticulture. In short, nothing has been said about spraying-programmes for specific crops.

Whether or not pesticides are applied to a particular crop depends upon the probability of their giving an economic return (see chapter 1). The increased yield resulting from the application must be worth substantially more than the costs of the material, the costs of the application including labour, and the depreciation of the spraying machinery. The possible damage to the crop by the material and the passage of the machinery must also be considered. For many pests this gain need only be gained in the year of application, but when 'insurance' spraying is carried out, for example against potato blight, the gains in the years when the spraying has proved to be helpful must also pay for the years when hindsight shows the spraying to have been unnecessary.

Much depends upon the value of the crop. An orchardist accepts that he will have to spray his apples several times during the year with fungicides, insecticides, and acaricides, and that he may possibly have to apply herbicides as well, if he is to get the high yield of the blemish-free fruit that his public demands, and he budgets accordingly. On the other hand, a crop that brings a comparatively low yearly income per acre – for example, cereals or timber – normally cannot 'afford' sprays, apart from herbicidal ones. An East Anglian farmer finds it economic to use organomercurials and insecticides in seed dressings against smuts and wheat bulb fly, for these are cheap and do not entail any extra operations, but he must depend upon methods other than chemical application for other forms of pest control. In between are such crops as sugar-beet or potatoes, where a comparatively small number of spray applications are feasible.

Sir Harold Sanders discussed some of the factors which have led to the wide acceptance of pesticides by British farmers (Sanders, 1966). The application of pesticides can replace many pest control practices which make heavy demands on labour. Between 1939 and 1966 the cost of labour on the land increased at least eight-fold, the price of land five-fold, but the prices received by the farmer for his produce only three-fold. The farmer has responded to this by increasing his yields by about one-half, by spending more on machinery, fertilizers, and pesticides. Since 1966 these processes have accelerated, though there may have been a small check in the rate of increase in the use of pesticides.

There is also a tendency towards larger farms, each of which concentrates on a few products, with the owner becoming a specialist in these crops. Pesticides have helped to bring about these changes by relieving the farmer of peak labour demands which formerly could be avoided only by suitable rotations.

A spraying programme for fruit will vary with the crop, and the way in which it is grown, the geographical location and local climate, and the insect, mite, and disease complex during the particular season. It will also depend upon the past history of spraying, for this alters the ecology of the crop drastically, and may also have initiated the development of resistant strains. The programme should also take into account the various biotic factors which regulate the pests; unfortunately, it rarely does so now, but, it is hoped, will in the future.

It will also vary with the biological understanding of the grower or the salesman who supplies his materials. If the grower or his advisor is able to recognize the pest organisms present and can accurately assess the danger that they represent, he will probably apply fewer sprays than a person who follows mechanically a chart prepared by a pesticide firm or an advisory headquarters.

For most pests and diseases there are usually several materials on the market, some, of course, better than others. It is rare that the grower will use the 'ideal' combination for most show 'brand loyalty', buying most of their materials from one company – and no company has a monopoly of the best materials. In short, a spray programme does not depend only upon the distribution of the pests, but also upon the distribution of salesmen, although some growers might not differentiate between the two!

For these reasons, and particularly because of the upsurge of resistance, it is impossible to outline a fruit programme that would be valid in, say, five or ten years' time. There is value, however, in a series of papers by the late Dr A. D. Hanna which appeared in the *World Review of Pest Control*, dealing with the application of insecticides to various crops throughout the world. If these are revised frequently to take into account the development of resistance, the introduction of new materials, the spread of pests and the appearance of new ones, and the introduction of integrated control methods, they will prove more valuable than any that can be given in such a text as this. Despite this we must have a basis for the discussion of integrated control methods later, and it is therefore necessary to consider some current spraying programmes.

Apple spraying programmes in south-eastern England
Kent and adjacent apple-growing areas have much the same problems as most apple-growing districts of the world, namely codling moth *Carpocapsa* (*Cydia*) *pomonella*, spider mites, especially *Panonychus ulmi*, aphids (green apple aphid, *Aphis pomi*, woolly apple aphid, *Eriosoma lanigerum*), apple scab, *Venturia inaequalis*, and apple powdery mildew, *Podosphaera leucotricha*. Routine applications are made against

these pests in most orchards. What is done in Kent, however, cannot be transposed unchanged to the United States or Australia, for the life cycles of the pests vary from region to region, and other pests become prominent. In warmer areas, for example, the San José scale, *Quadraspidiotus perniciosus*, trypetid (tephritid) fruit flies and the Oriental fruit moth, *Grapholitha* (*Cydia*) *molesta*, call for control measures. The codling moth passes through three generations a year in the Australian Capital Territory – in the region of Canberra, where the apple trees tend to have light frames and smooth bark (Clark *et al.*, 1967). In Kent there is only one generation, with sometimes a partial second brood: the trees are often heavier, with a rough bark. In America, South Africa, and Australia the woolly apple aphid establishes colonies on the roots of trees, and in the United States it may also winter on the elm, *Ulmus americana*. It is rarely found on the roots of apples, or on elm, in England (Massee, 1954).

In England, apple mildew overwinters as hyphae within infected buds and as perithecia (cleistocarps) on the twigs, but the latter appear to be sterile in Britain and New South Wales, however, they may be important elsewhere.

Most aphids, but not the woolly aphid, may be controlled by ovicidal winter washes or by spring sprays against the newly emerged insects. Tar oils may also be applied, preferably at high volume, during the winter when the buds are dormant, or DNOC-petroleum oil may be sprayed at the delayed dormant stage. Although these sprays are directed mainly against the aphids they also give control, good or partial, of a number of other pests, including winter moths and other Lepidoptera, and apple sucker eggs (*Psylla mali*). Alternatively, systemic insecticides such as dimethoate or demeton-S-methyl, or contact insecticides such as lindane, can be used later, after the nymphs have emerged.

Additional sprays later in the season may be needed, especially on young trees, but these often double up as controls for spider mites.

Codling moths begin to emerge at about the end of May, but most moths fly from late June to mid-July. Light traps set up in packing sheds can give the grower an indication of the times of emergence of the adults. Sprays are usually applied to kill the eggs or newly emerged larvae before they enter the fruit and, ideally, a protective deposit should be maintained from about mid-June to mid-August. This calls for two or three applications.

The most commonly used insecticides are now organophosphates such as azinphos-methyl, phosphamidon, or malathion, which also give some control of red spider mites, or compounds such as carbaryl or DDT which have no acaricidal activity. Lead arsenate, a stomach poison, is also sometimes used, but less than a few years ago. Carbaryl should not be used from pink bud till after petal fall as it may thin the fruit.

A wide range of chemicals kill red spider mites. This is fortunate as they rapidly develop resistance, especially if spraying is carried out routinely by the calendar rather than according to need. Resistance to ovicides which kill summer eggs has become very troublesome; resistance to organophosphates is also serious.

As a rule two sprays are necessary, the first being applied at petal·fall or when the hatching of the winter eggs is at its peak, and the second about three weeks later when, in a normal season, the populations are beginning to build up again. Various organo-phosphates can be used, such as azinphos-methyl, oxydemeton-methyl, dimethoate, or phosphamidon. Alternatively some of the bridged diphenyl acaricides such as chlorbenside or tetradifon will give a good control and it must not be forgotten that certain of the fungicides that may be used against powdery mildew will also be of value. These last two groups of chemicals will not destroy any of the insect predators of the mites.

Routine sprays against scab are absolutely necessary, but there are two distinct approaches. Either an attempt is made to keep a protective layer of fungicide constantly on susceptible surfaces throughout the normal infection period, or sprays are applied only when it is known that weather conditions, suitable for infection, have occurred. The fungicide is applied as soon as possible after a Mills period has passed. Infection can only take place at a given temperature if the leaves remain wet for a certain period, for example, 30 hours at 5·5 °C, 14 hours at 10 °C, and 9 hours at 16·5 °C.

In the first approach the first application is made at bud burst, and the spraying is repeated at ten-day intervals till late June. Some of these sprays may be combined with insecticidal or acaricidal sprays, provided that the chosen products are compatible. Suitable fungicides have been mentioned earlier. Most of the fungicides can be applied at high volume or at low volume, but the amount of active ingredient is increased at the low-volume rate. Dodine acetate and the organomercurials are usually chosen for spraying after a Mills period, but they can also be used for preventive spraying.

Mildew overwinters in the infected buds, which appear white and mealy when they break in spring. These primary infections can cause secondary infections in other shoots and blossoms, and in young buds which will continue the infection in the following year. The pruning out and destruction of the infected twigs during winter, when they can be recognized by their pale grey colour, is an excellent way of reducing mildew, but when this is impossible a DNOC-petroleum oil spray in mid-winter, or a lime sulphur spray at early pink bud stage, is sometimes substituted. These measures are followed by oragnic fungicidal sprays or sulphur sprays at seven- to ten-day intervals from blossom time onwards. Thorough wetting of the trees is essential, so high-volume applications are preferable. Whenever they are appropriate, insecticides or acaricides may be added to these fungicidal sprays.

This has been a brief outline of the spraying programmes used on apples in south-eastern England during the 1950s and the 1960s, but they are likely to change markedly during the next ten years. Fortunately, orchardists, in general, are willing to accept changes where necessary, as may be gathered from Jary's reminiscences on developments in pest control in orchards and hop gardens since the 1920s (Jary, 1965).

Citrus pest control in Australia

Most of the Australian citrus groves are found in New South Wales and Victoria, with three-quarters of the acreage being taken up by Navel and Valencia oranges. Queensland also produces a considerable tonnage, but its nickname, the banana State, shows where its real interests lie.

Citrus crops differ radically from the familiar fruits of northern Europe in that the trees are evergreen. In addition lemons, another important product, flower throughout the year so that the fruit may be harvested frequently.

As in most citrus-growing areas the main arthropod pests are scales, the most important species being the California red scale, *Aonidiella aurantii*. The common name is somewhat unfair: the pest was introduced into the United States from Australia in about 1879.

Other important armoured scales on citrus are the closely related yellow scale, *A. citrina*, the Florida red scale, *Chrysomphalus aonidum*, and the purple scale, *Lepidosaphes beckii*.

The commonest and most troublesome unarmoured scales are black scale, *Saissetia oleae*, two or three species of wax scales, *Ceroplastes* spp., brown soft scale, *Coccus hesperidum*, and the infamous cottony cushion scale, *Icerya purchasi*. The last was also introduced into California from Australia, but was soon controlled biologically in a most dramatic fashion.

Australian citrus also suffers from trypetid (tephritid) fruit flies, as do many other citrus industries. In Western Australia the chief culprit is the introduced Mediterranean fruit fly, *Ceratitis capitata*. Although it was also once present in eastern Australia it is now absent, partly because of the strict quarantine measures, and possibly because it has been displaced by the Queensland fruit fly, *Dacus tryoni* and the closely related *D. humeralis*. These two fruit flies have extended their range southwards in the last twenty or thirty years.

Two aphid species are troublesome, namely *Toxoptera aurantii*, the vector of the destructive Tristeza virus, and *Aphis citricidus*. Another important sap-sucking pest is the bronze orange bug, *Musgraveia sulciventris*.

A number of other pests occur but usually do not call for routine sprays. Spider mites are often found, but appear to be less troublesome than in many other countries.

It will be noted that only a few of these pests are native to Australia, and of those that are at least one, the cottony cushion scale, has spread to most other citrus areas. For this reason Bodenheimer's admirable *Citrus Entomology*, though devoted to Israel and the Middle East, serves well for Australia and other citrus areas (Bodenheimer, 1951).

The scales are generally controlled by white oil sprays, or white oil with insecticides such as malathion, azinphos-methyl, or dimethoate. These are applied in the summer months to kill the crawlers as they are liberated from the female scale. This is usually early in December or late in February. The exact time depends upon the season and the district. Some scales are better controlled by lime sulphur, though there is danger of plant damage if the spray is applied too soon after an oil spray. The aphids are usually controlled by an organophosphate such as fenthion or malathion.

There are several chemical approaches to fruit fly control. Cover sprays to kill the adult flies may be applied at fortnightly intervals during the flight period. Such compounds as DDT and malathion are among the chemicals which have been used. Alternatively poison baits may be applied to the foliage at weekly intervals. A typical bait contains sugar, protein hydrolysate of yeast, and a stomach insecticide such as malathion. These materials are scattered on the foliage, an all-over cover being unnecessary. Finally, penetrating organophosphates such as fenthion and dimethoate can be used to kill the larvae as they emerge from the eggs which have been laid just below the surface of the fruit. Chemical control measures start about five weeks before harvest, and continue for at least a month after it is finished.

The main Australian citrus diseases include black spot, *Guignardia citricarpa*, which causes spotting of the fruit, melanose, *Diaporthe citri*, another blemishing disease, and scab, *Spaceloma fawcetti*, especially on lemons. These diseases can be controlled by copper sprays which usually contain white oil. The oil content may be increased when the scales are troublesome.

Citrus trees are also susceptible to a number of diseases of the roots and stems which, of course, threaten the life of the tree. Chief of these is the root and collar rot caused by *Phytophthora parasitica*, a fungus which also attacks a number of other plants. Control is usually by cultural methods such as the use of resistant root stocks, the avoidance of infected soil and so on, but there has been some success with soil fumigants such as Vapam ®.

Cotton pest control

The crops of the American cotton belt are attacked by well over one hundred different pests and diseases, several of which, unchecked, can lead to severe losses. Chemical pesticides are by no means the only control agents used, and integrated methods are being increasingly employed. Cotton was also one of the first crops for which selective breeding gave strains resistant to certain insect pests.

There are several species of *Gossypium* which are cultivated for their fibre, but the most important is *G. hirsutum*, the Upland type. Despite its shrubby appearance the plant is grown as an annual. In fairly dry climates the plantations are irrigated.

Without doubt the boll weevil, *Anthonomus grandis*, is the most famous of all cotton pests, but though it always brings to the European mind visions of the old slave plantations it was not known in the United States before 1892. It is now troublesome in the whole of the cotton belt, apart from California, New Mexico, and Arizona.

Boll weevil is difficult to control with insecticides because the chief damage is caused by the grubs developing within the squares (flower buds) and bolls (developing fruits). The adults which emerge from hibernation in debris, under bark, and so on, also feed on the foliage and puncture the squares and bolls.

Damage may be reduced by any measures which encourage rapid growth and early fruiting, followed by the destruction of the stalks after picking, so that the adults are deprived of their food before hibernation. Chemical control varies from district to district, for it is often found that a chemical which works well in one place is unsatisfactory in some others, Methyl parathion, azinphos-methyl, endrin, and carbaryl are among the most frequently used chemicals, but monocrotophos is also showing considerable promise against this pest. The first application is made when about 10 or 20 per cent of the squares are punctured, and further applications are then made at intervals of four or five days till the weevil is controlled. There are usually about seven generations in each season.

The bollworm, *Heliothis zea* is a general feeder which is also a severe pest of maize, when it is known as the corn earworm. In the cotton belt it passes through several generations each year, but in the northern United States there are only one or two.

The newly hatched larvae usually feed on the outside of the squares, but older ones tunnel into the squares and the bolls, completely hollowing them. The aim of insecticidal spraying is to kill the young larvae before they start tunnelling, so the first application should be made when there are numerous eggs and approximately four larvae per hundred plants. Carbaryl, DDT, endrin, malathion, and methyl parathion are among the insecticides used, although DDT sometimes leads to an increase in the severity of aphid attacks. Deep, winter ploughing and cultivations are used to destroy the pupae in the soil.

Another troublesome noctuid on cotton is *Alabama argillacea*, the cotton leaf worm. The moth cannot overwinter in the United States, so it seems that it migrates northwards from tropical America each year. Generation after generation follow, while the migration continues, so that in favourable years the moth may reach as far as Canada. Normally, this moth is killed by the sprays directed against the boll weevil. Late in the season its damage may be welcomed as it robs the last generation of adult boll weevils of their food.

The larvae of the cosmopolitan pink bollworm, *Pectinophora gossypiella*, feed on the seed of cotton, but they also reduce the quantity and the quality of the lint. Quarantine regulations have restricted the areas of infestation in the United States, but favourable weather has led to the increase in populations in infested areas in recent years, with a consequent spread to new districts, especially in Texas.

Larvae hibernate in old bolls and other cotton debris, and in gin trash. Many also overwinter within seeds – either a single seed, or two fastened together with silk. Control measures therefore include fumigation of seed and gins with methyl bromide, heat sterilization of the seed, the destruction of cotton stalks, the burning of gin rubbish, and so on. Dusts and sprays are also applied to fields at weekly intervals until most of the bolls are open, or the infestation is controlled. DDT, carbaryl, and DDT/methyl parathion mixtures have been used.

DDT tends to lead to a build-up of aphid populations on cotton through the destruction of lady-birds, syrphids, and other natural enemies. The commonest aphid species is the polyphagous *Aphis gossypii* or melon aphid, but *A. craccivora* and *Anuraphis maidi-radicis* can also cause trouble. These pests are controlled by the non-DDT insecticides used for the control of other cotton insects, especially good results being given by the organophosphates. Disulfoton or phorate can also be applied in the furrow at planting time.

The tiny cotton leaf-hopper, *Psallus seriatus*, a mirid, is often overlooked because of its small size, but it can cause considerable damage, including a characteristic spindly plant growth, the death of very young squares, and the prevention of blooming. A wide range of pesticides can be used for its control, including chlorinated hydrocarbons, organophosphates, and carbaryl.

Cotton is also the favoured host of several tetranychid spider mites which have increased in importance with the growing use of organic compounds on the crop. Bridged diphenyl compounds, organophosphates and sulphur are all used for their control, the choice depending upon the species present and on the control measures used for other pests.

The most important cotton disease is probably bacterial blight or angular leaf spot whose causal organism is *Xanthomonas malvacearum*. Fortunately, upland cotton is relatively resistant, but Egyptian varieties are very susceptible. The disease severely limited cotton cultivation in Sudan till the introduction of resistant varieties such as 1730.

Organomercurial seed dressings are used to control this disease, and some other soil- and seed-borne diseases of cotton, and this is combined with the destruction of infected material. Bacterial blight can spread from plantation to plantation, so it is important that seed dressings should be used uniformly over wide areas. The dressing of cotton seed is, however, such a simple and economic process that little untreated seed is ever planted.

Other soil- or seed-borne diseases that are controlled by copper or organomercurial seed dressings are cotton anthracnose, *Glomerella gossypii*, damping off, *Rhizoctonia* spp. and *Macrophomina phaseoli*.

Black root, *Fusarium vasinfectum*, is best controlled by rotations and soil fumigation measures taken against nematodes. Verticillium wilt, caused by *V. alboatrum* and *V. dahliae*, is particularly troublesome in wet areas or irrigated plantations, and is very difficult to control chemically. The best approach at the moment appears to be fairly long crop rotations and the use of resistant varieties.

The economics of pesticide production

There has been, during the last five or ten years, a marked slackening of the rate at which new pesticides are introduced on the market. It is becoming increasingly expensive to carry out all the necessary steps in the research and development which come before the marketing of a new product. In the 1950s and 1960s it was sufficient to carry out chronic and acute toxicity tests on a few mammals and birds, and to study the residue behaviour of the compound on those crops for which it was proposed. Sometimes, tests would also be carried out on the toxicity to fish and pheasants to try to assess the danger to wild life. Now new tests are added yearly to check, for example, possible carcinogenic and mutagenic dangers. Far greater efforts are made to assess the dangers to wild life. This process is, of course, 'open-ended'. It is obviously impossible to test the effects of the pesticide on all the many different species with which it could come into contact in the field, or even to select a number of species which would serve to indicate the danger to the fauna and flora in general.

Table 3.1 (reproduced by kind permission of the Agricultural and Veterinary Chemicals Association of Australia) outlines the R. and D. stages in the early history of a new pesticide and the various costs (in Australian dollars) involved. *Chemical Week* (Anon, 1969) gives comparable figures for the United States of America. These figures are already out of date as costs are increasing monthly.

These costs include those of the various pesticides which never get to the marketing stage – for the successful products must also finance the work on these. The odds against a chemical that passes the initial screening stage passing to the next stage, toxicity testing, are about 100 to 1. The probability of it being finally marketed is about 1 : 36 000 (Anon, 1969).

The chances that a particular manufacturer will be able to market a chosen microbial pesticide are far better – about 1 in 144 (Dulmage, 1971). Furthermore, the initial screening will have been carried out by universities or government agencies, whereas chemical pesticides are usually entirely the products of their manufacturers. Nevertheless, the total costs to marketing are not much lower. The producer of a microbial agent cannot, of course, patent a particular organism, although he can patent a new manufacturing process. The chemical pesticide manufacturer, on the other hand, can enjoy complete protection for his product for several years. It is doubtful, however, whether this period is long enough for him to get a sufficient return on his costs before rival companies can manufacture the product themselves. It must be remembered also that even when a product eventually reaches the market it will not necessarily remain in use for long. Dangers that were not detected during the R. and D. period may become evident during widespread use, and the compound may have to be withdrawn prematurely. Furthermore, most chemical pesticides have only a limited useful life; when they are intensively used against a particular pest resistance often appears rapidly and the compound has to be replaced by a new and more effective one.

There is a demand from many biologists for compounds which are more selective

Table 3.1 Agricultural and veterinary chemicals. Schematic representation of a research and development programme

Time	Number of compounds	Research and development	Toxicology	Manu-facturing	Marketing	Legal	Corporate planning and management	Cost
	1000s (2000 to 10 000 per company per annum).	Candidate compounds: synthesis, intermediates, by products. Biological activity against standard insects, plants, disease organisms. Laboratory–Greenhouse–Field.	Range finding rodents.			Basic patent filed		
6–12 months								
	100s	Physio-chemical properties. Formulation. Analysis. Radio tagged compound. Mode of action.	acute, cumu-lative } oral, dermal, inhala-tion. Range finding fish, birds.	Preliminary process studies. Raw materials availability.	General appraisal of prospects.	Patent: foreign filing.		
Australian participation commences 1–2 years		Trials: greenhouse, field, representa-tive of climatic zones.	Preliminary metabolism animals.	Cost estimates.	Definition of outlets.	Patent: infringe-ment studies.		$500 000
	10s	Field evaluation world-wide, including residue studies. Metabolites analysed.	Pharmacology. Metabolism. 90 day and/or 2 year feeding studies to 2 or more species (of compound and/or metabolites). Genetic, carcinogenicity, wild life studies.	Process development studies. Pilot plant production.	Continuing assessment of competition, potential sales	Infringe-ment studies on preferred manu-facturing route.	Progressive mobilization of necessary resources. Review progress at end of each stage and decide whether to proceed. Fit product into total company operation.	
2–4 years								$1 000 000
	1–5	Field adaptation trials to fit into farm management programmes. Packaging, compatibility and storage stability.	Continue, and supplement where necessary.	Manu-facture and formulation cost estimates and profitability studies.	and selling prices.	Trade mark. Common name.		
4–6 years								$2 000 000
	1–3	Round off experimental work. Establish with holding periods/ residues/ tolerances. Issue product specifications. Finalize analysed method for: product, formulations, metabolites, residues. Collaborative soil fauna studies.	Collaborative study on wild life in field. Complete programme. Develop procedures for safe handling. Develop tolerances.	Profitability evaluation. Design and erect commercial plant. Quality control procedures finalized.	Policy and objectives. Integrate sales estimates. Advertising policy and literature. Technical literature. Train: technical service, salesmen, dealers.			
6–8 years								
	1–3	Clearance (Agriculture and Health). Registration.		Production, Packaging, Ware-housing.	Distribution Sales. Technical Service.			$3 000 000 to $6 000 000

Reproduced by kind permission of the Agricultural and Veterinary Chemicals Association of Australia (*AVCA Bulletin*, July 1970)

122 Pest Control

than the polytoxic compounds, such as DDT and parathion, which have been popular so far. They are needed, of course, for integrated control programmes. The discovery and development of a monotoxic compound would be extremely expensive if present screening methods are used, for these would have to be expanded to include a far greater number of test species. Even if such a compound reached the marketing stage it is very unlikely that the demand would be great enough to cover the costs. Unterstenhöfer (1970) discusses this aspect of integrated pest control from the agricultural chemical company's point of view.

It is not surprising therefore that many smaller companies have left the industry, and that the larger companies are producing fewer new chemical pesticides than they did ten years ago.

4

The drawbacks of chemical control

In many ways, the discovery of DDT was the most important medical discovery of the century. In the warmer parts of the world DDT and its allies have probably saved more lives, by the destruction of disease-transmitting insects, than penicillin. It almost certainly spared war-torn Europe from the ravages of typhus when the outbreak was checked in Naples. Insecticides, fungicides, and weed-killers have played an essential role in the provision of food and fibre for a starving world. Despite these successes the insecticides have fallen out of favour, at least among the better fed and better clothed nations. Unfortunately, this disfavour is not unmerited.

When DDT was first introduced there were great hopes that all insect pest problems were near to being solved. This optimism, based on the chemical's high toxicity to insects, and on its great stability, brought the neglect of long-established techniques such as cultural control. The ideal was a completely pest-free environment, achieved simply by the use of the spray gun. As time passed, it became clear that DDT was losing its effectiveness against many pest populations, especially those, such as the house-fly and malaria-carrying mosquito populations, which were subjected to the most intensive spraying. It was not only DDT which lost much of its potency. Insects were soon to show marked resistance to dieldrin and other cyclodienes, and a little later to the organophosphates as well.

Resistance

Resistance to insecticides was defined by the WHO Expert Committee on Insecticides (1957) as 'The development of an ability in a strain of insects to tolerate doses of toxicants which would prove lethal to the majority of individuals in a normal population of the same species'. This definition does not set any quantitative level but it avoids the confusion that can arise when an attempt is made to differentiate between a 'tolerance' and a 'resistance' on the basis of a two- or three-fold increase in the LD 50. Such a small increase is often of little practical importance except as a hint of possible trouble to come. Far more important were the levels actually reached in some populations, such as several hundred-fold to the organochlorines after thirty generations or less. In general, the increase in resistance to the organophosphates was less, but it was still serious enough to cause difficulties in control.

In 1946, one public health insect, the house-fly, was known to be resistant in some populations, but by 1950 between five and eleven species had developed resistance to one or more insecticides. By 1969, the number had increased to 102 species – 55 to DDT, 84 to dieldrin, and 17 to organophosphate compounds (some insects had developed resistance to all three kinds of pesticides). Resistance also appeared in agricultural pests. In 1958, at least thirty plant-feeding insects were resistant to various pesticides (Brown and Pal, 1971). Some authorities give even higher figures. Georghiou (1965) states that by 1964 there were 166 species of insects and 19 species of mites and ticks that were known to be resistant to various chemicals. At the time of writing the only groups of public health insects which do not contain resistant species are the tsetse flies, *Phlebotomus* sandflies and reduviid bugs, while resistance has only recently appeared among the *Simulium* blackflies.

It is unfair to suggest that no one foresaw these difficulties as resistance to chemicals was not unknown among organisms before the introduction of DDT. It was recognized in bactericides and with drugs designed to retard cancerous growth. It was even known with insecticides, at least eight species having developed resistance to such materials as hydrogen cyanide and arsenates before 1940. The first recorded case was in 1908. The unexpected aspect was the rapidity with which the resistance towards the new synthetics developed. The reason for this lay largely in the extreme initial toxicity of these chemicals. They quickly eliminated the susceptible individuals, allowing the scarce resistant forms to reproduce explosively with the minimum of intra-specific competition.

This brings us to the paradox that pesticides do not really cause resistance. They merely select resistant individuals already present in the population. Thus insecticide-resistant individuals existed before insecticides were applied although, of course, they were very rare, and completely absent from some populations. There is no evidence that insecticides induce mutations which confer resistance, and, indeed, there have been many experiments in which pure-bred lines have been subjected to insecticidal pressure for many generations without specific resistance appearing. A possible exception (though with a rodenticide rather than with an insecticide) is warfarin, which is believed to have induced a resistant mutation among rats in Wales (Mellanby, 1967).

The inheritance of specific resistance is usually quite simple, being, in many cases, monofactorial, although secondary modifying genes may enhance the effect of the principal gene. In a review of the genetics of insecticide resistance in a variety of insects and acarines, Georghiou (1965) tabulated data on DDT, dieldrin, and organophosphate compounds. In 13 studies out of 17 the inheritance of resistance to DDT was monofactorial; the figures for dieldrin and the organophosphates were 20 out of 23, and 3 out of 7. Brown and Pal (1971) give similar results: DDT resistance was monofactorial in 13 species, dieldrin resistance in 16 species, organophosphate resistance in 5 species, and carbamate resistance in 2 species.

In the above species DDT resistance was dominant in 3 species, intermediate in 2, and recessive in 8; dieldrin resistance was intermediate in 15 species and dominant in the tick *Boophilus*; organophosphate resistance was dominant in 4 species and intermediate in 1; carbamate resistance was dominant in both species. This relatively simple mode of inheritance is a further reason for the rapid build-up of resistance in field populations. It is, however, surprising that resistance to dieldrin and DDT develops more quickly than resistance to organophosphates when it is considered that the genes concerned in DDT and dieldrin resistance are usually either recessive or intermediate in effect, whereas those of the organophosphate resistance are dominant. This is partly explained by an examination of the respective log-dosage/response curves of the various chemicals on susceptible populations. Those of the organochlorines are shallow, indicating a greater variability in the susceptibility among the individuals of the population compared with the susceptibility to organophosphates, which generally produce steep response curves.

An insect can show resistance to several insecticides at once. When a single detoxification mechanism confers resistance to two compounds the phenomenon is called cross resistance. This, of course, involves the same genes for the two chemicals. Resistance to DDT usually brings cross resistance to TDE and methoxychlor, but not to lindane and the cyclodienes which fall into another cross-resistance group. The organophosphates can be subdivided into a further two cross-resistance groups typified by parathion and malathion, but even more subdivision is possible. A fifth resistance group is represented by the pyrethrins although resistance to this pesticide is, so far, of less practical importance.

Where different mechanisms for resistance to toxicants belonging to different groups exist in the same individual, it is said to show multiple resistance. A further possibility is the existence of two or more mechanisms in the one individual which confer resistance to one kind of toxicant. If these mechanisms work serially, the first, perhaps, restricting absorption, the second being detoxifying, their combined effect is multiplicative. If the two work in parallel, to use Busvine's phrase, their effect will be additive (Busvine, 1971). Multiplicative resistance (again using a phrase of Busvine) of this kind will be difficult to overcome completely by the addition of a synergist to the original insecticide, for these compounds often work by interfering with a detoxification mechanism.

When a population is first subjected to insecticidal pressure from, say, DDT, it also becomes less susceptible to other pesticides which are not in the DDT group. This initial increase in tolerance is probably due to the elimination of the most susceptible individuals, those which are the least capable of dealing with a toxicological insult of any kind. The mode of inheritance involved in these cases is polyfactorial, but the mechanism is incapable of producing such high levels of resistance as the more specific kinds.

Those individuals which were pre-adapted for insecticide resistance were, as stated above, extremely rare in their populations. They were, possibly, at some disadvantage compared with their fellows; there certainly seemed to be no reason for the mutation to spread through the population in the absence of insecticides. When subjected to an insecticidal environment, the resistant forms survived and reproduced and, at the same time, there was a general selection for a genotype better fitted to the environment as a whole. Because of this, the development of resistance in a population when first subjected to insecticides is initially quite slow, but accelerates later when the general fitness of the survivors improves. Milani (1956, 1959) has pointed out that the rapid reappearance of resistance in a strain that had reverted to the susceptible status because of the extended withdrawal of insecticides may be explained by the prior development of a genotype compatible with insecticide resistance, and, generally, vigorous.

The development of resistance can be studied in the laboratory, but its development in the field is modified by various factors which are usually absent from controlled experiments. A further reason for the rapid development of resistance to the organochlorines, for example, is the extreme persistence of these compounds in the field. An insecticide used in the form of a wall deposit, for instance, can remain concentrated enough to maintain a selection pressure for several months. It has been suggested that an insecticide used for residual deposits in anti-malarial work should be designed to disappear at the end of the transmission period. Some workers, on the other hand, insist that insecticidal pressure should be maintained continuously in an attempt to interrupt the transmission of the disease before resistance becomes troublesome.

If parts of an infested area only are treated it is possible that significant resistance will be greatly delayed, or not even occur, provided that the surrounding untreated populations are mobile enough for susceptible individuals to 'dilute' the resistant survivors.

Field and laboratory trials have shown that the use of a pesticide as a larvicide is likely to bring resistance much sooner than is its use as an adulticide, since a greater part of the population is subjected to the material when it is used in this way. An extremely rapid development occurs when the material is used in both ways, and many workers have insisted that this should never be done.

The study of the physiological mechanisms of resistance is just as important as the study of its inheritance. It might be thought, considering how little we know about the mechanisms of action of many pesticides, that most of this work would be abortive but this, fortunately, is not so. DDT-resistant insects, for example, often have a high titre of the enzyme DDT-dehydrochlorinase which aids in the conversion of DDT to the less toxic DDE:

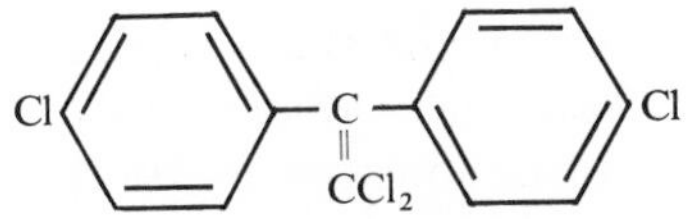

$$Cl-\!\!\left\langle\bigcirc\right\rangle\!\!-\underset{\substack{\parallel\\CCl_2}}{C}-\!\!\left\langle\bigcirc\right\rangle\!\!-Cl$$

This compound, incidentally, is the principal storage material in the human body after the ingestion of DDT. The enzyme is also present in normal, susceptible flies but not in sufficient quantities to detoxify the DDT before the insect is killed. DDT-dehydrochlorinase will also detoxify some of the analogues of DDT, but it is characteristically less effective when deuterium is substituted for hydrogen on the secondary carbon of DDT. Another possible detoxification pathway leads, in some species, to dicofol,

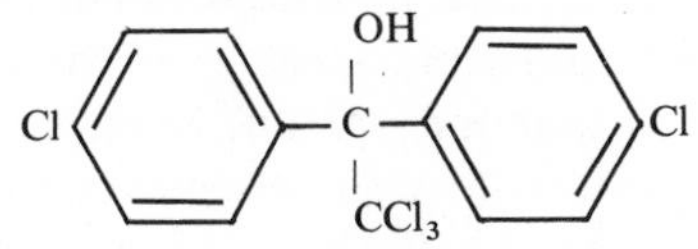

$$Cl-\!\!\left\langle\bigcirc\right\rangle\!\!-\underset{\substack{|\\CCl_3}}{\overset{\substack{OH\\|}}{C}}-\!\!\left\langle\bigcirc\right\rangle\!\!-Cl$$

which, oddly enough, is a very effective acaricide.

Organophosphate resistance is more complicated. It will be remembered from the previous chapter that most organophosphates must be activated *in vivo* before they can exert their toxic effects. They may also be detoxified by phosphatases and carboxyesterases. The toxicity of the chemical thus depends upon the balance of toxifying and detoxifying enzymes. Normal flies contain large quantities of an aliesterase which is present in only small amounts in some organophosphate-resistant strains. These, on the other hand, contain comparatively large quantities of a new phosphatase enzyme which degrades organophosphate insecticides. It has been shown that the gene which produces the aliesterase is an allele of that which brings about the synthesis of the resistance-conferring phosphatase. Thus, in the house-fly at least, one gene, producing an enzyme of obscure function, has mutated to one producing an enzyme which endows the fly with organophosphate resistance. The actual mechanism is more complicated as modifying genes are also involved.

A microsomal mode of detoxification, involving NADP and oxygen, has been shown to be one possible cause of DDT resistance in some insects. Busvine (1971) has suggested that this may explain the cross-resistance between DDT and pyrethrins which he first observed in the early 1950s. Yet another resistance mechanism may be found in the reduced penetration of the pesticide through the cuticle of the resistant insect. There is little known of this mechanism as yet, as the study of the penetration of insecticides has proved difficult. It is not likely that it will confer a high level of resistance, but it may magnify the effects of other forms of resistance. Finally, there may be differences between resistant and susceptible strains in the nature of the target of the insecticide, but as we are still ignorant of the mechanisms of action of most pesticides such differences cannot be demonstrated (Oppenoorth, 1965).

Populations of pests can become resistant to pesticides by mechanisms other than those described above. Differences in the behaviour of individuals can lead to their picking up different quantities of the pesticide, or even to their avoiding all contact with it. Certain strains of anopheline mosquitoes will not settle for more than a few seconds on DDT-treated surfaces, while others will not come into buildings (where the treated surfaces are) for blood meals. Among crop pests larvae of various populations of codling moth have evolved the habit of discarding the first bite of the fruit when first boring into the flesh, and thus escaping the effects of the stomach poisons with which the apples had been treated.

The study of this kind of resistance is, by its nature, more difficult than the study of biochemical resistance, and it is even more difficult to circumvent such resistance problems by the substitution of another chemical.

The development of resistance, and particularly of multiple resistance, has led to great difficulties in both vector control and agriculture. The two fields are not separate. In many rural areas heavy pesticide pollution following crop spraying has produced resistance in vectors towards insecticides to which they were not directly subjected.

DDT resistance became intolerable in many populations of house-flies within a year of its introduction, and lindane was substituted, followed, when that failed, by chlordane or dieldrin. Within two or three years these chemicals no longer gave satisfactory control, and the organophosphates, malathion and diazinon, had to be used in their place. These performed well enough for four or five years, and then yet other chemicals had to be found. Carbamates have been useful, but flies which are organophosphate resistant often show a high level of resistance to carbamates as well. Populations of mosquitoes, bed bugs, fleas, and German cockroaches have also gone through the same stages. Many agricultural pests have developed multiple resistance in a similar way, and have become increasingly difficult to control. Red spider mites in many parts of the world, and on a variety of crops, have been particularly troublesome.

The development of resistance in anopheline mosquitoes has led to the resurgence of malaria in many areas where the disease had been almost eradicated. There was, for example, an outbreak in western India in 1958 when *Anopheles culicifacies* developed resistance to dieldrin. Multiple resistance in *A. albimanus* has prevented the eradication of the disease in El Salvador, and the problem has also halted the attempted eradication of the yellow fever mosquito, *Aedes aegypti*, in the Americas. Parts of Central America and Brazil which had been freed from this pest are now being re-infested by strains which are resistant to DDT and the cyclodienes. Resistance has blocked the Indian filariasis campaign which depended upon the control of the vector *Culex pipiens fatigans* (Hamon and Pal, 1968; Busvine, 1968; Brown and Pal, 1971).

Many anti-malarial campaigns have been hampered by the development of resistance in bed bugs. In many countries bed bugs became more troublesome after a few years when organochlorines had been used as a spray against mosquitoes. The bugs had apparently developed a resistance to the pesticides, and possibly their mobility was increased by the irritating DDT. Whatever the reason, many owners of houses objected to the continuance of the spraying, and even resorted to white-washing freshly sprayed walls.

There is, of course, always the possibility that a resistant population will undergo reversal when the insecticide or group of insecticides concerned is no longer used. Partial or complete reversal is known to have occurred in laboratory stocks, even when these were not diluted with susceptible individuals, and similar reversals have been observed in field populations. Dieldrin resistance in a population of *Anopheles culicifacies* in India disappeared within three years of the substitution of DDT. Unfortunately, when the original pesticide is again used resistance reappears very rapidly. In Denmark, for example, house-flies lost their resistance to DDT and to the cyclodienes when organophosphates were used in their place, but trials showed that this susceptibility disappeared within one month when DDT and dieldrin were used again.

The substitution of a second insecticide for the one to which resistance has developed could accelerate reversal if the substituted insecticide is more toxic to the resistant forms than to the susceptible forms (negative cross resistance). While this phenomenon is known for certain *Drosophila* strains it does not seem to have had any practical application so far. Positive cross resistance appears to be much more likely.

Brown and Pal (1971) list some of the measures that can be taken in public health work, using current insecticides, when resistance problems arise. DDT resistance may be played against dieldrin resistance, and if multiple resistance develops, then DDT may be used at shorter intervals, as resistance to this insecticide rarely becomes so high that it becomes completely useless. House-fly resistance to the organophosphates may be combated by using baits containing trichlorfon; the insects acquire a very high dosage in this way.

A natural suggestion is to use two kinds of pesticides alternately. When this method was tried against house-flies, using DDT in the first generation, malathion in the next, and continuing in this way for several generations, both DDT and malathion resistance developed to the same high levels as those in the two control populations, one of which received DDT only, and the other malathion only. Similar results were obtained in trials with cockroaches subjected to chlordane/malathion mixtures, and with bed bugs treated either with a DDT/malathion mixture, or with the two chemicals alternately, in consecutive generations. There does, however, seem to be some merit in changing alternately from one chemical to the other at intervals of five or six generations.

Synergists do give some relief from resistance problems. A number of DDT-dehydrochlorinase inhibitors, such as WARF Antiresistant

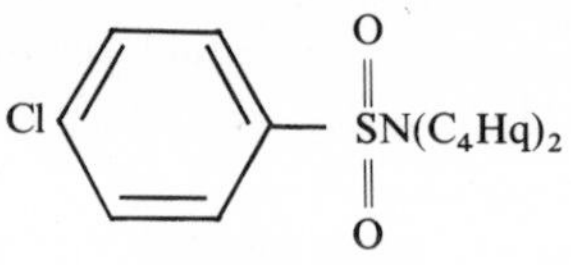

$$Cl-\!\!\bigcirc\!\!-\overset{\displaystyle O}{\underset{\displaystyle O}{\overset{\|}{\underset{\|}{S}}}}N(C_4H_9)_2$$

and chlorfenethol have restored the toxicity of DDT to some resistant fly populations, and synergists such as sesamin and piperonyl butoxide which inhibit the microsomal enzymes have also been useful against some kinds of DDT and carbamate resistance. Unfortunately, many populations rapidly develop resistance to the combined pesticide and synergist.

Until satisfactory non-chemical methods of control are developed resistance will have to be countered with new insecticides. It has sometimes been suggested that resistant flies are, in some way, 'super flies', and that any new chemicals which control them will have to be even more toxic than those now in use, the implication being that they will be thus more dangerous to mammals. This, of course, need not be so; they will merely have to exploit other biochemical pathways in the target insect. Even if a wide range of insecticide types is developed resistance will sooner or later develop towards each of them, but there will be longer periods during which reversal may take place, and this may facilitate control.

Resistance is not, of course, confined to arthropods and insecticides, but, in general, other kinds of pesticides have not suffered to the same extent. Chickweed, *Stellaria media*, has been known to have become resistant to 2,4-D by the third generation; in California the green mould on citrus, *Penicillium digitatum* has developed resistance to biphenyl. Penicillin resistance is common among micro-organisms, and many pathogenic protozoa can no longer be controlled with drugs which formerly killed them.

At present there is concern about the appearance of resistance to anticoagulant rodenticides in rats. Normally the brown rat, *Rattus norvegicus*, is the most susceptible of the common pest rodents to anticoagulant rodenticides. The LD 50 of warfarin is roughly one day's feeding on a 0·025 per cent bait. The black rat, *R. rattus*, gives an LD 50 of three to four days' feeding on a bait of the same concentration. The house-mouse, *Mus musculus*, is normally even less susceptible. Resistance to anticoagulants in the house-mouse was first reported in Britain in 1965. The physiological basis of this resistance is not known, but probably depends upon several biochemical factors. The inheritance of the resistance is almost certainly polyfactorial. In the brown rat, however, the resistance originating in the Welsh outbreak strain was found to be controlled by a single autosomal gene, and the same mechanism seems to operate in south-east England. Interestingly, in the Welsh resistant strain there is a linkage between the gene for resistance and genes for factors affecting the coat colour of the rats. The physiological basis of the resistance appears to be the production of a protein involved in the synthesis of blood clotting factors dependent upon vitamin K. Warfarin and vitamin K compete for these factors, but in the resistant strain the affinity of the factors for vitamin K is slightly reduced, but that for warfarin is greatly reduced. Most of the physiological work has been carried out on the Welsh strain and it is possible that other mechanisms are involved elsewhere (Greaves, 1970).

Brown rat strains resistant to the anticoagulant rodenticides have been reported from various parts of the United Kingdom, where they appear to be spreading, and also in Denmark, Holland, West Germany, and the USA. Recently resistance has been reported in populations of the black or ship rat, *R. rattus* in the docks of Liverpool, England. This is a serious development because of the likelihood of the transport of the resistant rats to other ports, despite the efforts of the British authorities to contain the population. The World Health Organization has advised all Port Authorities to be on their guard for the appearance of resistant black rats, and it has made test-kits available from its Geneva headquarters.

Rodents have developed resistance to other control agents. A later chapter describes the use of cultures of *Salmonella* spp. for the biological control of house-mice and other rodents. The technique was first used in the nineteenth century. Resistant strains of the house-mouse have been recognized since the early 1930s, the resistance being inherited more or less monofactorially, with a single main autosomal dominant gene. Rabbits (not, of course, rodents, but lagomorphs) have developed resistance to the virus of myxamotosis.

Effects of pesticides on the agroecosystem

A pest in a crop is only a part of a complicated ecosystem. A crop, though apparently simple, contains a surprising number of different species. Boness (1953) examined over 100 000 animals from German lucerne fields and found among them representatives of about 1500 species, mostly insects. Of course not all species occurred in any one field (the maximum was about 800), many were rare and about one-quarter were casual visitors. As well as the animals a number of flowering plants are to be found in such fields, and a variety of fungi and bacteria. The whole forms an integrated community with interactions between the various species which reduce the possibility of any of them increasing explosively. Even the pest species will be regulated by natural enemies to some extent, so that economically important outbreaks will be fairly infrequent. Furthermore there are probably several potential pests which have never caused important damage because their populations have always been kept low by density-dependent mechanisms.

It is impossible to predict all the consequences of the applications of pesticides to such an ecosystem. Here it is well worth stressing, as does Moore (1967), that pesticides are always applied to ecosystems, and not merely to pests. We can be fairly sure that there will be a big fall in the numbers of the target pest, but a number of other organisms will also be destroyed, or damaged in some other way. Among these will be some of the species which help to regulate the pest's population density. Indeed, parasites and predators of insect and mite pests are, in general, more easily killed by pesticides than are the pests themselves. Furthermore, predators at least are usually far scarcer than their prey at the best of times. The poison will also affect the parasites and predators in another way, namely by depriving them of their prey so that if they are not killed, they must either move out of the area or starve. The net result just after spraying is a smaller number of pests and a greatly reduced parasite–predator population. When the pesticide finally disappears from the environment (possibly to reappear somewhere else), or its quantity is so reduced that it no longer affects the pest, the survivors can increase rapidly in the absence of significant competition, predation, and parasitism. Thus an all too common phenomenon, a pest 'flare-back' or resurgence occurs. Furthermore, these survivors may well show a developed resistance towards the pesticide. The pesticide has not affected the pest's final population density in the way we wanted. It may even have made matters worse, except that, for a time at least, it protected the crop. This situation is very common with insect pests, but it is not unknown with other organisms. Weed-killers are often applied to kill a specific kind of weed which may be growing in association with other, less noxious species. The weed-killer may well destroy all the broad-leaved weeds, many of which were in fierce competition with the target species. If this species is an efficient colonizer of bare or partially covered ground, it could become far more troublesome than it was before spraying. Egler (1964) cites ragweed in America as an excellent example. This plant releases a pollen which is notorious for causing allergic responses in sensitive people, so medical authorities often call loudly for its chemical control. Medical men, unfortunately, are often far worse ecologists than even chemical salesmen.

A further complication may arise when a pesticide is applied at some time before some important density-dependent mechanism exerts its effects upon the pest population. If the pesticide reduces the population to a very low level then the density-dependent factor acting later will destroy proportionately less. We have, of course, already considered this in part when we discussed the reduced intraspecific competition. A chemical pesticide which produces a large but temporary drop in the numbers of a pest, negates, to a large extent, the effect of density-dependent factors acting subsequently. This mechanism will occur even if the factors themselves (predators, parasites, and so on) are unharmed by the pesticide. It is not only a drawback of chemical pesticides but can occur when any control or regulatory agent produces a catastrophic fall in pest numbers. In any case the grower is more concerned about the stage of the pest which causes the damage, and he usually concentrates his control measures against this. The chemical control of a caterpillar could, for example, allow the subsequent survival of a larger number of moths than would have survived if the caterpillars had been left alone, but the adult moths do not damage the crop, and there are often further factors which will reduce *their* numbers.

A pest can start to 'come back' when there are still enough residues to kill parasites and predators. This has been shown to happen with the carabid predators of the cabbage root fly, *Erioischia* (*Chortophila*) *brassicae*, a pest which became more troublesome after the use of aldrin and dieldrin (Wright, Hughes, and Worrall, 1960; Coaker, 1965).

Red spider mites (Tetranychidae) have long been pests under glass, but their spectacular rise to being the most important pests of fruit and many other crops throughout the world has only taken place since the Second World War. Before the use of DDT on agricultural and horticultural crops they were merely potential pests, probably not even noticed by the majority of farmers and growers. Unfortunately, while they are not susceptible to DDT, many of their predators, such as the black-kneed capsid bug, *Blepharidopterus angulatus*, are easily killed by this insecticide. Furthermore, the predators are much slower breeders than the mites whose rate of reproduction may even be stimulated by DDT. Massee (1954), is of the opinion that DDT is not the only culprit; he puts part of the blame on the use of tar oil winter washes, used for the control of aphids.

Some pests, as said in the second chapter, are poorly regulated by density-dependent mechanisms. They are characterized by violent fluctuations from year to year. Many of the sawfly and moth pests of forest trees fall into this group. Since their numbers are largely determined by climatic factors, and since they often 'escape' from their natural enemies, it would not seem that using pesticides would aggravate the situation. But even in these cases the pesticide may merely protect the foliage and have little effect on the population density of the following season. Watt (1968), for example, examined the records on the aerial spraying against the gypsy moth, *Porthetria dispar*, in New York State from 1945 to 1961, and on its annual damage, in terms of acres defoliated, from 1925. Using the sum of the mean monthly temperatures for July, August, and September as a measure of the climatic conditions each year, and employing multiple regression techniques, he showed that changes in damage from year to year were much more closely correlated with climatic conditions than with the acreage sprayed with insecticides. In fact, in some years the damage rose considerably, despite extensive spraying the year before.

The unhappy conclusion is that when we start to use insecticides (and some other pesticides as well) we are, in fact, treating the symptoms of the trouble, and having little effect on the cause. To make matters worse, the longer we keep on using them the harder it is to stop because it could take several seasons for the natural enemies to re-establish themselves, especially if treatments have been carried out over a wide area. If, as is often the case, the development of resistance aggravates the problem, the amounts of pesticide used must be continuously increased until they reach intolerable levels.

Possible effects on man
This leads us back to the question of the effects of pesticide pollution on other non-target species, and, in particular, on man. The danger of residues in the food that we eat has been, till recently, the most publicized aspect of pest control. Town dwellers – and in the technically advanced countries that means nearly everyone – have little appreciation of the damage that uncontrolled pests can cause, but they do have a healthy objection to being poisoned. As a result, there has been a flood of publications on the danger, ranging from emotional predictions of mass sickness to soberly written, technically executed reports of biologists and medical men. Fortunately, governments have chosen to listen to the latter and have refrained from banning the use of pesticides entirely. Instead, they have chosen to regulate them.

This control is usually mandatory, but in the United Kingdom a voluntary scheme, depending upon the cooperation of manufacturers, importers, and government agencies, has worked well for several years. The scheme controls the application of pesticides by providing for their registration and by giving recommendations for such things as minimum periods between application and harvest, maximum rates, and so on. Recently, the Advisory Committee on Pesticides and other Toxic Chemicals has recommended the introduction of various mandatory measures (Dept. of Education and Science, 1969). Unlike most other countries which use pesticides in large quantities, the United Kingdom has not legislated on tolerable residues in food, although there are regulations to cover food contamination in less specific ways. The USA and several European and Commonwealth countries depend heavily upon the use of tolerances to control pesticide usage. Unfortunately, the setting up of 'zero tolerances' for some products has made the use of some valuable chemicals almost impossible. The advance of analytical chemistry, now capable of detecting a few parts in a thousand million, an unimportant quantity in practice, has far outstripped the slow march of legislation.

In the opinion of the above committee the lack of tolerance levels for home-grown food does not present any threat to the British public, for they found 'no substantiated evidence that the consumption of foodstuffs bearing residues of active ingredients from accepted agricultural practices in the United Kingdom has led to injury in man'. In addition, they point out that the setting up of tolerance limits is still largely guesswork, mainly because we lack the basic knowledge of what is an acceptable daily intake. Such data are being collected by the WHO and the FAO, but the work is, by its nature, slow, and only a few pesticides have been dealt with so far. There may be justification for setting up tolerance limits for imported foodstuffs if only to curb any temptation to 'dump' in one country food which is unacceptable in other countries.

Deaths and illnesses from pesticides have occurred through accidents and carelessness. Insecticides, for example, have been stored in soft-drink bottles and swallowed by children. Empty tins have been dumped in drinking-water supplies. Parathion was often used in post-war Germany as a means of suicide, though it is difficult to imagine a more unpleasant way of poisoning oneself. These deaths and illnesses cannot, however, be held against pesticides any more than road fatalities can be held against the motor car. As in road safety the accidents could be reduced by education. Many countries, including the United Kingdom, are tightening up the regulations against, and increasing the penalties for, the dumping of waste toxic materials in potentially dangerous places.

Foodstuffs can also become contaminated by accident and by carelessness, as well as during routine agricultural operations. This usually happens during storage or transportation, especially when food is carried just after, or at the same time as, pesticides. The 1956 outbreak of endrin poisoning in Wales, when at least fifty-nine people were affected, was traced to flour contamination by spillage in a railway truck. Two separate outbreaks in 1967, resulting in the deaths of twenty-six people and the illness of at least 1000 more in Saudi Arabia and Qatar, were much more serious. One outbreak was traced to flour contamination by endrin leaking from a deck cargo to a hold below, the other to endrin escaping from the lower 'tween-deck to flour stored in the hold beneath. Parathion, contaminating foodstuffs stored in the same hold, killed 106 people in Kerala, India, in 1958. There is a number of similar cases, but most accidents could have been avoided by better cargo storage, and a keener appreciation of the danger of such products by the masters of vessels, and transport officials. There is almost no excuse for the use of empty pesticide sacks for food storage, as happened, for example in the Dominican Republic. In poorer countries, however, empty bags are valuable, and readily resold as containers. It should be remembered that in such areas the level of literacy is very low, and few people would be able to read warnings printed on containers. In the Dominican Republic it is also the custom for the poorer families to eat in large communal groups, a habit which accounts for the large number of cases of parathion poisoning on this occasion – nineteen of illness, and thirteen of death.

Literacy is not, however, a guarantee of safety. It is surprising the scant attention some people will give to the warnings printed on a pesticide container. In a recent case in Western Germany an elderly forestry worker, who had long 'treated' a persistent skin condition with petroleum, decided to use parathion in its place. He diluted the contents of the bottle with 250 ml of water and rubbed about 70 ml of the resulting emulsion on his chest, stomach, and legs. It was estimated that he applied about 2 g of the active ingredient to his skin. Fortunately, he recovered sufficiently to be released from hospital within a few days, but he still suffered from hepatitis several months later.

The residues of pesticides in food resulting from routine agricultural operations, even when all the proper precautions are taken, and the pesticides present in the environment in general, present another kind of problem. Here we are concerned with their possible long-term effects on human health, whereas in the cases mentioned above we were considering acute poisoning. Despite all the regulations on pesticide usage, whether they be mandatory or voluntary, concerned with application or with residues, most people carry some pesticides in their tissues. The most common are the stable chlorinated hydrocarbons and their metabolites which are stored in the fatty tissues of the body. It is fairly simple to take samples of fat and other tissues from members of the general public by biopsy, or during a post mortem, and to analyse these for chlorinated hydrocarbons and their metabolites, or to measure the depression of blood cholinesterase to assess the effects of organophosphate and car-bamate insecticides. It has, however, proved to be very difficult to relate the content of a given pesticide with any specific intake, and impossible to link the results cate-gorically with any particular illness or symptoms. We are not, of course, considering here the symptoms of acute or chronic poisoning resulting from exceptionally high exposures.

There is a common misconception that as long as organochlorine residues are ingested with food, or taken in from the general environment in some way, the amounts of these compounds or their metabolites will continue to rise within the body. Some of the material is, in fact, eliminated from the body and, after a time, the level of these products reaches a plateau which is related to the daily intake. In those countries where the compounds have been widely used for the last two decades there has not been any significant recent increase in the mean quantities of DDT and similar compounds stored in the bodies of members of the general public. Hunter and his colleagues (Hunter *et al.*, 1967), for example, reported that there had been no significant increase in dieldrin residues in the general public in the United Kingdom and in the USA during the preceding five years.

Table 4.1 surveys the levels of DDT and DDE, expressed as total DDT, found in samples of fatty tissue taken from the general public in various countries at various times. It also gives some details of dieldrin levels. In the United Kingdom and the USA it can be seen that the total DDT levels have reached a plateau and, indeed, show some signs of falling in the USA. As many countries are now restricting the use of chlorinated hydrocarbon insecticides it is probable that the levels will decline even more.

In many surveys there were no significant differences between various age groups, and between various occupational and racial groups. In the USA, however, non-white groups often had higher levels of total DDT than did white groups. The ratio was almost two to one in one American survey in 1968. There were no differences, however, in the dieldrin levels in the 1968 Florida survey. In South Africa, on the other hand, the samples from Bantu people contained less total DDT than did the samples from white subjects. The higher levels found in the non-white Americans

Table 4.1 Mean concentration of DDT and DDE as 'Total' DDT and dieldrin in human body fat and other tissues. Various sources

Population	Year	No. in sample	DDT (p.p.m.)	Dieldrin (p.p.m.)	Remarks
USA	1942	10	0	—	
	1950	75	5·3	—	
	1955	49	19·9	—	
	1961–2	130	12·6	—	
	1961–2	30	5·3	0·15	
	1962–3	64	—	0·11	
	1964	64	7·6	0·31	
	1964	25	10·3	0·29	
	1966	71	10·2	0·22	
	1962–6 }	994	10·4	—	
	1962–6 }	221	—	0·14	Not detected in 103 of the 221. In rest, range 0·01 to 1·39
	1968 {	90	8·4	—	Whites, >6 years
	{	35	16·7	—	Non-whites, >6 years
	1966–8	70	6·65	0·14	
	1968	146	—	0·22±0·16	
United Kingdom	1961–2	131	2·2*	0·21*	
	1963–4	65	3·3	0·26	
	1964	100	3·9	0·21	
	1965–6	101	2·8	0·23	
	1965–7	248	3·0	0·21	
Australia (Melbourne)	1965	53	1·81	0·046	
(Western)	1968	46	9·5	0·67	Dieldrin: 12 samples
Canada	1959–60	70	6·63	0·14	
	1966	27	4·1	0·22	
New Zealand	1966	52	5·7	0·27	
Israel	1963–4	254	19·2	—	
	1965–6	133	18·1±12·6	—	10–89 years
		71	10·2±9·2	—	0–9 years
DDR	1967	100	13·2	—	
Germany	1958–9	60	2·3	—	
Denmark	1966	17	3·4	0·2	
Holland	1968	29	1·75*	0·17*	Population in vicinity of a dieldrin/aldrin manufacturing plant
India (Delhi)	1964	67	26·0	—	
South Africa	1970	123	6·38	—	African < white Women < men

* Geometric mean.

probably reflect socio-economic differences rather than racial differences. Ehrlich and Harriman (1971) suggest that American Negroes have a diet which contains more fatty foods than the diets of wealthier whites. There were no differences in the levels when groups of poor whites were compared with poor non-whites.

In the Israeli survey of 1965–6 (Wassermann *et al.*, 1967) there were no significant differences in total DDT levels between the age groups within the 10 to 89 years, but the levels in the 0 to 10 years group were significantly lower than those in the older group. The 0 to 10 years group included 12 still-born children, 24 babies aged between 0 and 30 days, and 25 older children. There were no significant differences between these three sub-groups. In this, and in several other surveys, there was clear evidence of the transmission of chlorinated hydrocarbons across the placenta. Human milk also commonly contains these pesticides. Polishuk and his colleagues (1970) compared the levels of total DDT and other chlorinated hydrocarbons in the tissues of pregnant and non-pregnant women. The levels were lower in pregnant women, and the authors suggested that in pregnancy the metabolism of these insecticides is increased, with the foetus participating in the metabolism of some of them.

Not surprisingly, the levels of DDT and its derivatives are higher in populations in warmer countries, where agricultural spraying is carried out during many months of the year, and where public health pests are important. Furthermore, rural populations have higher burdens of insecticides than do urban populations, especially in areas of intensive agriculture and fruit growing.

Unfortunately, we do not know how important these pesticide burdens are. There is no direct evidence that they cause disease, and in most studies there is little or no association with morbidity or pathological conditions. There is, however, a disquieting association between lindane and blood dyscrasias, and West (1967) has called for intensive research into the possible relationship. On the other hand Hoffman and his colleagues (1967) studied the contents of DDT, lindane, and related compounds in 994 specimens of abdominal wall fat taken during autopsies, and found no significant correlation between levels of insecticide and any pathological condition. In a later paper Hoffman (1968) surveyed the findings of various workers on pesticide residues in human tissues and their acute and chronic effects on man, and concluded that there is no evidence available that the pesticides, in the concentrations found in the body, have any harmful effect.

Hayes (1971) discusses, in a WHO document, the degree of hazard to man when DDT is used against malaria. He points out that in the few cases where numerous small doses have produced symptoms over an extended period, these symptoms were qualitatively similar to those produced by a small number of large doses. No new kinds of disorder have been detected. He stresses the remarkable safety record in malaria campaigns. Thousands of workers have applied large quantities of DDT; the homes of millions of people have been sprayed annually, a dose of 2 g/m^2 being applied to the interior walls. In yellow fever campaigns DDT has been added to community water supplies at the rate of 1 p.p.m. and to some private water supplies at rates up to 5 p.p.m. He also examines the evidence for DDT being a carcinogen. While it is impossible to say definitely that DDT is not a carcinogen in man, no evidence has been put forward to indicate that it is. Hayes also doubts that DDT is a mutagen, and, in fact, he claims there is considerable doubt that any heritable change has been produced in a mammal by a chemical. Alkylating agents can produce dominant lethals resulting in the death of the embryo, or the sterility or semi-sterility of the offspring, but these characteristics, by their nature, cannot be inherited. They are not even recognized as a problem by people who are working with true alkylating agents, and DDT does not fall into this class of compounds.

An indication of the possible long-term dangers of pesticides can be obtained from a study of people who are exposed to them as part of their work. It is, of course, difficult to disentangle acute effects resulting from carelessness, and chronic effects caused by a constant low-level exposure. In general, the safety record in pesticide manufacture is good. In one study thirty-five men had been working in a DDT manufacturing and formulation factory for periods ranging from eleven to nineteen years. Their total DDT content in body fat ranged from 38 to 647 p.p.m., with a mean of 204 p.p.m., while that of the general population at that time was 7·67 p.p.m.

Twenty of the men, those most exposed to the insecticide, had an estimated daily intake of about 18 mg per man, compared to an average of 0·04 mg per person for the general population. The medical histories of the men examined, and routine medical and clinical checks, including chest X-rays, did not reveal any disorder which could be attributed to DDT.

Paramonchik and Platanova (1968) report various effects on the liver which might be attributed to DDT and related compounds. Paramonchik (1968) also discussed abnormalities of both the liver and the nervous system among workers in the pest control industry, but Hayes (1971) considers the second report to be difficult to interpret as it gives no indication of storage or excretory levels, and as the various complaints listed do not correspond with those known to occur in cases of DDT poisoning. Furthermore, these complaints also occur in the same country, Russia, among workers on compounds unrelated to these insecticides. Hayes also treats with reserve two reports by Schüttmann (1966, 1968) concerning eight cases of chronic liver disorder and a few cases of peripheral neuritis among an unspecified number of people occupationally exposed to DDT and lindane. He points out that similar complaints have not been found, even though looked for, in countries where DDT is manufactured and used on a large scale.

Certain effects on the liver are, however, well established. DDT, in common with many other substances such as ethanol, oxidized fats, several antioxidants, and many components of food which are not themselves nutrients, causes an increase in microsomal enzymes in the liver, although Hayes (1971) reports that workers with heavy occupational exposure to DDT appear to have normal liver enzyme activity.

Ehrlich and Ehrlich (1970) doubt, however, that we can take much comfort from observations on people exposed to DDT and similar compounds during the course of their work, or from trials with volunteers. They are concerned with the possible effects of a lifetime's exposure to these residues, with the first intake occurring at birth, or even before. The people who have been exposed in this way are still less than thirty years old. Workers who have been exposed to the chemicals for periods of ten or twenty years would have received their first intake during adult life. Nevertheless, these workers carry burdens which are often hundreds of times greater than those of the general population so, on present evidence, there seems to be an ample safety margin for the general public.

There are other facts to be considered. DDT and related compounds, as we have noted before, are stored in fatty tissues. These tissues often contain the food reserves of the body, to be drawn upon in times of need. If this should happen the pesticide content would be released with possible harmful effects. This mechanism is known to have killed birds in trials. These had a much greater burden of insecticide than humans carry, but something similar could happen to people in times of severe hardship, or during enforced slimming, especially if they were originally grossly overweight. Carson (1962), for example, cites a New Zealand case of a man who was being treated for obesity and who showed symptoms of dieldrin poisoning (Paul, 1959).

The problem of food contamination and environmental pollution is further complicated by the presence of residues of more than one kind of chemical. The number of possible combinations is large, and this increases the chance of harmful effects, through synergistic action.

Obviously, any insecticide, or any pesticide for that matter, is undesirable in human tissue, but the possible disadvantages in their use must be balanced against the known advantages. The main hazard lies in the possible chronic effects of the extremely stable chlorinated hydrocarbons, so any restriction of their use, when a satisfactory substitute is available, is to be welcomed. Unfortunately, none has yet been found for malaria control where a relatively cheap and persistent material is needed. There will be, no doubt, a reduction in the use of chlorinated hydrocarbons, and of fungicides containing heavy metals, for crop and food pest control, in at least the cooler countries, and in those which can afford the more expensive, but far less stable substitutes.

It is, unfortunately, impossible to gain any clear understanding of the dangers, if any, posed by pesticides by a study of the literature, or, at least, by a study of the more popular literature. We have already noted the intra-tribal squabbles of population ecologists, but these are mild compared with the quarrels of those writers who are opposed to widespread use of pesticides, and those who defend them. Those who oppose pesticides, or who consider that their use should be greatly restricted, frequently accuse the defenders of being either in the pay of chemical companies, or at least dependent upon them for research funds. The defenders, on the other hand, often claim that their opponents are far too concerned about conservation, and have insufficient regard for, or appreciation of the value of, pesticides in the protection of food and health. Ehrlich and Ehrlich (1970), for example, list in their bibliography J. L. Whitten's *That We May Live*, a rather optimistic book, describing it as 'A clever piece of pro-pesticide propoganda by a U.S. Congressman'. They suggest that we should 'Read this if you want to know your enemy.' The citation closes with a quotation from *Playboy* attacking other activities of the Congressman.

Carson's *Silent Spring* was, of course, the most important stimulus for the present investigation into the undesirable effects of pesticides, and should be consulted for historical reasons. It did bring to public notice many of the undoubted abuses of pesticides in the United States. It has been rightly described as an advocate's plea rather than a work of science, for it is, unfortunately, inaccurate in several places. The author, for example, tends to confuse a statistical correlation with a clear-cut demonstration of cause and effect, and she also, at times, derives undemonstrated conclusions from established facts.

A reading of *Silent Spring* should be complemented by a study of Rudd (1964), Moore (1966, 1967), Whitten (1966), Mellanby (1967), and Slater (1967). The 1966 National Academy of Sciences Symposium on Scientific Aspects of Pest Control (National Academy of Sciences, 1966) should also be consulted. There is, in addition, a number of texts dealing with the environmental crisis which devote some space to pesticide problems. These include Arthur (1969), Ehrlich and Ehrlich (1970), and Wagner (1971). There are several more, and their omission from this survey should not be interpreted as a criticism of them. There have been, at the same time, several publications in which it is clear that the authors have little knowledge of the subject; one example describes aldrin and dieldrin as the 'terrible organophosphate twins'.

Possibly the present situation is best summarized by Moore when he wrote: 'When honest men and women can disagree so fundamentally about the value of chemicals which save life and increase food production, it is probable that the issues at stake are both real and complex.' Clearly there is an urgent need of even more research into the possible effects of pesticides on human health. Unfortunately, the failure to demonstrate any cause-and-effect relationship between any specific pesticide and a specific disease, other than acute poisoning, will never prove that pesticides are incapable of causing illness. Some doubts will always remain.

Effects on the general environment

It is natural that man's first concern about pesticides is for their possible effects on his health. It is also natural that he should take precautions to avoid both acute and chronic poisoning. It now seems that there is little direct danger to man in the residues in his food, but he is beginning to realize that there may be greater dangers elsewhere.

Once again, the main concern is with chlorinated hydrocarbon insecticides, because of their extreme stability, and their affinity for organic matter, a consequence of their fat solubility. Many other pesticides can have a catastrophic effect on an ecosystem, but most of them are comparatively unstable, so that their effects are relatively transitory. The most important exceptions are the various substances which contain metallic ions, some of which are persistent and can be troublesome. Most of the following discussion will be about chlorinated hydrocarbon insecticides, but other pesticides will be mentioned where appropriate.

A very large number of papers have appeared on the acute, chronic, and sublethal toxicities of pesticides to various kinds of wild life. Most of these studies were, however, carried out under laboratory conditions, and it is difficult to relate these findings to dangers in the field. Nevertheless, they are basic for any studies on the effects of pesticides on organisms in the wild. Often, of course, animals found dead in the field are analysed to determine their contents of chlorinated hydrocarbon insecticides and their derivatives. It is then sometimes possible to say that the content was sufficient to cause death, but frequently there will be doubt, especially if the carcases are so old that it is impossible to rule out other possible causes of death.

The widespread death of fish is, for example, common in many rivers and estuaries. Insecticidal run-off from agricultural land is often blamed, but sometimes the death may be caused by disease organisms, by general industrial effluent, or by deoxygenation of the water following the entry of large masses of decaying organic matter during flooding. In such cases a large sample of dead and dying fish should be analysed before any firm conclusions are made.

The chlorinated hydrocarbons are very toxic to most fish. The LC 50 for 48 hours exposure to DDT is, for example, 0·0006 p.p.m. for the fish *Mugil curema*. It is clear, therefore, that any direct application of these pesticides to water for the control of mosquitoes or rice pests, or even the application to surrounding farm land and forest, is a grave threat to these animals. The low concentration mentioned above should be contrasted with that required to control mosquito larvae, namely 0·03 p.p.m. The organophosphate insecticides, as a group, are far less toxic to fish than are the chlorinated hydrocarbons, and herbicides, fungicides, and algicides are, in general, still less toxic.

Fish, and many other aquatic organisms, are exposed to pesticides both in the food which they eat, and also in the water that has to be passed over their gills. The intake of suspended water-insoluble compounds which are fat soluble is likely to be greater than that of substances dissolved in the water.

A number of factors can influence the toxicity of a pesticide to fish in natural waters. The formulation of the product is important, as emulsified and oil solutions are far more harmful than granular formulations, wettable powders, and dusts.

The toxicity also varies with the temperature and chemistry of the water. Bottom muds with a high organic content will, for example, bind a large proportion of the insecticide mixed with the water. A high pH seems to reduce the toxicity of DDT to salmon and related fish. Cope (1971) discusses these factors and others which affect the toxicity of pesticides to wild life.

There are several reports of fish populations developing resistance to pesticides in the field. In the Mississippi region *Gambusia affinis* and bullheads, *Ictalurus* spp., sometimes show a hundred-fold resistance to DDT, and they often display a cross resistance which is similar to that found in insects. Fish, in general, can concentrate insecticides from the surrounding water, and in resistant forms this ability is enhanced. Some Mississippi *Gambusia* were found to contain up to 214 p.p.m. of endrin in their body fat after a two-week exposure to a concentration of 0·05 p.p.m. Such high residues would, of course, threaten the health or life of predators of such resistant fish.

The fate and effects of pesticides in the soil after intentional application, or after contamination by drift, depend upon several factors in addition to the kind of pesticide and its formulation. Lichtenstein (1966) lists a further eight, namely soil type, moisture and temperature, wind or air movement, cropping and cultivations, the method of application to the soil, and soil micro-organisms.

Insecticides persist longer in soils with a high organic content, just as they tend to bind with silts and muds in natural waters. Although they are more persistent in such soils than in mineral soils, they are less available to the organisms present there. There is a much greater uptake of insecticides by plants from sandy soils than from organic soils.

The effect of the moisture depends upon the kind of pesticide. Water, for example, displaces aldrin from soil particles, allowing subsequent volatilization. It does not appear to have the same effect with DDT. Moisture is also necessary for the metabolism of micro-organisms, some of which attack pesticides. Temperature influences the rate of breakdown and the volatilization of the chemicals within the soil.

Cultivations expose the soil to the air and hasten the volatilization and disappearance of the pesticides. Working the pesticide into the soil, after a surface application, reduces, of course, the loss, as does a cover crop such as lucerne which cuts down the exposure of the soil.

The disappearance of a pesticide can often be correlated with the abundance of micro-organisms. Aldrin and parathion, for example, persist longer in dry and autoclaved samples of soil than in control samples, though some of the aldrin loss in the latter is due to microbial epoxidation to dieldrin.

Some crops extract surprisingly large amounts of pesticide from treated soil. Carrots translocated more soil residues than any other crop tested in Lichtenstein's experiments.

Newsom (1967) in his review of the consequences of insecticide use on non-target organisms reports that rates of application usually recommended do not lead to the accumulation of large amounts of residues in soils, and in most cases the amounts would not exceed the quantities applied in any one year. There are, however, examples of very high accumulations in salt marsh in New Jersey where chlorinated hydrocarbons have applied for many years in succession for the control of mosquitoes.

Edwards (1969) and his colleagues at Rothamsted Experimental Station have studied the effects of various soil pollutants on the small soil-dwelling invertebrates.

These animals, which range in size from earthworms to protozoa, play an important part in the cycling of plant debris, breaking it down into smaller particles which can be more rapidly attacked by soil micro-organisms such as bacteria and fungi. The total biomass of all these animals in the soil under one square metre can reach 500 g.

In some of the trials the insecticides were mixed as thoroughly as possible with the top few inches of soil. Perfect mixing was not, of course, possible, so the smallest animals had the greatest likelihood of avoiding contact with the pesticides. In general, predatory mites, pauropods, isotomid springtails, and larvae of Diptera were susceptible to most insecticides, while most other families of springtails, the symphilids, the earthworms, and the enchytraeid worms were unaffected by moderate concentrations. Chlordane, heptachlor, phorate, and carbaryl did, however, seriously decrease earthworm populations.

One common consequence of insecticide treatment was an increase in the numbers of some species of mites and of springtails, compared with numbers in untreated soils. There was also a decrease in the number of predatory mites. In one long-term experiment the population of predatory mites was only five per cent of that of untreated soil, four years after a single application of DDT. Other workers in Europe have obtained similar results.

Edwards suggests that the springtails are normally regulated by predation, rather than by competition for food, so that when the predators are removed by pesticides, a population explosion of springtails occurs. Edwards also noted a reduction in the number of species present in unploughed pasture after insecticidal treatment, though a similar reduction would follow after ploughing. Edwards concludes that, in general, even the most drastic effects of pollution cannot persist for more than a few years after the last traces of a pollutant have disappeared, and that even the most persistent pesticides are gradually degraded. He observes that the increases in the numbers of the smaller soil animals, such as springtails, after soil treatment, may compensate for the loss of the larger invertebrates that play a part in breaking down soil organic matter. In one experiment in which leaves were enclosed in nylon mesh bags to exclude earthworms, 73 per cent of the leaf material had disappeared after nine months in untreated soil, 90 per cent in DDT-treated soil, and 43 per cent in aldrin-treated soil, and these data could be correlated with the numbers of springtails present. Edwards considers that it is only in forest and woodland soil that insecticides are a potential hazard in that they may slow down soil formation and prevent the maintenance of soil fertility.

Fungicides containing heavy metals such as mercury, or copper may contaminate the soil, though often in some changed form. In one East Anglian orchard copper sprays had been used several times annually for a number of years, and analyses of the surface litter revealed a content of 0·2 per cent copper. Although this seemed to have no ill effects on the well-established Bramley apple trees, it is probable that young trees would be damaged. The leaves accumulated on the surface, forming a peaty layer, because of the absence of earthworms. It is also probable that the leaves remaining on the surface increased the likelihood of scab (*Venturia inaequalis*) infection of the growing trees (Mellanby, 1967).

It is true that even persistent chlorinated hydrocarbons disappear slowly from treated soil, but only a part of this disappearance can be attributed to degradation by micro-organisms and other agencies. Much of the material finds its way elsewhere.

Some is translocated into growing plants; some vaporizes or co-distils with evaporated water. Some is washed or blown into streams, rivers, and, eventually, the sea. During spraying or dusting, a large amount of the material never reaches the target surfaces, and drift may spread the chemicals to places far from the area of application. Finally, of course, some is taken up by animals, and enters into food webs.

Air and rain samples usually contain at least traces of chlorinated hydrocarbon pollutants. In the rural Midlands of Britain rain was shown in one survey to contain up to 0·0001 p.p.m. of lindane, and even smaller quantities of dieldrin. DDT was possibly present as well. Similar figures were obtained from rain-water collected in central London, but the dieldrin and DDT contents were somewhat higher than in the Midlands. Lindane is, of course, more volatile than most of the other chlorinated hydrocarbons. Air sampled over London contained 10 nanogrammes per cubic metre of DDT and its metabolites (a nanogram is $1·0 \times 10^{-9}$ grammes). Air sampling over California in 1963 also revealed DDT in trace amounts. Quantities were higher over New Mexico where 45 μg of DDT were recovered from the air collected in thirty minutes of flying time.

Observations on dust blown from Europe and Africa to the Barbados region led Riseborough and his colleagues (1968) to study the transatlantic movements of pesticides in the north-east trade winds. They calculated that the values for DDT and related compounds ranged from 6·5 to 233 femtogrammes per cubic metre (a femtogram is $1·0 \times 10^{-15}$ grammes). By making various assumptions about the amount of the pesticides absorbed on the dust and about the density of the dust it was estimated that about 600 kg of the insecticides would be deposited annually into the 19 400 000 km^2 (approximately 7 500 000 mile^2) of ocean under consideration (Galley, 1971).

As Galley points out it is difficult to make any generalizations about the extent of air contamination with pesticides. The amounts vary from place to place and from season to season. There is also a regrettable lack of data for tropical and sub-tropical areas where anti-malarial campaigns are in progress. The quantities are, of course, exceedingly small, and are not likely to have any harmful effects whatsoever. They are far lower than the concentrations in many foods of both man and wild animals.

Wild animals can accumulate pesticides from their food. Seed dressings were introduced into eastern England for the control of the wheat bulb fly, *Leptohylemyia coarctata*, during the late 1950s. Trials were also carried out in which the insecticides, mainly lindane, dieldrin, and aldrin, were added to the fertilizer, or broadcast in some other way, but seed dressings were recommended for winter-sown wheat because, at 420 g per ha, there would be a minimal effect on the important soil fauna and flora. The eggs of the fly are laid on bare soil in the late summer, oddly enough in the absence of their chief host crop, wheat, and they overwinter in the soil. The eggs hatch in January and February and the larvae bore into the young winter-sown wheat. Spring-sown wheat thus usually escapes any serious attack and seed dressing was not recommended. The method proved fairly successful against a very serious pest.

The first troubles arose in the spring of 1956 when a number of dead seed-eating birds were found in the cereal-growing areas, especially in the eastern counties. In the following springs more birds, wood-pigeons, rooks, stock doves, pheasants, and so on, were found dead, and in many cases dieldrin was found to be the most probable cause of death.

The poisoning extended beyond the seed-eating birds for, dieldrin being a very stable chemical, the dead and dying victims were themselves poisonous to predators, such as foxes, which ate them. In the winter of 1959–60 over 1000 dead foxes were found, as well as numerous dogs, cats, badgers, and predatory birds.

Seed-eating birds rarely take winter-sown seed grain from the ground, especially if it is drilled fairly deeply, as there is sufficient food which is more easily found. In the early spring alternative food is scarcer and spring-sown wheat is greedily eaten. The majority of deaths seem to have been caused by the seed-eating birds feeding on spring-sown wheat which had been treated with dieldrin seed dressing. As was said before, such a seed dressing is not needed. Some farmers were also quick to realize that wood-pigeons could be easily killed by leaving some treated grain on the surface and doubtless some of them used this knowledge.

It is worth mentioning that the author and his colleagues carried out many of the trials with seed-dressed winter wheat between 1953 and 1956 and although they inspected hundreds of acres of treated wheat they did not find any sick or dead birds. It is possible, of course, that poisoned birds had died elsewhere, and it is obvious that the acreage treated was small compared with that treated later, but there was certainly no evidence then that winter-sown wheat, treated with dieldrin, was a serious danger to birds.

The result of these deaths was an agreement that seed dressings containing aldrin, dieldrin, or heptachlor are only to be used on wheat sown before the first of January and then only in wheat bulb fly areas, in fields where attacks are likely and only during years when eggs are numerous in the fields.

Since this ban was imposed all the species concerned have shown population recoveries, but it is still too early to be certain about their ultimate fate (Moore, 1969). There are many other ecological factors which influence the population density of birds, one of the most important being the destruction of their normal environment by urbanization and such farming practices as the grubbing out of hedgerows to make larger fields. Many of the birds of lowland Britain really belong to a woodland fauna and have survived for so long largely because of the substantial acreage of hedgerows. Simms (1971) devotes several pages to the avifauna of British hedgerows and lists some of the losses following their destruction.

There is no doubt, however, that the use of chlorinated hydrocarbons has led to the disappearance of predatory birds over large parts of Britain and Western Europe. A kestrel, *Falco tinnunculus*, for example, can be killed by consuming seven mice which have fed on dressed seed corn, and a peregrine, *Falco peregrinus*, by eating three pigeons.

The death of predatory animals which have eaten the remains of the victims of persistent insecticides is an example of the disastrous passage of such pesticides along a food chain. Since a single predator consumes a large number of individuals of the prey species, there is often a concentration of the pesticide within the predator. One of the best known examples of this followed the spraying of Clear Lake in California with the chlorinated hydrocarbon DDD. Clear Lake lies in an important tourist area but it was made most unpleasant by clouds of mosquitoes, biting midges, and a *Chaoborus* gnat. DDD was chosen for their control, rather than DDT, because it is less toxic to fish. The technique of application was so designed that most of the insecticide sank to the bottom where the midge larvae were found. At first the operation appeared to be both successful and selective, the midges being destroyed while other invertebrates and fish were scarcely harmed. This was in 1949 and the one application gave satisfactory control for a number of years. Further applications were made later since the DDD seemed to have produced no harmful effects. In 1954, however, large numbers of dead western grebes, *Aechmophorus occidentalis*, were found around the lake.

It took several years to trace all the steps in the poisoning of the grebes (Hunt and Bischoff, 1960). The birds lie at one end of a food chain, the earlier links of which are phytoplankton, small fish, and larger predatory fish. The phytoplankton contained about 5 p.p.m. of DDD, the smaller herbivorous fish about 10 p.p.m., and the larger predatory fish up to about 200 p.p.m. In the visceral fat of the grebes the values ranged from about 700 to 1600 p.p.m. This and similar work on other aquatic environments in America and Europe has been summarized by Moore (1967) in a review which is an excellent and valuable analysis of the pesticide problem.

Pesticide concentration is likely to be more severe in aquatic environments than in terrestrial ones as many aquatic animals acquire pesticide both in their food and by the passage of large quantities of water over their gills and similar organs. It also appears that fish can store levels of pesticides which approach the LD 50's. The same phenomenon does, however, occur commonly in terrestrial environments. Woodwell (1967) tabulates DDT and DDT derivative levels in marine, freshwater, and terrestrial animals in various parts of the world, in a review of toxic substances and ecological cycles.

Rudd (1964) has introduced a number of technical terms to cover the ways in which an animal may be harmed by eating contaminated prey or carrion. In secondary poisoning one animal dies after eating another. Death may occur shortly afterwards, or it may be delayed (delayed toxicity). If the animal does not die itself, but is eaten by a third species which does die, then the phenomenon is called 'delayed expression'. Very often the intermediate animal has concentrated the pesticide in its tissues, the result of eating many meals dosed with the poison. Such an animal is called a biological concentrator; earthworms furnish an excellent example.

We have already noted that fish populations may develop resistance to pesticides, and that this can lead to their being still more efficient biological concentrators. There is no reason to think that other forms of wild life could not do the same, and thus become greater threats to their predators. It is, of course, impossible to say how often such resistance has developed among wild populations, for there are usually no records of the original susceptibility of the various species.

Chlorinated hydrocarbon insecticides have caused most of this kind of ecological damage, because of their persistence and solubility in fat. Pesticides containing heavy metals are just as persistent, or at least their metallic ions are. Of these the most widely used are the fungicides containing mercury. Mercury compounds were included in the seed dressings for wheat bulb fly, as a protection against certain seed-borne diseases. Mercury residues were found in many of the dead birds but the amounts were not thought to be lethal. Dieldrin remains the chief culprit. Some work from Sweden suggests that mercury alone can kill wild animals and can be concentrated in food chains (Borg *et al.*, 1966). The deaths were widespread in Sweden among seed-eating and predatory birds, and rodents, and the cause was largely traceable to ethyl and methyl mercury compounds. These alkyl compounds are eliminated from the body more slowly than are their main competitors, phenyl mercury compounds. In passing it should be noted that some mercury compounds are natural constituents of animal tissues, a fact which makes interpretation rather difficult.

Mercury is, of course, also an industrial pollutant arising from electrical and other industries. A second group of industrial pollutants has also confused the picture as far as the chlorinated hydrocarbon insecticides are concerned. These are the polychlorinated biphenyls (PCB's), chlorinated hydrocarbon compounds discovered in the late nineteenth century, and brought into industrial use from about 1930 onwards. They make excellent plasticizers for paints and adhesives, and because of their fire-resistant properties are used in heavy-duty electrical equipment and hydraulic machinery.

During the monitoring of chlorinated hydrocarbon insecticide residues by gas–liquid chromatography many workers in Europe and America noted peaks for unidentified compounds which they presumed were unknown metabolites of the insecticides. They were eventually identified in 1966 as peaks corresponding to PCB's. Since then they have been found in many marine and freshwater organisms throughout the world, and their identification in museum specimens collected in 1944 shows that they were pollutants even before the persistent pesticides were introduced. In many ways they seem to behave in a way similar to the chlorinated hydrocarbon insecticides, as would be expected from their chemical structure.

The Monsanto chemical company are the only manufacturers of PCB's in the USA and Britain, and they also supplied part of the European market. They withdrew these compounds from sale voluntarily from March 1971, for all uses which could lead to environmental contamination (Tinker, 1971).

In general, apart from the direct killing of plants, and the destruction of food plants of herbivores, weed-killers seem to have a comparatively small effect on wild life. Nettles are often controlled with weed-killers, and despite their unpleasant properties, they provide the food for a number of beautiful and harmless species of insects such as the peacock butterfly, *Nymphalis io*. Doubtless many attractive species of plants have become much rarer since the introduction of herbicides, especially when farmers have been eager to carry out 'cosmetic' farming. Doubtless, too, there has been too much indiscriminate spraying of roadside verges to control tall growth. These applications, like the blanket spraying of millions of acres of forest and crop land in South-East Asia for military reasons, must be regarded as abuses of weed-killers. It is impossible to forecast the ecological effects of this defoliation in warfare. It can certainly be said that it will be drastic. One consequence of the removal of vegetation will be an increase in soil laterization in which soil nutrients are leached out by heavy rains, and the remaining soil, rich in iron and aluminium oxides, takes on a rock-like consistency under the influence of strong sunlight and exposure to oxygen.

Wurster (1968) has reported that laboratory trials have shown that DDT in low concentrations can interfere with the photosynthetic activity of marine phytoplankton. It is difficult to assess the ecological significance of these findings. Ehrlich and Ehrlich (1970) write of Wurster's publication that 'this may turn out to be one of the most important scientific papers of all time', and although they do not predict a mass destruction of all marine phytoplankton they do suggest that DDT pollution could lead to significant changes in phytoplankton floral composition as the various species appear to differ in their susceptibility to the chemical. Other authors appear to put less importance on these findings.

We have discussed the possible acute and chronic poisoning of animals in the wild by persistent insecticides. Death is not, however, the inevitable result of an illness, nor do all poisoned animals die. Sublethal effects can be very important as they reduce the numbers of a species in other ways. This is particularly noticeable with birds in which reproduction is often impaired. Clutch size may be reduced, a smaller proportion of the eggs hatch, and the young are less viable. There is also evidence of an increase in egg breaking and eating by birds of prey. The breaking may be partly caused by changes in adult behaviour induced by sublethal doses of pesticides. Much of it is due, however, to a reduction in the average egg-shell thickness since the introduction of the chlorinated hydrocarbons. Such an effect has been found associated with diminishing populations of a number of predatory and other birds in Europe and North America. Species concerned include the American bald eagle, *Haliaeetus leucocephalus*, the osprey, *Pandion haliaetus*, the peregrine falcon, *Falco peregrinus*, and the herring gull, *Larus argentatus*. In some osprey populations the weight of the egg shell has been reduced by more than one-quarter.

Various mechanisms have been suggested for the reduction in the thickness of the egg shell. Among these are elevated steroid metabolism, increased thyroid weight with a reduction in colloid content of the follicles, and abnormal vitamin metabolism (Hickey and Anderson, 1968; Cope, 1971; Wagner, 1971).

Changes in behaviour, even small ones, can make a poisoned animal more conspicuous than its fellows, and more likely to be taken by a predator – an unlucky meeting for both. There is even a report of a dieldrin-poisoned fox wandering within reach of its chief predator – into the yard of the Master of Fox Hounds – although it was not, presumably, consumed.

Possibly the most valuable analysis of the pesticide problem which has yet appeared is Moore's 1967 review. In this he stresses the difference in attitudes between the ecologist, the agriculturalist, and pest controller trained in chemistry. The latter often speaks of the side-effects of pesticides, and regards the application of a pesticide as being an application to a pest population, other effects being incidental, and often regrettable. This is an oversimplification, as Moore points out. Pesticides are never applied to a pest population alone, except in the laboratory; they are applied to ecosystems, and the effects should be viewed in this way. To stress the point he suggests that pesticides, even when selective, should be regarded as biocides, rather than insecticides, herbicides, and so on.

One inevitable consequence of the application of a chemical to an ecosystem is that the ecosystem is, at the very least, simplified. There is a reduction in complexity, and, as has been pointed out before, simpler ecosystems are generally less stable than more complex ones.

The controversy over the undesirable effects of pesticides has been more heated in the USA than in Britain and Europe. There are good reasons for this. In Europe fields are generally smaller than in the United States so that applications of pesticides are, usually, much more localized. Any catastrophic effect is thus relatively restricted in area, and there is therefore a strong possibility of recolonization from surrounding areas when residues disappear. Furthermore, extensive spraying campaigns to control public health pests such as mosquitoes and the fire ant are far commoner in the USA than in at least the more northerly European countries. Probably many of these campaigns were taken up far too enthusiastically. There is much doubt, for example, about the necessity for the fire ant campaign. In addition, forestry is a far more important industry in North America than it is in Europe, and widespread aerial spraying with pesticides has been carried out frequently in Canada and the USA. In Britain, on the other hand, the latest report of the Research Committee on Toxic Chemicals (Agricultural Research Council, 1970) states that undesirable effects of pesticides occurring in nature have not been severe, widespread, or prolonged, although there have been a few cases of importance.

Nevertheless, the controversy has been of service. It has led to the cessation of much unnecessary spraying. It has led to valuable research, much of it of a basic ecological character. It has pointed out areas in which research is badly needed. On the other hand, it has discouraged chemical manufacturers from pursuing pesticide research. The development costs are so high that a company is reluctant to invest the huge sums necessary unless there is a strong possibility that a new compound will not be withdrawn precipitously within a few years of introduction. This is to be regretted since the need is for more pesticides, but of a less damaging nature.

There are still many pieces of information missing. It is very difficult, for example, to obtain reliable figures for pesticide usage. Companies are reluctant to make their sales figures known. Without such knowledge it is even more difficult to assess the ecological consequences of pest control by chemicals.

Before concluding this chapter we must pay a final 'tribute' to the stability and 'spreadability' of DDT and the other chlorinated hydrocarbons. These compounds are frequently reported as contaminants many thousands of miles from any possible area of use. Some of it must have been transported in organisms which are carried along by physical forces, or which travel by their own efforts, and yet more must be transported by water or air currents after drifting as a fine spray from the area of application, or after volatilization or co-distillation from soil and plant surfaces. Some, it is now known, is carried from near-by land masses in those other pollutants, oil slicks. Residues have been detected in fish in the far North Atlantic, and in the far South Pacific, and DDT has even been found in appreciable quantities on the Antarctic continent.

The quantities reported from the Antarctic are of little, if any, biological importance. Their discovery is, in fact, as great a compliment to the sensitivity of analytical techniques as it is to the physical properties of DDT. The analytical method used on snow and water was capable of detecting one millionth of a gramme of DDT in 2000 g of water. If 600 tonnes of DDT were spread evenly over all the earth's surface, terrestrial and marine, every two square metres would contain about one millionth of a gramme. Such a distribution is mere fancy, but the calculation gives some idea of the sensitivity of gas chromatographic techniques towards chlorinated hydrocarbons.

George and Frear (1966) took several samples from water, and from surface and deep snow at various sites in the Antarctic but, using apparatus with the above sensitivity, were unable to detect any insecticides. This does not rule out atmospheric contamination entirely, but if DDT was present it was at levels below 0·0005 p.p.m.

Similarly, no DDT residues could be found in any of the marine invertebrates sampled, which were drawn from four phyla, namely Nemertinea, Arthropoda, Mollusca, and Echinodermata. The limit of sensitivity in this case was 0·005 p.p.m. Three species of fish were also studied, and in one sample from four of the species

Rhigophila dearborni DDT was detected at 0·44 p.p.m. As far as is known, this fish is restricted to the Antarctic, and it must therefore have concentrated the chemical from some local source. It is, however, impossible to say what this source was as water samples containing algae, and all the marine invertebrates sampled, contained no detectable amounts.

DDT and DDE were commoner in the higher vertebrates. Fourteen adult Weddel seals, one immature male and one immature female were sampled, and four of these, all adults, contained from 0·042 to 0·12 p.p.m. of DDT in the body fat, but not in other tissues. No DDE or other chlorinated hydrocarbon was detected. Weddel seals, *Leptonychotes weddelli*, are predators of fish, from which they presumably derived the insecticide. The absence of the pesticides from several of the seals suggests a patchy distribution in the fish.

Four of the sixteen Adèlie penguins, *Pygoscelis adeliae*, examined contained DDT (body fat only) at concentrations ranging from 0·015 to 0·18 p.p.m. The Arctic skuas, however, carried the greatest burden of DDT. All of fourteen adults contained both DDT and DDE, and one of two chicks contained DDE. The DDT sometimes exceeded 1 p.p.m., and the DDE reached 2·8 p.p.m. in the kidney tissue of an adult male. The whole chick sample contained 0·03 p.p.m. of DDE.

Samples were also taken from sixteen skuas, *Catharacta skua maccormicki* (*C. maccormicki*), fourteen of which were adults. All the adults contained both DDT and DDE residues, and one of the chicks, DDE. One adult male contained as much as 2·8 p.p.m. in the kidney tissue.

The skuas, unlike the penguins, do leave the Antarctic, ranging, for example, to Australia. It is probable, therefore, that they acquire some of the DDT during their migrations. The other species have very small amounts. Mellanby (1967) has pointed out that if the figures for the penguins are typical of these birds, then the whole population of the Antarctic, estimated as about ten million, would contain about half a pound of the insecticide. Furthermore, if the penguins and seals are both typical, then the total amount of DDT and its metabolites in all the Antarctic birds and mammals would be less than the amount which would be routinely applied to a group of elms in an American suburb.

It is interesting to speculate about the way in which the insecticide was introduced into the region. One possible route would be in the food supplies for the scientists stationed there, but George and Frear (1966) consider that the residual amounts in such food would be insufficient to account for the levels observed, and, in any case, it would be expected that other chlorinated hydrocarbons would also be involved. The skuas have been seen feeding at the garbage tips, so some could have come in this way. Since the animals examined were collected near research stations Tatton and Ruzicka (1967) extended the sampling to animals remote from any station, and again demonstrated the presence of DDT and its metabolites, as well as other chlorinated hydrocarbon insecticides. Pesticides were also detected in krill (*Ephausia* sp.).

The DDT brought to the region by wind and water currents would be so diluted that it seems that no organism at the base of a food chain would be able to concentrate it to the levels required. It is possible, however, that some may have reached the Antarctic in oil slicks in which it had concentrated.

There is a small possibility that DDT was brought in as part of a sea or air cargo, and that some escaped, but it is difficult to see what its purpose would have been.

It is more probable that a ship or aircraft which had previously carried DDT, perhaps to some tropical military outpost, was the source of the contamination.

There is, of course, yet another possibility, as Robinson (1970) points out. Most of the residues were at levels which could only just be detected by the instruments used at the time, and this time was before the demonstration of the importance of PCB's as pollutants. Robinson concluded:

> Any attempt to assess the significance of the results obtained in Antarctic animals must include a critical re-apparaisal of the analytical procedures. The majority of concentrations are very small and it is probable . . . the analytical procedures . . . were not sufficiently sophisticated, particularly in regard to possible interference by organochlorine compounds such as the polychlorobiphenyls.

Galley (1971) points out that such criticisms could be applied to much of the earlier work on residues, particularly when these residues were very small. There are now reports, based on the use of gas–liquid chromatography and mass spectrometry in tandem, of PCB's in the eggs of various birds. There would be no difficulty to explain the presence of these industrial compounds in the Antarctic.

There is no doubt that chlorinated hydrocarbons and mercury residues are present in many wild animals at damaging levels, and that many animals can concentrate such substances from their environment, even when these chemicals are present in very small amounts. It must be remembered, at the same time, that modern analytical techniques are capable of detecting residues that are so small that it is difficult to believe that they have any biological effects. Galley (1971) has likened the identification of DDT in the air above Barbados to the selection of a single second in about three thousand million years. It was pointed out above that an estimated 600 kg of the insecticide was deposited annually into about forty-seven million square kilometres of ocean from the north-east trade winds. Graham (1970) compares this with the amount of insecticide deposited annually into the sea by a major river system. The Mississippi, for example, is thought to carry down about 10 000 kg of mixed pesticides each year into the Gulf of Mexico, and the input into San Francisco Bay is estimated at 1900 kg annually. The comparison is hardly meaningful, however, as the river pollution leads to much higher local concentrations than those arising from aerial deposition. Such local concentrations are likely to be harmful, especially to molluscs, crustacea, and fish, within the area affected directly, and possibly far away from the river mouth if the pesticides are carried by currents or by contaminated organisms.

Pesticides entering any environment, including the ocean, are eventually destroyed by chemical decomposition and microbial attack. Thus, if the input is constant, the levels of even persistent pesticides are likely to reach plateaus related to the input, just as they do in the human body. Organisms of various kinds can, however, concentrate the pesticides from the environment, and the pesticides can then enter into a food chain. In general, the concentration ratio (concentration in the organisms to concentration in the medium) is higher at lower concentrations in the medium. There is, probably, however, a threshold concentration below which an organism cannot accumulate the pesticide. This is likely to vary considerably from organism to organism, and it will also depend upon the uniformity of the dispersion of the pesticide in the medium. Unfortunately, despite the importance of such basic data, there seem to be few if any publications on such threshold concentrations, although they would be valuable for the assessment of ecological dangers of pesticides.

There is also a need for the continued and intensive monitoring of pesticide residues in environments, foodstuffs, human beings, and other organisms. This work has been started in the developed countries but little is being done, as yet, in the poorer countries, where DDT is widely used for vector control. The techniques will have to be sophisticated. Results are likely to be misinterpreted if gas–liquid chromatography alone is used. They should be confirmed by a different analytical method, especially when very small amounts are concerned (Galley, 1971).

Clearly, persistent pesticides should not be used without good reason. Considerable quantities are used by private gardeners, usually at rates much higher than those recommended. Gardeners rarely worry about the costs of pest control, whereas farmers must balance the costs against estimated gains. In the United States there have certainly been a number of unnecessary campaigns involving large quantities of DDT and its allies.

Nevertheless, insecticides, even the persistent ones, do bring undoubted benefits, and these must be balanced against the drawbacks in their use. Of course every effort must be made to minimize or eliminate these drawbacks. It is sometimes implied in the more emotionally written books and articles that because pests usually reappear after pesticide application, and often reappear at higher population levels, that the pesticides have served no purpose. This is, of course, nonsense; pesticides have saved millions of tons of food and fibre from destruction by pests. It is true, however, that it would have been better to sacrifice some of this material to the pests by using far less pesticide, and allowing the control of the pests to be carried out partly by other agents. This is, of course, the technique of integrated control which is discussed later.

The World Health Organization is adamant that DDT is essential for their work on vector control in warmer countries. At present DDT is used for the control of vectors of malaria, African sleeping sickness, onchocerciasis, bubonic plague, typhus, and leishmaniasis. BHC also is used against the body louse, and dieldrin and BHC against the reduviid vectors of Chagas disease. Malathion, abate, fenthion, and Dursban® are the chief weapons against the vectors of the other important insect-borne diseases.

In malaria control, DDT is the cheapest available satisfactory insecticide. The next cheapest, malathion, which is unsatisfactory on mud walls, would cost eight times as much as DDT to give equivalent control. The third cheapest is twenty times as expensive as DDT, but it does give a satisfactory residue on mud walls. Over one thousand million people are now living in areas freed from endemic malaria and about three hundred million live in areas where spraying with DDT is continued. Even when the disease is reduced to a very low level, as it was in Ceylon, there can be little relaxation. Ceylon is facing a return of the disease, and more than two and a half million cases were reported in 1968 and 1969. It is not surprising, therefore, that the authorities of developing countries, and the WHO, are uneasy about the hostility to DDT, most of which, incidentally, arises in countries which do not have these immense public health problems and in which diets are adequate.

Malaria control probably leads to less environmental pollution than does agricultural pest control, when comparable amounts of DDT are used. Very little – about one ton each year – is now used for larviciding. Almost all spraying is carried out indoors, the object being to kill any resting female *Anopheles* mosquitoes after they have taken a blood meal. A little is sprayed around the eaves of huts and buildings, but most of this would escape to compacted earth around the dwellings.

At the peak of the malaria control campaign, during the 1960s, 66 000 tonnes of DDT, 4000 tonnes of dieldrin and 500 tonnes of BHC were used each year. As *Anopheles* mosquitoes rapidly develop severe resistance to dieldrin, very little of this insecticide is now employed. The amount of DDT used has dropped to 46 000 tonnes each year, but the WHO considers that this level will have to be maintained if malaria is to be contained. The peak of DDT usage has obviously been passed, as it has in agriculture, and this drop has not been just the result of banning in various countries. Much of the fall is caused by the substitution of more effective compounds for certain pests, and by the development of DDT resistance. Nevertheless, there is an urgent need for monitoring in the countries with massive vector control programmes (World Health Organization, 1971; Galley, 1971).

During the last few years, there has been a steady stream of literature on the ecological problems of the world, and much of it condemns the use of pesticides. These are also attacked on the grounds of possible effects on human health. Often the publications are written by eminent scientists, skilled in the field of ecology. Sometimes the writing is emotional, mirroring the author's feeling of urgency. Recently, however, some equally eminent scientists have been reacting against some of the more extreme opinions; see, for example, Bawden's 1972 review of a book by Commoner. There are also some recent balanced official or semi-official reports on the use of chlorinated hydrocarbon insecticides in agriculture which reflect the reluctance of the authors to advocate the wholesale banning of these pesticides (Department of Education and Science, 1969; Australian Academy of Science, 1972).

This flood of articles and books has renewed interest in non-chemical methods of control. There have been demands from conservation bodies and similar groups for the replacement of chemical methods by these alternative techniques, but few people who are not directly involved with pest control seem to have any clear ideas about these methods, and the difficulties they entail. The remainder of this book deals with non-chemical methods. Some are already well established, some are in the experimental stage, and some are still merely ideas. Yet others, it is hoped, await discovery. The book concludes with a discussion of integrated control, the welding together of several methods so that satisfactory control is obtained, without the disadvantages of the unilateral approach.

This chapter concludes with a number of tables (4.2–4.8) showing residue levels in various organisms or sites, and the usage of pesticides in various countries. The data have been taken from a wide selection of sources, so they must be considered as only approximately comparable.

Table 4.2

(a) *Residues in soil and soil invertebrates (mean values, p.p.m.). Various sources*

Location	Year	Pesticide	No. of sites	Soil conc.	Species	Conc.	Storage: exposure ratio
USA	1958	Total DDT	1	9·3	*Lumbricus terrestris*	19·2	2·06
					L. rubellus	680·0	73·10
USA	1965	Total DDT	1	9·9	*L. terrestris*	140·6	14·2
					Allobophora caligonosa	140·6	14·2
USA	1967	Total DDT	67	0·98	Earthworms	9·64	9·8
United Kingdom	1968	Total DDT	10	0·3 (Arable)	Earthworms	Trace	0·06
			2	9·7 (Orchard)	Earthworms	19·6	2·10

(b) *Persistence in soil (to 95 per cent disappearance)*

> MCPA: 2 to 3 months
> 2,4-D: 4 to 6 weeks
> 2,4,5-T: 3 to 12 months
> Parathion: up to 6 months
> Dieldrin: 5 to 25 years (average 8)
> DDT: 4 to 30 years (average 10)
> Lindane: 3 to 10 years (average 6·5)

Table 4.3   Total DDT residues in water surveys. Mean values in parts per 10^{12}. Various sources

Location	Year	Type of water	Concentration
USA	1967	Mississippi Delta	8·2
USA	1962	Various, California	0·62
United Kingdom	1966	Sewage	36·0
United Kingdom	1971	Rivers	1·6

Table 4.4 Residues of pesticides in water and in aquatic animals and plankton. Mean values: water in parts per 10⁹. Organisms in p.p.m.

Location	Year	Pesticide	Water conc.	Species	Conc.	Storage: exposure ratio	Remarks
USA	1965	DDT	0·5	Shrimp (*Penaeus setiferus*)	0·14	280	
USA	1960	DDD	20·0	Plankton	5·3	250	Clear Lake, California
				Fish	1·0–196·0	—	
				Grebes	723–1600·0	—	
USA	1966	Total organochlorine	0·6	Invertebrates	0·4	667	Tule Lake, California
				Fish	2·5	—	
				Birds (feeding on invertebrates)	1·3–42·8	—	
				Birds (feeding on fish)	4·0–57·2	—	
USA	1966	Total organochlorine	—	Invertebrates	0·41	—	Lake Michigan
				Fish	3·35–5·6	—	
				Birds (feeding on invertebrates)	6·0–98·8	—	
United Kingdom	1963	Dieldrin	—	Trout (*Salmo trutta*)	0·1–0·71	—	
United Kingdom	1964	Dieldrin	—	Trout (*S. trutta*)	0·06–0·92	—	
Ireland	1966	DDT	—	Salmon (*Salmo salar*)	0·71	—	
Queensland, Australia	1970	DDT	—	Oyster (*Crassostrea commercialis*)	Range 0 to 0·94	—	µg/g whole meat (various rivers)
		DDE			Range 0 to 0·20	—	
		DDD			Range 0 to 0·51	—	
Northern Territory, Australia	1972	DDT	—	Barramundi (*Scleropages leichhardti*)	0·70	—	Darwin River (Agricultural Area). In isolated area, all less than 0·01 p.p.m.
		DDE			0·27	—	
		TDE			0·21	—	
Victoria, Australia		DDE	—	Crocodile (*Crocodylus sp.*)	0·10	—	Isolated area
		Total DDT		Rainbow trout (*S. gairdnerii*)	1·3	—	Geometric mean
				Trout (*S. trutta*)	0·22	—	

Table 4.5 Residues of pesticides in predatory birds. Mean values, p.p.m. Various sources. (Residues in liver, unless stated otherwise)

Location	Year	Species	No. in sample	Dieldrin	Total DDT	Remarks
United Kingdom	1964	Sparrow hawk *(Accipiter nisus)*	8	1·5	12·95	
	1966	Sparrow hawk *(Accipiter nisus)*	8	1·46	12·86	
	1964	Kestrel *(Falco tinnunculus)*	28	3·07	5·21	
	1966	Kestrel *(Falco tinnunculus)*	9	5·8	6·50	
	1964	Barn owl *(Tyto alba)*	23	2·36	5·39	
	1966	Barn owl *(Tyto alba)*	7	1·18	2·37	
	1963	Heron (Adult) *(Ardea cinera)*	6	1·05	46·85	
	1964	Heron (Adult) *(Ardea cinera)*	17	2·17	23·91	
	1965	Heron (Adult) *(Ardea cinera)*	4	4·82	11·84	
Australia (Victoria)	1972	Brown goshawk *(Accipiter fasciatus)*	2	—	2·0	Geometric mean
	1972	Peregrine falcon *(Falco peregrinus macropus)*	1	—	7·0	Fat
	1972	Kookaburra *(Dacelo gigas)*	8	—	0·45	Geometric mean
	1972	Nankeen kestrel *(F. cenchroides)*	2	—	0·71	Geometric mean
	1972	Boobook owl *(Ninox novaeseelandiae)*	5	—	0·20	Geometric mean
USA	1964	Bald eagle *(Haliaeetus leucocephalus)*	—	—	9·4	Muscle

Table 4.6   Residues of pesticides in various birds. Mean values p.p.m. Various sources

Location	Year	Species	No. in sample	Pesticide	Residues	Remarks
USA (California)	1962	Pheasant (*Phasianus colchicus*)	—	Dieldrin Total DDT	0 to 25·0 ⎤ 0 to 2930·0 ⎦	Fat tissue
USA (Lake Michigan)	1965	Herring gull (*Larus argentatus*)	—	Total DDT	2441·0	Fat tissue
USA	1965	Canada goose (*Branta canadensis*)	—	Total DDT	1·85	Muscle
USA	1966	Robin (*Turdus migratorius*)	35	Total DDT	64	Brain (Birds tremoring)
United Kingdom	1965–7	Blackbird (*Turdus merula*)	8	Total DDT	215·80	Muscle ⎤
			7	Total DDT	414·73	Liver ⎬ Orchard
	1965–7	Pheasant (*P. colchicus*)	6	Total DDT	9·42	Muscle ⎦
Australia (Victoria)	1972	Blackbird (*T. merula*)	1	Total DDT	0·76	Muscle
	1972	Little wattle bird (*Anthochaera chrysoptera*)	1	Total DDT	3·9	Fat tissue
					88	Liver
			2		25	Brain
					0·77	Muscle (Geometric Mean)

Table 4.7   Residues in food (p.p.m.)

	Date		No.	Dieldrin	DDT's
United Kingdom					
Beef, Kidney fat	1964		66	0·07	0·07
	1966		63	0·04	0·07
	1968		34	0·03	0·03
Butter	1964		18	0·03	0·05
	1966		13	0·02	0·05
	1968		16	0·03	0·06
Milk	1964		60	0·003	0·005
	1966		75	0·003	0·004
	1968		76	0·001	0·002
Whole diet	1966	4th quarter		0·005	0·03
	1967	1st quarter		0·0045	0·03
		2nd quarter		0·0035	0·025
		3rd quarter		0·0035	0·025
Australia					
Mutton, Kidney fat	1964		6	<0·01	0·06
	1966		11	0·02	0·09
	1968		16	0·01	0·17
Butter	1964		45	0·01	0·41
	1966		53	0·01	0·54
	1968		20	0·01	0·26

Australia 1970
    Total DDT's.   Daily intake from whole diet as mg/day/70 kg/youth.[a]

Brisbane	0·242 [b]
Adelaide	0·083 [c]
National Mean	0·119

FAO/WHO acceptable daily intake 0·350.

[a]15–18 years old. Male, *ca.* 4000 cal./day
[b]Greatest for a State Capital
[c]Lowest for a State Capital

Table 4.8   Usage of pesticides in various countries

(a) *England and Wales 1966 (Department of Education and Science, 1969)*
Insecticides:

Crop	Approximate acreage grown	Organo-chlorines seed dressing/dips	(%)	Organo-chlorines sprays	(%)	Organo-phosphates	(%)	Others	(%)
Farm crops	9 300 000	4 080 000	43·9	78 100	0·8	511 100	5·5	—	—
Vegetables	373 000	73 000	19·6	58 300	15·6	175 260	47·0	—	—
Fruit and hops	191 700	0	0	189 100	99·0	183 300	95·5	52 750	27·5
Outdoor bulbs and flowers	13 100	300	2·3	600	4·6	2 400	18·3	310	2·4
Forestry	844 400	16	—	1 220	—	—	—	—	—
Total	10 722 200	4 153 316	38·7	327 320	3·1	872 060	8·1	53 060	0·5

N.B.   Some areas treated more than once per annum (e.g. fruit), often with different insecticides

(b) *Total use of DDT (active ingredient in tons) in 1969*

Austria	27
Italy	2456
United Kingdom	166
West Germany	190
USSR	2500

(Partial bans in force in all cited countries in 1969)

(c) *Domestic production and imports of DDT into Australia* (domestic production ceased, 1971)

1966	968·8 tons
1967	968·3 tons
1968	917·9 tons
1969	929·9 tons

5

Biological control

Forty years ago every biologist knew the meaning of the term 'biological control'. Since then the definition has been stretched and twisted to include a number of techniques which are closely allied to those of the original concept. Unfortunately, different biologists have twisted it in different ways so now there is always some uncertainty whenever the phrase is used. In this text we shall use it to mean the deliberate introduction of living material (other than resistant varieties of the victim) into the environment of the pest so that its population density is reduced or the damage it causes lessened. In what might be called classical biological control the living material is a natural enemy of the pest – a parasite, predator, or pathogen of an animal pest, or a pathogen or other attacker of a plant pest. This chapter, and the two which follow, deal with this kind of biological control. Later chapters will cover the use of the pest species itself as the living material, and also with the use of a competitor of the pest. For present purposes viruses will be considered to be living material since they multiply within their hosts. We shall also discuss the use of spore preparations of certain bacilli which kill by the toxic action of crystalline protein particles formed at sporulation. This is an arbitary inclusion since the control is not regulatory, and the material is not persistent. In these respects the method differs from most others which use living agents.

The basic ideas of biological control are perhaps best presented by describing some of the well-known success stories. These have been told many times before but they are worth recounting again.

Some early successes of biological control

The control of cottony cushion scale

The cottony cushion scale, *Icerya purchasi* (Fig. 5.1), was accidently introduced into California sometime in the 1860s, possibly on wattle (*Acacia*) from Australia. By 1887 the pest was forcing many citrus growers out of business. C. V. Riley, who was the Chief of the Federal Division of Entomology at that time, turned to biological control for a solution.

Unfortunately, Riley found it difficult to obtain the necessary financial support that he needed from the government. Apparently, there had been too many excursions to study European agriculture first hand (mainly by Riley himself) and a ban had been placed on foreign travel at the state's expense. The Commissioner for Agriculture was luckily as much a politician as a scientist and found a way around the problem.

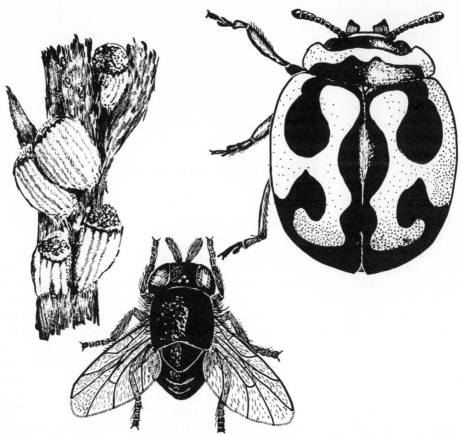

Fig. 5.1. Cottony cushion scale, *Icerya purchasi,* **and its biological control agents. Left: the scale; Right: the vedalia lady-bird,** *Rodolia cardinalis;* **Centre:** *Cryptochetum iceryae.* **(Lady-bird adapted from D. Miller (1971)** *Common Insects in New Zealand,* **A. H. and A. W. Reed, Wellington)**

An International Exposition was to be held in Melbourne in 1888 and Riley's colleague, Albert Koebele, was chosen to go, officially to represent the department, unofficially to search for natural enemies of the scale.

Koebele is usually given all the credit for the introduction of natural enemies of *Icerya,* but before he even left California an Australian, Frazer Crawford, sent parasitic flies for release in California (Fig. 5.1). He had found this species, *Cryptochaetum iceryae,* parasitizing the scale in Adelaide in 1886. It still plays a part in controlling the pest.

On arriving in Australia Koebele soon found many of these flies himself and shipped about 12 000 of them back to California. He also found the larvae of a small lady-bird beetle, commonly known as Vedalia, feeding on *Icerya* in an Adelaide garden. Koebele believed that the parasite would be a much more effective control agent than the predatory beetle, but wisely he sent samples of both species.

The first consignments of the Vedalia beetle, *Rodolia cardinalis* (Fig. 5.1), were released on a tented tree near Los Angeles during December 1888 and January 1889, and by the following April they had multiplied and cleared the tree of scale. The population was then allowed to expand on to surrounding trees, and sub-colonies were distributed to other parts of the state.

The progeny of these 129 beetles, and of a further 400 received a little later, have been sufficient to reduce the scale populations to insignificant levels throughout California. From about five years after its introduction till recently the scale has not been a pest anywhere in the state. In the last few years, however, insecticides which were applied against other pests have killed many of the Vedalias, allowing the *Icerya* to become troublesome again.

The probable reasons for the great success of Vedalia are as follows. Both the adult beetles and the larvae are active animals which easily search for and destroy the sedentary scales. The beetle usually completes two generations while the scale goes through one, and it is not itself greatly troubled by natural enemies.

The successful control of *Icerya purchasi* in California was naturally noted in many other countries, and in the following years the lady-bird was released, with similar results, in Florida, Hawaii, New Zealand, South Africa, and other areas.

Biological control of insect pests in Hawaii

Koebele did not rest on his laurels after his success with Vedalia. Soon after the Californian growers had presented him with a gold watch, and his wife with diamond ear-rings (Doutt, 1964a), he resigned and went to work for the Hawaiian government. In these islands sugar cane, an imported plant, is a major crop. It is attacked by numerous insect pests, most of which are also immigrants. Among these is *Perkinsiella saccharicida*, a leaf-hopper which is an important vector of virus diseases. In 1905 and 1906 Koebele and Perkins introduced two chalcid wasp parasites from Queensland and Fiji. Although the numbers of the leaf-hoppers quickly fell the control was insufficient and a search was made for other natural enemies. A suitable egg-predator, the mirid bug *Cyrtorhinus mundulus*, was found, also in Queensland and Fiji, and introduced by Muir in 1920. This bug withstands the heavy rains of the islands and has continued to give complete control ever since. It is apparently not so successful in Fiji where the crop continues to be ravaged by the Fiji disease virus, transmitted by *Perkinsiella*, although resistant varieties of cane are proving useful.

Muir continued his work with another pest of cane, the New Guinea sugar cane weevil, *Rhabdoscelus obscurus*. At the time the country of origin of this weevil was not known and, accordingly, Muir began a long search through Southern China, Malaysia, and Indonesia before he eventually found the weevil in Amboina, near Ceram. Here it was parasitized by a tachinid fly, *Ceromasia* (*Microceromasia*) *sphenophori*. Unfortunately, the life cycle of this fly is so short that it was impossible to ship it to Hawaii, so Muir extended his search to New Guinea, where once again he found the beetle and its parasite. Even then he had to establish intermediate breeding stations in Australia and Fiji before the parasite could be brought alive to Hawaii in 1910. By 1913 the fly had firmly established itself in the islands. It has given a useful degree of control of the weevil but is ineffective in windy, rainy regions, especially when the cane lodges.

The biological control of prickly pear in Australia

When the First Fleet arrived at the site of what is now Sydney in 1788 the ships carried a mixed cargo of convicts, soldiers, officials, and the smooth tree pear, *Opuntia monacantha*. Governor Phillip had collected these South American cacti at Rio de Janeiro with the intention of starting a cochineal industry. This red dye stuff is extracted from the mealy-bug *Dactylopius coccus* which feeds on cacti.

Other species of this cactus genus, namely *O. aurantiaca* (tiger pear), *O. imbricata* (Devil's rope), *O. stricta* (spiny prickly pear), and *O. inermis* (common prickly pear) were introduced later, though no one seems to know why. Possibly their bizarre shapes made them attractive as ornamentals. The cochineal industry was not established but the prickly pears were. Although graziers esteemed them at first as emergency fodder for drought years, by 1900 it was realized that the cacti were beginning to usurp the sheep and cattle. In 1900 40 500 km² were infested, and land continued to be lost at the rate of almost 10 000 km² each year. At the height of the infestation, in 1925, 243 000 km² in Queensland and New South Wales were covered with the pest. Over half of this area the stands were so thick that no grass or low plants could grow. The most important species by far were *O. inermis* and *O. stricta*.

Chemical or mechanical control was out of the question since most of the land was low-value grazing, even before the invasion, and carried little stock. Biological control was the only possible solution.

The first attempts were made in 1903 with the introduction of *Dactylopius indicus*. This failed, but a reintroduction in 1913 was more successful, giving a satisfactory control of *O. monacantha*. There was no respite however from *O. stricta* and *O. inermis*, so in 1920 the Commonwealth Prickly Pear Board was set up. Biologists were sent to the Americas to search for arthropods attacking cacti and to select those which promised to control the prickly pears. These were to be tested on a wide range of economically valuable plants to make sure that they would not turn to these in Australia. This work continued from 1920 till 1937 and resulted in the despatch of almost 60 species to Australia. Not all of these could be acclimatized, or bred in large numbers. Between 1921 and 1925 several species were released including a coreid bug, a boring moth, and the red spider mite *Tetranychus desertorum*. These insects and mites began to reduce the stands of prickly pear but their success was overshadowed after the introduction of the South American moth, *Cactoblastis cactorum*, in 1925. There was only one importation, a mere 2750 eggs, but it was enough.

The females of this moth, which can fly several miles in a single night, lay their eggs in long sticks on the cactus pads. Each stick contains about 75 eggs and resembles somewhat the spine of the cactus. Such sticks provide a convenient means of collecting and transporting the eggs from one district to another. The larvae which emerge feed gregariously within the pads and the destruction they cause is aggravated by various secondary fungi and bacteria. When fully grown the larvae emerge from the pads and pupate in rubbish and other sheltered spots on the ground.

The 2750 eggs received at Brisbane were used as breeding stock. After two generations 2 250 000 eggs were available for release at twenty sites in the infested area. Between 1927 and 1929 3 000 000 000 eggs were collected in the field and redistributed. By 1930 whole stands of prickly pear were being destroyed by the caterpillars. Hordes of hungry larvae were left without food but, fortunately, they did not turn to any other plant. There was, of course, a catastrophic fall in the numbers of the moth, followed by a resurgence of the cactus from 1932 to 1933, but the moth populations soon built up again. By 1935 the real danger was past. Now only scattered patches remain in Queensland, held well in check by *Cactoblastis*. The moth has not been so successful in the cooler State of New South Wales where its work has to be supplemented by chemical and mechanical control, and by the activities of a scale, *Dactylopius* sp.

The campaign in Australia caught the public's imagination, but people still occasionally grow prickly pears in their gardens though they are, of course, prohibited plants. The people of Dalby in Queensland, at the centre of the former prickly-pear country, still felt enough affection for the moth for them to erect a cairn in its honour in 1965. The inscription reads, in part, as follows:

IN 1925 PRICKLY PEAR, THE GREATEST EXAMPLE KNOWN TO MAN OF ANY NOXIOUS PLANT INVASION INFESTED FIFTY MILLION ACRES IN QUEENSLAND OF WHICH THIRTY MILLION REPRESENTED A COMPLETE COVERAGE. THE DALBY DISTRICT WAS THEN HEAVILY INFESTED. THE BIOLOGICAL CONTROL INVESTIGATION WAS UNDERTAKEN BY THE COMMONWEALTH PRICKLY PEAR BOARD, THE JOINT PROJECT OF THE COMMONWEALTH, QUEENSLAND AND NEW SOUTH WALES GOVERNMENTS.

EARLY IN 1925, A SMALL NUMBER OF CACTOBLASTIS CACTORUM INSECTS WERE INTRODUCED FROM THE ARGENTINE BY ALAN PARKHURST DODD, O.B.E. WHO WAS OFFICER-IN-CHARGE OF THIS SCIENTIFIC UNDERTAKING. THEY WERE BRED IN VERY LARGE NUMBERS AND LIBERATED THROUGHOUT THE PRICKLY PEAR COUNTRY. WITHIN TEN YEARS THE INSECTS HAD DESTROYED ALL THE DENSE MASSES OF PRICKLY PEAR.

(Dodd, 1936; Dodd, 1959; Holloway, 1964.)

The practice of biological control

The choice of suitable agents

The successful examples discussed above illustrate the ways in which biological control projects are carried out, using arthropodan natural enemies. The pest is usually an invader from some other country which has established itself, and whose population density is not kept below economically acceptable levels, by native parasites and predators. A search is made in the area in which the pest originated, or in which it is well regulated, to find suitable predators or parasites which may be shipped to the invaded country. Here they are held in quarantine, with stringent precautions against accidental escapes, while further studies are made. Those which are still considered promising are then cultured in large numbers and released in the infested areas. The fates of the pest and the natural enemy or enemies are followed in the field for several years until the project is deemed to be a success, a partial success, or a failure.

This procedure appears to be fairly straightforward but it actually involves the expertise of many different biological specialists, besides the ecological field worker. Not the least among these is the often forgotten museum taxonomist and systematist.

The area in which the pest is endemic is not always known, for this may be remote and infrequently visited by biologists, or it may be a region where the pest is scarce because of the attentions of its natural enemies. The specialist in the systematics of the group, borrowing from his knowledge of the evolution within the group, may be able to make a shrewd suggestion about the place of origin of the pest. Between 1918 and 1928 California and Federal entomologists searched Australia, Argentina, and Uruguay unsuccessfully for natural enemies of the plant bug which spreads curly top virus of beet. This bug was then known as *Eutettix tenellus*, but careful work by the systematist Oman showed that it really belonged to the Old World genus *Circulifer*. A subsequent search in Spain and North Africa revealed both the bug and suitable parasites for importation. Systematists also suggested that the best area to search for parasites of the olive scale, *Parlatoria oleae*, would be in, or near, Pakistan. Their predictions were correct and, as a result, efficient parasites have been introduced into the olive groves of California.

When the area of origin of the pest, or the area in which it is endemic, has been pin-pointed a search for natural enemies is carried out. A variety of predators and parasites is usually found and a decision must be made about which are the most suitable species. Unfortunately, we are unable to predict accurately the fate of any species which is introduced. There are, however, certain characteristics which indicate which species are most likely to be good biological control agents (see also the discussion of Holling's work in chapter 2).

The natural enemy that is the most abundant and obvious is often a poor biological control agent. Such an abundance is usually coupled with heavy infestation by the host species. A desirable species is one which can search out and destroy its host even when this is scarce. A suitable species is thus likely to be an uncommon species. The searching ability is considered by most authors to be far more important than fecundity; the latter is thought to be adequate in most parasites and predators. This probable scarcity of the most promising species is unfortunate because it makes the collection of large numbers difficult, and there is a danger in attempting to initiate a biological control programme with only a few imported individuals. Such a small number would carry only a small sample of the genetic variability present in the parent population. While the sample of genes may be favourable from the point of view of the project's organizers, it is also possible that they could be unfavourable and lead to a failure of the importation.

Remington (1968) has reviewed the influence of the nature of the source population on the success of colonization of individuals taken from it. His arguments deal with both the pest species and natural enemies. There is not space enough to discuss his views fully but, to summarize, he advocates the introduction of a large wild sample taken from a large source population. The individuals of such a source population should be numerous and spread continuously over a large geographical area. Those imported should be taken from an environment as similar as possible to the one infested or, better still, a number of samples should be introduced in close succession. Each of these should be taken from a different kind of environment, each of which should be similar to one or another of those infested. In this way there is a good chance of the habitats of the pest and of the natural enemy coinciding over the pest's range. The wide genetic variability in such introduced samples will make the successful adaptation of the species to the new conditions more probable. It should be remembered that Remington is concerned with the successful colonization of the organisms but a successful colonizer is not necessarily an efficient biological control agent, though the reverse must apply. Remington agrees that the scientist must avoid introducing a few individuals from a large source population, or releasing large numbers of laboratory progeny derived from a few founders. Such introductions are likely to fade away.

There is always the possibility that the genetics of the progeny of a few individuals could be modified by laboratory selection to make them more suitable as control agents. This would, however, take several seasons and the economics of the situation usually make this undesirable.

It has been noted that the most successful agents are usually fairly specific, if not completely specific to the pest. This specificity relates to the behaviour in the field, for in the artificial conditions of the laboratory an insect will sometimes parasitize or attack a host which would be disregarded in the wild. Host specificity is particularly important when arthropods are introduced for weed control, for there is no point in trading a weed pest for a new insect pest. It also seems to be an important feature in the biological control of arthropod pests. This is to be expected as a host-specific parasite or predator is highly adapted for dealing with its host or prey. Their life cycles are synchronized so that the natural enemy does not perish because of the temporary absence of a suitable stage of the host for attack. A specific enemy is also efficient at searching for the host. Often, however, the initial stages of the search are directed towards habitats in which the host is normally found; on a certain food plant or in a certain community of plants. If the habitat of the pest in its new home is somewhat different from that in its country of origin, an introduced parasite may not be as efficient.

This could occur, for example, if *Itoplectis conquisitor*, an ichneumonid endo-parasite of the pine shoot moth, *Rhyacionia buoliana*, were used for the biological control of the pest on red pine. Arthur (1962) showed that the odour of the food plant of the caterpillar is very important in the host-finding by the parasite. The ichneumonid showed a distinct preference for the odour of Scots pine to that of red pine in laboratory experiments, and, in the field, the parasite was much commoner in pupae that had developed from larvae feeding on Scots pine than in pupae derived from red pine.

In another series of experiments pupae of the fir budworm, *Choristoneura murinana*, were transferred from their natural host tree to near-by oak trees, where they were placed in rolled oak leaves so that they resembled the pupae formed by certain oak tortricids. The budworm pupae on the oak trees were readily attacked by the ichneumonid *Apechthis rufata* which ignored the pupae on the near-by trees. Clearly the oak trees were the over-riding attractant for the ichneumonids.

A natural enemy which is not completely specific may perform better as an intro-duced control agent than a truly specific species, since it can survive on alternative hosts when the target host becomes very rare. In such cases the natural enemy will be able to deal quickly with a resurgence of the target species, if this is the preferred host.

The question of synchronization raises, in passing, one difficulty which sometimes occurs. Natural enemies often have to be transferred from the Northern Hemisphere to the Southern, or transported in the other direction. The seasons are, of course, out of phase, and the biologists may have to spend a considerable time in bringing the two organisms into step. The difficulty increases the further the two areas are from the tropics.

The systematist plays an important part at this stage also. It is sometimes found that what appears to be a widespread parasite or predator attacking its host in a variety of habitats is, in fact, a complex of races or sibling species each with a restricted habitat. It is obvious that if the pest is to be controlled throughout its range it is not sufficient to introduce only one of these races or sibling species. Such studies bring in all the modern techniques of what has been called the 'New Systematics' (Huxley, 1940) for morphological studies alone are insufficient; by definition sibling species are morphologically indistinguishable. The difficulties posed to the field worker collecting in the area of origin or the region in which the pest is endemic are clearly great. It is also important that material and cultures should be preserved for reference throughout the project. With these samples for comparison with those taken from the field it is possible to determine which varieties or sibling species are bringing about the control, and which are contributing little.

A rapid rate of increase is an obviously desirable feature of a biological control agent. It should have a high fecundity and a short developmental period compared with that of the host so that a sudden increase in the population density of the latter is followed closely by one of the agent. One of the probable reasons for the success of the Vedalia beetle, as was pointed out before, is the comparative shortness of its life cycle.

In all biological control projects the imported material must be held for some time in quarantine so that it may be studied. It is essential, for example, to ensure that the stock is free from diseases and parasites. Usually a relatively small number of specimens of each species or strain is brought in. For both these reasons a way must be found to increase the numbers by laboratory or glasshouse breeding, and this may prove difficult or impossible with some species. Usually, of course, the host must also be cultured, though this does not often prove difficult.

The modes of life of adult parasites and their larvae often differ markedly. Consequently, their modes of nutrition also differ. The larvae are generally parasites which obtain all their needs from the host, while the adults are usually free-living, obtaining their food from flowers, honey dew, and similar sources. The roles are reversed in such forms as mosquitoes and fleas, but the adults of these insects are probably better regarded as specialized predators. As the adult's source of food may be remote from the places in which the hosts are found, the question of egg maturation becomes important. Some parasitic females lay their eggs very soon after emergence and mating. The materials from which the eggs are formed are obtained when the insect is feeding as a larva. Other parasitic females lay eggs over an extended period and must obtain food as adults for the maturation of the eggs. If these are to be effective control agents throughout the range of the pest, adult food supplies must always be obtainable near to the pest infestations, or the environment must be manipulated in some way so that suitable food is provided. These contrasting types of females are known technically as proovigenic and synovigenic respectively.

An efficient natural enemy should be able to distinguish between suitable and unsuitable hosts, and should avoid egg laying in one which will not allow the young to develop normally. Since the larvae of many parasitic insects need the exclusive occupancy of a host if they are to develop normally, a host individual which already contains the egg or larva of the species must be regarded as an unsuitable host. Sometimes, neither of the larvae matures, sometimes only one, or if both develop into adults these adults are somewhat stunted in size. Superparasitism, as it is called, can retard biological control, especially when the female lays her eggs completely randomly in host individuals.

Earlier workers considered that egg-laying females could not distinguish between healthy hosts and those already parasitized, either by individuals of the same species, or of other species (multiparasitism). It was calculated that if egg laying was purely random about 60 parasite eggs would be needed to achieve 50 per cent parasitization of 100 host individuals, and about 450 to achieve 99 per cent parasitization – clearly a great waste of biological control agents. Many later workers have shown, however, that many species avoid laying in already parasitized hosts. *Trichogramma evanescens* apparently recognizes eggs which are already parasitized and usually refrains from laying in them. If healthy hosts are few, superparasitism can occur. This, of course, makes an estimate of the probable percentage parasitization difficult. The parasite is able to detect the odour of other individuals which have previously walked over the eggs, though it is probable that this odour is removed by rain.

Several workers, using optical and electron microscopy, have demonstrated the presence of minute sense organs on the ovipositors of several species of endoparasitic Hymenoptera, and these are, presumably, used to test the suitability of a prospective host individual (Fisher, 1971). It is possible that such sense organs can detect changes in the haemolymph of the host when this has been parasitized. Other parasites may be able to detect clotted haemolymph on the surface of a parasitized host, or the presence of a parasite within the host by the sound produced when the latter is tapped with the antennae.

Usually, therefore, females are selective and will only lay in already parasitized hosts when the egg-laying drive is intense, and most of the hosts have already been attacked. Even in these circumstances the females of some species retain their eggs rather than lay them. Superparasitism is undesirable, but as it can occur in most parasite–host associations, at least to some extent, it should not be a reason for rejecting an otherwise promising biological control agent.

Apantales glomeratus, the commonly observed larval parasite of the large cabbage white butterfly, *Pieris brassicae*, is an excellent example of a large group of parasites which lay large numbers of eggs in a single host. The sulphur-yellow cocoons on the carcase of the caterpillar must be a familiar sight to most gardeners and growers in Europe. It is doubtful, however, that a single host individual could support two such broods, and such superparasitism, like that discussed above, must lead to wastage.

A single host individual may support several different species of parasites, a phenomenon known as multiparasitism. It is arguable whether such interspecific competition is disadvantageous for biological control. To consider the simplest case, the one in which only two parasitic species are involved, the question becomes would the pest population be kept at a lower level by a single parasite species than it is by both. The two species clearly compete. Furthermore, they also compete even if multiparasitism by the two species does not occur. The answer is of practical importance because it determines whether or not both species should be introduced for biological control. We have already mentioned in passing the effect of interspecific competition between predators and parasites on the stability of the population of the prey species (chapter 2).

Experience seems to show that multiparasitism (or interspecific competition without multiparasitism) is not disadvantageous. There are many examples where the successive introductions of different species can be correlated with increasing control of the pest. The competition may, of course, lead to the displacement of one or more of the competing species. This can also occur if several species are introduced at approximately the same time. In Hawaii, for example, *Opius longicaudatus*, a parasite of the Oriental fruit fly, *Dacus dorsalis*, increased rapidly after its release, but was soon displaced by *O. vandenboschi* which had been released at about the same time. This was, in its turn, displaced by a parasite introduced later, namely *O. oophilus*. During the whole period the control of the pest improved. Since 1950 *O. oophilus* has been the dominant species, and usually achieves a parasitization rate of about 70 per cent.

There are other examples in which, because of the differences in the preferred habitats, the two parasites have somewhat different ranges. When, however, they do occur in the same area they achieve the greatest control.

It is theoretically possible that when one species replaces another, the replaced species is the better biological control agent, capable of finding the host individuals when these are scarce. In general, however, most biological control experts appear to be in favour of multiple introductions, when suitable species are available.

When discussing multiple parasitism we have been considering parasites competing for the same stage of the host. It is well known, however, that different parasites specialize in different stages. Thus there are egg parasites such as *Trichogramma*, larval parasites such as *Apantales glomeratus*, pupal parasites, and adult parasites. Predators may also specialize. *Cyrtorhinus*, for example, feeds on the eggs of the leaf-hopper *Perkinsiella*. Howard and Fiske (1911) advocate the introduction of such a series of natural enemies whenever they are available. Their argument is that if the attack on one stage fails because of adverse weather conditions or some other reason, control may still be brought about by the other agents. Other workers reason that most of the highly successful cases of biological control have been accomplished by a single parasitic or predatory species, and consequently conclude that such a sequence is usually unnecessary. Doutt and DeBach (1964) point out that when a satisfactory control is not achieved, then the sequence approach is well worth trying.

We have discussed in the last few pages the criteria which may be used in choosing species for the biological control of arthropodan pests. The project scientists must be extremely careful that they introduce into the field only members of these species. It is always possible, indeed probable, that some of the individuals reaching the quarantine receiving area in the new country will, themselves, be parasitized. If such a hyperparasite or pathogen is released into the field along with the control agent, then the agent itself could be destroyed or hampered by the hyperparasite or disease. That the danger is a real one may be gathered from the experience of various early workers who, finding insects emerging from the dead body of a pest, took them to be parasites of that pest when, in reality, they were hyperparasites.

When a weed has been successfully controlled by an introduced insect care must be taken not to introduce a natural enemy of that insect. This is a real danger if the weed-controlling insect is related to some pest of crops. Australian entomologists view with suspicion any proposed pathogen or parasite which might attack *Cactoblastis cactorum*.

While the biological control of weeds resembles that of phytophagous pests in many ways, there are certain differences in approach which must be considered when selecting species for importation. The main concern is for the economically important plants grown in the area. There is always a risk that the introduced species will turn its attention to some valued crop. This danger has probably been overemphasized in the past and, consequently, some potentially valuable agents may have been discarded needlessly. The danger may be best estimated (or over-estimated) by starvation tests and by consideration of the botanical position of the pest plant. The prickly pears and the widespread St John's wort, *Hypericum perforatum*, are both suitable candidates for biological control, since they are both widely separated from any economically valuable plant. The blackberry, *Rubus fruticosus*, on the other hand, belongs to the immensely important Rosaceae with its apples, pears, loganberries, raspberries, and a host of ornamentals. The blackberry is possibly the most important perennial weed in New Zealand, but there seems to be little prospect of controlling it by an introduced biological control agent.

Acaena sanguisorbae, piri-piri, is a native weed of New Zealand and also belongs to the Rosaceae. It is, in fact, closely related to the strawberry. Sweetman (1958) has, however, pointed out that although the leaves resemble those of the strawberry the fruit is quite different. It is the fruit that is the main reason for the trouble caused by piri-piri for it is a bur that becomes entangled in sheep-wool. An insect which attacks the fruit of piri-piri is unlikely to attack that of the strawberry, but a leaf-feeder would be a possible pest of the strawberry. Two leaf-eating chrysomelid beetles, *Haltica* (*Altica*) spp., were rejected when found to feed on both plants. A sawfly which feeds on the leaves of piri-piri but not on those of strawberries was tried, but did not become established, possibly because of predation by birds.

Sweetman's suggestion of a fruit-attacking insect does not seem to have been followed, possibly because no such insect has, as yet, been found. The idea of highly specialized feeders is quite popular for two reasons. First, highly specialized feeders (bud-borers, stem-borers, root-borers) which are oligophagous are considered to be less likely to adapt to other plants. Second, it is thought that they will themselves be less subject to attack by predators and parasites. Seed-feeding insects are not usually favoured for biological control, as most projects are directed against perennial weeds growing in relatively undisturbed land – forest, rangeland, etc. Such perennial weeds do not depend upon seeds for survival from year to year. There is, however, an awakening of interest in seed-feeding insects for the possible control of annuals which are utterly dependent upon seeds to survive from one season to the next.

There seems to be little real danger that an oligophagous insect will, after introduction, change its feeding habits and attack a plant which it had ignored in starvation tests. The feeding habits of insects were established long ago, and show little change. Careful work by Huffaker (1957) and others has shown that most recorded changes of host plants in the literature really related to the recording of a food choice which had not been noted previously. It is evident, however, that biological control workers should be aware of the work being carried out on the mechanisms of host-plant selection, even if they do not have the facilities to carry out such research themselves. With this knowledge the dangers associated with the introduction of biological control agents can be assessed more scientifically than they are at present. The basis of host-plant selection has been reviewed by Schoonhoven (1968). A more recent review of the problem is presented by Zwölfer and Harris (1971) who also stress the importance of studying the chemical and physical basis of host-plant selection by an insect. Other pre-release studies which they recommend are an examination of the biology of the insect species, particular attention being given to any adaptations which might restrict its host range, and a survey of the host-plant range of related insect species They also discuss the possibilities of phytophagous insects transferring to new hosts.

There are very few cases known of an insect introduced for biological control of a weed becoming a pest. Two exceptions are *Thecla echion* attacking egg-plant in Hawaii, and certain insects attacking spineless cacti in South Africa. The second was expected to happen at the time of introduction, and the first occurred because starvation tests were not carried out.

A suitable biological control agent may look most unpromising whilst feeding on the weed in its native home. This performance cannot be taken as a guide for its probable effectiveness when imported into the invaded area. It is most probable that the agent itself is regulated by its own parasites and predators. In its new environment, in the absence of the parasites and predators, it may be much more destructive to the weed, just as an introduced and highly destructive pest insect is often of little consequence in its country of origin. It is also difficult to assess how widely the weed would be distributed in the absence of the proposed agent. It is possible that the weed's range has been restricted to areas which are comparatively unsuitable for the insect or pathogen. The recorded history of the control of St John's wort in California illustrates this. Formerly it was widespread throughout large areas of rangeland. After the introduction of the beetle *Chrysolina quadrigemina* it receded and now only flourishes in shaded places which are not favoured by the beetle. Without a knowledge of what had happened earlier, an observer would consider *Chrysolina* to be a poor candidate for biological control.

Remington's recommendations for choice of suitable source populations apply here too. Whenever possible, the environment of the source population should be similar to that in the invaded area. When this is very large, as was the case with prickly pear in Australia, sources may have to be found in a number of regions, although it did not prove necessary for *Cactoblastis*.

The search for suitable agents need not be restricted to the weed itself. It is useful to examine related species, especially when these occur in a region climatically similar to the infested one. *Apion antiquum*, introduced into Hawaii for the control of *Emex spinosa* (emex), was found in South Africa on *Emex australis*. This weevil was found to be specific to the genus, unlike the most promising insect found on *E. spinosa*, which also attacked broad beans. The success with *Apion antiquum* is especially encouraging since it is the first time biological control has shown promise with an annual weed.

Most attempts at biological control of weeds have been directed against exotic species, as would be expected. On the other hand, many workers have pointed out the destruction caused by introduced pests to native plants, the most frequently quoted examples being *Juniperus bermudiana* attacked by introduced scales in Bermuda, and native chestnuts destroyed by the fungus *Endothia parasitica* in North America. The fungus attacks chestnuts in Asia, but these are resistant and little damage is done. It would seem worthwhile, therefore, to pay more attention to the parasites and pathogens of foreign relatives of native weeds, or, indeed, enemies of the weed itself in other countries. It may lead to the effective biological control of some such pests.

Before concluding this discussion of the control of terrestrial weeds it should be pointed out that the introduced agent is not going to be the only biological control agent present. The weed is growing in fierce competition with other plants, and these will help to eliminate individuals weakened by insects or pathogens. In this respect the biological control of weeds differs from most cases of biological control of phytophagous pests where the competition from other species is not so intense. The replacement of a pest by a competing innocuous relative has occurred in public health work, and the concept is being investigated enthusiastically, but this will be discussed later.

Apart from the age-old practice of selective grazing of weeds by goats and other large herbivores, biological control projects against terrestrial weeds have all, so far, used arthropods as agents. In contrast insects have not, as yet, been used for the biological control of aquatic weeds. It has been suggested that insects feeding on aquatic plants may not be sufficiently oligophagous, and may spread to rice-growing districts. Biological control workers have turned to other animal groups such as the Mollusca, the fishes, and the mammals. Carp and fish belonging to the genus *Tilapia* are widely used and, of course, can also be cropped for protein. A South American snail. *Marisa cornuarietis* has also shown promise but cannot be used in rice-growing areas. But possibly the most fascinating agent is that aquatic mammal which may have given rise to the legends of mermaids, namely the manatee, *Trichechus manatus*. This ungainly Sirenian is used to keep channels and ponds clear of weeds in Guyana, and the practice may be extended to other tropical and sub-tropical coasts. Its recruitment to man's collection of 'domestic' animals should, at least, save it from its otherwise almost certain extinction.

Biological control of aquatic weeds is particularly desirable as chemical control is dangerous for the fish. There are many urgent problems. Imported species such as the water hyacinth, *Eichhornia crassipes* and the fern *Salvinia auriculata* are blocking the flow and passage in waterways throughout Africa and Asia. Weeds provide mosquito larvae with cover so that they escape from *Gambusia* and other insectivorous fish which are themselves important biological control agents in the campaigns against malaria.

A number of current biological control of weeds papers were given at an International Symposium in 1969 (Zwölfer, 1970; Dunn, 1970; Harris, 1969; Goeden, 1970).

It has been pointed out that biological control is, in general, out of the hands of the individual grower. This is stressed by a recent project which has been started by the United Nations Fund for Drug Abuse Control in cooperation with the Commonwealth Institute of Biological Control. The Institute has been provided with funds to investigate, over three years, possible biological control agents for the cultivated opium poppy, *Papaver somniferum* (the source of heroin and morphine) and Indian hemp, *Cannabis sativa*. Presumably the agents, if found, will be released to destroy the growing crops without the agreement of the growers (Ryder, 1972). Unfortunately, neither plant is grown merely to provide drugs: hemp is an important fibre crop, and the poppy is widely grown for its edible oil-rich seeds. Substitute crops must be found which will fit into the farming pattern of the peasants concerned. When 'ordinary' weeds are controlled there is no need to search for a substitute, when a crop is made impossible to grow a substitute or substitutes must be found which may be grown in rotation with the other, more socially acceptable crops.

This discussion has been confined almost entirely to the selection of arthropodan natural enemies which are to be used against phytophagous insects, disease vectors, and weeds. Other agents are important and their selection differs in various ways. These agents will be discussed in later chapters and consideration of their necessary characteristics will be postponed till then.

Biological control is not merely, of course, the selection and introduction of suitable natural enemies. Some pest problems are more likely to be solved by biological control methods than are others (at least at the present stage of development of the technique). The selection of problems which are likely to succumb to this approach is discussed in the next section.

The choice of suitable pest problems
In *Biological Control of Insect Pests and Weeds* DeBach (1964) tabulates the biological control projects in which imported entomophagous insects were used against crop and household insect pests. The degree of control is assessed in each case as 'complete', 'substantial', or 'partial'. There are, no doubt, many examples of attempted biological

Table 5.1

Pest	Complete	Substantial	Partial
Scale insects and mealy bugs	22	25	25
Aleyrodidae	3	3	1
Aphidae	1	3	3
Lepidoptera	3	14	16
Coleoptera	4	10	9
Diptera	0	2	5
Orthoptera and Dictyoptera	0	1	4
Hymenoptera	1	4	0
Dermaptera	0	0	2
Hemiptera Heteroptera	0	1	0

A summary of biological control projects employing imported entomophagous insects. Adapted from Table 12 of *Biological Control of Insect Pests and Weeds* edited by Paul DeBach; Chapman and Hall Ltd., 1964, by kind permission of the publishers.

control which failed and which were not reported in the literature: these are not, of course, listed in DeBach's table. His Table 12 is summarized in Table 5.1. Other tables in his paper list further successes against *Icerya purchasi* (in twenty-nine countries), *Eriosoma lanigerum*, the woolly apple aphid (in twenty-four countries), *Pseudaulacaspis pentagona*, white peach scale (in five countries), and *Aleurocanthus woglumi*, citrus black-fly (in six countries).

An examination of Table 5.1 shows the high proportion of cases that concern scale insects, mealy bugs, and aleyrodids. The number of such species controlled biologically is disproportionately large when one considers the whole range of pestiferous insects. There are a number of probable reasons for this.

Scales and mealy bugs are easily transportable on plant material and have therefore invaded numerous new areas. The biological control of insects is directed mainly against exotic insects. They are also difficult to kill by chemical means and, since they occur on high-value crops such as fruit, growers have been forced to look for other methods of control. Perennial crops such as fruit are possibly advantageous to natural enemies since they offer a very stable environment compared to that of short-lived crops such as vegetables and cereals. These insects are also probably biologically suited since they are, for a large part of their life cycle, sessile and thus easily found by their natural enemies. Also, they are usually in exposed positions on their host plants.

It must not be forgotten, however, that the most outstanding success of the biological control method (and one of the earliest) was against a scale insect, *Icerya purchasi*. It is very probable that many applied biologists followed the lead of Koebele, the scientist largely responsible for this work, and concentrated on similar problems before turning to others. In time this preponderance of successes with one group may disappear as biologists meet with success with other kinds of pests. It is worth noting in this context that only two species of parasites of the codling moth had been introduced into the USA before 1967 (Doutt, 1967) although this widespread insect is known to be attacked by at least 120 insect parasites, 22 predators, a fungus, and a granulosis virus. The codling moth is the insect that shares with the house-fly and mosquitoes the greatest 'coverage' in texts of applied entomology.

For the reasons discussed earlier, introduced pests are the most common subjects for biological control. This is likely to continue as far as insect pests are concerned, when control is attempted by the 'inoculation' of small numbers of arthropodan natural enemies into the environment. Nevertheless, the prospects of controlling native pests by biological control methods are good, especially when the environment can be inundated by natural enemies or pathogens. In these cases the goal is a rapid destruction of the pest, and is thus akin to chemical control. Egg parasites or viruses are often suitable agents as they can be mass produced and, in some cases, stored. The agents may well persist after the inundation, but this is not the main objective. The alfalfa butterfly, *Colias eurytheme*, is a native pest in the United States, where stocks of a virus are held in readiness for outbreaks (chapter 7). The biological control of native weeds has been discussed earlier.

One concept that arose early in the consideration of biological control is the 'Island Theory'. Many of the earlier successes were on islands such as Hawaii, Fiji, and New Zealand. Other successes were in areas, such as California, which could be regarded as ecological islands. An extreme example of the ecological island theory involved the control of *Trialeurodes vaporariorum*, a whitefly, in British glasshouses by the chalcid *Encarsia formosa*. The theory gained so much ground that many biologists were pessimistic about the value of the method in continental areas. It is now thought that the method was particularly successful in Hawaii and Fiji because the growers and authorities there decided to apply the method seriously, and because most of the pests were introduced species on introduced crops. Oceanic islands have a notoriously small native fauna and flora, offering little resistance to invading species. Their equable climates often favour introduced natural enemies but no more than do those of many continental areas. The subsequent successes in continental areas has pushed this theory into the background.

A survey of the literature shows that most cases of effective biological control have occurred in warmer climates. Such climates allow the host species to breed more or less continuously so that often the generations and stages overlap. An introduced natural enemy can synchronize its life cycle with such a host. There are, however, many pests which have life cycles broken by obligate diapauses and these prove difficult to control in this way. It is also probably true that it is easier for an introduced species to aclimatize to a climate somewhat warmer than its native one, than to a colder one, and this too helps to account for the success of the method in warmer countries.

This limitation must not be taken too seriously. As with the Island Theory there are exceptions. In Canada, for example, good control of the forest pest *Nygmia phaeorrhoea*, the brown tail moth, has been achieved by three introduced parasites. DeBach lists two other cases of complete control, nine of substantial control, and four of partial control in Canada.

It would seem, therefore, that rules are made to be broken in biological control. A limitation of the method is suggested, such as its restriction to warmer climates, but sooner or later the limitation is shown to be invalid. As DeBach remarks, the success of the method is limited, in the long run, only by the amount of research and importation carried out.

Possibly, lack of research is one of the reasons why there appear to be no cases of the biological control of plant pathogens by introduced natural enemies – or, at least, none known to the author. Fungi and bacteria attacking plants do have, in their turn, parasites which attack them, or species which compete with them. Soil pathogens, in particular, can be combated by encouraging the growth of other soil micro-organisms which are antagonists of the pest species. This method, however, is better left to a later chapter.

Measures taken after importation

After being introduced into their new country natural enemies are kept in quarantine for some time. During this period they are screened for hyperparasites and their numbers are increased by a suitable mass breeding programme. This gives the biologists an opportunity to learn more of the organism's biology, its relationships with the proposed host and other species, and something of its climatic and ecological requirements. During the breeding programme particularly strong and suitable strains may be produced by selection. The quarantine handling of entomophagous insects, their culture and nutrition and those of their hosts, are described by various authors in DeBach's *Biological Control of Weeds and Pests*. Another modern book devoted to the same topics, but stressing the mass production of pest species of insects, mites, and ticks is *Insect Colonization and Mass Production*, edited by Carroll N. Smith (C. N. Smith, 1966).

A successful breeding programme produces large numbers of the natural enemy. These are not merely released into the infested area and left to fend for themselves. Early in the breeding programme, as soon as the cultures can spare them, a small number may be released; experience has shown that sometimes this is all that is necessary. Usually, however, such attempts prove abortive and then great care and planning has to go into the successful colonization of the species. The aim of this is to establish permanent colonies of the natural enemy at various sites from where it can spread into surrounding infested areas.

These sites should be chosen so that the infant population is protected as much as possible from unexpected inclement weather. A number of sites is chosen, covering a range of different habitats and climatic conditions, for tests during the mass production stage cannot duplicate completely the field conditions. It is difficult to know how many individuals should be released at each site. With an insect such as the Vedalia beetle it is only necessary to release a few mated females at each suitable location, but the colonization of some Hymenopterous parasites calls for the release of many thousands of individuals.

Many of the released individuals will be destroyed by inclement weather or predators such as birds and spiders. They should therefore be at such a stage when they are set free that they will seek out their hosts or prey as soon as possible. For this reason adults, already containing eggs ripe for laying, are usually released in close proximity to the pest they are to control. They may be given further protection from predation, and from competition from other natural enemies, by enclosing them in cages or muslin sleeves together with their hosts. Such a method was used with the first release of Vedalia beetles in California.

One of the greatest technical difficulties is to ensure that the cultured natural enemies are at the correct developmental stage when the appropriate stage of the pest appears in the field. Because of the differences in climate at different candidate sites, it is usually possible to choose some of these with hosts at the right stage when the natural enemies are ready.

While good powers of dispersal are an advantage in an established natural enemy they can make initial colonization difficult. If the colony breaks up too quickly the chances of mating are lessened. This is a further reason for the common practice of confining the population during the early stages of the project.

There will, of course, be more than one release at each site. It must be decided, however, how long releases should be continued if the earlier ones are not particularly successful. Clausen (1951) has observed that, as a rule, if a natural enemy is going to be successful as a control agent it is showing good control by the end of three years. There have been some exceptions to this rule but, generally speaking, an ultimately successful natural enemy shows its potentialities quite soon. It is thus usually uneconomical to continue releasing an organism at a site for more than two or three years.

After the agent has been given a reasonable time in which to establish itself recovery surveys are made. These are usually qualitative. Their purpose is to discover, using simple sampling methods, whether the natural enemy is still present in the area, and, if so, to make a subjective assessment of the impact on the pest population. Usually, recoveries are quite good in the first generation – those individuals which develop from the eggs of the released ones. After this the population may increase, continue to hold its own, or disappear. Before the population can be considered to be permanently established it must be subjected to the extremes which the local climate can offer.

After a natural enemy has been successfully colonized the pest population may decline. It cannot be assumed that this decline is the result of the action of the natural enemy, for there may have been some concurrent change in the environment. Sampling can be carried out in the field to assess the population density of the host, the population density of the natural enemy, rates of parasitization or predation and so forth, but such work, as explained in the second chapter, is difficult and often inaccurate. There are other, more direct, ways of assessing the effects of natural enemies.

The best method is a series of field trials in which natural enemies are excluded from some of the 'plots' (using this word in its statistical sense). With sessile pests such as scale insects the plot could be a branch of a tree suitably enclosed by a muslin sleeve. The branch could be 'cleaned' by fumigation and pests re-introduced. A number of such plots could then be sealed to prevent entry of the natural enemies, while others are left with entrance holes. It is important to enclose both kinds of plots by sleeves as these may have a marked effect on the microclimate of the branch.

There are other possible methods of carrying out this so-called check method for the determination of the impact of natural enemies. Plots may be sprayed with a highly selective pesticide which destroys the controlling agent but which does not affect the pest.

Flanders (1971) summarizes two other methods which are sometimes valuable. Many authors have noted that populations of aphids which are attended by ants are often protected by the latter from the attentions of parasites and predators. In certain cases ants may be used to assess the importance of an introduced agent in the control of homopterous pests. The second method involves the release of large numbers of laboratory- or glasshouse-bred specimens of the pest, and observing if they are controlled by the agent.

In the past many biological control projects have been started without adequate thought being given to the methods of assessing the effectiveness of the campaign at a later date. In some cases (as with the cottony cushion scale) the impact of the agent was so catastrophic to the pest and so obvious that validation was unnecessary. In less spectacular cases it is often impossible to say, without extensive experimentation and sampling, if the decline in the population densities of the pest is due to the introduced agent or agents, or to some other cause, and, if due to the agents, which agent is mainly responsible. The ultimate application of validation methods should be planned at an early stage of the project.

When natural enemies have been successfully colonized throughout an infested area they can still be helped by suitable manipulation of the environment. These measures are, of course, equally valid for the encouragement of indigenous natural enemies of a pest and are thus not exclusively a part of classic biological control. They will be discussed in a later chapter.

6
Biological control agents: animals

Applied biologists have found biological control agents throughout the living world, but animals and micro-organisms have been the most frequently used organisms. For convenience the nematodes and protozoa are usually considered with the micro-organisms.

So far, the greatest successes have been achieved with parasitic and predatory animals, largely because of their mobility and ability to seek out their prey, even when these are scattered. Parasites have usually been more effective than predators as, in general, they are more specific in their choice of prey and have life cycles which are closely geared with those of the hosts.

The insects and vertebrates, especially fish and mammals, are outstanding among the animals, but other vertebrate groups, such as the birds, amphibians, and reptiles, as well as the arachnids, centipedes, and molluscs, have also been employed, or are regarded as potentially useful biological control agents.

Insects

There are probably more species of insects which are entomophagous than species which are not. Although they are usually classified as parasites or predators of other members of their class it is difficult to draw a clear dividing line between the two modes of life, for parasitic insects differ in many ways from other kinds of parasitic invertebrates. Insect parasites almost always kill their host insects, but the victim usually lives long enough for the parasite to complete its development. The parasite is usually quite large in relation to the host (hence the latter's eventual death), unlike most invertebrate parasites of vertebrates. Generally, only the larval stage is parasitic, and the adult insect is 'normal' in most respects – free living, winged, highly mobile, and feeding on such foods as pollen and nectar.

When the larvae feed within the body of the host insect few would quarrel with the use of the term parasitism, but often the larvae consume their host from the outside and this differs from predation only in the restriction of the attack to a single victim.

Both larval and adult insects show predatory habits. In some insects both stages are predatory, in others only one stage or the other. The adults of some parasitic insects are also predatory in that they feed on some individuals of the species in which their larvae develop. The predatory habit is more widespread than the parasitic one among the insects, most Orders containing at least some predatory species.

The chief Orders which contain species which parasitize insects are the Coleoptera,

Diptera, Hymenoptera, Lepidoptera, and Strepsiptera. All the members of the Strepsiptera are parasites but usually in insects which are not pests. They will not be considered further. The Diptera and the Hymenoptera are the two most important groups for biological control.

The Hymenoptera

This immense Order of insects, many of which still await discovery, contains a large number of predatory forms as well as a number of groups which are entirely parasitic. With the possible exception of *Orussus* none of the sawflies are parasites. The Suborder Apocrita, on the other hand, contains the ants, bees, and wasps (Aculeata), many of which are predatory or parasitic, and the Parasitica which, as the name indicates consists largely, though not entirely, of parasitic forms.

The Hymenoptera show many unusual reproductive features. These peculiarities are extremely important for ignoring them could ruin a culturing or establishment programme. A fertilized egg develops in the normal way, but almost always produces a female. An unfertilized egg, on the other hand, develops into a male. The ovipositing female can control the release of sperm from the spermatheca as the eggs pass down the oviduct, and thus determine whether or not eggs are fertilized. The males thus have a single set of chromosomes in the germ line, although many of the somatic tissues are diploid. Eggs are produced by a normal meiosis but this is modified in sperm production so that there is no reduction in the number of chromosomes. This is the reason why the eggs are checked for fertilization before the females are released, despite their ability to reproduce parthenogenetically. Males are, however, rare or unknown in some species and females are regularly produced by *diploid* parthenogenesis.

A complication in certain species is that the female and male larvae are parasites in different kinds of hosts. Among the Aphelinidae, for example, the mated females of some species lay eggs which develop in their hosts as primary parasites, while the unmated females lay eggs in hosts in which the larvae develop as hyperparasites, sometimes even devouring their sister larvae. It is essential, therefore, to arrange for some of the females to oviposit before mating if they are to be cultured in the laboratory.

Another reproductive variation found in the Hymenoptera and, as far as the insects are concerned, almost unique to this Order, is polyembryony. This is a form of asexual reproduction in which a single egg divides during development to give several embryos. *Platygaster hiemalis*, a parasite of the Hessian fly, *Mayetiola destructor*, lays eggs, each of which can give rise to one or two larvae. At the other extreme is a braconid parasite of the moth *Plusia gamma* which is said to produce 2000 individuals from a single egg. This form of reproduction could clearly increase reproductive potential, but it seems that this is offset by the laying of fewer eggs than are laid by females of closely related species.

The eggs of the parasitic Hymenoptera are usually laid on or within the host, but sometimes they are laid apart from it. The perilampids, for example, oviposit on leaves, and the young, emerging larvae seek out parasitized caterpillars to feed hyperparasitically.

As might be expected the larvae of parasitic Hymenoptera, especially the internal parasites, show marked adaptations for this mode of life. Egg parasites, for example, are tiny, apparently non-segmented larvae which absorb their food through the skin.

The first instar larvae of the parasitic Hymenoptera show the greatest range of forms, details of which may be found in a number of texts (Imms, revised by Richards and Davies, 1957; Clausen, 1940; Hagen, 1964). Fisher (1971) reviews many of the aspects of endoparasitic Hymenoptera which are important for their use of biological control agents. Both he and Salt (1970) discuss defence mechanisms of insects attacked by endoparasitic Hymenoptera.

Superfamily Ichneumonoidea (Fig. 6.1). All the members of this large superfamily are parasites of insects or other Arthropods. They are, mainly, solitary endo- or ecto-parasites of the caterpillars of moths and butterflies, but some attack the larvae of sawflies and other Hymenoptera, often as hyperparasites (some *Pimpla* spp.), and

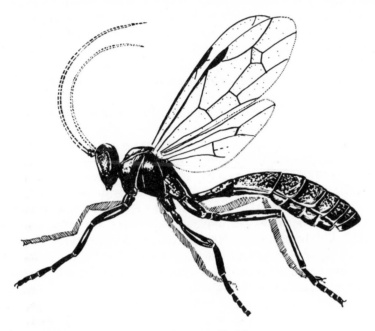

Fig. 6.1.   Ichneumonid adult

also those of beetles and flies. Some other kinds of insects and spiders are sometimes parasitized. The ectoparasitic type of feeding is usually associated with hosts which live in protected positions. Notable examples are *Thalessa* and *Rhyssa* whose long ovipositors can reach wood-wasp larvae (Siricidae) deep within a log. Wood-wasps have recently invaded the Australian mainland so that Australian biologists are now studying such parasites at a biological control unit in England.

Possibly the most familiar European ichneumonids are the large reddish-brown adults of *Ophion* which, being attracted by light, often come indoors during the summer months. Their larvae parasitize the caterpillars of the economically important noctuid moths. Another British ichneumonid, *Glypta haesitator*, has given substantial control of the pea moth, *Laspeyresia nigricana* in British Columbia.

In addition to their value for biological control projects ichneumonids, like the members of many of the families which follow, are of immense importance in the natural regulation of pests and potential pests, and care should always be taken to promote this activity. As the adults are often to be seen feeding on flowers, especially Umbellifers, it is clear that these should be conserved if possible.

The Braconidae rival their close relatives, the Ichneumonidae, in their importance in the natural regulation of insects. The main hosts are again caterpillars of butterflies and moths, a single one of which can nourish well over a hundred larvae of some braconid species. A heap of the sulphur-yellow cocoons of *Apantales glomeratus*, clinging to the husk of a *Pieris* caterpillar, may contain 150 individuals. This contrasts with the members of the braconid sub-family Aphidiinae which are parasites of aphids. A single aphid does not usually contain more than one of these parasites.

The association of these small braconids with their aphid hosts is so close that it is suggested that they have, over long periods of time, shown co-evolution. The association of the aphids with *their* hosts is likewise very close in certain species; because of this many workers believe that the parasites are influenced as much by the habitat as by the aphid species when searching for suitable victims. An aphid may thus have one series of parasites on its primary host and another series on its secondary host. A biological control programme for aphids should not be restricted to the introduction of parasites from one of these habitats.

The importance of the habitat in host finding by natural enemies is not restricted to this group of parasites. It is, in fact, so widespread that it must always be taken into account when carrying out a programme on a pest which attacks a number of hosts, or which is found in a variety of habitats.

An aphid may be attacked at any stage. This applies at least to *Toxoptera graminum* when parasitized by *Aphidius testaceipes*, and probably to many other associations. If the aphid has undergone the second ecdysis she may be able to produce some nymphs, but not as many as her unparasitized sisters. An aphid which carries a parasite which is in the pupal stage is easily recognized by its inflated appearance and its brown, yellow, or black colour. Such an aphid is usually known as a 'mummy'. Larvae of *Praon*, on the other hand, leave the body of the aphid to pupate in a tent-like cocoon beneath the cadaver (Fig. 6.2).

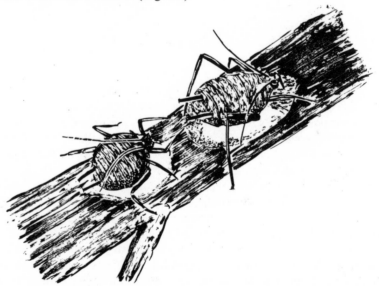

Fig. 6.2.   Aphid 'mummies' with pupae of *Praon* sp. below

Despite their undoubted importance in the natural regulation of aphids these tiny parasites have been used in very few biological control projects. As Hagen and van den Bosch (1968) pointed out in their review on the natural enemies of aphids biological control workers have doubted that the use of natural enemies could give sufficient control of pests which are such efficient vectors of plant virus diseases. Nevertheless, they cite one case where the method seems to have been adequate, namely the control of motley virus of carrots around Melbourne, Australia, after the introduction of *Aphidius* species from California. *A. matricariae* gave a good control of *Myzus persicae* when released in Californian glasshouses, and later spread to field populations of the aphid. Two members of this sub-family (*Praon exoletum* and *Trioxys complanatus*) and an aphelinid, *Aphelinus asychis* (see below) were introduced into California from the Old World and have contributed to the successful control of the spotted alfalfa aphid, *Therioaphis trifolii*. The ranges and seasonal peaks of activity of the parasites were, to a great extent, complementary.

Superfamily Chalcidoidea Some members of this large superfamily are plant feeders but the majority are parasites or hyperparasites of other insects (Fig. 6.3). The

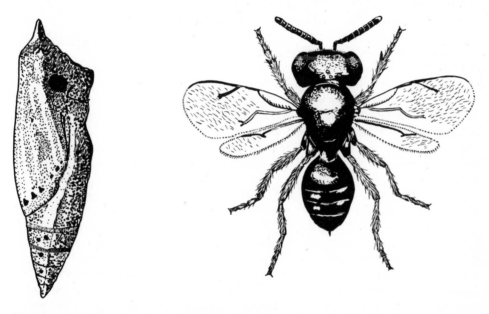

Fig. 6.3. *Pteromalus puparum*, **a pupal parasite of** *Pieris* **spp. Left: pierid pupa with exit hole of parasite; Right: adult parasite. (Adapted from USDA)**

main primary hosts are larval Lepidoptera, the eggs of Lepidoptera, Homoptera (especially coccids), and Diptera. As so many of them are parasites of scales and mealy bugs it follows that this has been one of the most widely used groups in biological control. Generally, chalcids are small insects and, indeed, the smallest of all insects are to be found among them. There are several families but only the most important will be discussed.

The Aphelinidae are primary and secondary parasites of scales, mealy bugs, aphids, and aleyrodids. The members of only three genera attack aphids (Hagen and van den Bosch, 1968). *Aphelinus mali*, imported successfully into many countries for the control of the woolly apple aphid, *Eriosoma lanigerum*, is one of the few natural enemies of pests used for biological control in the British Isles. It was introduced into Kent at various times between 1924 and 1937 without any great success, but later introductions have given better results. Woolly apple aphid is at least less troublesome than it was before the introduction of the parasite, although the results have not been as good as they have been in some other countries. They can be improved by keeping the parasitized aphids indoors during the winter.

Aphelinus mali females, and the adult females of other species of the genus, feed on the body fluids of the aphids after laying their eggs. While this ensures that the females obtain enough protein for the development of further eggs it sometimes leads to the death of the host and the parasite within it. The predation itself is often significant in the control of the pest. The adult females of some other species of *Aphelinus* feed on honey dew produced by aphids.

Although the woolly apple aphid is thought by many entomologists to be of North American origin (it was known in England as American blight in 1802, the year it was originally described in Germany) it was not observed on Vancouver Island and in lower British Columbia till 1892 where, in the absence of *A. mali*, it soon became a major pest. In the 1920s the parasite was introduced from Ontario and gave satisfactory control within a few years. There was a resurgence of the pest in the 1940s and 1950s, possibly because of the use of chlorinated hydrocarbon insecticides for the control of codling moth. The pest then decreased sharply in importance from 1953 (LeRoux, 1971). There was a similar sequence of events in South Africa. DeBach (1964b) lists twenty-three other introductions of the parasite, eight of which resulted in the complete control of the pest. Moreton (1969) summarizes the use of this and other biological control agents in the British Isles. Starý (1970) has recently published an extensive monograph on aphid parasites, and their use in integrated and biological control.

A second biological control agent which has been successful in the British Isles also belongs to this family. It was not intentionally introduced into the country, but appears to have found its own way there. This small chalcid, *Encarsia formosa*, blackens the nymphal stages of the glasshouse whitefly *Trialeurodes vaporariorum*. After it was first found by L. Hawkins, a Hertfordshire gardener, in 1926 it was distributed to many growers in Britain, and was also shipped to Canada and to South Australia. The distribution of parasitized nymphs was not continued in Britain after the Second World War, partly because of the danger of spreading other pests, and partly because of the advent of new chemical sprays for aphids and spider mites. There is, however, great interest at present in the biological control of spider mites under glass, and in integrated control of the pest complex (including the whitefly), so it is probable that the use of *Encarsia* will be revived commercially. Hussey, Read, and Hesling (1969), and Moreton (1969) discuss the use of these parasites in the United Kingdom (see also chapter 11).

Other aphelinids have been used for the control of coccids, important examples being *Coccophagus gurneyi* against the citrophilus mealy bug, *Pseudococcus gahani*, and *Aspidiotiphagus citrinus* against the coconut scale, *Aspidiotus destructor*.

The Encyrtidae is a further chalcid family widely used for the control of coccids although some of its members also attack other kinds of insects including, unfortunately, lady-bird beetles. Some are hyperparasites attacking chalcids within coccids and aphids. Polyembryony occurs in a few genera, all of which are restricted to butterfly and moth caterpillar hosts (Clausen, 1940). Among the many species used in biological control we may mention the British *Blastothrix sericea* introduced into British Columbia against *Eulecanium coryli* on forest trees, and *Metaphycus* spp. for black scale, *Saissetia oleae*, on citrus and olives in many parts of the world.

Fig. 6.4. *Trichogramma* **sp. ovipositing in moth egg. (Adapted from USDA)**

Egg parasites are common among the Hymenoptera but the habit is particularly associated with the Trichogrammatidae (Fig. 6.4). *Trichogramma* species (the taxonomy is very difficult) have been bred in huge numbers and released periodically in attempts to inundate the habitats of certain caterpillar pests. They can be reared very easily upon Angoumois grain moth, *Sitotroga cerealella*, and the parasitized moth eggs can be kept in cold storage until needed. Recent work has shown that the use of regularly fluctuating temperatures in the insectary produced parasites which performed better in the field than did those bred at a constant temperature.

Trichogramma species, used in this way, gave some control of codling moth in the United States, and substantial reduction in the USSR. In Russia results were better in low-lying orchards than in those on high ground. Its impact on the Oriental fruit moth, *Cydia molesta*, the cotton leaf worm, *Spodoptera* (*Prodenia*) *litura*, the European corn borer, *Ostrinia nubilalis*, and the sugar cane borer, *Diatraea saccharalis*, has been discouraging.

Yasumatsu and Torii (1968) summarize the use of *Trichogramma* spp. for biological control of rice pests, and describe the bionomics of the main species in Japan. *T. japonicum* and the scelionid *Telenomus dignus*, together with three larval parasites, were introduced into Hawaii in 1928 for the control of *Chilo suppressalis*, the rice stem borer. The pest had been accidentally introduced on rice straw from Japan. The following year the yields were back to normal, but the above authors say that other factors may have been important, although it was ascertained that the parasites became established. In 1929 a *Trichogramma* species was introduced into Japan from the Philippines, but it was found to be of little value compared with a native species, *T. japonicum*. Later it was pointed out that the introduced species was already present in Japan. Three species, including *T. japonicum*, have been established in the Philippines, but they do not appear to give much control of the stem borer.

In some areas of Japan *T. japonicum* passes through fifteen or sixteen generations each year, and one or two generations are passed in *Chilo suppressalis* which is a favoured host. At other times of the year other caterpillars are attacked. Although *Trichogramma* species are usually used as inundative biological control agents, and prolonged control is not considered essential, the presence of alternative hosts, provided they are of non-economic species, is useful.

In general, however, *Trichogramma* species have proved to be disappointing control agents. DeBach and Hagen (1964) suggest that the very ease with which the insects may be mass produced has led to their use against some insects for which they are not suitable, and they point out that in most cases no studies were made before release to ensure that the correct species or physiological strain of *Trichogramma* was chosen for the particular pest. Furthermore, use had not been made of fluctuating temperatures during breeding.

Superfamily Proctotrupoidea Although these insects are important in the natural regulation of pests they have been little used in biological control. A few have, however, been used successfully and doubtless, with further experience, others will also be of value.

Scelionids are small egg parasites. Some have been used for the control of horse-flies (tabanids) in Texas, a rare example of the use of a natural agent against a medically important pest. Others, including *Telonomus nakagawai* and *Trissolcus mitsukurii* have been used successfully against the cosmopolitan green vegetable bug, *Nezara viridula*, in Japan (Yasumatsu and Torii, 1968).

The Platygasteridae are also egg parasites, usually on gall midges (Cecidomyidae), though some species attack coccids and other Homoptera. Unfortunately, most species are univoltine, irrespective of the number of generations through which the hosts pass, and this, of course, reduces their value as biological control agents. *Misocyclops marchali* has, however, been used successfully against *Perrisia pyri*, a pear midge, in New Zealand.

Superfamily Vespoidea This group consists of the true wasps, the insects which are recognized, and usually feared as such, by the general public. Despite their notoriety, and the damage they sometimes cause to ripe fruit, hornets and wasps (*Vespa* spp.) are useful scavengers and predators. These social insects provision their nests with insects, especially larval Diptera and Lepidoptera. Their solitary relatives such as *Odynerus* are also predacious on moth larvae and similar insects which they seal into nests containing eggs.

One species of the social genus *Polistes*, colloquially known as 'Spanish Jack', has been credited with the partial control of various pests in the West Indies and has been transported from island to island for this purpose (Clausen, 1940).

Superfamily Formicoidea The most primitive sub-families of the Formicidae, those which contain such insects as the driver ants of the tropics and the bulldog ants of Australia, consist of savage carnivores which doubtless destroy large numbers of harmful insects and arachnids, but whose ferocity precludes their use as biological control agents. The higher ants usually have a more mixed diet or a vegetable one. Many species 'cultivate' aphids for honey dew and, in doing so, protect the aphids from parasitism and predation to some extent. *Formica rufa* (in the wide sense) is protected by law in German forests and is even provided with artificial nesting sites by the foresters. The most important species in Germany appears to be *F. polyctena*. Gösswald (1970) describes techniques for the establishment and protection of colonies of this species. Finnegan (1971), on the other hand, has recently studied the indigenous ants of Quebec and concluded that none are promising as biological control agents. The diet and behaviour of ants has been summarized by Sudd (1967).

Superfamilies Pompiloidea and Sphecoidea Almost all of these solitary wasps build underground nests or mud cells which they provision with paralysed prey for their larvae. The stinging of the prey and the way in which it is carried to the nest are complicated, so that the behaviour of these wasps has long been a favoured field of study by entomologists. It has been well illustrated in Olberg's *Das Verhalten der solitären Wespen Mitteleuropas* which, though written in German, has English captions to all the photographs (Olberg, 1967).

The pompilids are almost all predators of spiders so they are unlikely to be used in biological control except, possibly, against venomous species such as the black widow, *Latrodectus mactans*. The sphecoids are more catholic in their diet. Some attack spiders, while others prey on Heteroptera, Homoptera, Orthoptera, and Dictyoptera. Caterpillars are also favoured prey. *Larra luzonensis* was introduced into Hawaii to give useful control of the African mole cricket, *Gryllotalpa africana*, on cane. *Ampulex compressa* and *Dolichurus stantoni* have been used in the same islands in an attempt to control the common domestic cockroaches *Periplaneta americana* and *P. australasiae*.

The Diptera

The Hymenoptera is, without doubt, the most important group of biological control insects, but it is closely rivalled by the two-winged flies. Both predators and parasites are to be found throughout the Order, often in the most unexpected families. There are, for example, a number of mosquito larvae which prey on the larvae of the same family. *Megarhinus*, *Zeugnomyia*, *Psorophora*, *Eratmapodites*, and *Lutzia* are the most important genera and sub-genera (Bates, 1949).

Equally surprising is the way in which some species capture their prey, or handle it. The larvae of the New Zealand glow worm, *Arachnocampa luminosa*, use their light to attract midges to their sticky webs of fine threads. The larvae of *Vermileo vermileo* and related genera (worm lions) trap ants in conical pits dug in light soil, a method resembling that of the totally unrelated ant lions (Myrmeleontidae, Order Neuroptera). Finally, the males of many empids capture prey to offer, in a bundle, to the females during mating.

In an Order which contains such a diversity of species with such a wide range of predatory and parasitic habits it is only possible to deal with those families which are the most useful in biological control and natural regulation.

Syrphidae The larvae of the hover flies have taken up several different modes of life, ranging from scavenging in ant and bee nests, through an aquatic life in stagnant pools, breathing through a 'telescopic' tail (rat-tail maggots) to predation on aphids, scale insects, and other small arthropods.

The aphidovorous species (Fig. 6.5) have a marked impact on populations of these plant pests, rapidly devouring any colonies they chance upon. The adult flies search out colonies in order to lay their eggs very close to the aphids. The blind

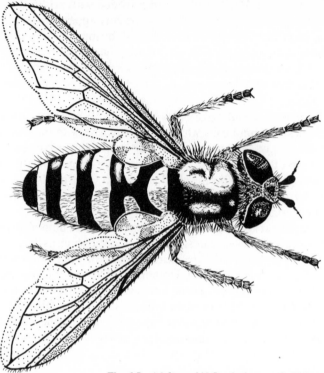

Fig. 6.5   Adult syrphid fly, the larvae of which are predators of aphids

maggots which hatch from the eggs are ignored by the aphids as they feed among them. The attacked aphid is pierced by the maggot's mouth parts and is sucked dry whilst being held aloft so that its struggling legs cannot get a purchase on the leaf. Meanwhile, its fellows continue their unconcerned feeding.

Before the slug-like maggot changes into a tough, tear-drop-shaped pupa, it disposes of some two hundred aphids in ten days or so. If the first colony is exhausted it starts to climb till it either encounters a colony or the tip of a branch. If the latter the tip is explored thoroughly. If no aphids are found it descends and repeats the process on another branch. As aphids are generally found on new growth the method is usually successful. Some larvae can, however, survive for short periods on plant sap or other vegetable material.

The adult syrphids are strong and persistent fliers which feed on flowers and honey dew. Pollen is essential for the maturation of the eggs and syrphids are, in fact, important pollinators. As Schneider (1969) points out in his review on the bionomics and physiology of the aphidovorous species, these requirements make their ecological manipulation difficult. They appear reluctant to oviposit in areas which are not rich in flowers, and will readily leave such areas. For this reason, and also because they are fairly non-specific, syrphids have been little used in biological control, although they play an important part in the natural regulation of aphids and other small insects (Hagen and van den Bosch, 1968). They are themselves heavily parasitized by ichneumonids and other natural enemies and this naturally reduces their impact upon the aphids.

A number of important papers on syrphids appear in a volume edited by Hodek (1966) on the ecology of aphidophagous insects.

Tachinidae (Larvaevoridae) Tachinid larvae are internal parasites of insects, spiders, woodlice, and centipedes. Lepidopterous caterpillars are the most common insect hosts, but the larvae of Hymenoptera and Diptera, and various stages of Orthoptera, Dermaptera, Hemiptera, and Coleoptera are often attacked. This large family is represented throughout the world.

Colyer and Hammond (1951) list four methods of oviposition in the family. Eggs, which may hatch immediately or after some time, are laid on the integument of the host and the emerging larvae bore directly into the body. Alternatively, the adult female pierces the host's integument with her ovipositor or with a special spine and lays the eggs within the tissue. Other species deposit their eggs away from the hosts, but in areas which they frequent. In some cases the eggs are eaten by the host and the larvae emerge in the intestine to bore their way to their feeding sites. Alternatively, the eggs hatch almost immediately to migratory larvae which search for, and bore into, suitable hosts. These migratory larvae are covered with plates which, when the maggots contract, prevent desiccation.

These methods of oviposition are sometimes modified in that many tachinids lay living young, the eggs undergoing full incubation within a uterus in the female's body. Needless to say, those species which deposit eggs or larvae apart from the host produce the largest number of progeny to make up for the loss of those which do not find a victim. In general, tachinids are solitary parasites, and hosts containing more than three or four tachinid larvae are uncommon.

Many species are fairly host specific, but others have an extremely wide range of suitable victims. *Compsilura concinnata*, a common British species introduced into Canada and the USA, attacks a greater number of Lepidoptera species than any other tachinid in Britain, and at least one hundred hosts in the USA, including representatives of eighteen families and three Orders.

Adult tachinids (Fig. 6.6), like ichneumonids, are often flower-haunting species, frequently taken at Compositae and Umbelliferae. Other species are to be found basking in the sun, head down, on fences and tree trunks.

Several tachinids have been used for the biological control of insect pests, especially defoliating caterpillars. *C. concinnata*, working together with the braconid *Meteorus versicolor*, has given complete control of the brown tail moth *Nygmia phaeorrhoea* in many parts of Canada. The fly was introduced into North America in 1906 and has since been carried from place to place within the continent to control such pests as the gypsy moth, *Porthetria dispar* and the satin moth, *Stilpnotia salicis*. According to Sweetman (1958) fertilization is not necessary for the development of the eggs of this species, but unfertilized eggs give rise only to males. In North America there are three or four generations each year, with the maggots hibernating within the hosts. Thus, the availability of overwintering host larvae controls the abundance of this parasite.

Compsilura uses an oviposition spine to place its larvae within the host's body, but *Hamaxia incongrua* deposits its larvae on the surface of its host, and *Prosena*

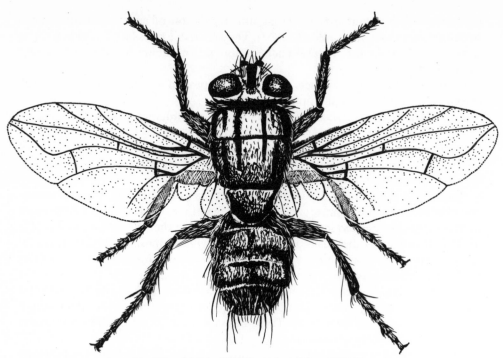

Fig. 6.6.   Adult tachinid fly. (Adapted from USDA)

sibirita chooses the soil as an egg-laying site (Sweetman, 1958). These tachinids were introduced into the USA from Europe and Asia in an attempt to control the subterranean larvae of the Japanese beetle, *Popillia japonica.*

Possibly the most successful use of a tachinid as a biological control agent was against the coconut moth, *Levuana iridescens*, in Fiji. The country of origin of this introduced pest was not known, but after a fruitless search in the Far East and Australasia it was decided to introduce parasites of a related species. A suitable one was found in *Ptychomyia* (*Bessa*) *remota*, which attacks *Artona catoxantha* in Malaysia. This strong-flying, quick-breeding fly has given complete control of the moth in Fiji.

Tachinids have been used in a number of other biological control projects, notably against the sugar cane borers (*Diatraea* spp.) in the Caribbean. Two particularly useful species which have helped to give substantial control of these caterpillars are *Lixophaga diatraea* and the Amazon fly, *Metagonistylum minense.*

Other families of the Diptera   The tachinids and the syrphids are the most important Diptera in biological control and natural regulation, but some other families have provided useful species. The Agromyzidae is an unexpected example.

Most agromyzids are plant feeders and their leaf mines are a familiar sight on many trees, shrubs, and plants. Yet this family has furnished one extremely valuable genus, namely *Cryptochaetum*. Indeed this genus is so atypical that many workers chose to place it in a distinct family, the Cryptochaetidae (e.g., Sweetman, 1958; Colless and McAlpine, 1970). These flies are parasites of the monophlebine coccids, and the best-known example is, of course, *Cryptochaetum iceryae* (Fig. 5.1), the Australian parasite of the cottony cushion scale. Its use against this pest has been overshadowed by the activities of the Vedalia beetle, but Essig (1931) is of the opinion that it is often the most important natural check of the scale and that 'Much of the credit given to Vedalia is due to this parasite'.

The adults are sluggish flies which feed on honey dew. They lay one or more eggs on each host and, it is reported, do not distinguish between healthy individuals and those already parasitized. The larval stages are quite different from those of other agromyzids. In all instars the caudal segment is extended to form two tails which, in later instars, contain tracheoles and blood. The first instar seems to be little more than a sac which absorbs food through the integument. There are five or six generations each year in California.

The related *C. grandicorne* is found in Europe where it parasitizes *Guerinia serratulae* and another species of the genus attacks *Drosicha* and *Iceryae* in Japan.

The Cecidomyidae is also a predominantly phytophagous family, but a few species are internal parasites of aphids and *Psylla* spp. while the larvae of other species are predatory on various small insects including aphids, scales, whiteflies, and thrips. *Schizobremia coffeae*, for example, attacks mealy bugs on coffee in Tanganyika. Cecidomyids, however, do not seem to have been used in biological control projects to any extent.

Locusts are frequently parasitized by Diptera and other insects, although it would seem that they are not very efficient in regulating these pests. A number of these parasites are found in families which are closely related to the Tachinidae. In the Sarcophagidae *Blaesoxipha lineata* and *B. filipjevi* are common parasites of nymphal and adult *Locusta migratoria* and *Schistocerca gregaria* in Asia and Africa. *B. lineata* often emerges from the still-living host which may even reproduce later, but *B. filipjevi* usually kills its victim within a few days and completes its development in the cadaver. *Oophagomyia* and *Wohlfahrtia*, which also belong to this family, prey on the egg capsules of locusts as do some species of the anthomyiid genera *Hylemya* (*Hylemyia*) and *Paregle*. An interesting example is the seed corn maggot *Hylemyia cilicrura* which is a widespread pest of beans and other crops. It is a general scavenger on soil organic matter, a category which includes locust egg pods. The bee fly family, *Bombylidae*, also includes numerous species which are predators of these eggs.

The larval anthomyiid *Phaonia mirabilis* is an aquatic maggot which preys on mosquito larvae, generally in 'container' breeding sites such as tree holes. Each larva can consume about one hundred larvae or pupae during its development.

Finally, many of the Calliphoridae, the family which contains the familiar blowflies, blue bottles, and green bottles, are parasitic during their larval stages in snails and earthworms (for example, *Melinda cognata* in the snail *Helicella virgata*). This family also contains a predator of the egg pods of the locusts *Locustana pardalina* and *Schistocerca gregaria*. In South Africa it is said to destroy between 50 and 90 per cent of the eggs capsules, each of which may contain about fifty eggs (Clausen, 1940).

These, and several other Dipterous families, are obviously important enemies of several pests and although tachinids have been the only ones so far used extensively in biological control, it is probable that other kinds of flies will be used commonly in the future.

The Coleoptera

This Order contains more known species than any other in the animal kingdom and a large proportion of them are entomophagous. Only eight families contain parasitic species, but the predatory habit is widespread. The most important families, for biological control, are the Carabidae or ground beetles, and the Coccinellidae or lady-bird beetles.

Carabidae The adult ground beetles, long-legged, swift and active insects, usually spend the day beneath stones and tree bark, and in other concealed places, but emerge at night to feed on other insects. The larvae are also active creatures which are usually found in the soil, but sometimes emerge to climb trees and shrubs in search of prey.

Their importance as enemies of pests has been recognized for centuries. Linnaeus, in a prize essay on the control of caterpillars in orchards, described *Calasoma sycophanta* as being 'like a wolf among the sheep, [it] creates terrible havoc among the caterpillars' (Ordish, 1967), and in about 1840 Boisgiraud used the same species in an attempt to control gypsy moth infesting the poplar trees in his native French village.

Calasoma sycophanta has been introduced and established in the New England states as a natural enemy of the gypsy moth, *Porthetria dispar*. It has been estimated that a single pair and their progeny can destroy about 6000 caterpillars and pupae in a year. The adults, which can have a life span of four years, climb the trees in search of prey while the larvae feed on caterpillars on the ground as well as in the trees.

The maggots of the cabbage root fly, *Erioischia brassicae*, are often attacked in Britain by the carabids *Bembidion lampros* and *Trechus quadristriatus*, as well as by the staphylinid parasites *Aleochara bipustulata* and *A. bilineata* when they reach the pupal stage, and it is thought that the increase in damage by this pest after the application of dieldrin (see chapter 4) is due to the destruction of these parasites and predators.

Staphylinidae The rove beetles are easily recognized by their short elytra which reveal the segments of the abdomen. The parasitic sub-family *Aleocharinae* contains the parasites of the cabbage root fly which have just been mentioned, as well as *A. curtula* which develops within the pupae of *Lucilia*. The adults of these beetles are predators on the species that their larvae attack and may be even more important as natural enemies.

Other rove beetles are important predators of pest insects. *Paederus fuscipes* is said to be one of the foremost natural enemies of the cotton leaf worm, *Spodoptera* (*Prodenia*) *litura* in Egypt where it may number up to 50 000 beetles in an acre. Another member of the genus has been reported to kill almost two-thirds of the larvae of the paddy pest *Schoenobius incertellus* in some Formosan infestations.

A few of these beetles have been used in biological control projects. *Philonthus aeneus* and *Creophilus erythrocephalus* were introduced into Hawaii as possible agents for the control of the horn fly, *Lyperosia irritans*. The adults of these flies suck blood from cattle and sheep, usually near the base of the horns, while the maggots develop in dung where the beetles prey upon them. Unfortunately, the first species did not establish itself in the islands.

An attempt was also made to establish *Thyreocephalus albertsi* in Hawaii, but this too apparently failed, although it is a common predator of fruit fly larvae in many parts of Oceania.

A tiny black staphylinid, *Oligota flavicornis*, is a common predator of the fruit tree red spider mite in Britain. A series of papers by Collyer (1952, 1953a, b, c) serve as a valuable guide to the predators of this mite.

Coccinellidae The lady-birds are the beetles most used as biological control agents and they include the most famous of all agents, namely the Vedalia beetle, *Rodolia* *cardinalis* (Fig. 5.1), which has been used so successfully against the cottony cushion scale. Although they have been used mostly against scale insects and mealy bugs they are possibly known better as predators of aphids.

Hagen and van den Bosch (1968) suggest that as lady-bird beetles feed on aphids both as larvae and as adults, they are more sensitive to changes in the population density of the prey than are syrphids and other predators which only feed upon them during the larval stage.

Several authors have studied the population dynamics and the effect of natural enemies of *Aphis fabae* on beans and sugar-beet, and on the primary host plant, the spindle (*Euonymus*). In general, it was found that parasites and fungi have little regulatory effect but that lady-birds and syrphid larvae sometimes prevented outbreaks on beet.

In Czechoslovakia, research has been carried out for several years on the effect of the two principal coccinellid predators of *A. fabae*, namely the seven-spot lady-bird, *Coccinella septempunctata*, and the two-spot lady-bird, *Adalia bipunctata*. When these natural enemies, and others, were present in bean fields early in the season they sometimes checked the build-up of the aphid populations, but sooner or later these 'escaped' with a subsequent population explosion. They also fed on the spring generations on spindle. In England it was found that the overwintering adults fed on other species of aphids that infested nettles and that the lady-birds migrated to the beet fields too late to affect the aphid populations significantly.

The Czechs also carried out experiments on beetle–aphid populations inside field cages in beet crops. In 1963 they found that the lady-birds suppressed or destroyed the aphid populations at a ratio of one beetle to seventy aphids, but did not prevent an eventual build-up at a 1 : 90 ratio. In 1964 the effective ratio was 1 : 200, the explanation being the higher temperatures and lower relative humidities of that season – conditions which favoured the beetles but which hampered the aphids. The workers recommended that insecticides should be applied about a fortnight after the aphids migrated to the fields whenever the ratio was unfavourable. They advised against the use of non-selective insecticides when the aphid populations reached the peak, for this would destroy the natural enemies which would reduce the overwintering population of aphids during the autumn.

The influence of natural enemies in the autumn has been put forward as a reason for the frequently noted alternation of bad aphid years and years when aphid populations are low, in temperate regions. In years of high population densities the predators and parasites make such heavy inroads on the aphids towards the end of the season that few eggs are laid, but in the following season this check is much less severe at the end of the season.

The Czech workers also suggested that suitable hibernation quarters should be provided near the beet fields for the beetles, and that plants should be grown which would carry non-economic species of aphids in the spring, and which would also provide nectar for the adults of.those natural enemies which need it.

A similar technique has also been tried to encourage the immigration of *Hippodamia convergens* and *H. tredecimpunctata* into infested crops in the United States. Corn fields were sprayed with sugar solutions and this was followed by an increase in the numbers of the beetles and a drop in the numbers of aphids. *Hippodamia convergens* is one of the species of lady-birds which diapauses in large groups and use has been made of this habit by their being collected and released into infested crops in California. Unfortunately, it is now thought that this long-established practice is largely worthless, for beetles collected during early summer do not disperse far enough, while those collected in winter and spring leave the infested area too quickly.

Despite their undoubted impact on aphid populations, lady-birds have not been very successful in classical biological control projects against these pests and their close relatives. Two species have been included, for example, among the thirteen natural enemies of the balsam woolly aphid that have been introduced into Canada, and they have become established, but without, so far, any worthwhile effect. Another species, *Coelophora inaequalis*, was introduced into Hawaii as an enemy of the sugar cane aphis, *Aphis sacchari*, in 1894, but was apparently of little value till 1930, although it is said to be now the most important control agent (Sweetman, 1958). Other predators and a parasite also play a part in the control of this pest.

An extensive bibliography and a useful summary of the impact of coccinellids on aphids will be found in the review by Hagen and van den Bosch mentioned before. Several papers appear in the symposium edited by Hodek (1966).

Although their role as natural enemies of aphids has been, so far, disappointing, the lady-birds redeem themselves as agents against scales and mealy bugs. The Vedalia beetle as a control of the cottony cushion scale has been mentioned several times earlier, but it is not the only Australian coccinellid which has proved to be a valuable ally of the grower. *Cryptolaemus montrouzieri* was introduced into California – again by Koebele – as an enemy of mealy bugs. It cannot, however, withstand the winters of most parts of the United States but it has become established in parts of California and Florida, and it is also bred and released in large numbers whenever necessary in other regions.

Cryptolaemus montrouzieri has also been used in several other countries where it has given substantial control of various mealy bugs. The beetle was imported from California into Chile, for example, where the pest was the citrus mealy bug, *Planococcus* (*Pseudococcus*) *citri*. In Western Australia also it was used against mealy bugs on citrus, and here it was brought from the other side of the continent, New South Wales. It has also given partial control of the green shield scale, *Pulvinaria psidii* on Puerto Rican shade trees.

Bodenheimer (1951) has reported on his disappointment with the performance of the beetle in Israel, Egypt, and Algeria, and related it to two unfavourable seasons, a humid and cool winter and a dry, very hot summer. He tentatively suggested that the optimal climatic zone for the beetle lay between 18 and 23 °C and 86 to 90 per cent relative humidity. He also found that native species of coccinellids, chiefly *Scymnus* spp., *Hyeraspis polita* and *Oxynychus marmottani*, were more beneficial than the introduced Australian species.

Although there is sometimes doubt about the effectiveness of *C. montrouzieri* there can be none about the West Indian species *Cryptognatha nodiceps* which, since its introduction into Fiji in 1928, has completely controlled the troublesome scale pest of coconuts and bananas, *Aspidiotus destructor*. It was just as successful in Portuguese West Africa after it had been brought there from Trinidad in 1955.

Finally, the genus *Stethorus* should be mentioned; the small beetles which constitute this genus are, as far as is known, all predators of mites, including the Tetranychidae. Mites are, apparently, necessary for reproduction, though the beetles can subsist on other kinds of food during shortages. Unfortunately, although they consume large numbers of mites during development and as adults, they are generally only abundant when the pests are exceedingly numerous, when the damage has become considerable. They do not appear to have been used for biological control as yet. Other lady-bird beetles and larvae will also feed on mites but never, it seems, as their main source of food (Huffaker, van de Vrie, and McMurtry, 1969).

Other families of the Coleoptera Various beetles, scattered through the remaining families of the Order, have been used in biological control, or are potentially useful agents. A few histerids have been introduced into various countries for the control of pests. These beetles are often found in decomposing organic matter where they prey upon insect eggs and larvae. The Javanese *Plaesius javanus* feeds upon the grubs of banana root borers such as *Cosmopolites sordidus* within their burrows and it has been introduced into Fiji, Australia, and other countries for this purpose. *Hister chinensis* was also taken to a number of Pacific islands from Java to feed upon muscid maggots in cow dung.

The colourful clerid beetles are usually predacious and often feed as adults upon wood-boring insects. *Thaneroclerus girodi* is a predator of the cigarette beetle, *Lasioderma serricorne*, a stored-product beetle which does not restrict its tastes to tobacco. The common stored-product beetle *Necrobia rufipes*, which is commonly known as the red-legged ham beetle, or the copra beetle, also belongs to this family. Although it attacks stored products with high fat contents it is also predacious on other pest insects such as the cheese skipper, *Piophila casei*. A third clerid, *Callimerus arcufer* was introduced into Fiji in 1925 as an enemy of the coconut moth, *Levuana iridescens*, but had little chance to prove its worth as the pest was already well under control by the tachinid fly *Ptychomyia* (*Bessa*) *remota*.

The Lampyridae or glow worm beetles may be of value in public health work for they are often predators of snails which are commonly intermediate hosts of flukes and other Platyhelminths parasitic in man and his animals. The adult males are usually winged and take little food, but the larvae and the often wingless, larviform adult females feed extensively on snails, earthworms, cutworms, and similar soft-bodied insects. An Asiatic species, *Luciola cruciata* is common in the mud and water of paddy fields and could be of great value to the public health biologist. The closely related Drilidae are also enemies of snails. The British glow worm, *Lampyris noctiluca*, was taken to New Zealand during the 1930s in an attempt to control the snail *Helix aspersa*.

The Neuroptera

The Order contains a number of families, all predatory and some aquatic. The most important families for biological control purposes or natural regulation are the Chrysopidae, or green lacewings (Fig. 6.7), the Hemerobiidae or brown lacewings, and the Coniopterygidae which are sometimes known as powdery lacewings.

Neuropteran larvae are usually active creatures with long, sickle-shaped mandibles, well adapted to the seizure of soft-bodied prey. One of the commonest European species is the chrysopid *Chrysopa carnea*, a large green insect which frequently comes indoors in the autumn to hibernate. The adults of this species apparently feed only rarely on aphids and, consequently, the larvae are not common in colonies of these insects. Most other adult green and brown lacewings feed on aphids and are thus more closely linked with aphid population changes than is *C. carnea* (Hagen and van den Bosch, 1968).

The American species *Chrysopa californica* was one of the most important natural

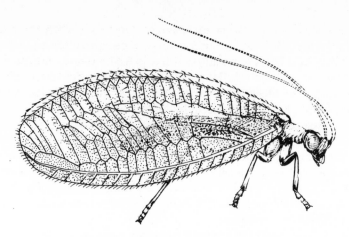

Fig. 6.7.   Adult green lacewing, *Chrysopa* sp.

enemies in the regulation of three *Pseudococcus* species and two tetranychid mites in glasshouses and citrus groves, and its activities have been augmented by the mass distribution of the eggs.

The cosmopolitan Hemerobiidae are, in general, smaller than the green lacewings. Their prey is usually aphids and other plant bugs, mealy bugs, and mites, and a single larva is said to be able to destroy several thousand prey (Sweetman, 1958). Some *Hemerobius* species have been introduced into Canada and Oceania as enemies of Chermidae on forest trees.

In the United Kingdom *Hemerobius humulinus* and *H. lutescens* commonly prey on the fruit tree red spider mite, *Panonychus ulmi.*

Bodenheimer (1951) and his colleagues claimed to have made effective use of the native *Sympherobius sanctus* in Israel against the citrus mealy bug, *Pseudococcus citri.* They reared and released large numbers of the lacewing in infested citrus groves. They attributed its success to its ability to build up to large populations during heavy infestations, its comparative immunity to attack by parasites and predators, and its adaptation to the local climate.

The powdery lacewings are tiny insects resembling white flies (Aleyrodidae) in appearance. *Conwentzia pineticola* has been observed feeding on red spider mites in the orchards of Essex and Kent. Despite their small size the larvae attack adult female mites readily and will dispose of some thirty daily. *C. psociformis* is associated with oaks in England, its prey being *Phylloxera* species, red spider mites, and coccids. It was unsuccessfully introduced into New Zealand in 1924 (Clausen, 1940).

The Hemiptera

Although the true bugs are usually thought of as plant pests a large number are predators of other insects. Often, a single family will contain plant parasites, predators, and species which take both kinds of food. Two outstanding examples are the Miridae and the Pentatomidae.

Miridae   It has been suggested that the reason that this family contains both plant-feeding and animal-feeding species is the ease with which a plant-sucking bug could turn its attentions to soft-bodied insects feeding on the same host (Miller, 1956). One particularly unusual association is that of *Cyrtopeltis droserae* with certain

Plate 1 (chapter 1)   Filariasis victim in Papeete, Tahiti, 1966. This disease is widespread in Polynesia. The causal nematodes are transmitted by *Culex* and *Aedes* mosquitoes

Plate 2 (chapter 2)   Adult winter moths, *Opheroptera brumata*, male on left. The population dynamics of this species have been studied in England by Varley and Gradwell (see chapter 2). The moth, which defoliates deciduous broad-leaved trees, has become established in Canada

Plate 3 (chapter 3)   Overwintering eggs of the green apple aphid, *Aphis pomi*, on apple twig. Greatly enlarged. Hydrocarbon oils may be used to destroy these eggs during the winter and early spring

Plate 4 (chapter 3)   Winter eggs of the fruit tree red spider mite, *Panonychus ulmi*, on apple bark. Greatly enlarged. This mite, and related species, have become more important with the destruction of their predators by some modern synthetic insecticides

Plate 5 (chapter 3)   Adult lappet moth, *Gastropacha quercifolia*. **The caterpillars of this large moth sometimes attack the leaves of apple trees in Britain, but rarely cause damage of economic importance. The species could be regarded as an example of a potential pest which, as a result of some environmental change, could become more important**

Plate 6 (chapter 3)   Adult codling moth (*Cydia pomonella***), a pest of apples which has spread to most apple-growing regions of the world**

Plate 7 (chapter 3)   Spraying apparatus used for laying down small trials with chemical pesticides. The biologist applies the spray with the lance held in the right hand. The spray mixture is held in the container on the right which is supplied with compressed air at a constant pressure from the cylinder on his left

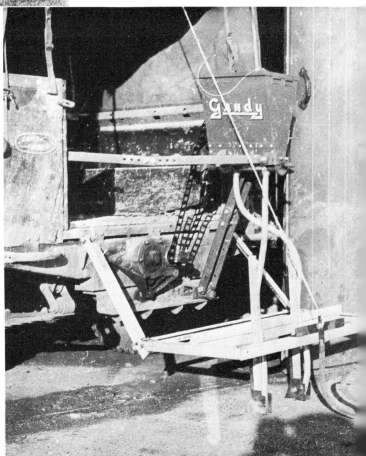

Plate 8 (chapter 3)   Apparatus for applying granulated insecticides to the soil, attached to the power take-off of a Land-Rover

Plate 9 (chapter 3) Merino sheep on an experimental station in New South Wales. The sheep in the centre is badly infested with the sheep louse, *Damalinia ovis*, and ked, *Melophagus ovinus*

10 (chapter 3) Dipping sheep in an insecti- ath to control ectoparasites

Plate 11 (chapter 3) Head and fangs of the Sydney funnel web spider, *Atrax robustus*, which has been responsible for several human deaths in New South Wales. The study of the mode of action of the venom of this spider and of other organisms could lead to the discovery of new pesticides

Plate 12 (chapter 3)   Hop garden in Kent, En▓
in which many of the vines have been destroy▓
Verticillium wilt. One possible method of c▓
involves the use of systemic fungicides

Plate 13 (chapter 6)   Agrotid pupa and the
ichneumonid adult which emerged from it

Plate 14 (chapter 6)   An Australian bu▓
ant (*Myrmecia gulosa*) attacking a meal▓
larva. These primitive ants destroy many in▓
but are too ferocious to be used for biol▓
control. Other species of ant may be of ▓
value

Plate 15 (chapter 6)   A healthy colony of *Aphis* *s* on milkweed bush

Plate 16 (chapter 6)   A colony close to that shown in Plate 15 but photographed two weeks later. The aphids have been attacked by aphelinid parasites which, after pupation in the 'mummies' have emerged through the tiny exit holes

Plate 17 (chapter 6)   Young syrphid larvae attacking ds

Plate 18 (chapter 6)   Adults and pupae of the cabbage root fly, *Erioischia brassicae*, and adults of pupal parasite, *Aleachara* sp.

Plate 19 (chapter 6)   Mosquito fish, *Gambusia affinis*, swarming around the author's foot in a public lake in Sydney, New South Wales

Plate 20 (chapter 6)   Larvae of an aphidophagous lady-bird emerging from a group of eggs

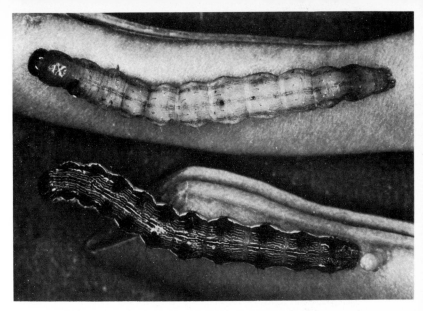

Plate 21 (chapter 7) *Heliothis* **larvae. The larva above is infected with a microsporidian. (By courtesy of the Queensland Department of Primary Industries)**

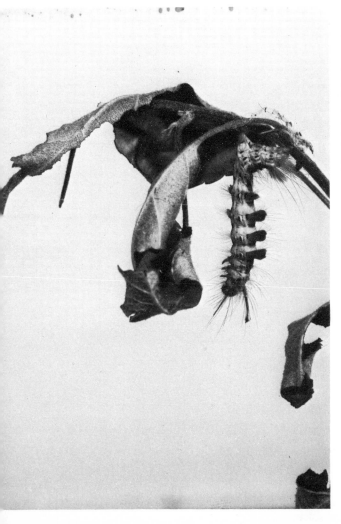

Plate 22 (chapter 7) Polyhedra of the *Orgyia anartoides* polyhedrosis disease. (By courtesy of the Queensland Department of Primary Industries)

Plate 23 (chapter 7) *Orgyia anartoides* larva killed by a polyhedrosis virus disease. (By courtesy of the Queensland Department of Primary Industries)

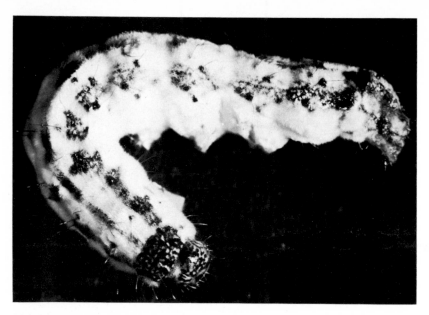

Plate 24 (chapter 7) *Heliothis* larva killed by a fungus disease. (By courtesy of the Queensland Department of Primary Industries)

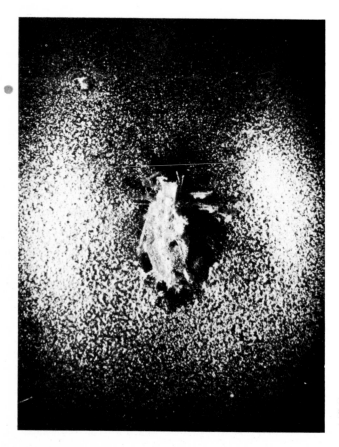

Plate 25 (chapter 7) Aphid destroyed by *Entomophthora* fungus. Note the discharged conidiospores surrounding the aphid. (By courtesy of the Queensland Department of Primary Industries)

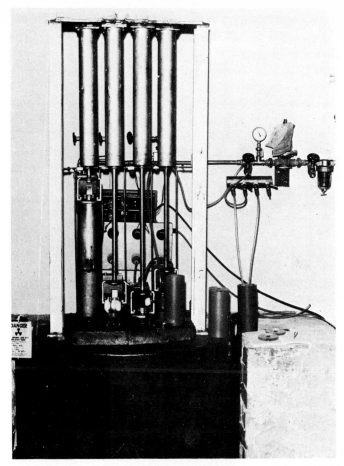

Plate 26 (chapter 8) A small Cobalt 60 source used in the experimental irradiation of insects. The source itself is surrounded by a thick, lead shield. The samples may be lowered into the field of the source pneumatically by the four units seen in the superstructure

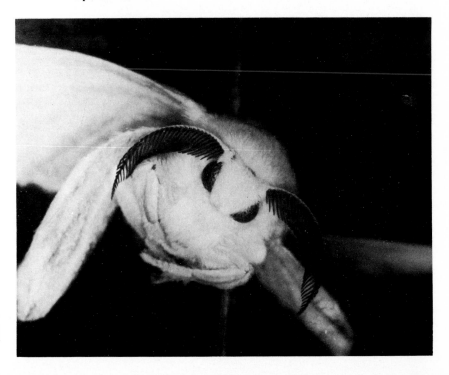

27 (chapter 8) Head of a male silk-
moth showing the large antennae
detect the sex pheromone released
female

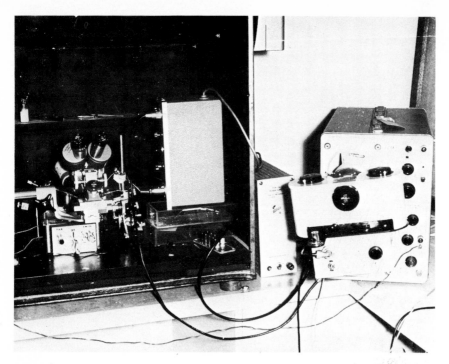

Plate 28 (chapter 9)   Neurophysiological apparatus used in neurological studies on the mode of action of insecticides, and for detecting responses to candidate attractants and repellents. The insect is placed on a stage beneath the microscopy, and electrodes applied to the nerve by micro-manipulators. The response of the nerve to the insecticide or to the repellent or attractant is displayed on the cathode-ray oscilloscope on the right, the screen of which is obscured by the camera which is used to obtain a permanent record

Plate 29 (chapter 10)   Umbelliferous pl and various other flowers, are importan the nutrition of adult parasitic hymeno and other beneficial insects

species of the insectivorous sundew plants in Western Australia. The bugs feed, unharmed, on insects captured by the sticky hairs of the plants.

The species of the genus *Cyrtorhinus* are predators on the eggs of homopterous insects such as leaf-hoppers. *C. mundulus* is an important natural enemy of the sugar cane leaf-hopper, *Perkinsiella saccharicida* in Hawaii, having been introduced from Australia in 1920. It has given complete control of the pest there but is apparently less successful in Fiji where the leaf-hopper is still a notorious vector of the Fiji disease of cane. Although the bug will attack the eggs of other leaf-hoppers on other plants it seems to be particularly attracted to cane fields. There can be as many as ten generations each year in Hawaii.

Blepharidopterus angulatus, the black kneed capsid bug is one of the most important predators of fruit tree red spider mites in Britain. There is only a single generation each year, but the bug can give a commercial control of the pest in some years, although it takes several seasons to build up to high numbers. While it is not affected by winter oil sprays it can be killed from hatching onwards by post-blossom sprays if these are not used with care.

In Collyer's studies, mentioned earlier in this chapter, another fifteen mirids were found to be of importance as natural enemies of pests in orchards. *Campylomma verbasci*, a bivoltine species, feeds on red spider mites and aphids, and was more abundant in cultivated orchards than in a derelict one which was also surveyed.

Pentatomidae   According to Clausen a member of this family, the European *Picromerus bidens*, shows a preference for bed bugs, *Cimex lectularius*, and was suggested as a predator for this then common pest in 1776. It was said that they were introduced into infested rooms which they cleared within a few weeks (Clausen, 1940). It will also feed on other slow-moving insects such as larvae of butterflies and moths, brown lacewings and beetles (Miller, 1956). Other members of the same sub-family, the Asopinae, are also carnivorous and they include *Cantheconidea furcellata*, preying on a defoliating moth of the coconut, *Artona catoxantha*; *Perilloides bioculatus*, a predator of the Colorado beetle in the United States and the Australian *Oechalia consocialis* which attacks the caterpillars of the day-flying moth *Phalaenoides glycine*, a common pest of grapes and ornamental climbing plants.

The Asopinae are not markedly adapted morphologically to a predatory life, but they capture their prey by stunning the insects with a paralysing saliva before grasping them with their fore legs.

Both *P. bioculatus* and another asopine, *Podiscus maculiventris*, have been introduced into France in attempts to control the Colorado beetle. The second bug also feeds on other chrysomelid beetles and various hairless caterpillars.

Reduviidae   The well-named assassin bugs are familiar predators of insects and of other animals, including man. The European *Reduvius personatus* attacks the bed bug as does, probably, *Vesbius purpureus* (Miller, 1956).

The venomous saliva of the reduviid bugs is very powerful and rapid in action, and is often capable of subduing prey much larger than the bug. *Physorhynchus linnaei*, for example, is said to paralyse giant millepedes some forty times its own length (Clausen, 1940).

Although the assassin bugs attack a wide range of insects, including many pests, they do not seem to have been used in biological control attempts.

Anthocoridae   The small bugs belonging to this family are well-known natural enemies of pests. *Anthocoris nemorum* is a very common British species which preys on a variety of insects and mites including aphids, coccids, caterpillars, and red spider mites. An adult will kill about fifty of the last named in a day, and during the spring and autumn, and in warmer periods in the winter, will also suck winter eggs. *Orius* species also are important predators in Europe and America.

A number of other families of the Hemiptera–Heteroptera contain natural enemies of pests. They include the Lygaeidae, the Nabidae, and the Coreidae.

In the last few pages the most important insect natural enemies and biological control agents of arthropod pests have been discussed. Without any doubt, the Hymenoptera, the Diptera, and the Coleoptera have been the most useful Orders, but the other groups will also be used in the future. There has not been space to consider other Orders of insects, apart from the Neuroptera and the Hemiptera, but many of these contain parasitic or predatory species which play some part in the natural regulation of Arthropod populations. The Odonata, for example, are predatory both in the aquatic nymphal stage and in the adult. Even the Lepidoptera has its zoophagous members; *Ereunetis miniuscula*, for example, has been recorded preying on the cottony cushion scale in Puerto Rico.

Arachnida

Apart from those mites which feed on plants, stored products, or dead organic matter, all Arachnids are carnivorous and undoubtedly play an important part in the natural regulation of animals, especially insects and mites.

The spiders (Order Araneae) are often thought of as the chief enemies of insects, especially flies. The orb-web spinners are, presumably, non-selective in their prey, the actual species depending mainly on the habitat, the habits of the prey, and the strength of the snare. Some of the web spinners do, however, specialize. According to Hagen and van den Bosch (1968) some species of *Erigone* subsist largely on aphids.

The hunting spiders, which use their silk more as life-lines than as snares, are probably greater specialists. *Dysdera crocata*, a species found in Britain and which is characterized by its huge chelicerae, was found by Bristowe (1958) to feed largely on woodlice, which may be classed as minor pests.

There appears to be few records of spiders being used for biological control. *Thannatus flavidus*, having been found to attack bed bugs in a Greek refugee camp in 1925, was introduced into an infested animal house in Germany where it successfully controlled the pest. *Ascyltus pterygotes* has been listed as one of the enemies of *Levuana iridescens* in Fiji.

Spiders are apparently quite important in the regulation of leaf-hoppers in paddy fields in some parts of Japan. From July to October each rice-plant hill supports one to seven spiders, a sufficient number to destroy most of the leaf-hoppers present. Their impact lessens from July onwards until, in October they are unable to keep the pests in check. Their importance may be gauged from the increased leaf-hopper densities experienced when lindane is used imprudently against *Chilo* species. Some Japanese workers believe that spiders are of considerable value in the control of tetranychid mites.

At present, spiders do not show promise of becoming biological control agents. As any collector of spiders knows, these animals are difficult to rear in large numbers since they must be kept separate from one another because of their cannibalistic habits. Sweetman (1958) points out, however, that little is known about the natural enemies of spiders, and the effect which these have upon them. It is possible that certain species of spiders, if introduced into new areas, would prove to be effective biological control agents.

The mites, on the other hand, are, without doubt, very important in the natural regulation of pest populations. They display both ectoparasitic and endoparasitic habits, and many are predators.

Collyer's work on the enemies of the fruit tree red spider mite, mentioned earlier in this chapter, showed that a number of mites, especially *Typhlodromus finlandicus*, *T. tiliae*, and *Phytoseius spoofi*, were to be found feeding on the pest in neglected orchards, but that their numbers were lower in commercial orchards. Nine *Typhlodromus* species are also known to attack this pest in Nova Scotia. All stages are taken but if animal prey is unavailable the predatory mites are reported to feed on plant sap as an alternative.

Tetranychid mites on out-of-door crops have become, as is well known, particularly troublesome since the introduction of chlorinated hydrocarbon insecticides. The various species are difficult to distinguish one from another as far as the layman is concerned (they also present difficulties to the expert). There has thus been a tendency to spray against spider mites on a 'rule of thumb' principle, treating all species as if they were equally damaging. The commonest species on grapes in certain parts of California, *Eotetranychus willamettei*, is far less damaging than *Tetranychus pacificus* which can be a very serious pest. Despite the difference between the two species it is likely, as Huffaker, van de Vrie, and McMurtry (1969) point out, that both species would be treated in the same way. It is possible that the less troublesome species could be best controlled by predators and by sprays only when absolutely necessary.

This integrated approach to mite control is being developed in Britain on glasshouse crops. Integrated control is, of course, particularly suitable for such environments which can be closely controlled.

Two tetranychid species, *Tetranychus urticae* and *T. cinnabarinus* are serious pests of a number of glasshouse crops in Britain. In 1959, the German entomologist Dosse, obtained specimens of the predatory mite *Phytoseiulus persimilis* in Chile. Experiments in Canada, the USA, Britain, and the Netherlands proved it to be an efficient predator of the tetranychid mites which it searched out at low densities. Hussey *et al.* (1969) state that the mite is capable of eliminating red spider mites under glass. Over a wide range of temperature (25–30 °C) its reproduction rate is twice that of its prey. When the mites were released at the rate of two mites on each cucumber plant they increased and were found on every infested leaf within nineteen days. If the mites are released while leaf damage is still slight they will control the red spider mites before the damage reaches the critical level at which losses are incurred. After the pests have been eliminated the predators can survive in their absence for about forty days, so good control may be expected for a total of about eight weeks after the initial introduction. The authors suggest that about one in a hundred cucumber plants could be set aside on which to rear the predator so that the second introduction that would be necessary to give complete control during the twenty-week cucumber 'season' could be dispensed with.

Unfortunately, insecticides have to be used to control other pests of glasshouse crops, so methods of integrated control are being developed which will conserve the predatory mite (see also chapter 11).

Many trombidiform mites are predators of insects and other mites. Among the most familiar (since they are so relatively large) are the scarlet mites such as *Allothrom-bidium* which prey on small insects in and out of the soil. *A. fuligonosum* is known to prey on the eggs of the European corn borer, *Ostrinia nubilalis*, in the United States, and on apple woolly aphid in Britain.

The trombidiform genus *Cheyletus* contains a number of species which are often found preying on tyroglyphid mites infesting stored products. At times *C. eruditus* is common enough to control the pests, at least during the summer. Descriptions and details of the biology of stored-product mites may be found in *The Mites of Stored Food* (Hughes, 1961); this bulletin is of wider use than its name suggests.

The sarcoptiform mite *Machrocheles muscaedomestica*, which feeds largely upon the eggs and first instar larvae of the house-fly, is also found in stored foods, or at least in those which are attractive to egg-laying house-flies. It is, of course, far commoner in manure heaps.

The adult female mite lays eggs which, if fertilized, develop into females and, if not, into males. At 27 °C the egg develops into an adult mite in about sixty hours. Each female mite lays, on the average, about five eggs each day, but in certain media this number may rise to over twenty. In one experiment, virgin females produced a total of about ninety eggs each; and fertilized females about sixty. They thus pass through three generations in the time taken for the house-fly to pass through one, but the rate of reproduction is, nevertheless, rather low.

The adults of this species feed preferentially upon house-fly eggs and larvae, but will also feed and reproduce, when necessary, upon nematodes. The nymphs, on the other hand, attack nematodes more readily than house-flies.

As dung ages there is a succession of mites, with parasitids appearing somewhat earlier than machrochelids. As the dung gets older the machrochelids are replaced by uropodids and, later, oribatids. Fresh dung is thus deficient in machrochelids, as is old dung. If advantage is to be taken of the machrochelid's predatory behaviour the dung must be managed in a suitable way. It should not, for example, be cleared away too soon. Similarly, fresh dung should be added periodically. There is also the danger that insecticides added to the livestock's food will kill the mites and, in fact, worsen the fly problem.

When the dung becomes unsuitable for machrochelids the female mites use house-flies and beetles for transport to new breeding sites – a phenomenon known as phoresy. This is common among the mites and their distant relatives, the pseudoscorpions. Usually, the association is not parasitic, though this may not be strictly true of that between *M. muscaedomesticae* and the house-fly.

The predation of house-flies by *M. muscaedomesticae* has been demonstrated by various workers in laboratory trials. Axtell (1963) found during a series of field trials that outdoor piles of dairy cattle dung produced about 30–45 per cent fewer house-flies than did similar piles treated with the specific acaricide dicofol, after 50 000 house-fly eggs were added to each pile. In experiments under cover the difference rose to over 90 per cent. The differences were attributed to the destruction of this mite.

Macrocheles muscaedomesticae seems to be most promising as a regulatory agent in indoor accumulations of dung, such as occur under the cages of poultry-batteries, especially when the control by the mite is properly integrated with insecticides used against the adult flies. The mite appears to be widespread and has not, as yet, been used as a biological control agent in the strict sense. An account of the biology of the mite and an assessment of its potentiality as a biological control agent was given by Axtell at the second International Congress of Acarology, and the account was later published by the WHO (Axtell, 1969).

Ectoparasitism is common among the mites (though phoretic associations are often confused with it), the usual victims being the insects and vertebrates. The Orthoptera, for example, are extensively parasitized by trombidiform mites. *Podapolipus grasii* is frequently found on locusts, the female mites remaining attached to their hosts throughout their life. *Pyemotes ventricosus* is a common larval parasite of the grain moth *Sitotroga cerealella* but, unfortunately, it can cause a troublesome dermatitis when the grain is handled. This so-called hay itch mite is also common in hay, so the usual victims are dockers and farm workers. The mite has been found in Britain in grain imported from Egypt and Morocco where, it is said, the indigenous people are immune but where 'White men never sit on the sacks of grain twice' (Alexander, 1972). The mite *Cheyletiella parasitovorax* has also been recorded as causing rashes in humans. Normally this mite is a predator of *Demodex* spp. in cats and dogs, and the owner becomes infested when handling the mangy pet (Alexander, 1972).

Among the endoparasitic mites is *Acarapis woodi* which infests the tracheae of the honey-bee. Potentially more useful are the species of the genus *Locustacarus* which attack locusts in a similar way.

A brief mention must be made of the water mites parasitizing mosquitoes. These brightly coloured hydrachnid mites are often found attached to adult mosquitoes, but it does not seem to be thought that they are of any great importance in natural regulation. It has been suggested, however, that they may serve as an indication of the breeding site of the mosquito because of their somewhat restricted distribution.

In summary, it cannot be said that mites have been of great value for the biological control of pests, although *Phytoseiulus persimilis* is promising in glasshouses in Britain, and on out-of-door crops in warmer areas. In addition, a tetranychid mite which escaped from quarantine made impressive inroads on prickly pear in Australia. In Australia too, the accidentally introduced *Biscirus lapidarius* proved to be an effective control agent for the lucerne 'flea', *Sminthurus viridis*, in some districts. Possibly, the lack of successful examples of biological control by mites reflects the shortage of acarologists more than the shortage of suitable mites. The Acarina, despite its great economic importance, is a comparatively neglected group. Enough is known, however, to contradict Savory's (1964) statement that 'Mites which perform actions which assist the human race . . . do not seem to have been evolved.' It is also significant that Moreton's 1958 Bulletin on beneficial insects was reissued in 1969 with the title *Beneficial Insects and Mites*. European readers will find this bulletin a valuable introduction (Moreton, 1958, 1969; Hughes, 1959; Savory, 1964).

Vertebrates

If we except those fishes used for the biological control of mosquito larvae, vertebrates, apart from a few mammals, seem to have been used little so far in the biological control of animal pests, and even their importance as agents of natural regulation is disputed. Vertebrates which attack other animals are, of course, predators; parasitism is very rare in the class. In general, their behaviour is far more complex and adaptable than that of invertebrate predators, so that they readily turn to other sources of food when one kind becomes scarce. They will, therefore, be poor control agents when the pest population is low. On the other hand, birds and mammals are warm-blooded animals with high metabolic rates and consequently with high energy demands. This applies particularly to small insectivores such as shrews which have high body surface area to volume ratios. Such animals must feed almost continuously, or starve.

Birds

Birds have long had their admirers, so they do not lack their supporters who proclaim their importance as natural regulators of insect populations. Unfortunately, there is little hard evidence to support their case.

Birds have often been introduced intentionally from one continent to another, but this has usually been done to allay feelings of homesickness or to provide sport (Elton, 1958; Laycock, 1966; Rolls, 1969), but, occasionally, with pest control in mind. The Indian mynah, *Acridotheres tristis*, for example, was taken to Mauritius in 1762 where it controlled a locust, *Nomadacris septemfasciata*. It is also said to have been released in Australia by the Victorian Acclimatization Society as a general predator. It is certainly common enough now in some Australian cities to be known as the starling, although the true starling, *Sturnus vulgaris*, is also numerous.

The mynah was also introduced into the Hawaiian Islands from India in 1865, but whatever good it may have done it became troublesome by spreading the seeds of the noxious shrub *Lantana camara*. Since then many attempts have been made to control this shrub by biological methods.

Another early introduction was that of the little owl, sometimes called the Dutchman, into the British Isles. The main introductions of this species, *Athene noctua*, took place between 1874 and 1900. Its economic status was subsequently so hotly disputed that the British Trust for Ornithology thoroughly investigated its biology during the 1930s. They came to the conclusion that, on balance, it is a beneficial species feeding largely on small rodents and insects, with the odd game bird or chicken.

The equivocal position of many birds can be best illustrated by the case of the starling, *Sturnus vulgaris*, in Britain (Murton, 1971). The species probably originated in the Asiatic steppes, but has settled in Britain long enough for a distinctive breeding biology to evolve. Nevertheless, it was not known as a breeding bird in Devon before 1830, nor in Wales till some time between 1830 and 1860. During the first half of this century its numbers increased enormously, and with this increase came a change in feeding habits. Before it became exceedingly common it was, on the whole, a useful bird, feeding on a variety of insects including such pests as cutworms and larvae of the diamondback moth, *Plutella maculipennis*. Its old Lancashire dialect name of shepster paid tribute to its useful habit of landing on the backs of sheep to remove ectoparasites. It also destroyed large numbers of the snail, *Limnaea trunculata*, the alternative host of the liver-fluke.

Now it is regarded as a serious pest, capable of destroying a young crop within a few hours. Murton (1971) records that on one large commercial poultry farm starlings were taking an estimated 1100 tonnes of animal food each year. In towns its populous roosts are an unmitigated nuisance. This urban roosting behaviour only arose at about the turn of this century, and has apparently occurred independently in several countries and towns. In 1962–4 there were 69 roosts in 39 conurbations in Britain, and it was estimated that the 13 roosts which were situated in trees contained about 400 000 birds in all.

In those countries into which it has been introduced it has developed, or is developing, into a pest. In the USA a few pairs released in New York Central Park in 1891 have populated the continent. Where they are still not too common they serve a useful purpose and, indeed, they are listed by Sweetman (1958) as insect destroyers, but in areas where their population density is high they are very troublesome. A similar situation exists in Australia where they are particularly disliked for introducing vermin into houses.

Some birds must be regarded as almost completely beneficial. The tits (Paridae), for example, are highly regarded by gardeners, horticulturalists, and foresters, who often provide artificial nesting sites. Seagulls of various kinds are well known for their habit of following the plough to pick up the insect larvae thus exposed. Sweetman (1958) tells of one Minnesota farmer who started his daily ploughing earlier than did his neighbours to ensure a lion's share of the birds. He almost completely eradicated white grubs, *Phyllophaga* sp., from his field.

This story stresses the nature of most of the evidence on the role of birds in the natural regulation of pests; it is anecdotal. Stories abound of huge flocks of birds appearing when insect outbreaks occur. The classic case is that of gulls and the Mormon crickets, *Anabrus simplex*, during the early settlement of Utah. There is an older story of an outbreak of voles in the Essex marshes, England, in 1580, but the disaster is said to have been checked by the arrival of owls in such numbers that 'All the shire was not able to yield.'

It is true that much research has been carried out on the feeding habits of birds, usually by the examination of gizzard contents, faeces, and owl pellets. This work has been, however, qualitative rather than quantitative. In order to assess the impact of a predatory bird it is necessary to know not only the quantity of food in the gut of a shot or netted bird, but also the rate at which it is digested and eliminated This was the approach of Mook and Marshall (1965) in their study of the relationship of the olive-backed thrush, *Hylochichla ustulata swainsoni* and the spruce budworm, *Choristoneura fumiferana*. They were able to estimate that the daily consumption of a thrush was about seventy-four pupae at a time when the budworm population density was some 49 000 per ha. A count of 0·33 singing males per ha gave an estimate of 1 to 1·2 birds per ha. As the budworms were only available as pupae for about two weeks, a simple calculation shows that the thrush population destroys only about 1 020 pupae per ha, a mere 2·1 per cent. The authors point out that their results can be applied only to this association, and that other species would display different rates of digestion. The paper gives a useful bibliography.

The feeding habits of birds may also be studied at the nest site by time-lapse photography. With practice, the identity of the prey can be determined.

Betts (1955) and her colleagues used several methods to determine the impact of various tit species on defoliating caterpillars on mature oak trees in the Forest of Dean in Gloucestershire, England. In one part of the forest it was estimated that there were approximately 3 100 000 defoliating caterpillars per ha, and in another part, where there were fewer oak trees, almost 2 500 000 caterpillars per ha. About three-quarters of the larvae were those of the winter moth, *Operophtera brumata*. During the nesting season it was found that the tits removed only about 1·4 per cent of the caterpillars at the first site, and 0·9 per cent at the other. There was only a slight increase in the predation by the tits when nest boxes were used to double the population. Some of these boxes were, however, used by other insectivorous birds which would have some impact on the caterpillars. This was not, however, measured.

Gibb and Betts (1963) obtained similar results in a pine plantation of the Forestry Commission at Thetford Chase in eastern England.

There are some records of higher predation rates by birds. In Hungary, for example, it was found that house-sparrows, *Passer domesticus*, killed 98 per cent of the fall webworm, *Hyphantria cunea*, before the moths could lay their eggs.

During their breeding season birds certainly destroy large numbers of insects, but their impact is lessened by the marked territorial habits of most species at this time. Their mobility allows them to respond numerically very rapidly to pest increases but, because of their relatively slow rate of reproduction, delayed numerical responses are probably less important. Their mobility, on the other hand, allows them to move away from an area when the prey population becomes scarce. In this respect they probably resemble insecticides and other 'catastrophes' which destroy a large proportion of a pest population quickly, leaving survivors which face a lessened intraspecific competition. Predators which are more closely linked with the infested area, and which thus extend their predation over a long period, and most insect parasites, exert a 'lag phenomenon' mechanism in which pests doomed to die without reproducing continue to compete for some time with those individuals which could survive to reproduce. This argument applies only to the rather special cases where flocks of birds move into an area for a short period, and not to such birds as tits, which apparently 'work' a district for a considerable time.

The arboreal habits of many insectivorous birds has led to their greater importance in forests, suburban gardens, orchards, and farming districts well endowed with hedgerows and copses, than in more open country. As we have said above, their presence is often encouraged by the provision of suitable nesting sites and cover, and by the supply of food during hard weather. Advising such a practice for pest control reasons is superfluous; many do it for purely aesthetic motives.

In general, however, many authorities consider that while birds consume large numbers of pests, and can have a considerable impact on localized outbreaks, they have little influence on widespread epizootics. Sometimes, indeed, they can hinder the regulation exerted by other agents, such as parasites. Tits, for example, took a disproportionately high percentage of *Thera* larvae parasitized by the encyrtid *Litomastix*, which changes the colour of the normally cryptic caterpillar.

We cannot close this section without mentioning some rather special uses for birds. Domestic ducks are employed for both weed control and keeping down noxious snails. Falcons and other birds of prey are being used again to hunt prey, or at least to drive them away from airports where birds, such as gulls, are a danger to modern jet aircraft.

Murton (1971) gives an excellent summary of the relationships between birds and man's economic affairs in Britain.

Mammals

Insectivorous mammals, other than bats, are rarely seen. Consequently their value in the regulation of pests is underestimated by the layman. To the townsman, for example, a shrew is some kind of mouse, and must be treated accordingly. To many countrymen it was a baleful creature, a bringer of disease to man and his cattle, a small animal to be incarcerated in the living tissue of an ash to cure the ills it caused. The bats are also unjustly feared.

The larger predators which prey on rodents may turn their attention to game birds, poultry, or even small lambs, and thus become troublesome if they are numerous. For such reasons, and because wild mammals cannot be studied easily from the comfort of an armchair or car seat, they do not call forth the same emotions as wild birds, and few claims are made about their values as beneficial species.

Shrews have a very high metabolic rate and a short, usually hectic, life. *Sorex cinereus*, a small species, has a metabolic rate four times greater than that of a mouse per gramme of tissue (Pearson, 1955). This rate is only exceeded by the even smaller humming birds, and even in these the rate falls dramatically during the night. To maintain this rate a shrew must eat almost continuously. One captive *Sorex* ate 3·3 times its own weight daily and, it is believed, wild shrews have a completely animal diet. The biology of the British species may be found described in *British Mammals* (Harrison Matthews, 1952).

Probably, the importance of shrews as natural enemies of pest insects, and as possible biological control agents, is best appreciated by the Canadian foresters. Several studies have been made of their impact on the population dynamics of sawflies attacking conifers. It should be mentioned that the Canadian conifer forests, unlike many European ones, are natural climax forests and consequently more complex ecosystems than their transatlantic counterparts. The role of vertebrates in such forests is reviewed by Buckner (1966).

Neilson and Morris (1964) showed by their key factor approach that small mammals, including shrews, made up one of the factors regulating the European spruce sawfly, *Diprion hercyniae*, but Buckner (1966) was unable to do the same with the larch sawfly, *Pristiphora erichsonii*. He points out that the sawfly population was then in the declining phase, and that many authors argue that at such times biotic factors are ineffective in determining populations.

Small mammals do not always consume all the insects that they find. Some are believed to be stored for later consumption, then 'mislaid'. If McLeod's (1966) work on the spatial distribution of *Neodiprion swainei* cocoons applies to other species of sawfly, this wastage may not occur with shrews and other small mammals. The distribution was found to be aggregated, the aggregations being the result of the collection by the shrews. All the aggregated cocoons had been destroyed, the animals having taken them to preferred feeding stations in their tunnels for consumption.

Compared with birds the small mammals are not very mobile, so immediate numerical responses are unlikely and, in fact, rarely observed. Conversely, they do not migrate with the seasons, as do many birds, nor move rapidly from the area when a particular prey species becomes scarce. Functional responses are difficult to assess because vertebrates, in general, take a wider range of foods than most invertebrate predators. Buckner does give some data on the mean seasonal potential 'kill' for certain species feeding on *Pristiphora erichsonii*. These range from 5000 (*Sorex cinereus cinereus*) to 9000 (*Blarina brevicauda*).

Shrews have been used successfully in at least one biological control project. Newfoundland, separated as it is from continental North America, has few small fossorial mammals. The only ones present are the Newfoundland meadow vole, *Microtus pennsylvanicus terraenovae*, the brown rat and the house-mouse. None of these have any worthwhile effect on sawfly populations. It was therefore decided to introduce shrews into the island.

This was a surprisingly difficult undertaking. A zoologist might prefer to call the 'rat-race' a 'shrew-race', for these tiny mammals live their life at a hectic pace. Fights to the death between males are common. It is even said that they can die of shock, for they can often be found dead, beside a sprung trap, but with no visible injuries, as though the sudden noise had frightened them to death. They are therefore difficult to capture alive, to keep alive in captivity for more than a few hours, and to transport. In 1957, several hundred were captured in Manitoba, but few survived. In 1958, a group of biologists captured sixty-nine masked shrews in New Brunswick and managed to deliver twenty-two of them, alive, to the release site at St George's, Newfoundland, fifteen hours later. There they were released into a prepared bog-land site partly isolated by barriers to prevent premature dispersal.

At least half of these shrews survived the winter to breed the following year. In 1961, shrews were taken from the well-established colony and released at two inland sites. Subsequently, the multiplying shrews spread outwards from their release sites, with clearly observable advance fronts. It is estimated that if the advance continues at the present rate the whole of the island will be occupied within a few years, and it is expected that, in the absence of competing fossorial mammals, the shrews will fill a wider range of niches and attack a greater variety of pests than they do in their country of origin. At the time of Buckner's review they were having a smaller impact on larch sawfly populations than they do in Manitoba, but they were still considered to be the most important biotic agent in the regulation of this pest.

It is natural that the larger vertebrates should be considered as control agents for vertebrate pests. Unfortunately, many authorities consider that far from reducing their prey populations the predators of many vertebrates maintain them at higher densities (Howard, 1967). The predator usually removes the surplus population, leaving the breeding stock untouched and free from the competition of individuals which would not have reproduced. Thus Errington (1946, 1967) discovered that mink (*Mustela vison*) preyed mainly on those muskrats (*Ondatra zibethica*) which had not been able to establish home ranges. These homeless individuals suffer injuries from the established ones and move to poorer habitats, especially along the banks of lakes and rivers frequented by minks. As Watt (1968) points out, these animals are also likely to be the weakest individuals and surplus in both a biological sense and in a genetic sense. Established muskrats are usually safe from predation within their burrows, except when drought exposes them. At such a time intraspecific competition is likely to be intense so that its reduction by increased predation benefits the population as a whole.

Other authors have made similar observations on other vertebrate–predator prey systems. Pocket gopher numbers, for example, appear to bear no relationship to the density of one of their chief predators, the coyote (Crissey and Darrow, 1949). The sparrow is heavily preyed upon by the sparrow hawk, *Accipiter nisus*, but the disappearance of the predator from many parts of Europe does not seem to have been followed by an increase in the numbers of sparrows.

For this reason and because of the disastrous results of some past introductions, there seem to have been very few recent cases of the use of the larger predators as biological control agents. Mongooses were released in a number of islands during the nineteenth century to control rats in agricultural land. This they often did successfully, but in some places the rat populations escaped the mongoose predation by adopting arboreal habits. When rats became scarce, or unreachable, the mongooses turned to other prey, including wild fowl and domestic poultry. In Puerto Rico the mongoose *Mungos birmanicus* reduced the numbers of a subterranean lizard, *Ameiva exsul* so that its prey, *Phyllophaga* larvae, became pests. This situation was corrected later by the introduction of the giant toad, *Bufo marinus*.

Mongooses, thanks partly to Rudyard Kipling, are popularly associated with snakes and, not surprisingly, they have been released in a few areas to control these reptiles. In Martinique and St Lucia, for example, they have almost eliminated the extremely venomous fer-de-lance, *Bothrops atrox*, from open country.

In the Netherlands the water vole, *Arvicola terrestris terrestris*, frequently causes damage, as might be expected in a country with so many waterways and meres (Van Wijngaarden, 1953, 1954). Between 1910 and 1930 they were particularly troublesome in a large plantation of young trees on the island of Terschelling. A few stoats, *Mustela erminea*, and about one hundred weasels, *Mustela nivealis*, were introduced to control them. While the weasels soon vanished, the stoats multiplied rapidly, reduced the water vole populations and that of the rabbits and then became a pest in their own right. Since 1939, however, it appears that although the stoats are still common they have become less troublesome.

The fox, *Vulpes vulpes*, may serve as a last example. Often enough it has been introduced not as a biological control agent, but as a species to be controlled – by a remarkably inefficient means, hunting. In Australia, however, it was released in an attempt to control rabbits but has since became an important nuisance itself.

The fox serves to illustrate another difficulty. Most people are in agreement when they designate some insect as a pest but there is often no such agreement when a vertebrate is considered. To some it *is* a pest, to others it is a source of sport or pleasure. Some weeds are regarded similarly. Even the prickly pear was valued by a few graziers, where the infestations were not too great, as a source of fodder in dry periods. This makes the decision of whether to carry out control measures or not difficult, unless they are applied as 'spot' treatments. This reduces the choice to chemical methods and trapping.

There is still interest, however, in the use of Mustelids as biological control agents. The World Health Organization is investigating the possibility of using the Japanese weasel, *Mustela sibricica itatsi*, to control rats in the Pacific Islands. The rats concerned are the introduced species *Rattus rattus* and *R. norvegicus*, for those brought by the Polynesians themselves, *R. exulans* and *R. hawaiensis*, have been greatly reduced in numbers since the coming of the Europeans. These small, comparatively harmless species were the only beasts of chase of the Polynesians who welcomed the new rats at first as bigger and better game. The Maoris even asked the captain of one early vessel to release his rats on shore for their sport. Now they are disliked, not only for the direct damage they do to the coconuts but also because the gnawed nuts make ideal breeding sites for the mosquito vectors of dengue and of Bancroft's filaria, the cause of elephantiasis. Hence the Organization's interest in the weasel. If the two diseases could be eliminated by the help of this predator it would amply compensate for any damage the weasels may do later to birds and poultry.

It is not thought that the weasels alone will have a sufficient impact on the rat populations so their use may be integrated with the rodenticide norbormide. This chemical is believed to be markedly toxic only to *Rattus* spp.

The use of the large aquatic mammal, the manatee (*Trichechus*), for weed control, was mentioned in the last chapter. Little appears to have been published recently on this project.

Amphibia and reptiles

Although they are predacious the reptiles seem to have found little use in biological control. The large monitor lizard, *Varanus indicus*, has been introduced into parts of Micronesia to control rats, but with disappointing results. Various studies have shown that many species of lizards destroy large numbers of insects, including pest species, and might merit consideration. *Ameiva exsul* has already been mentioned as a predator on white grubs, *Phyllophaga* spp. in Puerto Rico.

Surprisingly, the amphibia have proved to be more useful. The giant American toad, *Bufo marinus* has proved its worth in a number of warmer regions of the world, where it preys on a variety of pest insects. It has been particularly useful in sugar cane fields where it attacks *Phyllophaga* spp. and the sugar cane rhinoceros beetle, *Strategus barbigerus*. In urban areas it is, unfortunately, a frequent victim of road traffic. Dexter (1932) suggested that in Puerto Rico the waterways and reservoirs should be planted with bananas and other species which would attract *Phyllophaga* adults to the toads' habitat.

Gardeners often take the much smaller European toad, *Bufo bufo*, into glasshouses, where the toads can make a most effective contribution to insect control. They show a marked sense of location and usually return to a familiar spot, because of this they must be restrained for a time after capture, till they become used to their new surroundings. Sweetman even suggests that they could be released in kitchens at night-time to eradicate cockroaches, though he grants that most housewives would find the method abhorrent (Sweetman, 1958).

Fish

Fish, of course, can only be employed against pests which spend part or all of their lives in water. Their main use has so far been against larval mosquitoes and chironomids, but they could, possibly, also be used to control larval simuliids, the adults of which are important vectors of disease. These pests are all of public health importance and, so far, no use appears to have been made of predatory fish to control other kinds of pests, except in paddy fields (see below). One problem that may eventually yield to such an approach is the control of lampreys, *Petromyzon marinus*, in the Great Lakes system of North America, if a suitable predator can be found.

A large number of freshwater and shallow water coastal fish have long been recognized as insect feeders. Although some Central American tribes have been said to have introduced small fish into water containers to feed on mosquito larvae the technique was not carried out widely till this century, and even so was eclipsed for a time by chemical methods.

The most important Order for mosquito control is the Cyprinodontiformes, most species of which are small fish capable of reaching the larvae in shallow water or amidst weeds. The Order contains two useful families, Cyprinodontidae and Poeciliidae.

The well-known mosquito fish, *Gambusia affinis*, and *Lebistes reticulatus*, the Barbados millions or Guppy are highly prolific members of the Poeciliidae, and like all species of that family, viviparous. *Gambusia affinis* is a North American freshwater species while the guppy is endemic in the West Indies. Both species have been spread widely around the world and colonized in both surface water and containers of various kinds. It is now considered that much of this work has been unnecessary, for many areas into which they have been introduced contain suitable native fish, though not necessarily in the infested waters where control is needed. The fish may even be present in the infested waters, but may be hampered by weeds or other obstacles which could be cleared mechanically. Considerable research should be carried out before an alien fish is introduced for the native species may be eradicated by the newcomer. This would be undesirable if, for example, its eggs were capable of resisting drought in areas where this is common. This is the case with *Nothobranchius* spp. and *Barbus* spp. in East Africa. It also applies to Florida, where the salt marsh killifish, *Fundulus confluentus*, is a native. This species is not restricted to salt or brackish water and can be used in inland temporary waters. Despite the existence of this fish the University of California has received several requests from Florida for other species (Bay, 1968).

Gambusia affinis does not feed indiscriminately on all species of mosquito larvae present in its habitat. It tends to take active wrigglers, such as *Anopheles freeborni*, more readily than the sluggish ones such as *A. quadrimaculatus*. It also appears to prey more heavily upon culicine larvae than upon anopheline larvae, possibly because of the horizontal position of the latter when they are at the surface. Furthermore, they sometimes show a preference for chironomid larvae when these are present. It is clear that such differential feeding could favour a dangerous malarial vector when more than one species of mosquito is present.

Unfortunately, neither *Lebistes reticulatus* nor *Gambusia affinis* can be regarded as suitable food for humans, although they produce a large weight of protein for each acre of water. In fact Rolls (1969) claims that the name *Gambusia* is derived from an Asian word meaning worthless and that the fish is rejected as food on that continent. Furthermore, both species can be a threat to native food fish. Larger fish such as *Tilapia* spp. and carp, *Cyprinus carpio* have been used as control agents for mosquitoes and weeds, and, at the same time, as a source of supplementary protein. *Tilapia*, however, tends to breed so quickly that the population eventually consists of very small individuals, unless it is cropped heavily. This problem has been circumvented by the use of an artificially developed sterile male hybrid of *Tilapia mossambica*, although, of course, regular restocking is necessary (Bay, 1968). There are other dangers in the introduction of *Tilapia* spp. to serve in a dual role. They were used in Uganda, being introduced into artificial fish ponds, but the species used were found to feed on chironomids rather than mosquitoes so that the latter became more troublesome.

According to Grist and Lever (1969) it is generally considered that fish in rice fields can increase the yield by 4–10 per cent. The fish keep down algal growth, restrain weeds, and destroy many insect larvae which could be injurious to the crop. Young rice can be dislodged by the fish and these should not, therefore, be tolerated in rice nurseries. In established paddy fields, however, any damage that they cause is compensated for by their yield of high-grade protein.

The disturbance of bottom mud and silt by carp introduced into pools and lakes can make the water unsuitable for a number of game fish.

There have been a number of successes with mosquito-eating fish, ranging from their use at the turn of the century in Havana during the yellow fever eradication campaign, to current WHO control campaigns. In one trial in rice paddy fields in California, where *Culex tarsalis* bred profusely, and where the owners spent about one-third of their budget on mosquito control, fish proved promising. Some paddies were left as controls, others were stocked with *Gambusia affinis* at the rates of 200 and 1000 mature females per acre (half hectare). On later sampling the ratio of mosquito larvae in the three areas was 94:5:1.

Gambusia affinis is being used in the malaria eradication programme in Iran, but in Rangoon and Bangkok the guppy, *Poecilia reticulata*, is more valuable for it breeds readily in polluted waters. Before the Second World War the native Burmese living in Rangoon suffered little from filariasis, although the disease was common enough among immigrant Chinese workers. During the fighting, and during the civil war later, the drainage system of the city was so badly damaged that *Culex pipiens fatigans*, a species which is common in polluted water, became firmly established, and the incidence of the disease rose sharply.

In Hawaii, where insectivorous fish were first imported and established many years ago, several species of fish are being used to combat *C.p. fatigans* and *Aedes vexans* breeding in a large swamp on Oahu island (Wright *et al.*, 1972).

In the past insectivorous fish have been, perhaps, used too enthusiastically and uncritically. This trend has been reinforced by the widespread popularity of many of the fish concerned as aquarium specimens, a popularity which has ensured that they are readily available in many parts of the world. With the increasing resistance of adult mosquitoes to insecticides it has become necessary to turn to larval control again. There is, however, a reluctance to use chemical larvicides in large quantities, partly because of the danger of environmental pollution, and partly because this would merely aggravate the resistance problem already presented by the adults. It is likely, therefore, that larvivorous fish will find a place in the biological and integrated control of mosquitoes, but it is hoped that their introduction will be preceded by careful ecological studies.

Gerberich (1971) has published a bibliography key word index on the control of mosquitoes by the use of fish, covering the years 1901 to 1968.

7

Biological control agents: pathogens

If Louis Pasteur is to be regarded as the founder of modern microbiology, then insect pathology must be considered to be one of the earliest branches of this science, for much of his early biological work was on the diseases of silkworms. In a closely allied field, the resistance of organisms to infection, Metchnikoff made his pioneering observations of phagocytosis in crustacean haemocytes.

The first experimental demonstration, that a micro-organism can cause an infectious disease in insects, precedes even Pasteur's work. This achievement of Agostino Bassi, working with the fungus which now bears his name, *Beauveria bassiana*, came in 1834, five years before Schoenlein's discovery of the two human mycoses, favus and thrush and almost thirty years earlier than Davaine's experimental transmission of anthrax.

An acquaintance with disease in bees and silkworms stretches back to antiquity. Aristotle described bee diseases in his *Historia Animalium*, and they are also mentioned by Vergil and Pliny. The Chinese attribute the invention of sericulture and silk-weaving to the lady Hsi-ling Shih, a subject of the Yellow Emperor who is said to have reigned from 2698 to 2598 B.C., so the observation of silkworm diseases may date from soon after this (Sarton, 1952).

Despite these illustrious beginnings insect pathology languished in comparison with human and vertebrate microbiology. Le Conte did suggest in 1873 that insect diseases should be studied with a view to using them for insect control, and the Russians used the fungus *Metarrhizium anisopliae* in 1879 in attempts to control a chafer in wheat fields. There were also some early trials by the Americans with diseases of the chinch bug, *Blissus leucopterus*. Apart from these early developments insect pathology is almost entirely a product of the present century.

Before turning to their use as pest control agents it is necessary to describe briefly the main groups of pathogens and the diseases they cause. Although they belong to widely differing groups of organisms there are certain common features in the ways in which they are used. For the moment it should be noted that sometimes they are used as self-perpetuating control agents, similarly to the use of parasites discussed earlier; and at other times simply as microbial insecticides to control the pest population present at the time.

Bacteria

The biologist accustomed to the classification of plants and animals finds bacterial taxonomy confusing. The bacteriologist must rely largely on physiological and biochemical criteria rather than morphological ones, and his concept of a species differs from that of a zoologist. The biologist who intends to work in this field of pest control should consult Lysenko's paper on the topic (Lysenko, 1963).

Steinhaus (1949) grouped the bacteria associated with insects as follows:

(a) Bacteria which are normally present in the insect's environment, but which are not entomogenous.
(b) Bacteria regularly or occasionally present in the healthy insect's alimentary canal.
(c) Non-spore forming bacteria which are usually facultative pathogens.
(d) Spore forming pathogens (facultative, obligate, or stabilized).
(e) Crystalliferous spore forming pathogens.

The bacteria which offer the most promise so far for microbial control of insect pests are spore formers. These produce endospores which allow their persistence outside the host, and which germinate in the gut of a susceptible host after ingestion. Non-spore forming bacteria seldom cause a pathogenic infection because they cannot penetrate through the mid-gut wall except when the insect is under severe stress from some other cause.

The milky diseases

A small number of *Bacillus* species attack chafer grubs, turning the blood into a milky liquid which gives the larvae an opaque appearance. The best-known examples are those which cause Type A and Type B diseases in the Japanese beetle, *Popillia japonica*. The Type A bacterium, *B. popilliae*, is a slender, non-motile rod-shaped organism which forms a refractile body at the time of sporulation. *B. lentimorbus*, the causative organism of the Type B disease is similar, except that the refractile body is absent.

Spore powder preparations of these bacteria have been used successfully in the eastern United States for the control of the chafer in grassland. They were first used in Connecticut in the early 1940s where the beetle populations were greatly reduced, and have since remained at low levels. The pathogen penetrates the mid-gut cells after ingestion, and enters the haemocoel where, after multiplying for some time, it sporulates. The disease is slow in development, and although few larvae pupate they still cause some damage to the grass before dying.

These bacteria are not restricted to *P. japonica*, although they are confined to scarab hosts (Falcon, 1971). *B. popilliae* has a wide host range within the family, while the variety *B. lentimorbus* var. *australis* infects the Australian chafer *Sericesthis geminata* (*pruinosa*). The European cockchafer *Melolontha melolontha* suffers from a milky disease caused by *B. fribourgensis*.

Little is known of the nature of the refractile body or parasporal body in the milky disease organisms for, unlike that of the crystalliferous bacteria described later, it is retained tenaciously in the sporangium. Lüthy and Ettlinger (1967) have studied that of *B. fribourgensis*, using ultrasound to break up the sporangium, and they have shown it to be either a nucleic acid or a mixture of nucleic acids and proteins. They were not able to demonstrate by electron microscopy a microstructure similar to that of the *B. thuringiensis* parasporal body, although the geometric form of most of the bodies suggests a crystalline structure. Accounts of the milky disease bacteria are given by Dutky (1963), Steinhaus (1949, 1964), and Falcon (1971).

Crystalliferous bacteria

These bacteria form toxic protein crystals at the time of sporulation (parasporal bodies). These crystals are highly toxic to certain insects and can, indeed, be used as microbial insecticides against many Lepidoptera. They do not seem to be toxic to other organisms, apart from earthworms.

The taxonomy of the main crystalliferous species, *Bacillus thuringiensis*, is difficult, different varieties isolated at various times from different species of caterpillars having each been given its own specific name. Formerly, it was considered that there were at least two species, *B. thuringiensis* and *B. entomocidus*, each with more than one variety (Heimpel and Angus, 1963), but Heimpel (1967) has since grouped all crystalliferous bacteria as varieties of *B. thuringiensis*. Some authorities would include all these varieties within the species *B. cereus*, a common soil saprophyte which is sometimes entomogenous, but which is non-crystalliferous.

It is well established that the toxic crystal is responsible for the paralysis and ultimate death of infected caterpillars, for it can be used alone as an insecticide. It is known, however, that *B. thuringiensis* also produces other toxic substances, namely lecithinase C, a 'thermostable exotoxin', an endotoxin, a 'labile exotoxin', and an unidentified enzyme which may not be toxic.

The crystal or parasporal body is called an 'endotoxin' as it is formed within the cell during the sporulation. This, according to Lysenko and Kučera (1971), is a misnomer as a true endotoxin is a part of the bacterial cell wall. Furthermore, to be active it must be released from the cell and then activated. These authors regard the crystal as a waste product of the cell's metabolism.

Norris (1971a) surveys the biosynthesis and physical structure of the crystal. It is bi-pyramidal with prominent surface striations, and is composed of protein sub-units which are probably rod shaped. No unusual amino acids were detected in the analysis of the protein, but silicon has been detected in the crystal. Possibly the crystal has a siliceous framework.

The toxic parasporal body is broken by the alkaline mid-gut contents of susceptible insects, and the activated toxic material then affects the permeability of the epithelium, allowing the alkaline contents of the mid-gut to leak into the haemocoele, raising the pH of the blood. In some insects the protein causes an exfoliation of the mid-gut epithelium, with a paralysis of the gut. There are three main syndromes which have been designated Types I, II, and III by Heimpel and Angus (1959). In Type I the mid-gut is paralysed shortly after ingestion, and this is followed by a general body paralysis a few hours later, when the blood pH is found to have risen by about 1 or 1·5 units. In Type II there is again a gut paralysis, but no general paralysis and no rise in blood pH. The insects die within a few days. Type III is known in *Anagasta kühniella*, and both the toxic protein and the spores appear to be necessary. There is no general paralysis, and the caterpillars die within a few days of ingesting the pathogen. The mode of action is clearly complex, with several physiological mechanisms involved. The substance is not, however, toxic to mammals, probably because of the low pH optima of the proteases first involved in mammalian digestion.

The toxic protein is toxic only to Lepidoptera; the thermostable exotoxin, which was first detected about 1960, is, however, toxic to species from several orders of insects (Bond and Boyce, 1971). The effects develop slowly, and are only manifested at moulting or metamorphosis. At high doses puparia are deformed, or larvae die during moulting. The substance almost certainly contains a nucleotide, and there is strong evidence that it interferes with nucleic acid metabolism and protein synthesis. It is possible that this toxin will have some value in pest control, but more data are required on its toxicity to vertebrates. It is toxic to representatives of the Lepidoptera, the Hymenoptera, the Coleoptera, and the Diptera. One variant of the species, BA-068, may be useful against mosquito larvae (Wright *et al.*, 1972).

The labile exotoxin was found in two, as yet, unrepeatable batches of *B. thuringiensis* preparation. The aqueous extract was toxic to sawfly larvae feeding on treated foliage.

Other bacteria

A number of other bacterial diseases are known and some have been used in biological control attempts. One of the earliest was a coliform type now known as *Cloaca cloaca* var. *acridiorum* which causes a septicaemia and dysentery in grasshoppers and locusts. *Bacillus cereus*, the non-crystalliferous form closely allied to *B. thuringiensis*, is known to be pathogenic to many insects, including the larch sawfly, *Pristiphora erichsonii*, the codling moth and the house-fly. It is also one of a number of species which forms a scum on the surface of water, killing mosquito larvae, such as *Aedes aegypti*, living there. A number of bacteria are believed to be directly pathogenic to mosquitoes; they are listed in Jenkins's review of the natural enemies of medically important arthropods (Jenkins, 1964).

Viruses

Insects may be associated with viruses in two ways, as hosts or as vectors. This section is restricted to those which are parasitic within insects, and have pathogenic effects.

Most of the pathogenic insect viruses are quite distinct from almost all other known viruses in that the virus particles are enclosed within protein crystals, capsules, or membranes. The particles (virions) themselves are, in general, similar to those of other viruses. Besides these inclusion viruses, as they are called, there is a relatively small number of viruses whose particles lie free within the cells of the host.

Although the nature of viruses was not understood until this century, virus diseases have long been known in insects. The jaundice disease of silkworms was described, poetically, in 1527, and, according to Smith (1967), this is the earliest known reference to such a disease in insects.

The inclusion viruses – polyhedral viruses

The particles of these viruses are embedded in a protein matrix which is usually of a polyhedral shape. The polyhedra can be discerned readily by light microscopy, but the virus particles within them can be visualized only by the aid of the electron microscope.

There are two classes of polyhedral viruses, namely nuclear and cytoplasmic. These names refer to the site of multiplication within the cells.

Nuclear polyhedral viruses usually affect the cells of the epidermis, fat body, blood and, rarely, the silk glands. Most have been described from the larvae of Lepidoptera, but they are also known from sawfly larvae (*Diprion*, *Neodiprion*, and *Gilpinia*), from the larvae of the crane fly *Tipula paludosa*, and possibly from some mosquitoes and lacewings (Bibliography in Jenkins, 1964; Laird, 1971b).

The polyhedra are extremely resistant to water but are easily disrupted by weak acids and alkalis. Many of them can retain their infectivity for twenty-five years or more. Their sizes vary, even within a single host individual, but usually lie between 0·5 and 15 μ in diameter. Each polyhedral body contains a few to about one hundred virus particles, generally scattered randomly in the matrix.

Caterpillars are usually infected through the mouth or cuticle, but trans-ovarial infection also occurs, especially when the virus is latent in the population for a number of generations. When the disease is active there is an incubation period of a few days after which the larva becomes more and more sluggish till finally it stops feeding and dies. The dead or moribund larvae hang by their prolegs from the twigs and branches and the now fragile skin ruptures easily to release the polyhedron-laden blood and decomposing tissues. There may also be changes in the appearance of the integument as, for example, in the jaundice disease of silkworms.

There is often also an interesting change in the behaviour of the infected caterpillars shortly before death. They climb to the highest place available to them, an action which, of course, helps in the spread of the virus after death. This phenomenon is so common in the outbreaks of the disease among forest caterpillars such as *Lymantria monacha* and *Porthetria dispar* that the German foresters have called the disease Wipfelkrankheit (tree-top sickness).

The general symptoms of nuclear polyhedroses in sawfly larvae resemble those in caterpillars, but only the nuclei of the mid-gut epithelial cells are invaded, although the skin finally becomes fragile.

The symptomatology of the *Tipula* nuclear polyhedral disease is, however, quite different, for the integument is not affected by the disease. The larvae become paler and paler, till, finally, they take on a chalky white appearance. The progress of the disease is usually slow but the larvae finally die and rupture to release the crescent-shaped 'polyhedra'. A NPV has recently been isolated from *Anopheles sollicitans* in the USA (Wright *et al.*, 1972).

The cytoplasmic polyhedral diseases differ from the nuclear ones not only in the site of multiplication of the virus, but also in the shape of the virus particles which are sub-spherical rather than rod shaped. Furthermore, the nucleic acid present in cytoplasmic polyhedroses appears to be RNA while that of the nuclear polyhedroses is DNA.

Although the polyhedra are proteinaceous they are somewhat more susceptible to water, which etches them, and more resistant to alkalis which reduces them to a perforated matrix in which the particles remain embedded.

The recognition of the difference between the cytoplasmic and the nuclear poly-hedroses came only in 1950 but so many have since been discovered that Smith (1963) believes that they may outnumber the nuclear types. The diseases appear to be restricted to Lepidoptera and a few Neuroptera (*Hemerobius*, *Chrysopa*).

The internal pathology is restricted to the alimentary canal, with the mid-gut region succumbing first. In general, therefore, the diseases are less spectacular than the nuclear polyhedroses whose effects are more general. Affected larvae tend to be smaller than healthy ones, and slower in their development. There may be colour changes in the later stages of the disease when polyhedra may be regurgitated or passed out with the faeces.

The inclusion viruses – granuloses
The first granulosis disease was described in 1926 from *Pieris brassicae* by Paillot, and all those discovered subsequently have been found in the larvae and pupae of

Lepidoptera. The groups most affected appear to be noctuid moths and *Pieris* spp.. both of which include a number of important pests.

Generally, the virus particle is enclosed within two membranes and the whole unit embedded in a protein crystal which may itself be surrounded by a third membrane. The whole entity is from 0·2 to 0·5 μ in size and thus much smaller than the typical polyhedron. Each capsule (as it is called) contains only one virus particle. In many other respects the granules or capsules resemble the polyhedra of the diseases discussed earlier.

The fat body of the attacked insect is the main target organ, but sometimes other organs and tissues, including the skin, may be involved. The larvae often show colour changes, the larvae of the codling moth, for example, becoming a deeper pink. The virus multiplication usually starts in the nucleus of infected cells but later continues within the cytoplasmic area, but possibly utilizing the products of nuclear break-down.

Eventually, the disease kills the insect, which is often left hanging head downwards from its support, or in a characteristic inverted 'V' position so that the disease. in its gross symptomatology at least, resembles those caused by nuclear polyhedroses.

Non-inclusion viruses

The known diseases caused by non-inclusion viruses are fewer than those caused by the inclusion types. One of the best-known examples is the Tipula Iridescent Virus (TIV) which occurs naturally in the crane fly, *Tipula paludosa* but which can be transmitted experimentally to other insects including Diptera, Lepidoptera. and Coleoptera. The particles attack the fat body of the infected larva and their multi-plication gives rise to a microcrystal orientation which is the cause of the iridescent appearance of the fluid within the body. A similar iridescent virus has been noted in the Australian scarab *Sericesthis pruinosa* (SIV) and this has been transmitted experimentally to the yellow fever mosquito, *Aedes aegypti*.

A few non-inclusion viruses affect the honey-bee, Acute and Chronic Bee Paralysis Viruses (ABPV and CBPV) in the adults and Sac Brood Virus (SBV) in the larvae. Among the pest species of insects there are a few which are killed by potentially useful non-inclusion viruses. The Wassersucht Virus or Watery Degeneration Virus attacks cockchafers and there are reports of a disease in the armyworm, *Cirphis unipunctata*. Possibly the most interesting of this group of viruses is one which occurs commonly in wild populations of *Drosophila* species but whose presence is usually unsuspected. The so-called *Drosophila* Sigma Virus makes the infected fly extremely susceptible to carbon dioxide so that the insect cannot recover from even short exposures to strong concentrations of the gas; this is the only known symptom. Infected females pass the virus on to their progeny. In some cases all the offspring are affected, in other cases only about one-quarter. The males also can transmit the disease in their sperm (Seecoff, 1968).

Smith discusses the non-inclusion viruses at some length in his text *Insect Viruses* (Smith, 1967), as does Vago (1968) more recently.

There are a few viruses known to affect mites. The citrus red mite, *Panonychus citri*, rapidly succumbs to a paralysing non-inclusion virus which deposits, as a metabolic by-product of the disease, characteristic crystals in the tissues. It is not clear how effective it would be as a control agent. It is rather unstable so cannot be applied in a spray; it can only be spread effectively by the release of infected mites

(Stairs, 1971). A similar disease has been found in *P. ulmi* and a third condition, possibly caused by a virus, has been observed in *Tetranychus* spp.

Viruses, especially polyhedroses, have been used for insect control on a fairly substantial scale. The greatest successes have been with various sawflies attacking forest trees. The nuclear polyhedrosis of the European sawfly, *Gilpinia* (*Diprion*) *hercyniae* was accidentally introduced into North America with a parasite from Europe. It caused a massive epizootic, firmly established itself, and has been an important control factor of this alien pest ever since. The virus can be obtained easily in large quantities. After experimental spraying on spruce trees carrying a virus-free population it continued to give excellent control for many years.

The nuclear polyhedrosis of the European pine sawfly, *Neodiprion sertifer*, has also proved to be an effective control agent, and is now being produced commercially by several private organizations in America. The virus is applied by hand sprayers and portable mist blowers for good short-term control. Small trees need frequent applications as the foliage does not retain enough of the virus to give persistent control. Stairs (1971) gives details of experiments with field populations of four other sawfly species.

In general, experiments with the viruses of Lepidoptera have not been so successful as those with sawflies. A granulosis virus of the codling moth, *Cydia pomonella*, appeared among larvae in Central and North America in the mid-1960s, and trials have been carried out in California which indicate that it might be of value for short-term control. It is not known if it will be possible to initiate long-term epizootics.

A nuclear polyhedrosis virus of the alfalfa caterpillar *Colias philodice eurytheme* promises to be effective against this pest in America. It also attacks other pierids, including the small cabbage white butterfly, *Pieris rapae*. This butterfly, which is known as the imported cabbage worm in North America, has been the subject of trials with granulosis viruses in various parts of the world. Possibly, only one virus is concerned. The results were promising, though since the concentrations varied from trial to trial, it is impossible to say if the virus (or viruses) will be equally effective in all areas. A nuclear polyhedrosis virus has been used successfully against another brassica pest, the cabbage looper, *Trichoplusia ni*. Other caterpillars which show promise of being controlled by viruses include various *Heliothis* species, the army-worm, *Pseudaletia unipunctata*, the cotton leaf worm, *Spodoptera litura*, the African armyworm, *S. exempta*, and the forest tent caterpillar, *Malacosoma disstria* (Stairs, 1971). Wright *et al.*, (1972) discuss possible control of mosquitoes by viruses.

Rickettsiae and Rickettsiae-like organisms

Rickettsiae are minute organisms which combine some of the features of viruses with characteristics of bacteria. They have also been described as bacteria with extremely fastidious tastes (Ormsbee, 1969). As far as is known they are all obligate parasites or, possibly, commensals of arthropods or vertebrates. One species, *Rochalimaea quintana*, the causative organism of trench fever in man, has been propagated on a cell-free blood medium. Most species, however, are intracellular parasites. Typhus is caused by a member of this group, *Rickettsia typhi*, and although it is generally thought of as a disease of man it also shortens the life of its louse vector.

Rickettsiae of the tribes Rickettsieae and Ehrlichieae are important pathogens of vertebrates, and often have arthropods as alternate hosts. The members of the tribe Wolbachieae do not show an alternation between arthropods and vertebrates, and species of the genera *Enterella* and *Rickettsiella* are highly pathogenic towards arthropods. The former attack the epithelial cells of the gut, while most species of the second genus develop within the host's fat body cells.

Various *Rickettsiella* species attack beetles, including the Japanese beetle, and the European cockchafer and allied species. Unfortunately, *R. melolonthae* and *R. grylli*, and possibly other members of the genus, can produce fatal infections in mammals, although none have been reported in those regions of Western Germany where *R. melolonthae* causes epizootics in cockchafers (Krieg, 1971a).

Fungi

There is a wide variety of insect–fungus associations, though not all of these relationships are pathogenic. The fungus, for example, may be living as a saprophyte or as a secondary invader, or it may be a parasite with little pathogenic effect. Nevertheless, there are a number of species which can cause widespread epizootics. In general, however, fungus infections are probably somewhat more dependent on weather conditions than are those which are caused by other disease organisms, for the usual route of entry is through the integument, and only rarely through the mouth.

There is great interest at the moment in the members of the aquatic genus *Coelomyces* (Order Chytridiomycetes) for these attack, in the main, the larvae of mosquitoes, black-flies (Simuliidae), and sand-flies (Phlebotominae). Although the rate of infection is not usually very high they could be of use in integrated control techniques and, furthermore, it has been suggested that they may have more impact when introduced into new areas.

It is difficult to diagnose the disease in its early stages without a microscope, but when the disease is well advanced the sporangia can be seen as yellow, orange, or brown masses within the haemocoele. Infected larvae sometimes survive to the adult stage, and the fungus has been observed in the ovaries. Although this is probably uncommon it is important in the dissemination of the disease.

Species of *Coelomyces* have been used in field trials for the control of mosquitoes. In a particularly interesting study Laird (1967) introduced resting spores from Singapore into a large number of larval habitats on one of the islands of the Pacific Tokelau group. An earlier survey had shown that the fungus was absent from this group. On a second island he placed dieldrin cement briquettes in the larval habitats. A third island was used as a control area. Two years later the fungus had become well established on the first island and its incidence was five to seven times higher than in Singapore. The level of control reached 30 per cent. The mosquito populations were reduced when compared with those of the control island. Muspratt (1963) had previously produced up to 100 per cent mortality in *Anopheles gambiae* populations in Zambia, using *C. indicus*. Unfortunately, it is still difficult to obtain the fungus in more than experimental quantities; none of the species have so far been cultured on an artificial medium (Laird, 1971a; Wright *et al.*, 1972).

A more familiar order of entomogenous fungi is the Entomophthorales for some of its members are common pathogens of house-flies. The adult flies are often seen

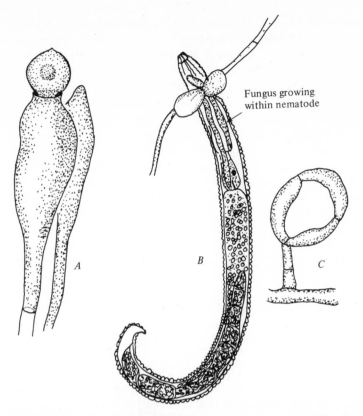

Fungus growing
within nematode

Fig. 7.1. Fungi-attacking pests. A: Conidiophores of *Entomophthora muscae*; **B and C: a 'lassoing' fungus, predatory upon soil nematodes. If a nematode passes through the ring of cells (C) these contract, trapping the fungus**

on window panes during damp weather, especially in autumn, their dead bodies surrounded by a halo of discharged spores (Fig. 7.1).

The nomenclature of the group has been confused in the past but there now seems to be a general agreement on two genera, *Entomophthora* (which now includes the former genus *Empusa*) and *Massospora*. The conidia of the *Entomophthora* species are extruded on conidiophores which protrude through the body wall of the dead host, and the conidia are discharged with some force (Fig. 7.1). The conidia of *Massospora* species, on the other hand, are formed within the host's body and released on its disintegration.

If the spore comes into contact with a susceptible host, and if the cuticle at least is damp enough, it puts out a germ tube which penetrates the integument. The mycelium eventually ramifies throughout the body, destroying the internal organs. A characteristic feature of many of the diseased insects is a restlessness which induces a tendency to climb shortly before death. This, of course, aids in the dissemination of the disease, just as it does with those virus diseases which bring about similar changes in behaviour.

It was once thought that the main victims of *Entomophthora* species were Diptera, but it is now known that many other kinds of insects are also attacked. Furthermore, one species, *E. coronata*, can cause disease in vertebrates, and thus cannot be used safely for microbial control. Members of this genus appear to be the only important pathogens of aphids which are capable of causing striking epizootics. About one dozen of the hundred or so species of the genus attack aphids. According to Hagen and van den Bosch (1968) they are not markedly host specific, some even attacking Diptera as well, but Gustafsson (1971) states that some species attack only certain aphids and that all species have specific host spectra. The latter author also observes that there is no definite correlation between the occurrence of these fungi and weather conditions. *E. fresenii*, for example, occurred as frequently, if not more frequently, in hot dry summers as in cold wet ones. It seems that high density of the host is important if the outbreak of the disease is to be significant, but epizootics have been noted at relatively low population densities of *Therioaphis trifolii*. Three species were introduced into California as part of the intensive campaign against the introduced spotted alfalfa aphid, *Therioaphis trifolii*. There was a slow spread from the centres of inoculation, with control following in many areas. Various species were also used against four aphid species on potatoes in Maine with outstanding success (Burges and Hussey, 1971b). Hagen and van den Bosch (1968) and Gustafsson (1971) cover the topic briefly in their reviews.

Other species have been recorded as pathogens of grasshoppers (*E. grylli*), the red-legged earth mite, *Halotydeus destructor* in Western Australia (*E. acaricida*), and the moth. *Plusia gamma*. In Canada the genus attacks insects in twenty different families, drawn from the Orthoptera, Homoptera, Coleoptera, Lepidoptera, Diptera, and Hymenoptera.

Epizootics, when they do occur, usually come towards the end of the season, but it may be possible to initiate them earlier, when they would be of more value, by placing cultures in the field, or by the release of artificially infected insects. The first method was used successfully against the spotted alfalfa aphid in California, and the second against the brown tail moth, *Nygmia phaeorrhoea* (*Euproctis chrysorrhoea*) in Canada. The pathogen was also successful against the brown tail moth in Maine and Massachusetts, where overwintering of the fungus occurred with a good spring control being achieved. Other species of the genus have given good control of the green apple bug, *Lygus communis*, in the Annapolis valley, the European apple sucker, *Psylla mali* in Nova Scotia, and the tiger moth, *Creatonotus gangis* in Taiwan.

The genus *Massospora* is recorded from cicadas and, so far, does not seem to be promising for microbial control.

The muscardine diseases are said to have taken their name from the French *muscadin*, a sweet whose texture is thought to resemble that of the insect killed by the fungi. The two most important Hyphomycetous fungi which cause these diseases are *Beauveria bassiana* (white muscardine disease) and *Metarrhizium anisopliae* (green muscardine disease). Related genera which are also sometimes entomogenous are *Isaria* and *Aspergillus*. All these fungi are facultative parasites and can be cultured easily on artificial media. The group has been reviewed by Madelin (1963) and various aspects of the biology and uses of its members were discussed more recently in an international symposium (van der Laan, 1967).

Most entomogenous Hyphomycetes are internal pathogens, and although superficial parasites also occur they appear to do little harm to their hosts. The usual route of entry is through the integument although infection may sometimes take place through the mouth. The hyphae of the fungus generally grow in the hypodermal region where they form pads which may become encased by the haemocytes of the host. After death the fungus continues to grow, till eventually the dead host is little more than a cuticle stuffed with a hard mass of the fungus. This mummy, with its enclosed so-called sclerotium, may persist for some time but if the conditions become humid sporulation occurs with the release of the conidia.

Most species have a wide host range, although there is marked specialization of fungal races within a species. The white muscardine disease has long been recognized as a pathogen of the silk worm. The causal organism, *Beauveria bassiana* also attacks a number of pests, including the codling moth and the European corn borer. It is also a significant organism in the natural regulation of the chinch bug *Blissus leucopterus* in North America.

The green muscardine fungus, *Metarrhizium anisopliae*, was first observed in 1879 by the Russian worker Metchnikoff infecting the larvae of a chafer, and was proposed by him as a microbial control agent. It has since been found to attack a large number of different kinds of insects, ranging over at least seven orders. Roberts (1967) has observed that several, possibly most, species of mosquitoes, including *Anopheles stephensi*, *Culex pipiens*, and *Aedes aegypti* are susceptible in the larval stage, and he has proposed its possible use for vector control. The viable spores were ingested by the larvae which often died before any fungal growth could be observed or before any invasion of the tissues took place. Roberts concluded therefore that the larvae succumbed to a toxin or toxins produced by the spores, and that these might be identical with the toxic Destruxin A and B which are known to be formed by the mycelium growing *in vitro*, and which have been shown to be toxic to both Lepidopterous larvae and, by Roberts, to mosquito larvae. Destruxin A has been shown to be a cyclic depsipeptide, and thus resembles in some ways the antibiotic Gramicidin S, and phalloidin, the toxic principle of the death cap fungus *Amanita phalloides*. Another toxin with some activity against mosquito larvae has been isolated from *B. bassiana*. It has been named 'beauvericin' and shown to be another depsipeptide (Lysenko and Kučera, 1971). *M. anisopliae* has been established in the West Indies where it has given some persistent control of the froghopper, *Tomaspis saccharina*. It is also applied in a spray for short-term control of the same pest. In Sweden the fungus has given reasonable control of *Agrotis segetum* in cold frames.

Although they have not been intensively tested as biological control agents the muscardine fungi are particularly promising because of their wide host range, their apparent harmlessness to vertebrates (apart from some allergic reactions), and the ease with which they can be cultured on artificial media. The related *Aspergillus* species, on the other hand, can cause severe diseases in vertebrates and are not likely to be used for microbial control.

This short survey has not exhausted the list of fungi which attack insects, although most of those remaining appear, at the moment, to hold less promise as microbial control agents. One genus, *Cordyceps*, cannot be ignored, for the species are often large and colourful. A hard sclerotium forms within the host's body and from this grows a long stem or stroma which bears, at the end, the fertile part. This stroma attains, in some species, a length of several centimetres. The larger varieties are highly valued in Chinese medicine and the fungi do, indeed, possess some interesting pharmaceutical properties. An antibiotic, cordycepin, for example, is produced which prevents the destruction of the dead host by bacterial rots. A *Cordyceps* species attacks scale insects, but better control has been achieved against these pests in several parts of the world by the imperfect fungus, *Cephalosporium lecanii*.

This discussion has only covered the pathogens of insects and other arthropods but possibly the relationships between the fungi and another group of animals, the nematodes, are biologically more interesting. As the pest species of nematodes are notoriously difficult to control by chemical methods the studies on such relationships are of great importance.

A number of fungi are predacious on nematodes which they capture by snares of various kinds. After the death of the trapped nematode fungal hyphae penetrate the body wall and extract the contents. Many of the fungi concerned had been known for many years before their predacious habit was discovered, for this behaviour is not obligatory, and the fungi rarely form their snares in the absence of nematodes. A chemical stimulus from nematodes causes a rapid production of the traps. These are of two kinds, adhesive snares and hyphal rings (Fig. 7.1). The rings act as traps into which the nematodes blunder and in which they are held fast till they die. In some species the rings contract, but not as a lasso as is often suggested, but by a swelling of the cells which form the ring.

Duddington has described the biology of these fungi, and that of certain endo-parasitic species, in a number of publications which the interested reader should consult (Duddington, 1957, 1962). He also describes a number of experiments in Hawaii, France, and England in which predacious fungi were tested as biological control agents. Although the fungi occur naturally in many habitats (for example, 82 records from 49 samples of English arable soil) a number of strains exist and the aggressiveness towards nematodes varies from strain to strain. It may be better, therefore, to inoculate the soil with strains of proved aggressiveness, although it must be recognized that this aggressiveness may disappear after a time. In the Hawaiian experiments, which were directed against a *Meloidogyne* species attacking pineapples, the best results were obtained when green manure was added to the soil. This led to a rapid increase in the numbers of free-living nematodes in the soil (but not of the *Meloidogyne*) which was quickly followed by a proliferation of the fungus. It cannot be definitely stated that the increased yields were due to the action of the fungi for much or all of it may have been the result of manuring.

In small-scale trials at Cambridge the addition of fungi and organic matter to the plots led to a great increase in the numbers of viable larvae and cysts of the potato root nematode, *Heterodera rostochiensis*. It has been suggested that during the early part of their growth the plants were protected by the fungi and stimulated to great root production by the manure, but that later the fungi lost their aggressiveness and allowed the nematodes to flourish on the augmented root system.

The French experiments, like the Hawaiian ones, were discontinued before really conclusive results were obtained, but they did give some evidence of the reduction of parasitism in sheep, the nematodes, of course, being killed during the free-living larval stage.

Protozoa

There are many protozoan insect associations, ranging from the commensal relationship of certain termites and cellulose digesting Mastigophora to the pathogenicity of the pébrine disease of silkworms. A few cause epizootics and many are present chronically in their host populations, but all are difficult or impossible to culture outside their hosts, and furthermore, the diseases which they cause are usually relatively slow in development. They have as yet found little use in biological control projects. They may be of value for inoculation methods, but it is not likely that they will be used for short-term control, as microbial pesticides.

The Class Sporozoa is particularly associated with insects which are frequently either hosts or vectors of the organisms. Unfortunately, as most zoological students know to their cost, these protozoa have complicated life cycles which have been well embellished with a confusing terminology. The reader should consult standard zoological or protozoon texts.

The sporozoons which are transmitted to vertebrates by insects, such as the *Plasmodium* species, show little pathogenicity to the vectors and, conversely, insect and invertebrate pathogens are usually harmless to vertebrates. Pathogenicity is associated with an invasion of the host's tissues; this familiar gregarines of the mid-gut lumen rarely cause their hosts any embarrassment.

Weiser and Briggs (1971) list the three sporozoon groups which contain insect parasites as the Sub-Order Haemosporina, the Order Neogregarinida (Schizo-gregarinida), and the Sub-Class Coccidia. Their stages in development include schizogony in which there is rapid asexual division which rapidly increases the number of individuals in the host, and sporogony which involves the production of gametes and spores. In the Sub-Order Haemosporina the schizogony does not occur in the insect hosts. As was stated above these are regarded primarily as parasites of vertebrates, and will not be considered further.

The neogregarines are considered by McLaughlin (1971) to be the only repre-sentatives of the Sub-Class Gregarinia which have potential for biological control. They infect various tissues of the host, especially the fat body, the Malpighian tubules, and the gut. Schizogony continues as long as there is suitable tissue left. Infection usually takes place through the mouth. Several species occur in both experimental and natural populations of stored-product insects, examples being *Lymphotropha tribolii* in *Tribolium castaneum* and *Mattesia trogodermae* in the khapra beetle, *Trogoderma granarium*. Other species are reported from the boll weevil, *Anthonomus grandis*, *Heliothis zea*, and from various other beetles, grasshoppers, and butterflies. The diseases take several weeks to develop fully and high population densities are necessary. They may be of value in controlling stored-product insects.

Coccidia are mainly parasites of vertebrates with only about five genera as true parasites of insects. In the genus *Adelina* the parasites emerge from the ingested cysts and penetrate the gut wall. A schizont stage develops and is spread to various tissues of the body in the blood. Those individuals which reach the fat body undergo further transformations with reproduction by both schizogony and sporogony. *A. tribolii* often causes epizootics in *Tribolium confusum*, the outbreaks occurring in cycles. *A. sericesthis* has been reported as a regulating agent of chafer grubs (*Sericesthis geminata* (*pruinosa*)) in Canberra. *Barrouxia* and *Chagasella* have similar life cycles but infest the gut epithelium, while *Legerella* is a parasite of the Malpighian tubules.

The Class Haplosporea are placed in the Sporozoa although in certain ways they resemble the Microsporidia (see below). *Haplosporidium* species are all host specific parasites which invade the host's tissues and can cause severe pathogenicity and death. One species, *H. typographi*, may be of value in the control of the bark beetle *Ips typographus*. McLaughlin (1971) suggests that it may be possible to infect the beetles at bait stations to which they might be attracted by pheromones. Other *Haplosporidium* species of interest are associated with aquatic invertebrates, and it is possible that *H. simulii* could be used for the control of the black-fly *Simulium venustum*.

Members of the Order Microsporidia are very common among insects, but their detection is difficult. They have, however, long been known as diseases of silkworms and honey-bees (pébrine and Nosema disease respectively) though even Pasteur failed to recognize the spores of the former as organisms. The diseases can be diagnosed in dissected insects, even some months after death, by the spores in the infected tissues. These spores contain long polar filaments (absent in haplosporidians) which are apparently used to attach the emerging organism to the gut wall after ingestion. A migratory stage, the planont, then migrates to the blood. The planonts are carried to various tissues of the body where two cycles of schizogony, followed by sexual reproduction and sporogony take place. The sexual reproduction is unusual in that it is autogamous, that is, there is a conjugation of two nuclei within a single cell.

Microsporidian infections are often associated with septicaemias caused by gut bacteria which, presumably, enter the haemocoele through the wounds made by the protozoa. Such a bacterial infection often inhibits the further development of the protozoa, obscuring the original cause of the sickness.

The biology and insect diseases caused by microsporidia have been reviewed by Weiser (1963, 1970) who gives useful bibliographies.

Species of *Nosema*, as already mentioned, attack silkworms and honey-bees, often causing severe losses. The species which attacks silkworms, namely *N. bombycis*, readily infects the pest species *Hyphantria cunea*, the fall webworm. Another potentially dangerous species is *N. cactoblastis* which attacked laboratory cultures of *Cactoblastis cactorum* in South Africa. Yet another species of *Nosema*, *N. carpocapsae*, is known from the codling moth in France. McLaughlin (1971) discusses the following microsporidian diseases which are potentially useful control agents: *Glugea pyraustae* in the European corn borer, *Ostrinia nubilalis*; *G. fumiferanae* in the spruce budworm, *Choristoneura fumiferana*; *G. mesnili* in *Pieris rapae*; *Nosema melolonthae* in the cockchafer *Melolontha melolontha* and *Octosporea muscaedomesticae* in various muscoid flies. Several microsporidians have been recorded from mosquito larvae which become inactive and sluggish, and which often die or fail to pupate. Other species infect the adult, curtailing egg production. *O. muscaedomesticae* has also been found to prevent female house-flies from laying eggs when they have been infected during their first meal after emergence from the pupae. Jenkins (1964) and Laird (1971b) give numerous references to microsporidial infections of public health insects. Wright *et al.* (1972) state that *Nosema stegomyiae* may be of use in the control of *Anopheles gambiae* in Nigeria.

The remaining protozoa can be disposed of quite briefly. A few diseases are caused by members of the Class Sarcodina, including the amoebic disease of honey-bees. The species concerned is *Malpighamoeba mellificae* and, true to its name, it attacks the cells of the Malpighian tubules by the insertion of pseudopodia, although it is actually situated extracellularly within the lumen. It is sometimes said to be an unimportant disease unless associated with some other disorder such as Nosema disease.

Malamoeba locustae also develops in the lumens of the tubules, but the mid-gut epithelium may also be attacked. The disease affects a wide range of grasshoppers and may be of some value in biological control. The infective stage is a cyst which is passed out with the faeces, and these have been used experimentally in the field to increase the incidence of the disease. Amoebic diseases are also known from fleas and other insects.

A number of ciliates (Class Ciliata) are of some importance in the natural regulation of chironomid and mosquito larvae. The most important genus appears to be *Tetrahymena*, the cysts of which are ingested by the larvae. Multiplication takes place within the body cavity and death often results, although some authors consider the ciliates to be merely facultative parasites. The small oval ciliates can often be seen through the transparent cuticle, especially in the anal gills and papillae.

Other ciliates are often associated with insects, but the relationship is often that of commensalism (for example, *Balantidium* spp. and *Nyctotherus* spp. in cockroaches).

The association of trypanosomatids and insects is notorious, for these members of the Class Mastigophora include the causal organisms of African sleeping sickness and Chaga's disease. It is less well known that the family also includes a large number of species which are restricted to insects and a few which are transmitted by insects to plants. *Leptomonas pyrrhocoris*, a species first described from *Pyrrhocoris apterus* has been transmitted experimentally to a number of other hosts and is sometimes pathogenic. Other *Leptomonas* species are known from other insects, including the dog flea *Ctenocephalides canis* and *Pulex* spp. Other trypanosomatids attack Diptera and Lepidoptera.

Lipa (1963) reviewed infections in insects caused by protozoa other than microsporidians and sporozoons.

Nematodes

Nematodes are frequently associated with insects, both internally and externally, but once again the association is often harmless or even beneficial to the insects. A number of species use insects as a mode of transport from one food source to another (phoresy) their mode of feeding being saprophagous. Many of the rhabditoid and oxyuroid species found living in the guts of insects are apparently commensals, although they may sometimes cause some small lesions. Insects are the vectors of a number of serious nematoid diseases of vertebrates.

True ectoparasitism is said to be displayed by only one genus, *Ektaphelenchus*, which attacks all stages of scolytid beetles, except the eggs, but internal parasitism, both facultative and obligate, is quite common.

Invasion of the host's body by nematodes may be by the ingestion of infective stages or by an active penetration through the cuticle or anus. The nematodes often do little damage but members of the family Mermithidae usually kill their hosts.

The mermithids are comparatively large nematodes which sometimes reach a length of 20 cm. All are parasites of invertebrates, mainly insects. The adults of the grasshopper parasite, *Agamermis decaudata*, live in small groups of one female and a few males in soil cavities. The infective juveniles, the newly hatched second stage larvae, emerge from the soil and climb wet vegetation where they search for grasshopper nymphs. After penetrating the cuticle they develop within the body cavity. The emergence of the fully grown nematode marks the death of the host. The final molt to the adult stage takes place outside the host. Most mermithids develop in a similar way but in the species *Mermis subnigrescens* the adult females lay their eggs on damp vegetation to be swallowed by grasshoppers. Otherwise the life cycle is similar to that of *A. decaudata*. Occasionally, these nematodes can attain high infection rates in grasshopper populations, and may be important in their natural regulation. Mermithids are also well-known parasites of ants, in which they often cause the development of intercaste characteristics. 'Mermithigates' of the genus *Pheidole*, for example, display a mixture of female, worker, and soldier features.

Mermithids are important parasites of public health insects (Culicidae, Simuliidae, Chironomidae), although their distribution is usually very patchy. Individual populations may be severely attacked, but most populations escape infection completely. Clearly, the mermithids could be distributed widely among uninfected populations if some suitable culturing process could be developed.

Most rhabditid nematodes associated with nematodes appear to cause their hosts little discomfort. The main exceptions are those belonging to the genus *Neoaplectana*. The first to be described was discovered by Glaser in 1929 within the dead larvae of *Popillia japonica*. Glaser soon developed a method of culturing the species, *N. glaseri*, on artificial media, and it was established experimentally in natural insect populations. In one trial against the larvae of Japanese beetle it brought high mortality. In a second trial with the same pest the population was reduced by about 40 per cent.

The closely related *N. carpocapsae* is also a form which can live saprophytically, but one strain, formerly known as DD-136, has developed a mutualistic relationship with a bacterium, *Achromobacter nematophilus*. Cells of this organism are carried by the infective stage nematodes within the gut, and are extruded through the anus when the nematode enters the body cavity of the host. The bacterium develops within the haemolymph of the insect which dies within a few days from a general septicaemia. The bacterium then serves as a food for the nematode which reproduces within the cadaver. This is preserved for some time, presumably by an antibiotic produced by the bacterium. Because of this mutualistic relationship the nematode is usually cultured for control and experimental purposes on the larvae of the wax moth, *Galleria mellonella*. An added advantage is that the nematode can be stored successfully for many months at low temperatures. Furthermore, it is not susceptible to many chemical pesticides, and thus it seems to be a promising candidate for biological and integrated control methods. It has a wide host range, which includes many pest Lepidoptera and weevils.

There have been many trials with the DD-136 strain against various Lepidoptera, Diptera, and Coleoptera. When suspensions of the nematode were sprayed on the trunks of apple trees a 60 per cent reduction of codling moth was achieved. Small reductions of Colorado beetle, *Leptinotarsa decemlineata*, cabbage root fly, *Erioischia brassicae*, European corn borer, *Ostrinia nubilalis*, and small cabbage white, *Pieris rapae* were obtained in other trials. Successful control is apparently dependent upon damp conditions. Details of other trials with this strain will be found in Poinar (1971).

Neoaplectana glaseri has not, apparently, been found again in wild populations of insects since Glaser's original discovery. It is not known, therefore, if this species is similarly associated with a bacterium although such relationships have been noted with other members of the genus.

Tylenchid nematodes are not normally considered to be associated with insect disease, but evidence is mounting that they are capable of regulating insect populations. *Oscinella frit* is often parasitized by *Howardula oscinellae* in Britain, and such frit flies are usually sterile. Similarly, a *Bradynema* species sterilizes a high proportion of the phorid *Megaselia halterata* in British mushroom houses in autumn. Nematodes of the genus *Parasitylenchus* apparently have a significant impact on populations of

Scolytus bark beetles in various parts of the world. Another potentially useful tylenchid is a species of *Heterotylenchus* infesting the face fly, *Musca autumnalis*.

Welch has reviewed the entomophilic nematodes (Welch, 1965) and discusses the infections they cause (Welch, 1963). Poinar's paper mentioned above discusses the use of nematodes for microbial control. He has also recently published a review of nematodes as facultative parasites of insects (Poinar, 1972). General surveys of insect pathogens will be found in the text by Steinhaus (1949), who may be regarded as the founder of modern insect pathology. A recent account of the 'state of the art' is given in the volume edited by Burges and Hussey (1971a).

Organisms associated with cockroaches as pathogens and commensals are surveyed in Roth and Willis (1960) and Jenkins's annotated bibliography on the parasites and pathogens of public health arthropods has been mentioned above. Laird (1971b) has provided a supplement to Jenkins's bibliography. Current accounts of research appear in the *Journal of Invertebrate Pathology*, formerly the *Journal of Insect Pathology*. Finally, it should be mentioned that the World Health Organization has established an International Reference Centre for the Diagnosis of Diseases of Vectors at the Ohio State University. Krieg (1971b) and Norris (1971b) give lists of key publications and guides to the literature. Weiser has published an illustrated *Atlas of Insect Diseases* which figures the chief pathogens of insects (Weiser, 1969).

Having surveyed the main pathogens of insects we can now examine some of the properties which characterize those which are efficient control agents, and also techniques of mass production and application.

Epizootiology

Epizootiology is the science which deals with the dynamics of disease in animal populations. It embraces the study of the spread and retreat of such diseases in space and time. The science has naturally borrowed heavily in the past from epidemiology but it is now rapidly forming concepts of its own, and accumulating a large body of facts and theory. It is not restricted to consideration of the pathogen and host alone, for environmental factors, biotic and abiotic, play a large part in the spread, inhibition, or waning of an epizootic.

The practical aim of most invertebrate epizootiologists is to seek ways of spreading virulent diseases among pest populations, although there are many practitioners who are more concerned with the control of diseases in useful insects.

An epizootic disease may be defined as a disease or a phase of a disease of animals, with high morbidity, and which is only irregularly present in a clinically recognizable form (Steinhaus, 1949). Morbidity refers to sickness, not necessarily to death, for which the term mortality is used. Some authors also use the terms pre-epizootic phase and post-epizootic phase for the periods between the severe outbreaks of the disease. An enzootic disease is one which is constantly present in a population of animals, but which has low incidence. The terms correspond to epidemic and endemic when human diseases are being considered.

Certain diseases can exist latently in insect hosts without producing any symptoms. This certainly occurs with many viruses and possibly with other kinds of pathogens also. The word 'latent' is best reserved to qualify 'infection', and the term 'latent virus' should be avoided. The phrase 'occult virus' is used to describe a virus present in a latent infection in which the virus particles cannot be detected. The terminology now used in the discussion of latent infections results from the decisions of two symposia held in 1958 in Wisconsin and in Stockholm, and the subject is discussed in Smith (1967).

Latent virus infections may be stimulated in various ways to produce overt infections. Steinhaus (1958a, b) has called the agents 'stressors' and he considers them to be any stimuli which tend to disrupt the homeostasis of an animal.

The chief stressors for latent virus infections are mainly physical factors, food quality, crowding, and other pathogens, either from other species or from the same species. As yet no universal stressor has been discovered but temperature and humidity seem to be the most important and common in the field (Franz, 1971).

Chemicals of various kinds can also act as stressors, and sublethal dosages of insecticides have been used as such in the field, but, as Franz (1971) points out, this has often been done with low rates of persistent organochlorines without regard to their possible effects on insect parasites and predators.

Bacillus thuringiensis applications have been found to induce the activation of latent virus infections in, for example, *Hyphantria cunea*, the fall webworm. The bacterium itself, of course, killed some of the caterpillars, especially the younger instars.

A virus of the same kind as that already present in the insect population may also act as a stressor (superinfection) so that in many cases virus applications are most effective when there is evidence that the disease is already present at a low infection rate in the target population (Franz, 1971).

If a stressor is acting over a large part of the area occupied by a latently infected population it can induce a widespread epizootic with numerous local foci of infection.

A pathogen which is capable of causing a severe epizootic is known as an epizootic strain. Two essential characteristics of such a strain are a high degree of infectivity and of virulence. These two terms are often confused but infectivity refers to the ease with which the pathogen spreads from one host individual to another when they come into contact, while virulence refers to the severity of the disease caused by the pathogen. The severity is usually expressed in terms of mortality, the LD 50 being the commonest measure.

The host population, the pathogen population, and the mode of transmission are the primary factors in the initiation and development of an epizootic, but these interact one with another and also with biotic and abiotic factors of the environment.

Persistence of the pathogen

The pathogen must be capable of surviving between outbreaks, either in the host's habitat or within the host population or associated organisms such as parasites or predators. The pathogens have, of course, evolved resistant stages in their life cycles which help to make this survival possible. The virions of polyhedral diseases, for example, persist within the resistant polyhedra; protozoa, fungi, and bacteria commonly form spores; nematodes can survive for long periods as ensheathed larvae or in the egg.

Such resistant stages can persist in the soil, on foliage, or in the faeces or cadaver of the host. The site depends upon the species of pathogen. Cadavers are of great importance in the persistence of many pathogens, such as the nematode DD-136, and it will be recalled that such cadavers often persist for many months by mummification or by antibiotic protection. The soil is a common site for there is usually enough moisture present to prevent the death of the resistant stages. Foliage is usually unsuitable because of the eroding action of the weather, but spores and so forth are often washed down into the soil or stubble and survive there. The milky disease organisms can survive for many months in the soil, as can microsporidia, spores of the white muscardine disease, virus polyhedra and nematodes. It is probable that facultative pathogens may also persist there saprophytically (Franz, 1971).

Persistence in the host population itself as a latent infection, or in an enzootic form, is common and important. Certain individual hosts also appear to function as carriers, never themselves succumbing to the disease, in much the same way as the human carrier of typhoid fever.

Pathogens may also survive in related alternate hosts or within parasites and predators of the normal host. The best documented example of persistence in an alternative host appears to be that of the microsporidian *Thelohania hyphantria* which kills overwintering pupae of the fall webworm, *Hyphantria cunea* but can survive in those of the brown tail moth, *Nygmia phaeorrhoea* (Weiser and Veber, 1957).

Various pathogens persist in the digestive tracts of parasitic animals, and are passed out with the faeces with no loss of infectivity and virulence. The nuclear polyhedrosis virus of the European pine sawfly, *Neodiprion sertifer*, has been shown to persist as it passes through the guts of the bug *Rhinocorus annulatus* and the European robin, *Erithacus annulatus* (Franz and Krieg, 1957), the cytoplasmic poly-hedrosis virus of the silkworm through the gut of the barnyard fowl, *Gallis g. domesticus*. The milky disease organisms and *Bacillus thuringiensis* will pass through various birds and small mammals (Hadley, 1948; Smirnoff and MacLeod, 1961). Such a passage of a pathogen through the gut of a non-susceptible predator aids, of course, in the dissemination of the pathogen in the host's habitat.

It seems probable that the persistence of a pathogen in the living host population or in associated animals is often more important than the persistence in the habitat of the host. It may be especially important when the host feeds on annual or deciduous plants.

The above discussion has considered only the natural persistence of pathogens. Various techniques have been developed to increase this persistence both in the field and in the laboratory. These will be described later.

The dispersal of the pathogen

The pathogen may be dispersed in the host's habitat by the host itself, by associated animals and by physical agencies such as wind and precipitation. They may also be spread by artificial means such as sprayers and dusters.

Many pathogens are age specific agents of mortality, attacking, usually, the larval or nymphal instars. Some infected insects do, however, survive to the adult stage and the movement of these adults is important in the dissemination of the pathogen, especially viruses. The younger stages, although usually less mobile, also play an important part. *Bacillus popilliae* may be spread by infected adult beetles and *Bacillus thuringiensis* var. *dendrolimus* (var. *sotto*) by the migration of the Siberian silk moth, *Dendrolimus sibiricus*. The granulosis virus of the large white butterfly, *Pieris brassicae*, is believed to have reached Britain from the European mainland during a mass migration of the adults in 1955 (Smith and Rivers, 1956). *Pieris rapae*, the small white butterfly, has been accidentally introduced into a number of countries and is also capable of migrating considerable distances. It has carried its granulosis virus to many of these new areas.

Louis Pasteur first demonstrated the transmission of an insect pathogen from generation to generation by infected eggs when studying the causal agent of pébrine, *Nosema bombycis*. This observation led Pasteur to stop the use of contaminated eggs and thus prevent the spread of the disease. The egg may carry the pathogen on the surface or within its substance. If the transmission takes place within the ovary the term used is 'trans-ovarian', while 'trans-ovum' is used when the transmission takes place outside the ovary.

Parasites are often efficient vectors of insect pathogens, transmission taking place during oviposition. Predators may also disseminate pathogens. These modes of dispersal have been reported for viruses, bacteria, protozoa, and fungi. Glaser considered that the polyhedrosis virus of the gypsy moth, *Porthetria dispar*, was introduced into the United States of America with one or other of the parasites imported for biological control (Franz, 1971).

Changes in the host's behaviour when diseased often aid in the spread of the pathogen. The tendency for virus and fungal infected larvae to climb shortly before death has been mentioned earlier. Among gregarious species diseased individuals will often wander away from the group and start new foci of infection.

Diseased individuals may infect others by regurgitation or defaecation of matter containing the pathogen, or by disruption of the body after death. Rivers (1967) noted that healthy larvae of butterflies and moths of various genera, including *Vanessa*, *Prodenia*, and *Mamestra* in captivity, and *Melanchra* in the wild, will drink the liquefied body contents of diseased larvae as soon as the skin ruptures. The behaviour was only noted with nuclear polyhedroses and granuloses. Rivers suggests that the stimulus is the moisture and the smell of concentrated food rather than true cannibalism. Cannibalism is, however, important in the dissemination of the disease when the habit is common in the host.

The main physical agencies which spread pathogens are streams of air and water, and rain. Irrigation is important in raising pathogens in the soil and surface litter to the foliage of growing plants. They can be spread further afield by flood waters, rivers, and streams. Air currents may spread pathogens to even greater distances, fungi having been recorded at a height of 11 000 m, bacteria and fungus spores at 6000 m, and yeasts and pollens at 5000 m and above (Rivers, 1967).

Rain will tend to wash pathogens downwards, although there will also be some lateral spread by splashing and wind. In Wipfelkrankheiten diseases, and others which produce similar changes in behaviour, rain is important in washing down the pathogen on to the foliage below which is being eaten by healthy larvae.

The host in epizootiology

The susceptibility of a host population depends upon the susceptibilities of the individuals comprising the population, and upon certain properties of the population itself. Thus a population could consist almost entirely of susceptibles, yet the population itself shows a resistance to the pathogen so that an epizootic does not occur.

Steinhaus (1949) classified the possible types of insects in a population subject to infection as follows:

(1) the typically diseased insect;
(2) the atypically diseased insect;

(3) the uninfected immune;

(4) the uninfected susceptible;

(5) the latently infected insect;

(6) the healthy carrier.

All six types have been shown to occur in one or another population but it is not known if they can all exist in a single one.

Resistance or immunity of individual insects to various pathogens has often been recorded, but little is known of mechanisms involved. There are also differences in the susceptibility of the various instars, for mortality caused by pathogens is markedly age specific. In general the earlier stages are the most susceptible to infection (maturation immunity), and some diseases are definitely restricted to one stage (e.g., foulbrood diseases affect only the larval stages of bees but, conversely, *Nosema* does not attack pre-adult bees).

The composition of the mid-gut fluids may prevent infection by ingested pathogens. Antiviral, bactericidal, bacteriostatic, and antifungal substances have been recorded from various species. The pH of the gut fluid is important in determining susceptibility to such pathogens as *Bacillus cereus* or the crystalliferous bacteria. A high pH (from about 9 to 10·4) confers resistance to the former but facilitates infection by the latter. Antagonistic micro-organisms present in the normal mid-gut microflora may also inhibit the development of bacterial pathogenicity. Antifungal substances in the waxy epicuticle shortly after moulting have been reported by a number of authors. Tanada (1967) suggests that fungal spore preparations used for microbial control should include abrasives and absorptive additives that will remove part of this waxy protective layer.

Both cellular and humoral immunities are known in insects but their relative importance is in dispute. Salt (1970) has recently reviewed the cellular defence reactions of insects, and this monograph is to be followed by another in the series which will cover humoral immunity.

Cellular immunity involves the ingestion of pathogens or foreign particles by phagocytic cells, the majority of which are found in the blood. In some cases the invading organisms or substances may be surrounded by a number of such cells which form an isolating capsule. Sometimes the invader survives; often it is killed. Frequently, some of the cells break down and deposit melanin on the foreign body. A third possibility is a combination of phagocytosis and encapsulation. This occurs, for example, with clumps of bacteria when some of the host's cells invade the spaces between the bacterial cells and begin to engulf the pathogen, while others form a capsule around the clump. This process is known as nodule formation.

Most kinds of micro-organisms can be destroyed by phagocytosis. Salt (1970) gives a representative list of viruses (polyhedroses, granuloses, irridescent viruses), bacteria, fungi, and protozoa. The presence of viruses in blood cells does not, however, prove that they have been engulfed by the cells, for many viruses attack the cells. Some micro-organisms, however, appear to be immune from phagocytosis, a well-documented example being *Pseudomonas aeruginosa* in grasshoppers.

It is difficult to judge the importance of phagocytosis in the actual defence of insects. It is known, however, that when the phagocytes are blocked with injected red blood corpuscles the insects succumb quickly to bacterial infections of the blood. Furthermore, the well-known sensitivity of first instar caterpillars to virus infections may be linked with the scarcity of phagocytic cells in their haemolymph.

Humoral activity is any immunity response due to factors in the haemolymph other than the blood cells. Such factors are well known in vertebrates; they are the antibodies which are formed in response to an antigenic stimulus. They are considered to be modified blood gamma globulin proteins which combine specifically with their corresponding antigens.

Gamma globulins apparently do not occur in insect blood, so that it is likely that if a similar mechanism does occur in insects then other substances must be involved. The study of mammalian antibodies has been facilitated by certain reactions which they can produce *in vitro* when a liquid containing the antibody is brought into contact with the correct antigen (Stephens, 1963). There appears to be no conclusive evidence that antibodies are formed by insects although humoral factors of other kinds, possibly analogous to the 'non-specific acquired immune factors' of vertebrates have been demonstrated, with activity against viruses and bacteria (Tanada, 1967).

As Salt (1970) points out, insects are at a great disadvantage with regard to massive infections by micro-organisms compared with vertebrates. The insect must respond by the production of a large number of additional cells whereas the vertebrate needs only to produce further molecules.

The biochemical study of the immunity of insects to disease organisms is clearly in a very confused state, but improvements in techniques may clarify the nature of humoral factors during the next few years.

Various environmental factors such as temperature, humidity, quality of food, and presence of other pathogens can alter an insect's susceptibility to a pathogen but the effect of the factor on the insect and the effect on the pathogen are often confounded, making interpretation difficult.

Immunity may be innate or acquired. Resistance can arise in laboratory stocks either by design or by accident. Pasteur, for example, produced a stock of silkworm larvae resistant to pébrine by careful selection, and thus helped to save the industry. David and Gardiner (1960) reported a resistance to the granulosis virus in his laboratory stocks of the large cabbage white butterfly, *Pieris brassicae*. The larvae were still resistant after thirty-six generations, spread over four years. The stock was equally resistant to its own granulosis virus and to that from another British stock, which suggests that resistance is involved rather than a loss of virulence of the virus.

House-flies have developed a resistance to *Bacillus thuringiensis* in laboratory trials, a fourteen-fold level being reached after fifty generations. The mechanism in this case is, however, probably very similar to normal insecticide resistance as the agent which kills the flies is the thermostable exotoxin known as the 'fly factor' (Harvey and Howell, 1965). Possibly, indeed probably, insects susceptible to the toxic crystal can also develop resistance to this agent if it is used intensively.

Acquired resistance is difficult to study in the field for it is not easy to distinguish between an innate resistance or immunity and that acquired earlier in the insect's life through subjection to a sublethal dose of the pathogen. In the laboratory it is possible to subject the insect to the pathogen or to a vaccine, and later to a challenging dose. In the tests that have been carried out it seems that immunity appears rapidly, often within a few hours, but disappears within a few days. The specificity of the immunity is also low. Active acquired immunity of this kind, and the passive acquired immunity derived from materials injected into the insect from other individuals subjected to the pathogen, are reviewed by Stephens (1963).

It is not known how important an inherited resistance, acquired through the subjection of the insects to the pathogen, is, but Burges (1971) does not believe that it has been of importance yet in the microbial control of insects. Resistance may be measured as a change in the LD 50, but Burges stresses the importance of studying the change in the complete log-dosage probit line. In many cases this simply becomes steeper, with the LD 99 remaining approximately the same. This is merely an indication that the more susceptible individuals have been weeded out from the population, but without the selection of extremely resistant forms, similar to those encountered in insecticide resistance. This would be indicated by a wholesale shift of the line to the right.

There is, however, no doubt that inherited resistance could arise in the field, but it must be remembered that the micro-organism itself can evolve and produce more effective strains. There is an alternative possibility. If the pathogen is so virulent that it kills the host so quickly that it endangers its own chances of survival, attenuated strains may be selected. This has apparently happened with the Myxoma virus of rabbits in Australia. The rabbits have acquired some resistance, but the virus has also become attenuated so now many rabbits survive attacks (see below).

The epidemiologist, recognizing that certain properties of the population at risk influence the development of an epidemic, speaks of 'herd infection' and 'herd immunity'. Steinhaus suggested the use of the terms 'population infection' and 'population immunity' when dealing with the interaction of an insect population and disease. Little is known of the factors involved in field populations, but population density and degree of aggregation are obviously important, for the more crowded the individuals the more likely is the spread of the disease. Crowding, as mentioned earlier, is also often a stressor in the induction of latent infections. The mobility of the individuals of the population, and the amount of immigration and emigration, will also affect the population susceptibility. A sedentary, highly aggregated population, with considerable distances between the groups, is likely to be resistant as a whole, although individual groups will be highly susceptible to the pathogen if it reaches them. It is clear that epizootiology is largely an ecological study. Just as the performance of an insecticide in the field cannot be judged from bioassays in the laboratory, the ability of a pathogen to cause an epizootic cannot be judged from its virulence to the host in laboratory trials.

The effect of environmental factors
Some of the factors which affect the epizootiology of insect pathogens have, of course, been mentioned above, but it is convenient to summarize them under a separate heading.

Physical factors   The influences of temperature and humidity have received more attention than those of other physical factors. There is little quantitative data available, but most workers agree that fungal diseases are most serious during periods of high humidity, and that temperature, sunlight, and wind have less effect on their incidence. This would be expected from their mode of infection: the germination of spores on the insect's integument. In laboratory conditions excess moisture often encourages the development of bacterial diseases.

There is disagreement about the effect of high humidity on virus diseases, although most workers recognize that rain helps to disperse the pathogens. Some workers have, however, associated virus outbreaks with wet and rainy weather, among populations of the gypsy moth, *Porthetria dispar*, and the cabbage looper, *Tricho-plusia ni*.

The free-living stages of nematodes are dependent upon a surface film of moisture for movement, and sometimes for survival, and thus require somewhat humid conditions. Paper and sacking collars have been used on the trunks of apple trees when using DD-136 nematode for the control of the codling moth. The collars serve as shelters for the larvae of the codling moth when they are about to pupate, but they also maintain a sufficiently high humidity for the nematodes.

Generally, the progress of a disease is accelerated by high temperatures, presumably because insects are poikilothermic animals, and at low temperatures there may be little or no development, or a reduced mortality rate. Furthermore, at low tempera-tures insect larvae may cease feeding, and most pathogens, apart from fungi, are ingested with the food. If the plant growth continues during this non-feeding phase the new growth will not be sufficiently protected with the applied pathogen (Franz, 1971). At higher temperatures insect virus diseases may not be as virulent as at lower temperatures, and this may make their use in tropical areas difficult.

Sunlight is an important factor, for it destroys many pathogens directly exposed to it. It is possible to use additives which screen out most of the ultra-violet light and thus protect the pathogen. Some workers apply pathogens at night-time to reduce this loss.

It should be stressed that the microclimatic conditions in which the insect is found are far more important than the macroclimatic conditions. The relative humidity at the height of a Stevenson screen or within a constant temperature/humidity chamber, may be very much lower than that of the air which is within a millimetre or two of the insect. The difference between the microclimatic and macroclimatic conditions probably accounts for the reports of the germination of fungal spores and invasion of insects at low ambient relative humidities.

Biotic factors The quality of the foliage on which the pathogen's host feeds can influence the development of the pathogen. The insect's susceptibility varies with its nutritional state, but the plant can also have a direct effect upon the pathogen for the foliage often contains bacteriostatic and bactericidal substances. The importance of these substances varies from plant species to species. The leaf sap of conifers, for example, is markedly inhibitory towards *Bacillus thuringiensis*.

Pathogens interact with other pathogens present in the host, and also with the microflora. The interaction may be mutualistic, synergistic, or antagonistic. Pathogens which damage the mid-gut epithelium of their hosts open a way to the haemocoele for gut bacteria and thus facilitate the development of a fatal septicaemia (Tanada, 1964, 1967).

Spatial and temporal development of an epizootic
A graph illustrating the progress of an epizootic in a restricted area may be obtained by plotting the number of deaths per unit of time against time. Needless to say, there are few such studies for insect epizootics, and we must turn to human epidemics to gain some insight into the process.

In a masterly study of the plague in London in 1665, Creighton gives the weekly bills of mortality for the disease (Creighton, 1891; summary in Watt, 1968). Over a period of thirty-five weeks the deaths rose rapidly to about 8000 per week in week 22, and then dropped rapidly to a few hundred per week at the end of the wave.

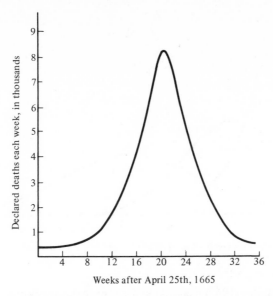

Fig. 7.2   The Black Death Epidemic, London, 1665. (Adapted from Watt, 1968)

The curve is roughly symmetrical (Fig. 7.2) and this is typical of many epizootics and epidemics. In other cases the curve may be skewed positively or negatively. The shape depends upon the virulence of the pathogen, the population susceptibility, environmental factors; in short, upon many or all of the influences discussed earlier in this chapter. The pre-epizootic phase of a milky disease outbreak, for example, is usually prolonged with the curve climbing slowly, because the pathogen is comparatively slow in action (Tanada, 1963).

The plague epidemic of 1665 was the third of that century in London. The series was unusual in that each killed approximately the same proportion of the population exposed to infection (13 to 15 per cent). Generally, the severity of a disease falls off with successive outbreaks because of the development of immunity within the population, and selection for reduced virulence in the pathogen (Watt, 1968). This fall in the severity has been noted, for example, in myxomatosis of rabbits in Australia, and in the series of plague outbreaks in fourteenth-century Europe, and in India during the 1950s.

Epidemics and epizootics are not restricted to one place, but spread from an outbreak centre or from a number of centres. The rate of spread depends upon many factors, some of which have been discussed earlier. The usual spatial–temporal progress of an epidemic or epizootic may be illustrated by the curves in Fig. 7.3, which show the incidence of the disease at successive times at various distances from the epicentre. It will be seen that at a given distance the incidence increases with time to a maximum, and then falls, and that the maximum incidence of the disease also decreases with distance from the epicentre. Curves resembling these idealized results were obtained by Bird and Burk (1961) when they disseminated a virus which controls the European spruce sawfly, *Diprion hercyniae*. The virus was sprayed on seven trees and the incidence of the disease measured over four years at various distances, up to approximately 900 m.

A number of workers have attempted to study epidemics and epizootics

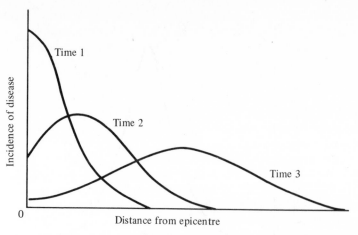

Fig. 7.3. Idealized character of an epidemic or epizootic wave. (Adapted from Watt, 1968)

mathematically. Bailey (1957) has treated the topic extensively and has also published a shorter account in his book *The Mathematical Approach to Biology and Medicine* (Bailey, 1967); Watt (1968) also discusses the subject.

The models used to simulate epizootics and epidemics resemble those of population dynamics in that they may be either deterministic or stochastic. An early deterministic model proposed by Kermack and McKendrick (1927) has been generalized by Watt (1968) to include immigration and emigration, variable infectivity rates, depending upon the weather, the interactions of various factors, and stochastic assumptions.

The original model envisaged a population of individuals more or less susceptible to the disease. One or more infected individuals enter the population and the disease spreads by contact infection. The infection spreads and the number of susceptible individuals falls till, eventually, the epidemic or epizootic fades away. The fate of infected individuals is either death or recovery with immunity. In either case the individual is said to be 'removed'. Unlike earlier models, it was not necessary for all the susceptibles to become infected before the epidemic or epizootic terminated. This was assumed to depend upon the relationship between the rates of infectivity, recovery, and death, and for each set of rates there is a critical population threshold density, below which the introduction of diseased individuals does not result in an epizootic or epidemic. Watt's modifications add realism to the model, but make the computations extremely burdensome. The work is lessened by the use of large digital computers with rapid memory access, and Watt outlines a suitable program using Fortran. The conversion to a stochastic model is achieved by Monte Carlo methods in which the mean values used for the variables in the deterministic model are replaced by suitable probability distributions. The computer then conducts sampling experiments which provide a simulation of the epidemic or epizootic.

Mass production of pathogens

However virulent a pathogen might be in the field it will have no value as a biological control agent if it cannot be produced easily in large quantities, except, possibly, for inoculation methods. Ignoffo (1967) estimated that 30 fungi, 33 viruses, 15 protozoa, and 14 bacteria had been mass produced and that, of these, 2 fungal species, 3 viruses, and 4 bacterial species were available for distribution.

The possible production pathways are, *in vivo*, by whole organism or tissue cell technologies, and, *in vitro*, by fermentation or synthetic culture technologies (Ignoffo, 1967).

Whole organism techniques

All obligate pathogens, by definition, must be cultured on living hosts, and even some facultative pathogens must be produced in this way, if enough of the infective stages are to be obtained. The method is usually expensive in time, space, and manpower and is therefore only used when no alternative method is possible – and then only if the pathogen is outstanding. It may be made less expensive if a suitable artificial diet for the mass rearing of the host can be found.

The living hosts may be healthy specimens collected in the field and artificially infected in the laboratory, specimens which are reared and infected in the laboratory, or diseased specimens collected in the field from naturally or artificially infected populations. In any case, the host may be the primary or the alternate host, but if the latter is used care must be taken that the pathogen does not lose its virulence towards the first. If this happens virulence may often be restored by passing the pathogen through a single generation of the primary host.

When the host is artificially infected with the pathogen the route of entry must be the one that occurs naturally in the field. If some other route is used there may be selection against strains which use the normal mode of entry.

Artificial and synthetic diets for the host insect simplify the mass production of their pathogens, and the viruses obtained from *Trichoplusia ni* and *Heliothis zea* reared in this way were as effective as viruses collected from the same species naturally infected in the field.

A number of useful pathogens have been produced in large quantities by these methods. Most of the control work against the Japanese beetle using *Bacillus popilliae* was carried out with pathogens cultured on healthy insects obtained in the field. *B. popilliae* is a facultative pathogen but does not sporulate readily on bacteriological media.

Rickettsiella popilliae, although not listed by Ignoffo, has also been mass produced on Japanese beetle larvae collected in the field. Probably, however, the technique will be most used in the production of insect viruses of various kinds. The nuclear polyhedrosis of *Heliothis* spp. has already been supplied commercially in the United States where it has been used on many thousands of acres of corn, sorghum, tomatoes, and cotton (Ignoffo, 1968).

Tissue culture techniques

Under this general heading are included methods of pathogen propagation using embryonated avian eggs, organ explants, dispersed cells, and established cell strains. Embryonated eggs have long been used for the culture of mammalian and avian viruses, often on a commercial scale, but there has been little success so far with insect viruses. Ignoffo suggests that some of the earlier attempts may have been unsatisfactory because the culturing was carried out at temperatures of 35 to 37 °C rather than at the usual temperatures of insect development, namely 25 to 30 °C. The embryonated egg is a suitable substrate for virus production because the chorioallantoic cavity is a large, irregular sack with an internal surface covered with a uniform layer of susceptible epithelial cells. Since a considerable amount of knowledge and technology has been built up using this approach in mammalian and avian virology, it seems worthwhile to carry out further work with insect viruses, especially since higher titres of viruses of vertebrates can be obtained in eggs than in the original hosts.

Organ culture, tissue culture, and dispersed cell culture are primary culturing techniques in which it is necessary to replenish the culture continuously with fresh host tissue if there is to be continuous survival, growth, and cell multiplication. There are, however, the so-called established cell strains available which are capable of continuous multiplication *in vitro*, without replenishment, and it is known that some obligate insect pathogens will multiply within such cells. Mass propagation may be feasible in the future, but there is a danger with all cell culture techniques that the pathogen will lose its ability to invade its natural host by the normal route, or that its virulence may be diminished in some way.

Fermentation techniques

If pathogens can be cultured on normal bacteriological media it is often possible to scale this up so that commercially useful quantities can be obtained. It is often possible to make the culture continuous, rather than a batch procedure. There are two approaches: surface culture, in which the pathogen grows on the surface of the medium; and submerged culture, in which the pathogen grows within the medium. The first technique naturally requires much more space than the second, so every attempt is made to use submerged culture methods.

Micro-organisms are now mass produced by both methods for a wide variety of purposes, ranging from the production of antibiotics to attempts to produce acceptable high-protein foods for famine areas. A number of universities have also established bio-engineering laboratories with pilot plants to investigate new techniques of batch and continuous culture of micro-organisms. There is thus sufficient background knowledge for these methods to be used successfully for the production of insect pathogens.

In a typical plant an incubating culture is introduced into a 500 gallon (2272 litres) jacketed vessel which is charged with a sterilized and cooled water–protein–carbohydrate mixture, and which is also aerated with filtered air. The mixture is then led into a 12 000 gallon (54 551 litres) aerated fermentation tank in which the actual multiplication of the pathogen takes place. After a suitable period the culture is screened and centrifuged, and the product assayed and standardized.

Krassilstschik used the surface culture technique in the 1880s to produce large quantities of *Beauveria bassiana* and it has since been used commercially for the production of *Bacillus thuringiensis* and *Metarrhizium anisopliae*. *B. thuringiensis* is now also produced by the submerged technique. Unfortunately, fungi do not sporulate readily when submerged, but it may be possible to produce strains that will do so. Ignoffo (1967) lists 25 fungi, 10 bacteria, and 3 protozoa, all insect pathogens, which have been propagated by fermentation methods.

Synthetic culture techniques

The final refinement in mass propagation of pathogens will be the use of chemically defined diets for then the manufacturer will have all the variables under his control and will be able to produce a more uniform product. The technique does not seem to have been applied successfully to the culture of insect pathogens as yet, but the technique seems to be at least possible.

A new terminology has arisen to describe the various kinds of culture media used for the propagation of micro-organisms, and since these terms are now used frequently in the microbiological literature (and, indeed, in publications describing the rearing of insects on artificial diets), they are defined here.

An holidic medium is one whose constituents, apart from sterile purified inert materials, are chemically defined before compounding. A meridic medium is an holidic one to which is added one or more substances of unknown structure or doubtful purity. In oligidic media most of the nutritional requirements are crude organic substances.

Axenic cultivation is the rearing of individuals of a single species on a non-living medium, not necessarily holidic. If there are a number of species, all known, present, the culture is said to be synxenic, but if an unknown species is present, the culture is xenic (Martignoni, 1964).

Reviews dealing with the mass production of insect pathogens by the various techniques outlined above will be found in Martignoni, 1964; Ignoffo, 1967 and 1968; Dulmage, 1967; Dulmage and Rhodes, 1971; Ignoffo and Hink, 1971.

Standardization of preparations

It is almost a truism to say that there are more variables to contend with in the mass production of a pathogen than there are in the manufacture of an insecticide such as DDT. The problem of the standardization of the final product is thus of great importance. A grower is accustomed to reasonably standard results from chemical insecticides, and he expects the same from a microbial insecticide. If he does not get such good results he regards the material as unreliable, and reverts to chemical insecticides. This happened to early preparations of *Bacillus thuringiensis*, and it is only in the last ten years that this material has come back into favour. Its early faults were aggravated by the failure to recognize the importance of the parasporal body in the toxic action of this microbe.

One of the earlier methods of standardization of preparations was to count the number of infective units (spores, polyhedra, etc.) in a unit volume of the preparation. This method has not been satisfactory, especially with those organisms, such as *B. thuringiensis*, which produce a number of toxins. Attempts are now being made to develop bioassay methods using standard strains of susceptible insects. The problem was thoroughly discussed at a symposium held at Wageningen in 1966 (Appendix to van der Laan, 1967). A more recent review is that of Burges and Thomson (1971).

The utilization of microbial control agents

Pathogens can be used for pest control in two ways. They can be applied to the pest population to give a rapid kill, but no lasting effect, or they may be introduced and colonized so that permanent control is achieved. The nature of the pest, the pathogen and the environment determine which method is chosen.

A pest that has a low threshold of economic damage must be controlled quickly so, in general, the short-term method would be chosen. This resembles the use of chemical insecticides and, indeed, the methods of application are very similar, although special care must be taken to ensure that the application technique does not reduce the virulence of the preparation. Formulation of the product is an important aspect. Water is often the diluent, and the product itself may simply be a suspension of cells or particles in water. Polyhedral bodies and endospores of various *Bacillus* species can be stored for long periods in this form, but it is not suitable for most pathogens. Some of these can, however, be stored as wettable powders with various fillers, wetting agents, etc., added to the dry cells. It may be possible to employ some pathogens in oil formulations for it is a common microbiological practice to store some micro-organisms under oil. Oil emulsion formulations can be used with various sprayers that have been developed for conventional chemical pesticides. Some organisms lend themselves to formulation in emulsions because the aerated broth method of culturing results in a preparation which is a wet paste. Immediate formulation as an emulsion eliminates the need to dry this paste. Finally, pathogens may be formulated as dry dusts. This has the advantage that the diluent dust is often abrasive, and helps the organism to penetrate the cuticle of the pest.

Various additives are used in these formulations. Even greater care is needed in the choice of these than is necessary for chemical pesticides. Emulsifiers that are used in the food and cosmetic industries have often proved to be suitable. Additives which are not usually found in chemical formulations are substances to maintain the correct osmotic pressure for the organism, and protectants which help to maintain the organism's viability after spraying.

Despite these difficulties microbial pesticides have certain advantages over chemical pesticides, some of them being as follows:

(1) Insect pathogens are usually harmless to other forms of life, so that toxic residues are not a problem. This is a particularly valuable property of the *B. thuringiensis* preparations which can be sprayed up to harvest.

(2) They are usually highly specific, so that beneficial insects escape injury. Even when the pathogen attacks several species, these species are often restricted to a single taxonomic group such as the Lepidoptera. Thus *B. thuringiensis* preparations are toxic to caterpillars but have no effects upon parasitic and predatory Hymenoptera and adult Diptera (Bailey, 1971).

(3) Pathogens are often compatible with chemical pesticides and therefore show promise for integrated control. In some cases there is a synergistic effect. There are exceptions: fungicides applied for the control of plant diseases often damage microbial agents (Benz, 1971).

(4) The apparently slow development of resistance in the host. It is still too early to say that resistance in the field will not develop, as microbial pathogens have not yet been used on a scale comparable with that of chemical pesticides, but laboratory trials, and experience with the diseases of silkworms and honey-bees, indicate that any such resistance would be slow in developing. In addition, pathogens, unlike chemical pesticides, can adapt to changes in their hosts but, conversely, a virulent strain can also lose its virulence (Burges, 1971).

(5) Even when a microbial pesticide is applied for short-term control there is often the possibility that the disease will persist within the residual population, and prevent resurgence.

There are, of course, some disadvantages:

(1) The timing of the application is difficult to judge, and is critical, because of

the time required for the disease to develop (incubation period), and because of the marked age-specificity of some pathogens.

(2) When a complex of pests has to be dealt with the specificity of most pathogens can be troublesome. A mixture of pathogens may have to be used to control a complex which might succumb to a single chemical insecticide.

(3) A pathogen is often difficult to store in a virulent and viable condition.

(4) The efficiency of many pathogens varies with the climatic conditions.

(5) The tendency for some diseases to leave insect-remains on the protected surface. This, for example, can make lucerne unpalatable to cattle, or brassicas difficult to sell unless carefully cleaned. On the other hand, it does help the pathogen to survive in the host's environment, and may make respraying for multivoltine pests unnecessary (Steinhaus, 1956; Cameron, 1967).

The necessity for an incubation period of possibly several days before death of the host, has led many biologists to doubt that microbial pesticides could be of value in the control of pests which cause appreciable damage at low population densities. The incubation period of a particular disease can, however, vary, and is often quite short when the infected insect is young. Thus, if the pathogen is applied as soon as possible after the attack starts, and if a thorough coverage is achieved, the chances of success are high. Furthermore, some pathogens, such as *B. thuringiensis*, interrupt feeding activity some time before the death of the larva so that, though still alive, it causes no damage. It is, however, sometimes difficult to convince a grower of this, so that there is sometimes resistance to the preparation on this score.

Good results have been obtained in many trials with *B. thuringiensis* preparations and the nuclear polyhedrosis virus against the corn earworm, *Heliothis zea*, on sweet corn and cotton, but in other trials the former was not effective enough against the larvae on sweet corn. The bacillus has also been used with mixed success against the European corn borer, *Ostrinia nubilalis*, and the codling moth. A less conventional use was to add the spore preparation to the food of birds and cows to control housefly larvae in the faeces. The insecticidal activity of the bacillus preparations resides in the toxic crystals, and the action is thus short-lived so that this pathogen cannot, as yet, be used for permanent control.

Ignoffo (1968) counted over thirty examples of the use of arthropod viruses for pest control. Although most kinds of viruses have been used, the commonest were the nuclear polyhedroses and the granuloses.

Viruses offer most promise for long-term control of pests with high thresholds of economic damage, but they can also be used for short-term control of low-threshold pests. Viruses are present in most environments but usually the epizootics which they cause occur after a considerable amount of damage has been done. In such cases it is worthwhile to apply the virus preparation early in the season and thus bring forward the epizootic. Spraying and dusting with nuclear polyhedrosis viruses has given good control of the forest sawflies, the alfalfa caterpillar, *Colias eurytheme*, the cabbage looper, *Trichoplusia ni*, and budworms, *Heliothis* spp. The *Pieris* granulosis virus gives excellent control of the cabbage caterpillars *Pieris rapae* and *P. brassicae*. A cytoplasmic polyhedrosis was used successfully against the pine processionary moth, *Thaumetopoea pityocampa*. Stairs (1972) summarizes studies on the impact of viruses on forest insects.

Apart from the early work with muscardine diseases there has been little success with fungal preparations. Czech workers have, however, claimed to have obtained

100 per cent mortality of Colorado beetle larvae, *Leptinotarsa decemlineata*, by the application of 5 per cent *Beauveria bassiana* dust. When the same species was tested against the chinch bug, *Blissus leucopterus*, in the USA, the results were disappointing, and the project was eventually abandoned. Careful surveys showed that the fungus was present naturally in most areas where the chinch bug occurred in Kansas, and was, in fact, the pest's chief enemy. The apparently rapid spread which occurred during the periodic epizootics was due to favourable conditions bringing the already present fungus into activity over large areas more or less simultaneously. Various workers have warned against the danger of extrapolating these findings from a restricted area to other districts and situations. More research on the ecology of the fungus and its host, they suggest, may lead to more success with fungal control agents in the future.

A pathogen applied to a pest population for long-term control is expected to act in a density-dependent manner. Ideally, it should regulate the host population at a low level, and respond rapidly to an increase in the host population density. In most ways, the pathogen acts like the other kinds of biological control agents mentioned earlier, but it differs from most of these in that it is usually distributed passively.

In short-term control the preparation can be applied with little or no regard for epizootiological principles, but in long-term control their importance is paramount.

The pathogen may be introduced into the host population in a number of ways. With virus diseases an aqueous suspension made from a few infected caterpillars can be sprayed over a fairly large area (five to ten caterpillars per acre has proved to be a useful rule of thumb). Other diseases may be introduced by placing infected but still active insects into the habitat, by the release of predators and parasites contaminated with the pathogen, or by the spraying or dusting of a mass-produced pathogen as in short-term control.

The success of a pathogen is judged from the occurrence of epizootics and enzootics with the eventual regulation of the pest at a lower population density. Infection does not have to bring immediate death for the pathogen to be of value. Some pathogens may have a debilitating effect on the pupae and adults, and they may sterilize the adults. These effects may be more important in the long-term control of the population than the immediate insecticidal effect on the larvae.

One desirable goal is the ability to predict outbreaks of a disease so that decisions may be made about augmentation of the pathogen, or the application of conventional insecticides. There have been few successes in such predictions so far, but techniques will certainly improve when methods of early detection of the disease in the population become reliable and rapid. Fluorescent antibody techniques can now be used for the detection of early stages of infection by nuclear polyhedrosis viruses, and biotic stressors can be applied to alfalfa caterpillars to reveal cytoplasmic polyhedrosis virus.

Viruses, fungi, and the milky-disease organisms have proved to be the most valuable long-term control agents so far. The nuclear polyhedrosis virus of the European pine sawfly, *Neodiprion sertifer*, sprayed on pine forests in Canada and the USA, has given excellent control of the pest, and that of the European spruce sawfly, *Diprion hercyniae*, has also given long-term control when introduced into disease-free populations (Stairs, 1972).

The efficiency of these diseases is apparently due to their high virulence and efficient means of dispersal and transmission, which is carried out mainly by infected females which transmit the viruses through their eggs, and by parasites and predators.

The strains of *Bacillus popilliae* and *B. lentimorbus* which have given long-term control of the Japanese beetle in parts of the eastern states of the USA are not highly virulent. These bacteria were successful because they persisted so well in their host's environment, the soil. The rate of dispersal is slow, but this is overcome by artificial spread. Highly virulent strains would not be successful because they kill their victims so rapidly that the bacterium is prevented from multiplying.

The spotted alfalfa aphid, *Therioaphis trifolii*, was accidentally introduced into the United States in 1954. It spread rapidly and, naturally, became the object of intensive research. This included the release of a number of fungi which spread gradually from the introduction areas and persisted despite periods of adverse weather. The complex of fungi has given a permanent control of the pest in many areas. A number of other fungi have been successfully established in various parts of the world and have contributed to the biological control of pests. *Entomophthora aulicae* was employed against the brown tail moth, *Nygmia phaeorrhoea*, in Massachusetts, just before the First World War, and related species against apple pests in Nova Scotia just after it.

There is yet another way of utilizing pathogens for insect pest control, although it does not seem to have been used on a field scale. This is the activation of a chronic infection by stressors or incitants. Laboratory manipulation of various factors such as the physical conditions during rearing, food quality, and infection with other viruses from the same or other host species, has led to the activation of various occult viruses, and it may be possible to adapt some of these techniques to field use. Sublethal doses of DDT, for example, have been found to activate viral and bacterial diseases of the gypsy moth, *Porthetria dispar*, and it is possible that less persistent insecticides or other chemicals will be found which will do the same, but which will not have adverse effects on beneficial insects. Russian workers have reported that one of their *B. thuringiensis* preparations provokes outbreaks of virus disease in Lepidoptera.

Interaction of parasites, predators, and pathogens

Most pest populations are attacked by a complex of natural enemies which interact one with another. Generally, the pathogens are more effective at the higher host densities, especially when these are above the 'escape' level. It is possible that a chemical insecticide might kill a higher proportion of the pest population than would a pathogen, but because the pathogen is usually compatible with the parasites and predators, and most insecticides are not, the pathogen may give better results in the long run.

Incompatabilities between pathogens, parasites, and predators can occur, as would be expected with such complex situations. The nuclear-polyhedrosis virus of the *Diprion hercyniae* sawfly, for example, so reduced some populations of the pest that certain of the parasites and predators no longer survived. Fortunately, other parasites and predators were introduced which were capable of surviving at the reduced host population density. Conversely, diseased insects are often taken more easily by predators than are healthy ones. This can either aid in the dissemination of the pathogen if this has multiplied and reached its resistant stage, or it may reduce the total infection. Infections by more than one pathogen can lead to a synergistic effect, or inhibition of one pathogen by the other.

Compatability with other methods of control

Many pathogens appear to be unaffected by chemical pesticides, and can be used in conjunction with them. This form of integrated control will be discussed in a later chapter.

Future trends in the control of insect pests by pathogens

The microbial control of insect pests has only got under way since about 1950, and it is obvious that much basic knowledge is yet to be won if the method is to be fully exploited. We do not know how to forecast epizootics, and this must be done before we can reliably start them at will. We have not yet discovered all the pathogens which attack insects, and foreign explorations, similar to those carried out in the search for predators and parasites, should bring to light a number of potentially useful micro-organisms. The yield may not be as rich as it has been with parasites and predators, for many pathogens have already been introduced accidentally with parasites and predators, or with host organisms.

A probably fruitful field of research will be the separation and study of the toxins produced by pathogens. This is in progress with the crystalliferous bacteria, and work of this kind should lead to the discovery of a number of highly specific insecticides.

Pathogens in vertebrate pest control

There is an understandable reluctance to use vertebrate pathogens for pest control. Man has suffered, and is suffering, from many diseases which are transmitted to him from mammals and birds; bacterial diseases such as bubonic plague, viral diseases such as various forms of encephalitis, and psittacosis, protozoon diseases such as African sleeping sickness and Chagas's disease, and parasitism caused by helminths such as the hydatid tape worm. The list is a long one, and contrasts sharply with that for diseases caused by arthropod pathogens of the kinds described earlier.

There is even disquiet about pathogens which have not been implicated in human disease, for these are often closely related to human pathogens, and it is feared that some comparatively small mutation could arise which would bestow pathogenicity towards man. The dangers are intensified when the pests live in close proximity to man, as is so often the case with mice and rats.

Louis Pasteur was the first biologist to attempt to control a mammalian pest by a microbial agent. A Madame Pommery, widow of the famous *vigneron*, had introduced rabbits into the walled enclosure above her cellars to provide sport for her grandchildren. The rabbits multiplied so quickly that they soon threatened her stocks of wine. Having tried, fruitlessly, to dissuade the rabbits from burrowing by feeding them with ample lucerne, she wrote in desparation to Pasteur. Pasteur dispatched his nephew Adrien Loir with cultures of *Pasteurella* which, when sprinkled on the lucerne, induced an epizootic among the pests. Meanwhile the Australian government had offered a prize of £25 000 for a solution for their rabbit problem. Graziers had been spreading rabbits suffering from a disorder known as Tintinallogy disease, named after the station in New South Wales where it had first been observed. This was, probably, a coccidiosis infection which was enzootic, but only occasionally epizootic.

Loir, and other French biologists, arrived in Australia and set to work in a small laboratory built for them on Rodd Island, in Sydney Harbour. The pathogen was never released on the mainland for it was believed that it was identical with, or closely related to, the causal organism of fowl cholera. This was already present among Australian poultry, but had failed to produce any epizootics among the numerous rabbits. The story of the bureaucratic manoeuvres, complete with contemporary quotations, has recently been retold by Rolls (1969), who also deals with the history of the rabbit in Australia.

The French were also leaders in the use of pathogens for the control of rodent pests. The German Loeffler had, in 1892, used cultures of *Salmonella typhimurium* against mice in Greece, and claimed success, though many observers did not agree. The year 1892 was also a vole plague year in France, as was the following year. During these outbreaks Danysz isolated a strain of a bacterium from diseased mice and voles which is now known as *Salmonella enteritidis* var. *danysz*. This organism, and that of Loeffler, cause diseases in mice and other rodents which are called mouse typhoid, for they are closely related to, but apparently distinct from, the causal organisms of human typhoid and paratyphoid, although they may cause human disease. Despite this relationship they were soon to be used, under the misnomer Rat Viruses, for rodent control.

Elton, in his work *Voles, Mice and Lemmings* (Elton, 1942), describes the early work of Danysz and his colleagues. The initial trials were carried out against a known number of rats in an enclosed sewer where, after several introductions of the culture, an almost complete kill was achieved. The original strain had not been very effective against rats, but its virulence was increased by several passages through white mice.

The severity of the vole plague soon forced Danysz to extend his trials to country areas. These trials were not as well conducted as the original sewer experiment. Control areas were not included, and few counts to estimate population density were made before baiting. Deaths from the disease were considered to be the consequence of the baiting but, as Elton points out, it was only a few months earlier that Danysz had isolated his organism from a wild population, and it is possible that the disease was already present in the treated populations. It must also be remembered that vole plagues are known to collapse suddenly as the result of illnesses caused by the stresses of overcrowding.

The 'viruses' were used widely in country areas both during this plague period and the following one in the first decade of this century. The contemporary discovery of the role of rats in bubonic plague epidemics led to the extension of the method to large cities.

In Denmark Bahr introduced his 'combined deratization procedure' in which epizootics were initiated with the product 'Ratin I', containing *S. enteritidis*. Three weeks later a preparation of red squills was used as a bait to complete the kill. This is probably one of the first examples of integrated control.

Eventually, cases of gastric enteritis in humans associated with the use of 'Ratin I' were reported from the USA, Denmark, Germany, and elsewhere, which led to the banning of 'Rat Viruses' in many countries. In passing, it is surprising to note that Danysz and his colleagues, skilful bacteriologists as they were, failed to notice any such side-effects during their massive rodent control campaigns. Gallons of the culture were mixed with bread or oats on the floors of school-houses, court-rooms, and public halls throughout France. Hundreds of people literally handled the material as they spread it in the fields, throwing a piece of infected bread with each step they took. Presumably the strains used were remarkably specific.

Despite their disadvantages the bacteria were used in Poland up to the outbreak of the war. Just after the war, in 1945 and 1946, Poland suffered great losses from rodents, and Danysz's organism was again put to use. Some fifty tonnes of the material were applied, but only after the local human populations had been inoculated against the disease. The rodents were reduced to a level at which chemical rodenticides could be used, and no cases of human illness were reported that could be linked with the campaign.

There are reports of a *Salmonella* strain from France which is pathogenic to rats, but harmless to humans. Strains of *typhimurium* – like *Salmonellas* have been used recently in the USSR where they gave 65–100 per cent mortality among small rodents at high population densities, and somewhat lower mortalities among rats. Obviously, the method is worth further consideration, especially since anticoagulant rodenticides may soon lose their value with the appearance of resistant populations, and because most of the alternative rodenticides have high toxicity to humans and domestic animals.

The best-known vertebrate pathogen which has been used for biological control is, of course, the myxoma virus, the cause of myxomatosis. As the disease and its impact have been well described elsewhere (Fenner and Ratcliffe, 1965; Thompson, 1961; Rolls, 1969), the topic will be discussed only briefly.

The disease was first noted among laboratory rabbits (*Oryctolagus cuniculus*) in Montevideo in 1896. There were further outbreaks among domestic rabbits in North and South America, which indicated that the disease was present in some native animal, but probably with no symptoms. In 1942 Aragao in Rio de Janeiro established that the disease occurred among the tapetis of Brazil, a wild rabbit belonging to a genus different from that of the introduced European rabbit, namely *Sylvilagus*. Infected tapetis sometimes bore a single tumour at the site of entry of the virus, but never the widespread tumours displayed by diseased European rabbits.

The virus infects other species of *Sylvilagus*, and sometimes the hares *Lepus europaeus* and *L. timidus*, but no hosts are known outside the Lagomorphs. In the European rabbit there is an incubation period of five to seven days, after which the symptoms begin to manifest themselves. These include swollen and permanently fused eyelids, and connective tissue tumours on various parts of the body. Death occurs eleven to eighteen days after infection.

The disease is transmitted by fleas and mosquitoes, the transmission being mechanical without any development of the virus within the insect. The disease can also be spread by contact.

The first releases in Australia were made in semi-arid pastoral land between 1936 and 1942, but the disease did not spread, probably because of predation of sick rabbits by foxes. In 1950 diseased rabbits were again released, this time at seven sites in the Murray valley. The disease spread rapidly from one of these, probably because of the far-ranging flights of infective mosquitoes.

Since 1952 the disease has been permanently established in Australia, but the rabbits are now 'coming back' rapidly. In 1953, although the mortality was still high, some rabbits were observed which recovered from the disease. In the same year Mykytowycz described an attenuated strain of the virus from a diseased rabbit caught in the Australian Capital Territory.

An attenuated virus which kills only slowly, or not at all, has, of course, a greater chance of being transmitted throughout a population than one which kills its host within a few days. The outbreaks were characterized by violent epizootics during the summer months, with high mortality. Attenuated mutant strains were more likely to survive during the winter months, and to spread rapidly in the spring.

Attempts have been made to introduce more virulent strains of the virus from France, and work is now being carried out on the possibility of introducing European rabbit fleas as vectors of the disease. The life cycle of this flea, *Spilopsyllus cuniculi*, is intimately bound up with that of its host, its breeding cycle being controlled by the sex hormones of its victim (Rothschild, 1967).

Myxomatosis spread rapidly throughout Western Europe after a French physician inoculated two rabbits on his estate in 1952. In France it is believed that mosquitoes were important vectors as many domestic rabbits were infected. The disease reached the British Isles in 1953, being first observed in Kent in the October of that year. Despite efforts to contain it the disease spread rapidly throughout England, Scotland, and Wales with profound ecological effects. The epizootiology differs markedly from that in Australia and France. The main vector is the rabbit flea, and mosquitoes are of little importance. Domestic rabbits are only rarely affected. The disease is much less seasonal than in Australia and France where summer epizootics are the rule.

Myxomatosis was supported by the government in Australia, but there was public and official opposition to the artificial spread of the disease in France and Britain. The French regard the rabbit as an important game animal, and the breeding of domestic rabbits is an important industry. The official policy since the war has been to regard the rabbit as a pest in the United Kingdom, and the commercial breeding of rabbits is only a small industry. The rabbit is, however, regarded with affection by the British public, and the disease is peculiarly horrible, with many of the victims dying in the open. It has been illegal to spread the disease knowingly so the appearance of attenuated strains has been of less concern than it has in Australia.

Although myxomatosis is a distressing disease it did have a massive effect on the rabbit populations of those countries in which it spread. Unfortunately, its effect is lessening so that rabbits are becoming troublesome again, and may eventually reach their former numbers. Every attempt should be made to use the advantage gained by the disease to keep the populations at low levels by other methods of control.

Pathogens do not appear to have been used for the control of bird pests, although they do suffer from a large variety of parasites and diseases (Rothschild and Clay, 1952). They may offer a solution, possibly the only solution, to what has become one of the main pest problems in Africa during the last twenty years. The quelea, *Quelea quelea*, a small weaver finch, is probably the most destructive and numerous bird in the world with a population in Africa which is believed to lie between 10^9 and 10^{11}. Huge flocks attack cereal crops in twenty-five or more African countries, and it is doubtful if the present control measures, aerial spraying, flame throwers, and explosives in the roosts, can ever do more than remove some of the biological surplus. While they may allow part of the cereal crop to be harvested before it is completely destroyed, the current control measures may be beneficial to the bird population as a whole.

A few diseases are known which cause epizootics among bird populations. Botulism is common among water-fowl, the bacteria growing anaerobically in rotting organic matter, and the disease sometimes kills shore-birds, game birds, and poultry. Starlings in the USA sometimes suffer from *Salmonella* infections, and although no widespread epizootics have been reported, the disease is potentially useful. It is doubtful, however, that public opinion would tolerate the widespread use of bird pathogens (Howard, 1967; Crook and Ward. 1968).

Pathogens for the control of plant diseases

The application of antibiotics, discussed in the third chapter, is probably the most convenient way of using bacterial antagonism. It is feasible, however, to use antagonistic micro-organisms themselves, and it has been suggested that this approach would be of value in developing countries. Most research has been directed towards the control of pathogens which are found in the soil, and if antagonistic strains of micro-organisms can be established in the soil, it may be possible to reduce the infection of crops planted subsequently.

Many workers doubt that antibiotics produced under natural conditions could exert a significant effect (Baker and Snyder, 1965; Baker, 1968). Nevertheless, some reductions in infection have been reported which may be due to antagonism. A strain of *Bacillus subtilis* has reduced the incidence of cotton wilt, *Verticillium* spp., and of *Erwinia amylovora*, the fire blight of pears. The actinomycete *Trichoderma* was found to reduce attacks by pathogenic bacteria and *Verticillium*, while a spray prepared from Bacterium strain A 180 reduced anthracnose of cucumber, *Colletotrichum lagenarium*. A liquid bacterial preparation has also been used to reduce grey *Botrytis* mould of strawberries.

One possible method of biological control of soil pathogens is to encourage other microbes which attack the pathogens and bring about their lysis, usually by means of enzymes such as chitinase, β-D-1,3-glucanase, and laminarinase. Work at Rothamsted Experimental Station and elsewhere has shown that the addition of chitin, in the form of prawn and shrimp 'shells', encourages the growth in the soil of bacteria which produce these enzymes, and that these inhibited the development of wilt disease in peas, *Fusarium albo-atrum*. Similar results have been obtained in Canada with *Rhizoctonia* stem rot of beans. Russian workers are also testing the effects of the addition of a variety of organic substances, and composts containing micro-organisms, on the development of beneficial micro-organisms in the soil. Baker (1968) reviews these and other possible biological control mechanisms in the soil.

Pathogens for the biological control of weeds

The application of plant pathogens for weed control has lagged far behind the use of insect pathogens, although the disastrous results of Dutch elm disease in America shows that introduced pathogens could be used for pest control if highly specific and stable forms could be found. It should also be remembered that much of the destruction of *Opuntia* species in Australia was caused by plant pathogens which invaded the tissues after the initial attack by *Cactoblastis cactorum*. The main species concerned are believed to be *Gloeosporium lunatum* and various bacterial soft rot organisms. It is thought that the moth would not have eradicated the cactus so quickly were it not for the secondary attack by these pathogens.

Wilson has reviewed the use of plant pathogens in weed control (Wilson, 1969). He discusses the following examples, among others.

Bathurst Burr, Xanthium spinosum, in Australia
This weed was found to be attacked by *Colletotrichum xanthii* in certain areas of
New South Wales. The disease was successfully established in other areas with
an annual rainfall of over twenty-five inches.

Crofton weed, Eupatorium adenophorum, in Australia
Crofton weed has been attacked by a leaf spot disease, *Cercospora eupatorii*, since
about 1952. It seems likely that this pathogen was accidentally introduced with a
gallfly, imported from Hawaii for the control of the plant. The gallfly established itself
for a time but was prevented from controlling the weed by certain native parasitic
Hymenoptera. The weed is not now spreading or increasing, being held in check by
the disease and a native cerambycid beetle.

Weed trees
An interesting but controversial use of a pathogen concerns the turning of oak
forest into more profitable soft-wood forest in certain parts of the United States of
America. French and Schroeder (1969) used the oak wilt fungus *Ceratocystis fagacearum* to kill the oak trees as they found it to be cheaper, more effective, and far more
specific than chemical controls.

Persimmon trees, *Diospyrus virginiana*, are troublesome weeds of pastures in
some areas in the USA. They are difficult to kill with chemicals or by mechanical
methods, but if they are cut off at ground level and the stumps inoculated with the
spores of *Cephalosporium diospyri* any regrowth is quickly killed. Under these
circumstances no spores are produced as they are by a standing tree that is attacked.
It seems, therefore, that the method is remarkably specific.

Inman (1967) has pointed out that most long-established plants and weeds have
developed resistance to the various native pathogens in their environment, but are
often susceptible to pathogens which have evolved recently in other countries on their
close relatives. He is currently importing weeds from America into Europe, under
quarantine restrictions, in order to inoculate them with suitable European pathogens.

Wilson discusses Inman's approach in more detail, and also other aspects such as
interaction of plant pathogens with chemical control, and various genetical considerations.

Higher plants in biological control

Insectivorous plants are well-known biological curiosities. They do not appear to
have a great potential in biological control, although *Utricularia vulgaris*, the greater
bladderwort, has been planted in a garden pool for the successful control of mosquito
larvae. A single plant destroyed 1800 small larvae.

Various aquatic plants have been used to cover the water surface to prevent the
development of mosquito larvae. The genera used include *Azolla, Brasenia, Lemna,
Nymphaea,* and *Wolffia. Brasenia schreberi*, the water-shield, a hardy plant which
grows in shallow ponds and lakes, inhibits the production of anopheline mosquito
larvae, even when the plants are growing sparsely. The white water lily, *Nymphaea
odorata*, may have a similar effect, but not when submerged aquatic plants are
present. Duckweed, *Lemna minor*, on the other hand, gives excellent control when
dense, but provides ideal cover for the mosquitoes when sparse. It is not likely

however, after the misfortunes caused by the invasion of water-hyacinth into various countries, that water plants will be introduced into new areas, but their growth could be usefully encouraged where they already occur. Jenkins's bibliography gives references to the control or regulation of mosquitoes by higher plants (Jenkins, 1964).

8

Autocidal methods of control

Possibly because man had previously been the only animal which indulged in the wholesale slaughter of its own species, he has found great satisfaction in manipulating pest insects so that they destroy their fellows. The employment of an organism, or some characteristic of an organism, to destroy other members of the same species, has been called the autocidal technique. The advantages of this method over chemical control will appear when the various methods which are in use, or which are potentially useful, are discussed. For the moment it is sufficient to say that the technique is highly specific, and should be far more economical than chemical control. Furthermore, the complete eradication of a pest from a wide area has been shown to be possible.

The sterile-male technique

Briefly, the wild population of the pest, which must reproduce bisexually if the technique is to be effective, is subject to large numbers of sterile males. If these are numerous enough most of the matings of the wild females prove fruitless, and the wild population begins to drop. If the pressure is maintained over several generations, and if there is no immigration of normal insects, the pest population disappears.

There are two possible ways of subjecting the wild population to the attentions of the sterile males. They may either be mass produced and sterilized in the laboratory, then released, or they may be sterilized in the field where they are members of the wild population. These are two quite different approaches with distinct techniques.

Although the first papers on the sterile-male technique were not published until 1955, the idea originated with E. F. Knipling in 1937. Having noted that the female screw-worm fly, *Cochliomyia hominovorax*, an important parasite of cattle, only mated once, he suggested that sterile males, with normal mating behaviour, could be used to eradicate isolated populations of the pest. One area was particularly promising, the south-eastern United States. Unfortunately, a suitable means of sterilizing the males was not available till much later, but Knipling pressed ahead with the construction of mathematical models.

Knipling (1955) suggested that a five-fold increase in population size in each generation would be typical of many insects during the exponential phase of growth. He considered the fate of an initial population of one million such insects in a given area when it is subjected to various kinds of control methods, or to none at all. The initial population he called the parental population, and the subsequent populations the first filial population (F_1), the second filial population (F_2), and so on, in the manner of the geneticists.

If the population is treated with an insecticide so that every member is subjected to it, and the insecticide gives 90 per cent kill, then the population would drop to 500 000 in F_1, and to 62 000 in F_4. As the same amount of insecticide would have to be used in each generation to give the same proportionate kill, it gives a diminishing return as the population density falls.

If, instead of using insecticides, 9 000 000 fully competitive sterile males are introduced, giving a 9 to 1 ratio of sterile insects to fertile ones, and a probability of a mating being fertile of 1 in 10. The expected number of fertile matings is thus $10^6/10$, and with a five-fold increase, the F_1 population size will be 500 000, as it was with the insecticidal treatment. If 9 000 000 sterile insects are again released the ratio of sterile insects to fertile ones becomes 18 to 1, the probability of a mating being fertile 1 in 19, and the expected number of fertile matings $5 \times 10^5/19$. The F_2 population will therefore contain 131 625 normal individuals. By F_4 the number of normal insects will drop to 50, the ratio will rise to 180 000 to 1, and the chance of a fertile mating will be almost zero. The pest will be eradicated from the area. The sterile-male technique thus becomes progressively more efficient as the wild population falls.

If, however, the number of sterile insects that is released is too small to cause a downward trend, then the population will either remain steady, or increase. As the population increases the efficiency of the method decreases, as Knipling shows in the following model, using the same hypothetical population. Starting again with 1 000 000 insects, an insecticide which gives 75 per cent kill will slow down the rate of population increase so that the F_1 contains 1 250 000 individuals. The uncontrolled population would contain 5 000 000. The introduction of 3 000 000 sterile insects would also give a F_1 population size of 1 250 000. If the treatments were continued in the same way in subsequent generations, the sterile method would become less and less effective, so that the natural population F_4 would consist of 9 294 115 individuals, whereas after the insecticidal treatments it would consist of only 2 441 405. Either method is superior to no control at all, but if we ignore the possible side-effects of the insecticidal method the sterile-male technique is clearly inferior.

It is, therefore, essential to overwhelm the natural population with so many sterile insects that a downward trend is certain from the start. This may be done by waiting until the natural population is at a low level, as happened with the sugar cane borer, *Diatraea saccharalis*, in Louisiana in the spring of 1962, or by reducing the natural population with an insecticide before the first sterile insects are released. If, for example, the population of 1 000 000 is reduced to 100 000 by the insecticide, and 900 000 sterile insects are released shortly afterwards (during the parental generation), and during the subsequent generations, the population will drop to 945 in F_3, and the number of reproducing insects to zero. The elegance of this approach lies in making use of insecticides when they are most economical, that is, when the population is high, and of sterile insects when they are most efficient, namely, when the population is low.

The choice and timing of insecticidal treatments are very important. A non-persistent insecticide is desirable, but not always essential. Frequently, the larval stages of the pest, and the adult stages, occupy different habitats, as is the case with mosquitoes, and it would be feasible to reduce the larval population with a persistent insecticide if no better one is available. Another possibility is to choose a strain of the pest which is highly resistant to insecticides for sterilization, as there would be no danger of the genetical factors for resistance being passed on to the progeny.

These mathematical models are greatly simplified, being deterministic rather than stochastic, although this restriction is comparatively unimportant when the populations are large, but it may have to be considered when the natural population becomes small. There are other simplifying assumptions which are probably more important, such as the full competitiveness in mating of the sterile insects, with no reduction in their life span, and monandry of the females. They serve, however, to illustrate the advantages of the method and, with suitable refinements, they can be used to plan actual control campaigns (Berryman, 1967).

Formerly, it was thought essential that a wild female should mate only once if the sterile-male technique is to be successful. This is not now considered to be so important, and tropical fruit flies (Tephritidae: Trypetidae), with polyandrous habits, have been reduced in various Pacific islands by the sterile insect release method. It does seem to be essential, however, that if polyandry does occur, the sterile males produce active sperm which is reasonably competitive with normal sperm, and which can fertilize the female's eggs. The sterilization technique should, in short, produce sperm, in normal quantities, which have dominant lethal effects which are manifested in the zygote.

It has been noted in some insects that a single mating is sufficient to fill the spermatheca of the female, and that even if further matings follow, the new sperm does not contribute to the female's sperm reserve. It is possible, in such cases, that sperm which is incapable of fertilizing eggs would be satisfactory. If, on the other hand, fertile sperm and infertile but competitive sperm are mixed in the spermatheca, and are randomly available for fertilization, the effect on the population as a whole would be the same as single matings.

The question whether or not the release should be restricted to males naturally arises in sterile insect release programmes. This is of great practical importance, as the released insects must be reared in extremely large numbers, and if the two sexes cannot be separated by some behavioural means it may be impossible to do it in any other way. The effects of introducing both sexes depends, of course, upon the mating behaviour of the species. If there are more matings available to the males in the whole population than there are to the females, then the reduction in reproductive potential is unaffected by the number of sterile females present. If not, then the mating habit of the female determines the effect that the sterile insects will have. If the female is polyandrous the biotic potential decreases with increasing numbers of sterile males or females, but if the female is monandrous the potential is independent of the number of sterile males, and decreases as the number of sterile females is increased (Proverbs, 1969).

Though monandry is no longer considered to be essential as long as the sterile sperm is competitive, the species and its wild population should have certain other properties for success to be achieved. The released adults should be harmless or at least the damage they are capable of causing should be outweighed by the advantages they bring. Adult screw-worm flies are virtually harmless as long as they are sterile, as are male mosquitoes. Sometimes, the act of sterilization reduces the danger posed by the released organism. Galun and Warburg (1968), for example, have shown that the soft tick, *Ornithodoros tholozani*, which normally feeds exclusively on blood, but only at long intervals, would rarely, if ever, take more than a single meal after irradiation. As adult males depend very little upon blood meals for survival and successful mating, this single meal ensures full sexual competitiveness in the sterile males.

The released males should be mobile enough to seek out wild mates, and to disperse sufficiently from the points of release. On the other hand, a species which can disperse over long distances may be very difficult to control because of the continual immigration of normal insects into the area, unless the treated population is very isolated. The first field studies undertaken with the screw-worm fly were unsuccessful because the site chosen, Sanibel Island in Florida, was too close to near-by infested areas.

Other desiderata are a means of sterilizing the insect without undue effect upon its mating behaviour and longevity, enough basic knowledge of the biology and ecology of the insect in the area in which it is to be controlled, and, perhaps most important of all, a means of mass-producing the insect. At the height of the Florida screw-worm campaign in 1958 more than 50 000 000 flies were reared, irradiated, and released each week, giving a total of 2 000 000 000 flies or more in eighteen months (Knipling, 1960). Smaller numbers are needed, of course, for the initial field trials, or for the control of incipient populations of introduced pests. It is essential, however, that any breeding programme is adaptable enough to be 'scaled-up' several-fold if the necessity arises. During the breeding the insects should be disturbed as little as possible, as excessive handling can have adverse effects on longevity and competitiveness.

The economics of the campaign must also be considered. If the complete eradication of a very harmful pest is the object, the costs may be high, yet acceptable; if annual treatment is envisaged then the mean annual costs of treatment plus uncontrolled damage must be less than those of the alternative method of control, unless this has some important disadvantage such as environmental pollution. Costs are naturally much more difficult to assess when the pest concerned is an important vector of human disease.

The technique in which a part of the population is sterilized in the field must be regarded as a completely different method from the introduction of, and overflooding with, sterilized insects. The method of sterilization itself is quite different as it is usually impossible to treat a substantial part of the field population by irradiation. Chemosterilants have to be used instead. Nor are the wild individuals subjected to the intense competition of the overflooding sterile insects which may force some of the wild population into unsuitable habitats.

The method resembles the use of insecticides in that the affected animals are dead to the population, as far as producing progeny is concerned. They continue to compete with fertile individuals, however, for such requisites as space and food, but there is an even more important bonus effect, as Knipling has called it. This is, of course, the failure to produce viable progeny when an infertile individual mates with a fertile one. Knipling constructed a model in which he made the usual assumptions of an initial population of 1 000 000 insects, capable of a five-fold increase. If each generation is exposed to a chemosterilant which sterilizes 90 per cent of the population, without reducing the competitiveness of the insects, the F_1 will contain 50 000 insects, the F_2 2500, and the probability of any insects appearing in the F_5 will be almost zero. This projection should be compared with the trends expected in a similar population subjected to an insecticide with a 90 per cent kill.

Unfortunately, most of the chemosterilants available at the moment are too dangerous to spray in the same way as insecticides, so they must be restricted in the field to bait stations. These can be charged with an attractant, and chemosterilant treated foods or surfaces. The successful use of this technique depends upon the discovery of a substance such as a pheromone which attracts members of the species from a considerable distance, although physical attractants are also worth considering. Pheromones and other attractants are discussed later.

Methods of sterilization and types of sterility

Partial or complete sterilization of organisms may be produced in several ways, some of which are artificial, while others occur naturally. Parasites, for example, often castrate their hosts, as is the case with Strepsipteron parasitization of certain Hymenoptera and Homoptera. The method which is the most useful to the economic biologist is the use of ionizing radiation or chemosterilants. Radiant energy, produced by photoflash equipment, has been shown recently to be a feasible technique, as are many of the methods of genetic manipulation of the pest population which have been proposed.

X-rays, neutrons and the alpha, beta, and gamma radiations of radioactive materials can all produce sterility. Beta rays, however, penetrate poorly as they are particulate (electrons), while alpha rays, which are also particulate (helium nuclei) have a short range, and are also readily absorbed by matter. X-rays and gamma rays, on the other hand, are electromagnetic radiations and can penetrate living matter easily. They have thus been used more than any other kind of radiation for sterilization. Neutrons may become important in the future.

X-rays are easily generated in the laboratory, so they have been much used in experimental work. X-ray machines cannot, however, be run for long periods continuously without damage to the expensive tubes so, despite the well-known difficulties of working with radioactive materials, gamma radiation from ^{60}Co sources is usually used in large-scale applications.

The unit of radiation which is most commonly used in these studies is the rad (or the kilorad, for convenience). A substance has absorbed a dose of one rad when the material has absorbed 100 ergs of energy per gram. The rad is roughly the same quantity as the older unit, the röntgen.

Doses of gamma radiation which have been shown to produce 97·5 per cent sterility or more in populations of various species are as follows:

European corn borer, *Ostrinia nubilalis*,	32 krad
Pink bollworm, *Pectinophora gossypiella*,	60 krad
Gypsy moth, *Porthetria dispar*,	20 krad
Yellow fever mosquito, *Aedes aegypti*,	6 krad
Melon fly, *Dacus curcubitae*,	10 krad
Mediterranean fruit fly, *Ceratitis capitata*,	10 krad
Sheep blowfly, *Lucilia sericata*,	6 krad
House-fly, *Musca domestica*,	2·85 krad
Screw-worm fly, *Cochliomyia hominovorax*,	2·5 krad
German cockroach, *Blattella germanica*,	3·2 krad
Boll weevil, *Anthonomus grandis*,	15 krad

The longevity and sexual vigour, when noted, varied from medium to good. These data are extracted from a table published by White (1967).

The difference between the Lepidoptera and the Diptera is striking, and a similar pattern is found with many of the chemosterilants. The difference is probably related to the method of attachment of the chromosomes to the spindle, and this will be discussed in more detail later.

Many chemicals reduce or completely inhibit the reproduction of insects and other organisms. Nevertheless, numerous compounds were tested in the United States between 1947 and 1950 without success, although in the same period British geneticists accidentally sterilized *Drosophila* spp. with chemical mutagens, but in ignorance of the American interest. Some later American successes led the United States Department of Agriculture to establish a screening programme in 1958, with house-flies and mosquitoes as the chief test insects. Now well over two hundred chemicals are known which will sterilize these or other insects. Some may be used as contact poisons, others as food additives, and many in both ways. While some chemicals are active against both sexes, many are more effective on one sex than the other. Sometimes, however, especially with female sterilants, the effects are not permanent.

The potency of chemosterilants may be expressed in terms of the ED 50 (effective dose 50), a measure analogous to the LD 50. Chemosterilants may also be insecticidal if given in large enough quantities, but as the aim is sterilization without any other reduction in fitness, it is essential that the ratio of the LD 50 to the ED 50 be large. Some insecticides, incidentally, produce sterility or reduced reproductive powers as a part of their poisoning syndrome, but such chemicals are, of course, of no value for the sterile-insect technique.

For convenience, chemosterilants are usually grouped into three major categories, namely alkylating agents, antimetabolites, and miscellaneous chemicals.

Alkylating agents are extremely active compounds which react readily with a number of substances. Their action is to replace a hydrogen atom with an alkyl group or a substituted alkyl group. Many of the most promising alkylating agents contain aziridinyl groups:

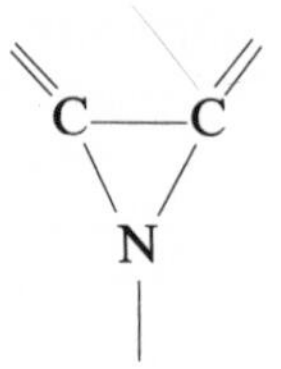

with apholate (1), TEPA (2), thio-TEPA (3), and metepa (4) as typical examples.

Most of these compounds are at least moderately soluble in water and many organic solvents, and all appear to be very unstable in acidic conditions. The exact mode of action at the molecular level is unknown, but it has been noted that DNA and lactic dehydrogenase synthesis do not occur in eggs laid by house-flies which have been sterilized by apholate.

The ED 50 of apholate to adult male house-flies by injection is 0·404 μg/fly, and that of metepa, 1·31 μg/fly. The normal LD 50 to ED 50 ratio of apholate in *Aedes aegypti* is 7·7, but some strains have developed resistance to sterilization, the ratio for the larvae falling to 1·6. The ratio for adult insects with TEPA is high, about 416, but the larval stages of insects are apparently much more sensitive to the chemical, presumably because of the high metabolic activity of the larval somatic tissues.

There are a number of other alkylating agents of promise, including those related

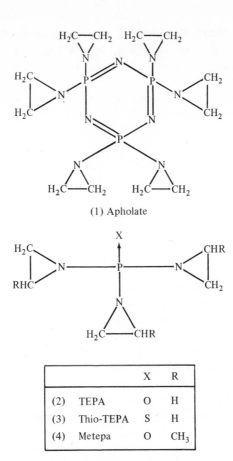

(1) Apholate

		X	R
(2)	TEPA	O	H
(3)	Thio-TEPA	S	H
(4)	Metepa	O	CH$_3$

Chemosterilants

to nitrogen mustard (5), and the sulphonic acid esters, such as busulfan (6). Nitrogen mustard is so-called because of its resemblance, structurally and pharmacologically, to the First World War weapon, mustard gas.

$$(ClCH_2CH_2)_2NHCH_3^+$$

(5) Nitrogen mustard

$$CH_3 \cdot SO_2 \cdot O \cdot (CH)_4 \cdot O \cdot SO_2 \cdot CH_3$$

(6) Busulfan

The nitrogen mustards and the aziridinyl (ethylenimine) alkylating agents are sometimes called radiomimetic chemicals because their gross physiological effects are similar to those induced by ionizing radiation. Several important differences have been found, however. Radiation, for example, does not cause crosslinking in DNA of living cells to any great extent, while this is marked with radiomimetic chemicals.

Alkylating agents are dangerous to mammals, especially during pregnancy. They are readily absorbed through the skin, and are mutagenic. This limits their application to closely controlled laboratory work, or to bait stations which are inaccessible to animals other than the insect. The mutagenic effect could also increase the risk of the rapid appearance of resistance in the pest insect or mite.

Antimetabolites are chemicals which are structurally similar to biologically active metabolites, and which may take their place in a biological reaction. Many of these have been tested for sterilizing activity but often only the females are affected, and even then not permanently.

An antimetabolite of uracil is 5-fluoro-uracil (7). It is an example of the sterilizing

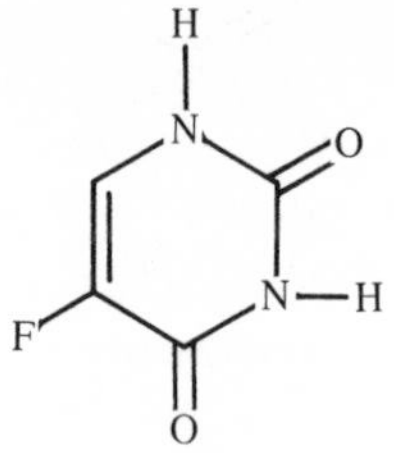

(7) 5-fluoro-uracil

compounds which are analogues of the purines and pyrimidines, several of which are important constituents of nucleic acids. It may compete with the normal pyrimidine for an enzyme which incorporates it into the nucleic acid, or its presence may prevent the synthesis of nucleotides by a feedback mechanism. Alternatively, it may take the place of its counterpart in RNA, thus disrupting its normal functioning.

The miscellaneous group contains a wide variety of compounds, and includes antibiotics such as methyl mitomycin and cycloheximide, the alkaloid colchicine and non-alkylating derivatives of s-triazine. Some of the organic tin compounds have sterilizing properties, as does DMF, a synthetic analogue of the juvenile hormone (Masner et al., 1968).

Summaries of work on the chemosterilants will be found in the publication edited by Labrecque and Smith (1968), and that written by Bořkovec (1966). These cover the chemistry and biochemistry of the chemicals, methods of testing and application, and the general theory of their use in sterile insect control programmes. They also display the wide variety of chemicals which show such activity. Despite this profusion it seems that most authorities consider that the alkylating aziridinyl compounds show the most promise at present.

If sterility is to be taken to mean the prevention of an F_1 individual from reaching maturity, it can be seen that this can happen in various ways. Treatment of the P_1 generation may result in the non-production of eggs, the non-hatching of eggs, the non-pupation of larvae, or the failure of adults to emerge from the pupae. There is a further possibility. The F_1 generation could survive and mate, but without producing viable progeny. If the damage caused by such an F_1 generation could be tolerated, the delay might be advantageous because of the natural increase in the number of sterile insects produced.

Sterilizing agents can affect the gametes of the treated insects in a number of different ways but the only effects which have been of practical use so far have been the production of dominant lethal factors in the sperm, and of infecundity in the females released at the same time as sterile males.

Ionizing radiation and some chemosterilants can produce dominant lethal mutations in sperm, by their actions on the chromosomes. At the doses used in practice the agents do not prevent the sperm from fertilizing eggs, but the resulting zygotes go through very few mitotic divisions before dying, because of increasing genetic imbalance in the cells.

The following discussion demands an understanding of the behaviour of chromosomes during normal cell divisions. The non-biologist will find adequate descriptions in most introductory biology textbooks.

While there is still much to be learned about the effects at the molecular level the general cytological effects of ionizing radiation and some radiomimetic chemicals are well known (Newcombe, 1971). A common form of damage is a break or breaks in one or more chromosomes. The resulting broken ends are 'sticky' and readily rejoin with other broken ends, but never with normal chromosome terminations. Usually, the union is restitutional with a reformation of the original chromosome, or with an exact interchange of parts between sister chromatids. These are the most

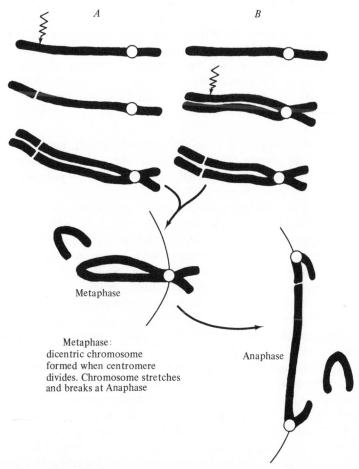

Metaphase

Metaphase:
dicentric chromosome
formed when centromere
divides. Chromosome stretches
and breaks at Anaphase

Anaphase

Fig. 8.1.   Non-restitutional breakage in a chromosome, leading to the unequal separation of genetic material during mitosis. Ionizing radiation breaks the chromosome. A: before replication; or B: after replication. The chromatids join to form one acentric segment, and one continuous chromatid with the centromere. At metaphase the acentric fragment does not attach to the spindle and will probably be lost. The chromatid attaches to the spindle but when the centromere divides to two centromeres which separate at anaphase, the chromatid is stretched, either preventing the complete separation of the daughter nuclei, or breaking unequally. (See text)

Autocidal Methods of Control   **257**

likely forms of union because of the proximity of the broken ends, but often the unions are non-restitutional. These are the more probable the longer the period between breaking and joining, and this can be prolonged by replacing air or oxygen with nitrogen immediately after treatment.

If the agent produces a single break through a chromosome (i.e., through two chromatids), there are two possible ways in which the chromosome can restitute, although one of them involves the exchange of material between sister chromatids.

A third kind of union, the non-restitutional one, would result in a dicentric chromosome (with two centromeres at metaphase), and one or more acentric fragments without centromeres. At metaphase the acentric fragments cannot attach to a spindle, and are usually lost to the daughter nuclei. A dicentric chromosome, on the other hand, attaches to the spindle, but, at anaphase, as the two centromeres migrate towards the poles, the chromosome is stretched between them and forms a bridge. This may either prevent the separation of the two daughter nuclei, or, more commonly, the chromosome breaks unequally so that the two daughter nuclei each receive a part (Fig. 8.1). The resulting 'sticky' ends may rejoin in the daughter nuclei to form new dicentric chromosomes, so that a *bridge–breakage–fusion–bridge cycle* occurs in a series of cell divisions. The cycle compounds the genetic imbalance between the numerous resulting cells, and, in a zygote, leads to an early embryonic death.

If the radiation or the chemical damages a chromosome after it has replicated it may happen that only one chromatid is broken. This will often restitute as the sister chromatid serves as a splint, but, if not, some chromosomal material is lost at the next cell division.

At higher doses multiple breaks in chromosomes become more common. If there are two breaks they may be both on the same side of the centromere (paracentric) or on opposite sides (pericentric). If the chromosomes do not restitute then deficiencies, inversions, or replications follow.

A broken chromosome that loses a part will produce one or more acentric fragments, and these will be lost at cell division. If the breaks are paracentric the two ends of the acentric portion may join to form a ring, and the terminal piece may make a union with the part with the centromere. With pericentric breaks the two 'sticky' ends of the centric portion may join one with the other to form a ring chromosome but in later divisions this often results in bridge formation.

In inversions the central part of the broken chromosome twists round, as it were, before joining with the terminal parts, so that the order of the genes in the central portion is reversed. Inverted chromosomes usually survive though synapsis is made more complicated.

A part of one chromatid may be incorporated in the sister chromatid, leaving the first with a deficiency and the second with a replication which may involve either a sequence of genes or a sequence of genes and the centromere. Some sections may be lost in that they become acentric fragments, but often all parts of the two original chromatids are retained in one or the other of the resulting centric chromosomes. At division, one of the daughter nuclei will contain excess genetic material (hyperploid nucleus) while the other will be deficient (hypoploid nucleus), although the parent nucleus has genes in the normal ratio (euploid nucleus). Chromosomes with replicated regions are likely to survive unless a large block has been replicated, or the replicated region contains a centromere so that the chromosome is dicentric.

If several chromosomes are broken in a single cell unions may occur which result in the transfer of material from one chromosome to a non-homologous one. Such translocations may be reciprocal, with an exchange of material between the two chromosomes; alternatively, the transfer may be one-sided. The translocation can result in two monocentric chromosomes (eucentric type) or in one dicentric chromosome with one or two acentric pieces (aneucentric type). When the translocation is aneucentric the chromosomes rarely survive, but eucentric chromosomes often do. Heterozygotes, however, are of low fertility as they produce gametes with deficiencies and duplications. Their possible use in pest control will be discussed below (Muller, 1954; White, 1961; Herskowitz, 1967).

Chromosome breaks can occur in both somatic cells or in germ line cells as the result of irradiation or chemical treatment (or, of course, by some spontaneous mechanism). Their effects depend upon the subsequent actions of the cells. Active somatic tissues which are undergoing cycles of mitosis, as in certain periods of larval life, are particularly susceptible. In the male germ line the diploid spermatogonia, which repeatedly divide mitotically, and the diploid spermatocytes, which divide meiotically, are all easily killed by sterilizing techniques, the early meiotic phase being particularly sensitive. Spermatids and spermatozoa (both haploid) are resistant, apart from the chromosomal breaks, and these show an effect only during the division of the zygote.

The time of spermatogenesis varies in the insects. In many species spermatogonia are present between hatching and the penultimate instar, spermatogonia and spermatocytes in the prepupal stage, and spermatogonia, spermatocytes, spermatids, and spermatozoa in the pupal and adult stages. In the emerged adult silkworm however, and probably in many other moths, only mature spermatozoa are present. The time taken for the development of mature sperm from spermatogonia also varies; in *Drosophila* the whole process can be completed in five or six days, whereas it takes ten days in the boll weevil, *Anthonomus grandis*, for an early primary spermatocyte to develop into a mature sperm. If it is desired to achieve a large number of dominant lethal factors in the sperm, with little somatic damage or reduction in sperm numbers, the time of treatment becomes critical, and this can be determined only by careful preparatory work (LaChance, North, and Klassen, 1968).

It was mentioned before that Lepidoptera, in general, require higher dosages of radiation or radiomimetic chemicals than the Diptera to produce comparable effects. This may be related to the mode of attachment of the chromosomes to the mitotic spindle. Diptera have monocentric chromosomes, each with a single centromere (normally) for attachment to the spindle; Lepidopteron chromosomes are apparently often holocentric, being attached over most of their length. It is not known if the attachment is due to numerous discrete centromeres, or to a diffuse activity, and it is probably incorrect, therefore, to refer to holocentric chromosomes as possessors of diffuse centromeres until this question has been resolved. The difference between the two types of chromosomes is seen most clearly at mitotic anaphase when the monocentric chromosomes are clearly seen to be bent as they are pulled towards the poles.

If a chromosome attaches at several places, or throughout its length, it becomes unlikely that chromosomal breakages will result in acentric fragments when restitution does not occur. It is also possible that single chromatid breaks at least will be even more likely to restitute than to make some other union. White (1970) lists the following orders of insects in which monocentric chromosomes occur: Neuroptera,

Coleoptera, Mecoptera, Diptera, Hymenoptera, and the orthopteroid orders. Holocentric chromosomes occur in the Hemiptera, and probably also in most Lepidoptera and Phthiraptera. If their presence is confirmed in the majority of moths it would go far to explaining the differences in sensitivity between the Diptera and the Lepidoptera.

Radiation and chemosterilants of various kinds can also produce point mutations in chromosomes, but these cannot be detected microscopically. It has not been established that point mutations can act as dominant lethal mutations. Chromosome breaks followed by non-restitutional unions are certainly sufficient to account for the dominant lethal mutation kind of sterility, as numerous chromosomal bridges can be detected in the dividing cells of affected gametes.

Sterilizing agents can also produce dominant lethal mutations in eggs, but this phenomenon must not be confused with the frequently observed infecundity. This is a lowered production of eggs. Dominant lethal mutations do not lower the egg production; they prevent their development after fertilization.

Oogonia, like spermatogonia, are very sensitive to radiation and to chemosterilants, and damage to them can result in permanent infecundity. Although the course of spermatogenesis and of oogenesis are similar in some respects, they do differ in that oogonia divide mitotically to give oocytes and nurse cells which have a nutritive function. Panoistic ovaries, that is to say, ovaries lacking nurse cells, do occur in the more primitive insect orders but the ovaries of the more advanced orders are meroistic, being well provided with such cells.

Nurse cells are extremely sensitive to sterilizing agents at certain stages of oogenesis for before they can service the growing oocyte they must undergo a process of endomitosis, during which the chromosomes divide several times, but without a division of the cell. Once a high degree of ploidy has been reached they become resistant, so that a dose sufficient to produce dominant lethal mutations in the ova has little or no further effect on the numbers of them produced.

Sterility in the males can be brought about in other ways, but these are of limited use in pest control programmes. Mistimed treatment, for example, can result in aspermia, or severely depleted sperm. Alternatively, sperm may be produced in normal amounts but be incapable of fertilizing ova because of the loss of motility or of the power to penetrate eggs. This type of sterility could be of value if the female is monandrous, and if the act of mating without fertilization is sufficient to satisfy the mating urge.

In general, the nuclei of sperm are much more sensitive to sterilizing agents than is their cytoplasm, so most workers agree that complete dominant lethality is obtained at lower dosages than sperm inactivation. Sterilizing agents can also have somatic effects which can alter the morphology, physiology, or behaviour of the insect in such a way that it becomes unable to mate. This is not sterility in the sense that has been discussed above, but it has been confused with it experimentally. It is not likely that an inability to mate would be of much value in insect-release programmes, but it could have an effect similar to that of a conventional insecticide when the agent is used to treat the wild population. Possibly, in release programmes, when the wild population is overflooded with such abnormal insects, the number of wild–wild matings would be reduced through the interference of the numerous treated individuals, and intra-specific competition would be much more severe, but this would be a very inefficient method of control compared with the release of truly sterile insects.

Applications of the sterile-male technique

It was fortunate that the first attempt to control a pest by the release of sterile males was made with the screw-worm fly. This in no way belittles Knipling's work on the pest. He and his colleagues showed remarkable insight when they brought together this method and this pest, for the screw-worm fly, as will be explained below, was well suited for the technique. Their success, however, led to the application of the method to other pests, sometimes uncritically. Unfortunately, the results have not always been satisfactory, and certainly none have been as outstanding as that obtained with the screw-worm fly in the United States. The initial outburst of work has led, however, to a greater understanding of the technique, and of its advantages and disadvantages, and to the accumulation of knowledge of the biology of various agricultural and medical pests. This will lead, no doubt, to further successes, possibly even to the eradication of some pests from wide areas.

The screw-worm eradication programmes

The larval screw-worm, *Cochliomyia hominovorax*, is an obligatory parasite of live-stock in the southern states of the USA, and in various Central and South American countries. In the United States it can regularly overwinter only in the southern parts of Arizona, California, New Mexico, Texas, and Florida. During the summer the infestation spreads northwards, partly aided by the shipment of carcases.

In 1937, while working in Texas, Knipling noted the suitability of the screw-worm fly for control by genetic means. An unsuccessful search for chemosterilants followed, but the discovery by Muller that ionizing radiation could produce dominant lethal mutations led to trials with X-rays and gamma rays.

It was established that the most efficient treatment was the irradiation of the pupae two days before the emergence of the adults. The males were found to be sterilized by a dose of 2·5 krad, and the females by 5 krad. Such females did produce a few non-fertile eggs and as these would make the field assessment more difficult, the pupae were irradiated at 7·5 krad for trials and eradication programmes. The sterility proved to be permanent and, in laboratory trials, the treated males were competitive with normal males.

The first field trials were carried out on the small island of Sanibel, near Fort Myers, Florida. The population was brought to vanishing point, but the island was not isolated enough to prevent the immigration of a few fertilized females from the mainland. The results were, nevertheless, so encouraging that trials were continued on the larger and more isolated island of Curaçao in the Netherlands Antilles. The release, every week, of 400 sterile males (together with sterile females) per square mile brought complete eradication by the fourth generation.

Funds were then raised to attempt the control of the pest in Florida. Work began on a rearing facility at Sebring which was to produce 50 000 000 flies weekly but an unexpected cold spell during the winter of 1957/8 so reduced the overwintering area of the fly that releases were started before the plant was completed. The research laboratory at Orlando had been producing 2 000 000 sterile flies weekly so its output was used, at the rate of 1 000 000 a week, from January 1958. The production was built up rapidly until it had reached 14 000 000 each week by July, when the Sebring plant was completed. The Orlando flies had not been sufficient to treat the whole of the infested area so they had been used to create a barrier across the peninsula of Florida, north of Orlando. This restricted the population to be treated by the Sebring production to the more southerly parts of the State. By 1959 the screw-worm had been eradicated from Florida, but a small outbreak occurred in 1961 at Pensacola which was checked by insecticides and local release of sterile flies.

After the success in Florida the work was extended to the south-west of the United States so that by 1964 outbreaks in Texas could be explained as the result of immigrations of fertilized females from areas in Mexico which were known to be infested. It has, however, proved possible to establish a barrier across the south-west, varying in width from 80 to 500 km, in which sterile males are released to patrol for immigrant females or their progeny. It has also been suggested that sterile males could be released so that prevailing winds carry them along the routes travelled by the females. It is hoped to push the population farther and farther southwards till the barrier is eventually established across the relatively narrow Isthmus of Panama. The annual cost of maintaining the present 1930-km-long barrier is about $5 000 000, and although this is only some one-twentieth of the costs of the screw-worm infestations before eradication, it could be reduced even more by using the more southerly barrier.

There is evidence that the flushing effect postulated by Monro (1966) has also been important during the screw-worm eradication campaigns. This is the forcing of a part of the wild population into sub-optimal habitats by the released insects. It also seems that many of the sterile males were abnormally aggressive and pursued the females till they died from exhaustion (Knipling, 1960; LaChance, Schmidt, and Bushland, 1967).

Fruit fly trials

The success of the sterile-male technique led to a search for other applications. Eradication is most useful when the pest is destructive even when the population density is low (for example, fruit-miners, certain vectors of plant and animal diseases), and which are difficult to control by chemical means. Fruit flies (Tephritidae; Trypetidae) are such insects and, furthermore, they can be bred in the laboratory in large numbers without great difficulty. In addition, they are widespread pests, troublesome in many countries.

Field trials with the Mediterranean fruit fly, *Ceratitis capitata*, started in a highland Hawaiian valley in 1959. Over the thirteen-month-long period nearly 19 000 000 sterile flies were released, and these achieved a reduction of the wild population to 10 per cent towards the end of the period. The area was not isolated enough to prevent the immigration of normal flies from other parts of the island, so that soon after the trial finished the wild population regained its normal density. It was considered that the experiment had shown the feasibility of using the technique for the eradication of this cosmopolitan pest from many areas.

Other trials were carried out in Costa Rica in 1963 and later in Israel on citrus. Encouraged by the results, a United Nations Special Fund Project began in 1965 in an attempt to eradicate the Mediterranean fruit fly from several Central American countries. Rhode and his colleagues (Rhode *et al.*, 1971) describe a trial in a large coffee and citrus area in Nicaragua in which forty million flies, sterilized by gamma irradiation, were released weekly from September 1968 to May 1969. A 2-km-wide buffer zone was established round the area by the aerial application of bait sprays. The numbers of eggs and pupae were reduced to less than 10 per cent of those found in adjacent control areas and, during the experimental period, the wild population increased only four-fold, compared with a fifteen- and a twenty-eight-fold increase in near-by control areas.

Promising trials have also been carried out with the melon fly, *Dacus cucurbitae*, on the small Pacific island of Rota, and with *Dacus tryoni*, the Queensland fruit fly, in New South Wales. In the Australian trials the flushing effect seems to have been of importance (Monro, 1966).

Sheep blowfly trials

Sheep blowflies, because of their similarities to the screw-worm fly, are obvious candidates for the sterile-male release technique. The eradication of the chief primary fly in Australia, *Lucilia cuprina*, would be exceedingly difficult and expensive because of the vast areas infested, but the control of sheep blowflies in Highland Britain might be possible. The most important species in Britain, *Lucilia sericata*, can reach a population density of about five adults to the acre in sheep country, which is a sufficiently low density for the method to work. A trial was carried out on Holy Island, off the Northumbrian coast, during 1956 and 1957. In the first year about 175 000 irradiated pupae were placed at three different sites on the two-square-mile island, and the adults allowed to emerge. Unfortunately, the proportion of adults which successfully emerged was low, partly, perhaps, as a result of predation, so releases were continued during the following year, but without any noticeable reduction of the wild population. The competitiveness of the males was reduced (one or two matings compared with the six or more of normal males) and it is also possible that the irradiated flies succumbed quickly to bad weather. Shipp (personal communication) has pointed out that larval blowflies are subjected to intense density-dependent mechanisms, so that a reduction in the number of eggs which are fertile may not be followed by a reduction in the number of adult flies emerging. The failure of this trial led to a loss of interest in the technique in Britain, but White (1967) has suggested that the 'phasing-out' of dieldrin and similar insecticides may revive interest, but with an emphasis on the use of chemosterilants.

House-fly trials

Musca domestica has been one of the most used test insects for the screening of chemosterilants so it was natural that several field trials were carried out with some of the successful chemicals. Among the earliest of these trials was one in which baits containing 0·5 per cent TEPA were scattered on a refuse dump, and on droppings in a poultry farm in Florida. The fly populations were greatly reduced within a few weeks and the remaining flies had a lowered fertility. Some Mexican trials using an apholate sugar bait and impregnated cords were less successful.

There have been some interesting experimental methods of dosing the flies and they may be of value in practice. Adult males and females emerging from the pupal medium were obliged to pass through expanded polystyrene foam which had been treated with 5 per cent TEPA. All the emerged flies proved to be sterile. In another trial female flies were 'booby-trapped' with pads of a chemosterilant which sterilized the males confined with them (Fye, 1968; Meifert, Morgan, and LaBrecque, 1967).

Field trials using irradiated flies have been less common. One was started in 1961 in a rural area of Italy, and during the four months of release the population density of wild females was greatly reduced. The area was not, however, isolated enough, and the radiation dosage was probably too low (2 krad instead of 2·85 krad or above). The populations increased rapidly soon after the trial stopped.

A composite trial with chemosterilants, insecticides, and the release of irradiated flies was carried out in 1965 on Grand Turk in the Bahamas. The chemosterilants, placed in the earth closets, reduced the fertility of the wild flies by only one half, so flies sterilized in Florida were flown in and released at the rate of 500000 to 1000000 each week. The island flies were difficult to colonize in the laboratory, and the Florida strain did not mate readily with the native females, but despite this 95 per cent sterility was eventually obtained. Unfortunately, the regular plane service from Florida closed down before eradication could be accomplished. Similar encouraging results were achieved with hempa (1 per cent) and dimethoate (1 g/m^2) in battery cage chicken houses in Florida (Meifert *et al.*, 1971).

Mosquito trials

The results of field trials with mosquitoes have been discouraging. Wright *et al.* (1972) consider that the logistical processes involved in the release of irradiated or chemically sterilized insects probably make it impractical for anopheline control. Other workers claim that a lack of basic knowledge about the power of dispersal of the males, and of mosquito behaviour in general, is partly to blame for the lack of success so far. Sterilizing techniques often reduce the mating competitiveness and longevity of the males and, furthermore, laboratory strains often differ from wild strains in their mating behaviour. Rajagopalan, Yasuno, and LaBrecque (1972) have reported recently, however, that in a trial with *Culex fatigans* chemically sterilized and irradiated males compared favourably with untreated males as far as dispersal and survival in the field is concerned.

Chemosterilants are effective against larval, pupal, and adult stages in the field but, as Weidhaas (1968), points out, the chemicals available at present are too toxic for broadcast application, and the lures which have been discovered so far do not attract a large enough proportion of most wild populations. Treating the larval breeding areas is also out of the question at the moment because relatively high concentrations – 10 p.p.m. or more – are needed to sterilize a worthwhile proportion of a wild mosquito population. Flood-water species would be particularly difficult to control because the eggs often remain dormant for many months so that it would be impossible to eradicate such a species over large areas.

Laboratory-bred strains of mosquitoes which are to be used for sterile-insect release programmes are unlikely to carry malarial parasites, pathogenic nematodes, or arboviruses, but many wild females do. It is thus important to know the effects of chemosterilants on such pathogens. There is evidence that *Brugia patei*, *Plasmodium gallinaceum*, and *P. cynomolgi bastianelli* are decreased in numbers by chemosterilants, although transmission can still occur. There remains the possibility, however, that such mutagenic agents could increase the pathogenicity of these organisms. Judson (1967) has indicated a related danger. Apholate can induce an increased biting rate in *Aedes aegypti* females. Chemosterilants may, therefore, increase the vector efficiency of this and other vectors during the course of an eradication programme.

Surprisingly high ratios of sterile males to fertile wild males have been attained in some trials without any marked effect on the population density of the wild mosquitoes. In field trials against semi-isolated populations of *Anopheles quadrimaculatus* and *Aedes aegypti* in Florida the ratio reached 941 to 1 in places, but no effects on the wild populations could be detected. In other trials involving irradiated males the wild populations were underestimated so that too few sterile males were released.

A small field trial in a California desert oasis, presumably well isolated, was more successful. Apholate was added to pot-holes on three occasions so that the concentration each time reached 75 p.p.m. Some time later no eggs of the mosquito, *Culex tarsalis*, could be found but three weeks after the trial finished the larval populations had reached their normal levels. It was suggested that the resurgence was due to the overlapping of the generations of the mosquito.

The limited dispersal of the released males may be a reason for many of the failures. In one small trial in Florida resting 'privies' and boxes were provided for *A. quadrimaculatus*. The inner surfaces were treated with TEPA at 5380 mg/m^2. The effects on the population were limited to small areas around the boxes.

Patterson and his colleagues (1970) have succeeded in eliminating a small island population of *Culex pipiens fatigans* by the sterile-male release technique. In 1968 males sterilized with TEPA were released on the island of Sea Horse Key (1·5 km long by one-eighth wide) which lies 3 km off the Florida coast. After eight weeks 85 per cent of the egg rafts collected were sterile, although the males were released at some distance from the breeding sites. The experiment was discontinued at this stage because of the coming of autumn, but work was resumed in 1969. As the males normally pupate before the females the authors collected only the first 30–40 per cent of the pupae, and further separation of the females was achieved by an apparatus described by Fay and Morlan (1959). This ensured that at least 98 per cent of the pupae retained were males. These pupae were then sterilized with 0·75 per cent TEPA and transported by car and boat to the island where they were allowed to emerge from pans kept under the Marine Laboratory. From the second generation onwards 8400 to 18 000 were released each day. By various sampling methods it was estimated that the ratio of sterile males to wild males ranged from 3:1 in the second generation to 100:1 in the fifth and sixth generations. The experiment lasted twelve weeks, and during the last two weeks no larvae could be found in the natural breeding sites of the mosquito. It is not expected, however, that the elimination will be permanent as the female mosquito is known to be able to migrate over two miles of sea and, furthermore, they are often found in the cabins of fishing boats which anchor off the island overnight. It is interesting to note that as the population density of the mosquito declined that of another species, *Culex nigripalpus*, increased.

It is clear that, despite this last success, it is going to be difficult, if not impossible, to eliminate large mosquito populations by the sterile-male technique. Other allied methods are, however, more promising, and these will be discussed later.

Tsetse fly trials
The last group of the Diptera to be considered, the tsetse flies, *Glossina* spp., would seem to be ideal candidates for control by this technique. Population densities are quite low, often no more than a few hundred a square mile, and such populations can often be reduced still more by insecticides although, because of the terrain, this is very expensive. Their reproductive potential is low, each female producing only a few fully grown larvae during her life. Immediately after extrusion these burrow into the ground to pupate so that immature losses are low, compared with those of insects with the usual types of life cycle. Knipling (1967) estimates that a doubling

of the population in a year (several generations) would be the maximum rate of increase achieved by most species in the field. Theoretically, therefore, a matching population of competitive sterile males would eradicate an isolated population quickly, if the numbers were kept high by sustained releases of sterile insects throughout the campaign. This could be integrated with preliminary insecticidal treatments.

The main difficulty is that tsetse flies cannot be easily reared in large numbers and, in fact, much of the experimental work has been carried out with pupae collected laboriously from the field. Techniques are being developed which should overcome this difficulty. Knipling has also suggested that it may be necessary to separate the females from the males so that they may be retained for breeding purposes.

A different approach is the replacement of the sterile-male release technique by the release of insects homozygous for a translocation, and it has been shown theoretically that this might be more economical as far as numbers released is concerned. This will be discussed in a later section.

Knipling has presented informal reports to the World Health Organization on the feasibility of controlling tsetse flies by the release of sterile males, and he summarizes his conclusions in his 1967 paper (Knipling, 1967). There is one factor which he mentions which is highly significant. Entomologists in Africa, spurred by the ravages caused by tsetse flies, have accumulated more details about the biology, ecology, population dynamics, and behaviour of these insects in the field, than have been acquired for any other species.

The eradication of the cockchafer, Melolontha vulgaris

The grubs of this beetle damage the roots of grasses and certain crops in Switzerland. The larval development is slow, and the flights of the adults occur in the spring of every third year, in any one district. The males emerge before the females and a nine-year study of the biology of the species enabled the investigators to predict the time of the flight accurately. Adult males were therefore collected in light-traps, sterilized with X-rays (3·3 krad), marked with a white dye, and released at the time of female emergence. In the 1959 trial the larval population was reduced to 20 per cent of that of the control area, and in the 1962 trial the larvae were completely eradicated. These trials are significant in that all the released males had been captured, so that breeding was unnecessary, and because the trial was completely successful despite the incomplete isolation of the area.

Trials with other pests

The codling moth, because of its low population densities, appears to be a suitable insect for control by this means, and field trials have been promising. A great advance has been made in the development of a breeding technique, using an artificial medium, which gives a yield of more than 5000 adults for each square metre of rearing tray (Proverbs, 1969; LaChance et al., 1967). There is also interest in the control of various cotton pests, such as the bollworms, Pectinophora gossypiella and Diparopsis castanea, and the boll weevil Anthonomus grandis. Sex-attractants may be of value in the integrated control of moths (White, 1967; Weidhaas, 1968).

Insects such as aphids, which reproduce largely parthenogenetically, do not, at first sight, appear to be promising subjects for control by this method. Steffan (1972) points out that many species have a sexual phase wherein males and females are produced. These mate and eggs are produced which are laid on the winter host, usually a woody plant. It may be possible to control such species as long as they do not also overwinter as parthenogenetic females. The numbers of sexual forms are comparatively small, and if the winter host plant is restricted in distribution it may be possible to swamp the wild population with laboratory-bred sterilized males. These could be produced in the laboratory or glasshouse a little earlier in the season than they occur in the wild by suitable adjustment of the day length. There is, of course, no difficulty in the mass production of most species of aphids.

Methods of genetic control

The production of dominant lethal mutations in eggs or sperm, and the use of these to control pest organisms, are closely related to the techniques of genetic control which has been defined by the World Health Organization as the use of any condition or treatment that can reduce the reproductive potential of noxious forms by altering or replacing the hereditary material (WHO Scientific Group on Genetics and Insecticide Resistance, 1964).

The techniques to be described now are usually considered separately from those in which insects are sterilized by radiation or chemicals, for they either do not depend upon the production of fully sterile individuals (whose sterility is not, in any case, inherited), or they utilize some genetic trait already present in the pest population, or in one closely allied to it. It must be stressed that these methods are so far unproved in any large-scale trials, but small field trials, laboratory experiments, and theoretical considerations are often encouraging. Furthermore, the genetics of pests, with the exception of *Drosophila* and mice, is a comparatively recent study, so that it is probable that future research will disclose further usable genetic characteristics, and possibly even completely new techniques. One important advantage of using genetic traits already existing in populations is that the insects concerned are likely to be fully sexually competitive with the insects in the target population.

Success is completely dependent upon a thorough understanding of the behaviour, ecology, and biology of the pest populations in the field, just as it is, of course, in the sterile-insect methods. Such knowledge is acquired slowly, so the application of the techniques will not be rapid. Initial costs are likely to be high, but the maintenance costs are likely to be lower than those when insecticides are used at regular intervals.

Public health entomologists currently show the greatest interest in genetic control methods. There are various reasons, the chief being the development of resistance by many important vectors towards insecticides, the necessity of reducing vector populations to a very low level, or of eradicating them completely, and the usual failure so far of the sterile-male technique in mosquito and tsetse fly control. The agricultural entomologist, on the other hand, deals mainly with pests with higher thresholds of economic damage and though, no doubt, he would welcome the eradication of many of them, his problems do not often demand so drastic a solution.

Among the genetic mechanisms which have been suggested, the following have been tested, or are being investigated: chromosomal translocations; cytoplasmic incompatibility; hybrid sterility; sex ratio distorters; conditional lethal genes. Population replacement is often listed here. When the pest population is replaced by a less noxious one of the same species few would quarrel with the classification, but when the replacing population is of a different species many would wish to class the technique differently.

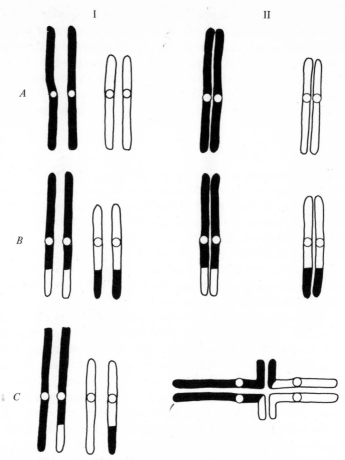

Fig. 8.2. Translocations. A: 'standard chromosomes'; B: translocation homozygote; C: translocation heterozygote. I: The chromosomes. II: Pairing during meiosis. Translocation homozygotes pair normally but translocation heterozygotes pair as shown, with subsequent difficulties which lead to abnormal distribution of genetic material in the gametes. (See text)

Genetic control by the introduction of chromosomal translocations

The production of translocations by ionizing radiation has already been mentioned. Curtis (1968) has suggested that homozygotes bearing suitable reciprocal translocations could be used more economically than sterile males in release programmes for tsetse flies.

The initial step would be the successful mating of two heterozygotes to produce homozygotes, and to culture from these large numbers of viable, fully competitive translocation homozygotes. Such individuals produce gametes with the full gene complement, but when mated to wild-type individuals they produce offspring with reduced fertility. During the first meiotic division of gametogenesis in these heterozygous offspring the chromosomes involved in the translocation each tend to pair with one chromosome along part of their length, and with another along the remainder. This abnormal synapsis results in cross-shaped figures which are visible during pachytene (Fig. 8.2). The segregation which follows gives rise to a variety of gametes, half of which have all parts of each chromosome represented once (orthoploid), while the remainder are aneuploid, with certain parts of the chromosomes duplicated, and some missing.

Half of the orthoploid gametes are similar to wild gametes, but the others carry the translocation. When they fertilize, or are fertilized by, wild-type gametes, the resulting zygotes are all viable, but half of them are heterozygous for the trans-location. The zygotes formed by the union of an aneuploid gamete and a normal gamete are all inviable. The net result of a mating of a wild-type individual with a heterozygote is a reduction of the fertility to one-half, with one-half of the viable offspring wild type, and the rest heterozygous for the translocation.

Such a result is not in accord with normal Mendelian segregation which would give the following ratios: one wild-type gamete to one orthoploid translocation gamete to four aneuploid gametes. The discrepancy arises from interaction between the chromosomes when in the cross configuration. The frequencies with which the various gametes are formed depends, to some extent, upon the nature of the translocation, but observations on a large number of different translocations in mice, fruit flies, and maize indicated that orthoploid and aneuploid gametes are usually produced in roughly equal numbers.

A mating of a translocation homozygote with a translocation heterozygote results in a reduction of the fertility to about one-half, with half of the viable progeny translocation heterozygotes, and the rest homozygotes. Finally, the mating of two translocation heterozygotes results in zygotes in the following proportions: inviable, 10/16 to 11/16; translocation homozygotes, 1/16; translocation heterozygotes, 3/16 to 4/16; wild type, 1/16. The theoretical proportions, based on Mendelian segregation, would be: 32/36; 1/36; 2/36; 1/36.

Curtis (1968) summarizes a mathematical model based on the above proportions which describes the results of introducing reciprocal translocation homozygotes into wild populations of tsetse flies, and compares them with those predicted for the sterile-male release technique. The essential difference between the two techniques is that the effects of the introduced homozygotes do not disappear with their death, as is the case with introduced sterile males, because the translocation characteristic is handed on from generation to generation. Since the model is mathematically complicated even after the usual simplifying assumptions have been made the computations were carried out on a digital computer.

A complicating factor is that the optimum number for release is not the maximum number which can be released. Wright (1941) investigated the problem of the fixation of translocations in a population and concluded that this would only occur when the incidence was 50 per cent. If the frequency of the translocation chromosome sets is less than this they ultimately disappear. If greater, the wild-type chromosomes disappear with translocation chromosomes being present only in homozygotes. Since these are fully fertile (Fig. 8.2), the population would stabilize.

It is essential, therefore, to release the insects so that a 50 per cent frequency of the translocation chromosomes is attained in the F_1 generation. This cannot be achieved by the release of homozygous translocation males alone as, in a mating, the female partner contributes half the chromosome set of the progeny. As there will always be some wild male–wild female matings it follows that the 50 per cent frequency of translocations cannot be reached by the release of a finite number of males alone. It can be reached in the F_1 generation if equal numbers of male and female homo-zygotes, in total equal to the wild population, are released. The population density will then fall, and asymptotically approach zero. In the sterile-male release technique, on the other hand, the population can only be eradicated if releases are made in each generation.

The main difficulty, as Curtis points out, will be the determination of the size of the field population so that the optimum release number can be estimated. As present sampling methods are not sufficiently accurate for this Curtis suggests that cytological techniques be developed so that the frequency of the translocation can be monitored in the campaign area and releases made when necessary to bring it to 50 per cent.

Curtis's calculations indicate that twice as many sterile males as translocation homozygotes would have to be released into a population to achieve the same level of control, and as the translocation releases should be of equal numbers of males and females it follows that the sterile-male technique would require a breeding facility four times as large as that needed for the translocation method. This is a very important consideration when culturing such slowly breeding animals as tsetse flies.

The search for suitable translocations is continuing. The demands of the technique will be difficult to satisfy as the translocation homozygote must be viable and competitive in mating, as must the heterozygote. Their behaviour and powers of dispersal must be similar to those of the target population individuals. Unfortunately, the stock of homozygotes will have to be derived from very few individuals and will differ from the field populations in many important respects.

In a more recent paper Curtis (1969) describes attempts in his laboratory to produce suitable translocations in *Glossina austeni*. The progeny of irradiated males were examined for manifestations of inherited partial sterility which would indicate that translocations had occurred. Thirty-four per cent of the viable male progeny of males treated with 5–7 krad displayed the desired characteristics. In two cases the mode of inheritance showed that the Y chromosome was probably involved, but in all others autosomal chromosomes underwent translocations.

The introduction of translocations for the control of mosquitoes is also being studied. Lorimer and her colleagues (1972) have induced numerous reciprocal translocations in the yellow fever mosquito, *Aedes aegypti*, but only two of these have yielded viable homozygotes which may be of value in the control of this pest.

It would, of course, be an advantage if the insects released carried multiple reciprocal translocations. McDonald and Rai (1971) simulated the release of single translocation individuals and double translocation individuals on a computer. Their results suggested that six introductions of single translocation individuals at a ratio of four to one will eradicate a wild population, whereas if double translocation insects are released, only five introductions need be made. It appears from their assumptions that they do not consider that Curtis's restrictions about the exact number of individuals to be released is valid. Lorimer, Halliman, and Rai (1972), however, believe that it would be very difficult to rear double translocation homozygotes in large numbers.

Translocations have been obtained in a number of pest species, but very few homozygotes have been produced. Lorimer *et al.* (1972) list the following data. The number of translocations obtained is given first, and then the number of homozygous translocations: *Culex pipiens*: 44, 2; *Glossina austeni*: 22, 1; *Musca domestica*: 193, 1; *Culex tritaeniorhynchus*: 46, 0; *Lucilia cuprina*: 6, 0; *Aedes aegypti*: 40, 2.

Although Curtis is one of the most active proponents of this method of vector control, he points out that it had been suggested by the Russian worker Serebrovsky in 1940, but it must be stressed that Curtis has developed the theory independently. The following papers are also of interest: Laven (1969), Rai and Asman (1969), Whitten (1971), Wehrhahn and Klassen (1971).

Genetic control by the use of cytoplasmic incompatibility

The topic is best introduced by considering the *Culex pipiens* complex as the phenomenon is common in this group of closely related mosquitoes. Mattingly recognized seven taxa in his 1967 paper on the systematics of the group, but concludes that the complex is best treated as a single polytypic species. The taxa are as follows: *C.p. pipiens*; *C.p. fatigans*; *C.p.* var. *molestus*; *C.p.* var. *pallens*; *C.p.* var. *comitatus*; *C.p. australicus*; *C. globocoxitus*. *C.p. pipiens* is mainly confined to the north temperate region of the world, but with some incursions into more southerly latitudes. *C.p. fatigans* is a tropical and sub-tropical insect, and a very important vector of filariasis. The other sub-species and varieties are not so widespread although *C.p.* var. *molestus*, an autogenic variety, is found in Europe, North Africa, the Near East, the United States, Japan, and Australia. An autogenic female mosquito is one that can produce eggs without first having taken a blood meal.

Many of the sub-species and varieties of *C. pipiens* will interbreed successfully, and *C. globocoxitus* will hybridize with *C.p. fatigans* in the laboratory. It has been found, however, that another breeding system exists, largely cutting across these taxonomic boundaries, which determines whether or not the mating produces viable offspring. A large number of strains have been identified and can be divided into a number of crossing types (seventeen when Laven reviewed the situation in 1967). Crosses between members of two different crossing groups are incompatible. In some cases the incompatibility is unidirectional so that matings between the males of one strain and the females of the other are successful, while matings in the opposite direction fail. Sometimes, however, the incompatibility is bidirectional with few or no offspring from either kind of cross. It has been shown cytologically, and by the use of genetic markers, that the male sperm enters the egg, stimulating meiosis, but does not fuse with the egg nucleus. Some cell division usually follows, but the eggs fail to hatch, apart from a very small proportion which develop into adult females. These are produced pathenogenetically, the sperm merely stimulating the egg. The factor which blocks the sperm is extrachromosomal, as may be shown by continuous back-crossing, but it is inherited with the crossing type for generation after generation.

Laven (1967c) discounts the theory that the factor which blocks the sperm is a viral agent or parasite, though this mechanism has been demonstrated for similar phenomena in other organisms. He argues that the complexity and the number of crossing types would necessitate an unreasonable number of different parasitic agents. He offers an alternative hypothesis to describe the factors which, it should be noted, are capable of self-replication since they persist, undiluted, through at least fifty generations. It is known that early in the embryological development of the mosquito certain nuclei together with cytoplasm separate from the rest of the embryo to form the pole cells, the future germ cells of the mosquito. The cytoplasmic factor responsible for the incompatibility must be carried in the germ cells from generation to generation, and as the cytoplasm of the pole cells is rich in RNA Laven suggests that the factor resides in a part of this nucleic acid. It is not unreasonable to suggest that this RNA is self replicating for the RNA of certain viruses is known to do just this. Egg and sperm will both carry such RNA and incompatibility between two kinds of RNA will result in incompatibility between the two gametes.

The geographical distribution of the crossing types is of interest. It is well known that in many kinds of animals sexual isolation, based on the genomes, exists between individuals from the extremes of a wide geographical distribution. This is often not the case with the cytoplasmic incompatibility of the *C. pipiens* complex. Strains from relatively close areas are often incompatible, while certain populations from the Philippines and North America, and from other widely separated areas, have been found to be completely interfertile. The phenomenon raises interesting speculations on modes of speciation differing from those described by Mayr (1942). These are discussed by Laven in his reviews (Laven, 1967a, b, and c).

The cytoplasmic incompatibility principle has already been used successfully in Burma to eradicate a *C.p. fatigans* population. The small village of Okpo, about sixteen kilometres north of Rangoon, was chosen for a pilot experiment because it is entirely surrounded by dry paddy fields during the winter and is thus, as far as the mosquito is concerned, an ecological island. Cage experiments had shown that a strain, D1, from the University of Mainz, was incompatible with the wild population, and that the wild females did not differentiate between wild males and D1 males. The wild population of mosquitoes was found to vary between 4000 and 20 000 individuals at the beginning of the experiment in February 1967. During the first month the rate of release of D1 males was too low but as the breeding techniques techniques were improved the rate was increased to reach an optimum of 5000 males each day. In the twelfth week of the experiment all the eggs found in the area were non-hatching, and the experiment was thus considered to be successful. It had been shown that the method can be used to eradicate *C. pipiens* populations from small areas, and that it is feasible to extend the technique to larger areas. *C. pipiens* is, however, a prolific breeder and it will be necessary to reduce the wild populations drastically by chemical methods before releases can begin. Knipling *et al.* (1968) present theoretical operational procedures for such control campaigns.

Cytoplasmic incompatibility has been found in other insects including *Drosophila*, the chironomid *Clunio*, the parasite *Mormoniella*, and the mosquito *Aedes scutellaris*. *A. scutellaris* is the most important vector of filariasis in many parts of the Pacific. As it is difficult to control by insecticides and since it occurs on many small isolated islands, it seems to be an ideal subject for genetic control. Mass-production methods have been developed for related species so it should not be difficult to raise this one in sufficient numbers. The existence of the phenomenon in this member of the genus raises hopes that it may also be found in the yellow fever mosquito, *A. aegypti*, but the search, so far, has been unsuccessful.

Before closing this section it should be mentioned that alternative hypotheses have been put forward to account for the sterility, especially in the case of *A. scutellaris*. These are discussed by McClelland in his paper on speciation and evolution in *Aedes* (McClelland, 1967).

Genetic control by the use of hybrid sterility
The crossing of certain strains of insects produces progeny which are infertile. Often the female progeny is normal, and the male sterile. The chief vector of malaria in Africa, *Anopheles gambiae*, furnishes an excellent example.

The *A. gambiae* complex is now thought to consist of five separate species, two of which develop in salt water, and three in fresh water. Attempts are being made to find morphological criteria which may be used to distinguish between these sibling species, but the only reliable methods at present appear to be crossing with standard laboratory populations, and the examination of the giant chromosomes. The five species differ in their efficiency as vectors of malaria.

The two salt-water species are *A. melas*, from the west coast, and *A. merus*, from the east. The fresh-water species are known tentatively as species *A*, *B*, and *C*. There are twenty ways of crossing these five species and in all cases the male hybrids are sterile and the females more or less normal. In six of the crosses, however, there are very few or no females produced (male *A* or *B* to female *A. melas*, *A. merus* or species *C*). Studies of the giant chromosomes in an $A \times B$ cross suggest that chromosomal aberrations in the autosomal chromosomes account at least partly for the sterility, but there is also evidence for a genic effect.

Although sterile, the hybrid males are fully competitive and, indeed, show hybrid vigour. Cage experiments show that they can be used for sterile-male release campaigns with good chances of success. The existence of crosses which produce nothing but males is particularly useful as their use avoids the difficult task of separating the two sexes. A less obvious advantage is that it makes it unnecessary to wait till the adults have emerged before the sterile insects are released. Larvae or even eggs may be seeded in the breeding sites so that the adult sterile males will emerge at the same time and place as the wild females. The tendency of females to mate with males from the same breeding site has been suggested as one reason for the failure of release trials in the past. A further refinement, easily accomplished, would be the incorporation of resistance to dieldrin or other insecticides in the released males.

The members of this groups are prolific breeders, and very difficult to control with insecticides. Releases will have to be made when the population density is low, and this would be in the dry season. At this time the number of breeding sites for the mosquito will be reduced and this will facilitate their seeding with eggs or larvae.

In anticipation of field trials using this technique Cuellar (1969) has investigated, by means of a mathematical model, the probable results of adding various numbers of sterile male producing eggs into breeding sites. He concluded that the optimum ratio of introduced to wild eggs would be one to one, and that this would give eradication in about nine weeks. Davidson *et al.* (1967) and Davidson (1969) review the *A. gambiae* complex, and the possibility of genetical control.

Sterile hybrids produced by the mating of individuals of different species are common enough in the laboratory, but in the field it is usually prevented by some mode of isolation, even when the two species occur in the same area. Hybrids between *Drosophila pseudoobscura* and *D. persimilis*, for example, can be obtained experimentally quite easily, but in the field *D. persimilis* tends to be active in the morning, and *D. pseudoobscura* in the afternoon. They also differ somewhat in their habitat preferences. There must, however, be further isolating mechanisms for when given the choice of partners in cage experiments the flies tend to mate conspecifically. It is therefore surprising that two distinct species of tsetse fly, *Glossina swynnertoni* and *G. morsitans* will mate apparently indiscriminately, even though the hybrid crossing produces very few progeny, and all of these are sterile. The two species are allopatric – that is, they occur in different geographical regions, so it has been suggested that *G. swynnertoni* might be controlled by massive introductions of male *G. morsitans*, if and when mass rearing is accomplished.

The progeny of the *Drosophila* cross mentioned previously resemble those of the *A. gambiae* crosses in that the males are sterile, and the females fertile. The females do, however, show hybrid break-down, for the progeny resulting from back-crosses between these females and males from the parent species are far less viable than pure strain insects. Such a mechanism may be of value in genetic control.

Doubtless, further research will uncover many potentially useful cases of hybrid sterility, especially in cosmopolitan pests, but a final example may be taken from forest entomology. There are two important forms of the gypsy moth, *Porthetria dispar*, namely the strong race and the weak race. Crossing the two produces intersex sterile female offspring. It may be possible to release males of the strong race into north-east North America where the weak race is particularly troublesome. This is regarded as the classical example of intersexuality in hybrids, and it was analysed by Goldschmidt in the 1930s. A lucid description of this type of hybrid sterility, and of many other kinds, will be found in Dobzhansky (1951).

Genetic control by the introduction of lethal and deleterious genes

On first consideration the introduction of deleterious or fully lethal genes into a population does not seem to be a promising way of controlling it. It would be expected that the processes of natural selection would eliminate the offending genotypes before the population was seriously depressed, unless a constant pressure was maintained by sustained releases. Various devices have been suggested which could tip the balance in favour of eradication or control. Knipling has presented mathematical models which show that a constant low level of mortality, coupled with the mortality caused by normal environmental factors, can bring about a drastic reduction in population density. He suggests that such factors as the inability to fly or to diapause could be of value, as could factors which bring special nutritional requirements, or sensitivity to certain temperature ranges. It is important that the factors should not make laboratory rearing difficult, or reduce mating competitiveness.

Factors which act after a delay or at particular times, killing, for example, the immature progeny or the adult during hibernation or aestivation, would have the greatest effect on the population. Factors whose expressions are markedly density-dependent so that they spread through the population when the density is low, and express themselves when the population increased, would fit in well with integrated programmes in which the population is first reduced as much as possible by chemical or other means.

There are several devices which could be used to help to propagate the deleterious genes in the wild population. Two of these are heterosis or hybrid vigour, and meiotic drive.

Heterosis is discussed by Dobzhansky in the text mentioned above. It is an adaptative superiority of heterozygotes over homozygotes, and it is, in fact, an essential condition for the establishment of balanced polymorphism in a population. Natural selection will retain a mutant in a population, if the heterozygote shows heterosis, even if the homozygote is poorly adapted.

There is, for example, a recessive lethal mutation in *Drosophila melanogaster* which increases the fecundity of female heterozygotes by 30 per cent.

The second possible mechanism is meiotic drive. In its absence a heterozygote Aa forms the two types of gamete, A and a, in equal numbers. Meiotic drive brings about a systematic deviation from this Mendelian segregation so that the two kinds of gametes are not produced in equal numbers. When the sex chromosomes are involved the sex ratio is seriously distorted; autosomes can also be involved in meiotic drive mechanisms. Craig (1963) points out that the phenomenon is probably widely spread because it had then already been found in insects, mammals, and plants. He cited from the literature sex ratio distorters in six species of *Drosophila* and in *Aedes aegypti*, and recessive tail-less alleles in the house-mouse, the pollen killer gene in *Nicotinia* and the knob chromosome in maize, all favoured by meiotic drive mechanisms. Theoretically, meiotic drive could increase the frequency of a factor, even a deleterious one, and spread it through a population. This has been tested in *D. melanogaster*, using the Segregation Distorter (SD). Wild-type/SD male heterozygotes produce many more gametes with SD chromosomes than with wild-type chromosomes, so that the SD frequency in cage populations increased, even though SD homozygotes show a marked reduction in larval viability. SD is also common in some wild populations of the fly.

Populations can, however, develop protective mechanisms against the effects of meiotic drive. An inversion in the chromosome homologous to the SD chromosome, for example, counteracts the SD effect. It may be possible to eradicate some populations of pests before such mechanisms develop.

According to Craig (1967) the first biologists to suggest the use of meiotic drive in pest control were van Borstel and Buzzati-Traverso, at a congress in Bombay in 1960. They proposed that if a meiotic drive chromosome carrying genes for female sterility were to be introduced, it would sweep through the population, with sterile homozygous females being produced, and with the proportion of males increasing steadily. Eventually, all the females would be sterile, and the population would disappear.

For full use to be made of this mechanism it will be necessary to develop methods of incorporating genes in meiotic drive chromosomes, or of introducing the factors conferring the meiotic drive into chromosomes carrying lethal genes; at present we must depend largely upon finding the two factors already linked.

Interest in meiotic drive mechanisms is, at the moment, declining for, as Whitten (1971) points out, there is evidence that they may invariably cause recessive sterility and this would prevent them, and the susceptibility genes which the chromosomes carry, from being fixed in a population.

Whitten suggests, as an alternative, that the desired genes could be carried by translocation homozygotes so that the genes would spread through the population and, at the same time, the population would also decline because of the sterility effects described earlier. He proposes that a suitable gene would be one that confers susceptibility to an insecticide for which the wild population has become resistant. The periods of release of the translocation homozygotes could then be alternated with periods of insecticidal treatment. Insecticides would be applied until resistance became troublesome, after which the susceptible translocation homozygotes could again be released. In his paper Whitten presents theoretical calculations on the efficiency of the method.

Werhahn and Klassen (1971), on the other hand, suggest that the desirable genes could be driven through a population by linking the desired gene or genes to one which confers a high degree of resistance to a suitable insecticide. If the releases are made at the same time as insecticides are applied, the carriers of the lethal gene will eventually become commoner than the wild types.

The object of using deleterious genes is, of course, to destroy the wild population, and even if a driving mechanism is at work it seems, at first sight, that the phenotypic expression of the deleterious genes would nullify the drive mechanism. If, however, the expression of the gene were delayed it might be possible for the gene to become widespread in the population before its effects were felt. In *Aedes aegypti*, for example, the mutant *intersex* makes the males infertile. When the larvae are reared at 27 °C the males are normal, but when reared at 30 °C the males are feminized and thus sterile. In suitable climatic regions the gene could be driven through the population during the cool season, and would become effective during the warmer periods. In *Drosophila*, several mutants are also known which make the insects more sensitive to heat or to cold than are the wild insects.

Certain widespread species or groups of closely related species have populations living in areas which differ one from another climatically. In such cases it often happens that the strains living in cooler regions are univoltine, and those in warmer regions multivoltine. Furthermore, some strains must undergo diapause to survive the annual periods of heat or cold. If voltinism or ability to diapause is genetically controlled, it may be possible to use the factor for the control or eradication of some populations. The manipulation of voltinism or diapause has been suggested for the control of the European corn borer, *Ostrinia nubilalis*, in North America, and work on the control of a field cricket in Victoria, Australia, by this means is in progress. The species present in Victoria, *Teleogryllus commodus*, survives the winter as a diapausing egg. The closely related Queensland cricket, *T. oceanicus* (till recently thought to be the same species), lays non-diapausing eggs and, furthermore, the eggs resulting from a reciprocal cross are also non-diapausing. This trait could be introduced into the southern populations as a conditional dominant lethal factor which would reduce the populations during the winter. A further safeguard is the almost complete sterility of the mature F_1 progeny which ensures the ultimate death of any individual which survives the winter. This hybrid sterility would limit the natural spread of the factor during the warmer months and, in fact, this would depend upon the breeding and dispersal of the introduced cricket.

Unfortunately, it seems that there is an intraspecific preference in mating, but it is possible that this difficulty can be circumvented by the use of a non-diapausing hybrid of the two species which has been cultured from a few viable eggs.

Whitten (1971) is sanguine about the possibilities of using such methods for the control of the Australian sheep blowfly by the release of suitable multiple transloca-tion homozygotes carrying conditional lethal genes. It would require smaller numbers of released insects than would be demanded by the classic sterile-male release tech-nique. Werhahn and Klassen (1971) also present theoretical calculations to support their suggestions in which it is estimated that a boll weevil population could be eradicated by the release of a number of suitable insects which is less than one-hundredth of those needed for other methods.

Pest control by population replacement

The underlying idea in this method is that the noxious population should be replaced by an innocuous or a less harmful one. The technique ranges from the complete replacement of the pest population without intervention by man, other than the introduction of the replacing species, to the eradication or near eradication of the pest by chemical or other means, followed by the introduction of the replacement.

In the latter case the aim is largely to fill the gap left by the pest to prevent its resurgence after immigration, or to prevent its replacement by another harmful species.

In practice, the wild population would be reduced by chemical means before the introduction of the new species, but no attempt would be made to eradicate it completely.

The phenomenon of population replacement depends upon the principle that two genetically distinct, and sexually isolated populations, cannot coexist indefinitely in the same habitat if the individuals of the two populations occupy the same niche (that is, if they are ecological homologues). The population that will remain in sole occupancy after a sufficient period will be the one which produces the highest number of reproducing females per individual.

The ecological niche must not be confused with the habitat. The habitat is the place where the organism is found: the niche is its role in the habitat – the food it eats, its other requisites, and so forth. In practice, the two organisms need to occupy the same niche for a part of their life, provided that they do so simultaneously; alternatively, the niches may be only partly identical. The two species may, for example, have distinct oviposition sites, but identical nutritional requirements. In such a case the organism with the larger rate of reproduction would eventually replace the other.

Although Darwin seems to have been familiar with the concept, one of the first biologists to formulate it clearly was Gause: hence it is often referred to as Gause's principle. The topic has been reviewed by DeBach (1966).

Population replacement as a means of control of pests has been suggested by a number of authors, but it does not appear to have been intentionally applied on a large scale as yet. There are, however, a number of records of one organism replacing another with a similar ecological niche. In parts of southern California the Californian red scale, *Aonidiella aurantii* has completely replaced the closely allied yellow scale, *A. citrina*, over a period of years, although there was apparently abundant food for both species. Similarly, the small cabbage white butterfly, *Pieris rapae*, accidentally introduced from Europe, has completely replaced the native *P. oleracea* over large areas of North America. In the Sydney district of New South Wales the introduced Mediterranean fruit fly has not been recorded for many years, although it was formerly abundant; its place has been taken by the Queensland fruit fly, *Dacus tryoni*, which has colonized the area from the north.

Examples may be found in groups other than the insects. Gause's original experiments, for example, were carried out with closely related species of Protozoa. At the other extreme there is a strong case for postulating competitive displacement in the disappearance of the British red squirrel, *Sciurus vulgaris leucourus*, from many parts of the British Isles, and the spread of the introduced grey squirrel, *Sciurus carolinensis*. Ecologists are not entirely agreed, however, that the disappearance of the red squirrel can be blamed entirely upon the grey; their niches are not identical in that the red squirrel shows a preference for coniferous woodland, and, furthermore, the native species was suffering from widespread disease at a time when the grey was spreading (Middleton, 1931; Shorten, 1954; Elton, 1966).

In most of the cases cited above a noxious species was displaced by another which was also harmful – sometimes even more troublesome. It would seem that this would usually occur when an agricultural or forestry pest is displaced by a species whose ecological niche is sufficiently similar. The component of the niches of the two species which is coincident usually involves that stage of the life cycle which is most damaging – most often the larval stage. The situation is quite different with many public health pests. The larval stages of the house-fly are of little economic importance, other than that they are the necessary precursors of the adults. A larval ecological homologue could, therefore, be of great value, provided that the adult stage is harmless: a candidate species has been found in the stratiomyid fly, *Hermetia illucens*, which has performed well in restricted tests.

There are several possible approaches with mosquitoes. It may be possible to replace a strain which is an efficient vector of a disease by a strain or a species which is not, or by a species which does not attack man. If this is not possible it may be feasible to replace a pest species which cannot be controlled by chemical or other means by a species which can. Thus exophilic strains and species, which feed out of doors, might be displaced by an endophilic one which could then be controlled by the widespread indoor spraying practised in anti-malaria campaigns. A similar technique might be used for the control of some agricultural and similar pests which are difficult to destroy by conventional methods. Such an approach would, of course, be tried only when all other methods had failed.

The replacement of a disease vector by a non-vector has already occurred on a large scale, though apparently not by design. Conventional mosquito control techniques against *Anopheles labranchiae* during the Sardinian anti-malaria campaign had greatly reduced the population density of the vector when its place was taken by a previously rare species, *A. hispaniola*, a non-vector. A similar result was obtained in East Africa when dieldrin spraying against *A. funestus* was followed by its replacement by *A. rivulorum*, a species which attacks cattle. It is significant that no attempt was made to control the vector in its larval breeding sites.

Finally, it should be pointed out that the competitive displacement principle might be used to displace an unsatisfactory parasite or predator of a pest by a more efficient one. DeBach lists a number of cases of the replacement of one parasitic insect by another. It would be necessary to make a thorough study of the biology of the two species before such a substitution is attempted, probably using the techniques pioneered by Holling (chapter 2). It is important, for example, that the introduced species should be more efficient than the established species at seeking out the pest at low population densities.

Pheromones and their use in pest control

Pheromones are chemical substances produced and released by organisms which affect in some specific way other individuals of the same species. As they resemble hormones in that they have an effect at some distance from the point of release they were originally called ectohormones: the term pheromone was proposed by Karlson and Butenandt in 1959 (Greek, *pherein*, to carry and *horman*, to excite or stimulate), and this term has virtually replaced the original name.

The study of pheromones has advanced rapidly in recent years. Its earlier neglect was possibly partly due to the relatively unimportant part played by the chemical senses in man, with a consequent lack of appreciation of their importance in many other organisms. Field naturalists have long been familiar, however, with the ability of the nubile females of certain species of moths to 'call' males from considerable distances.

It is obvious that at least some pheromones could be employed in the autocidal control of pests. The results have been, so far, disappointing, but this is probably largely due to a lack of knowledge of the biological mechanisms involved.

The classification of pheromones

Pheromones have been classified in a number of ways by various authors (Butler, 1967). Karlson (1960), for example, divided them into 'olfactorily acting' pheromones and 'orally acting' pheromones, whereas Wilson (1963) prefers to speak of 'releaser' substances and 'primer' substances. Releaser substances produce an immediate change in the behaviour of the recipient, apparently through some action on the central nervous system. The change in behaviour is reversible. Primer substances, on the other hand, trigger off a chain of physiological changes in the recipient, without any immediate change in behaviour.

Butler attempted to classify pheromones on the basis of their biological function, as far as this is known. The following outline follows Butler's plan.

Trail-marking pheromones   These are releaser pheromones which are used to form a chemical trail which is perceived by other individuals whose behaviour is thereby changed. There are two categories of trails: aerial and terrestrial.

Aerial trail pheromones are wind-borne, and they stimulate individuals for considerable distances downwind from the source. The trails are continuously renewed for as long as they are needed, and the pheromones are released by either stationary or moving individuals. Most are olfactory sex attractants which facilitate the meeting of the two sexes. Terrestrial trails are laid more or less continuously on the substrate, or at specific stations.

Aphrodisiacs   These pheromones are produced by one or both of the sexes as part of the courtship behaviour once the two sexes have come together. Some of the sex attractant pheromones may have an effect of this nature, but there are certain pheromones which are apparently specific aphrodisiacs. They may be olfactorily acting (those produced by certain male butterflies, for example), or at least partially gustatory (certain cockroaches).

Aggregation pheromones   Aggregation occurs in insects for a number of reasons, including mutual protection and mating. The aggregation may be temporary, as in overwintering lady-bird beetles, or persistent, as in the social insects. Aerial-borne and other pheromones often play a part in this aggregation.

Alerting pheromones   Many social insects, when alarmed, produce substances which influence the behaviour of their fellows. Sometimes, as is the case with the formic acid produced by many ants, the substance is also defensive. Alerting pheromones have been demonstrated in a number of ants, in some wasps, and in the honey-bee, and Butler suggests that they may also be used by gregarious insects such as locusts. Some odour trail pheromones elicit defensive behaviour at high concentrations.

Alerting pheromones are better known chemically than any other pheromones, as more have been isolated and identified. Most are very volatile terpenes with low molecular weights which, once released, are very short-lived so that their effects do not linger for long after the disturbance has passed.

Primer pheromones   In some social and gregarious insects the development of sexual maturity is partly controlled in some individuals by pheromones released by others. The sexually mature male desert locust, *Schistocerca gregaria*, releases a pheromone which accelerates the maturation of immature forms of both sexes. There is also some evidence that immature females may slow down the male maturation.

These pheromones are probably most effective at high population densities, and they probably serve to synchronize the maturation of large groups of this migratory species. It is probable that a similar mechanism exists in the African migratory locust, *Locusta migratoria*. Primer pheromones also control the development of many social insects so that sexual females do not develop when a mated queen is present.

Two examples of communication by pheromones have been reported in laboratory mice. The female oestrus cycle tends to become irregular when the females are kept in the absence of males, but the cycles tend to become synchronized a few days after the introduction of males ('Whitten effect'). The second example, the 'Bruce effect' is potentially more valuable in pest control. Mated female mice which are exposed for three days to 'foreign' males within five days of copulation frequently fail to become pregnant, apparently from the failure of the blastocysts to implant.

In passing, it should be pointed out that pheromones have not yet been confirmed in human beings, but it is interesting to note that musk and civet are important ingredients of many perfumes. These are believed to be aphrodisiacs or sex attractants in the mammals which produce them, the musk deer and the civet cat.

Although most, if not all, kinds of pheromones produced by pest species are potentially valuable in pest control, the sex attractants and aggregation pheromones have been given the most attention, and will therefore be discussed more fully below. Details of these and the other kinds of pheromones will be found in the following review articles and textbooks: Karlson and Butenandt, 1959; Beroza and Jacobson, 1963; Jacobson, 1965a, b; Butler, 1967; Callow, 1967; Shorey, Gaston, and Jefferson, 1968; Blum, 1969; Chapman, 1969; Downes, 1969; Ebling and Highnam, 1969; Highnam and Hill, 1969. These works should also be consulted for details on the sites of glands and so forth, as the following account is mainly on practical aspects in pest control.

Insect sex pheromones

These may be regarded as chemicals which directly facilitate mating (Shorey *et al.*, 1968). The recipient may be stimulated to approach the releasing insect from some distance ('attraction') or it may be stimulated to perform some close-range behaviour concerned with mating ('courtship'). Typically, the pheromone is produced by one sex only, and affects the other, but sometimes both sexes are affected. In other species both sexes release the pheromone.

Attraction by pheromones has been reported from the following groups of insects: Lepidoptera, Dictyoptera, Diptera, Coleoptera, Hymenoptera, Neuroptera, Mecoptera, and Hemiptera Homoptera. Shorey and his colleagues consider the case for the Hemiptera Heteroptera to be unproved, being based, in their opinion, on non-behavioural circumstantial evidence alone.

The sex attractants of the moths have received the greatest attention, partly because of the economic importance of the group, and partly because of the early observations of collectors. A few have been isolated and identified, but recent reports indicate that at least one, that of the gypsy moth, *Porthetria dispar*, was originally incorrectly described. This would account for much of the disappointment which had been experienced in attempts to control this pest with the aid of the synthetic homologue.

The gypsy moth sex attractant was thought to be 10-acetoxy-*cis*-7-hexadecenol (gyptol, 8), and the synthetic homologue, prepared from ricinoleyl alcohol, differs from it only in having two more carbon atoms in the skeleton. Gyptol is, without doubt, produced by the female moth, and released at the same site as the pheromone, but as neither pure gyptol nor the homologue, gyplure, appears to have any sexual activity towards male moths, gyptol cannot be the pheromone. Inconsistencies in trials on the control of the moth using gyptol and gyplure had been attributed to the masking of the compounds by substances such as the ricinoleyl alcohol; it now seems that the cause was variation in the quantities of 'contaminants' with pheromonic activity in the samples. It follows that the true pheromone must have remarkably high biological activity. This material has now been identified and synthesized as cis-7,8-epoxy-2-methyloctadecane (8a), and it has been given the name disparlure. In simulated field trials its efficiency was found to be very high, and the material holds great promise for the control of this pest (Jacobson, Schwarz, and Waters, 1970; Beroza et al., 1971).

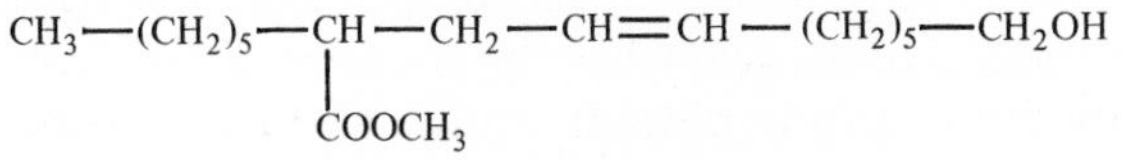

$$CH_3-(CH_2)_5-\underset{\underset{COOCH_3}{|}}{CH}-CH_2-CH=CH-(CH_2)_5-CH_2OH$$

(8) Gyptol

$$CH_3-CH.(CH_3)-(CH_2)_4-\underset{\underset{H}{|}}{C}\overset{\overset{\displaystyle O}{\diagup\diagdown}}{}\underset{\underset{H}{|}}{C}-(CH_2)_9-CH_3 \quad \text{disparlure}$$

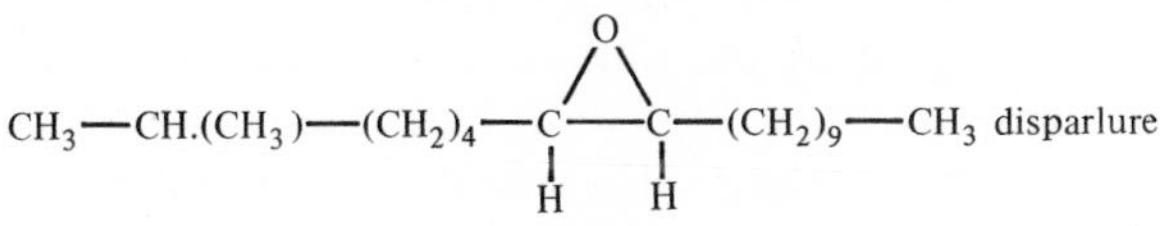

(8a) Disparlure

The silkworm attractant has been identified as trans-10-cis-12-hexadecadienol (bombykol, 9), that of the pink bollworm, Pectinophora gossypiella, as 10-propyl-trans-5,9-tridecadienyl acetate (10), and that of the cabbage looper, Trichoplusia ni, as cis-7-dodecenyl acetate (11). A pheromone from the honey-bee queen is reported to be 9-keto-trans-2-decenoic acid (12).

$$CH_3-CH_2-CH_2-CH=CH-CH=CH-(CH_2)_8-CH_2OH$$

(9) Bombykol

$$(CH_3CH_2CH_2)_2C=CH\cdot(CH_2)_2\cdot CH=CH\cdot(CH_2)_4O\cdot CO\cdot CH_3$$

(10) Pink bollworm

$$CH_3(CH_2)_3CH=CH(CH_2)_6O-CO\cdot CH_3$$

(11) Cabbage looper

$$CH_3\cdot CO\cdot(CH_2)_5 CH=CH\cdot COOH$$

(12) Honey-bee queen

It will be seen that these molecules are small compared with those of many natural products. They must be reasonably volatile in order to function; they must, at the same time, be reasonably specific. Total specificity is not, however, essential, for even if a female moth does attract a male of another species to her there are many other sexually isolating mechanisms which will make mating unlikely. The lack of specificity not only often extends beyond the species but sometimes beyond the limits of the genus. Schneider (1962), for example, showed that the pheromone released by a female of a species of *Saturnia* attracted males of the following genera within the same family, Saturniidae, as well as males of the genus *Brahmaea*, of the family Brahmaeidae: *Antheraea, Rothschildia, Samia, Hyalophora, Automeris, Aglia, Saturnia*. Conversely *Antherea* and *Rothschildia* females produced pheromones which were very attractive to *Saturnia* males, and *Samia* females one which produced some response. In passing, it may be noted that many ant alarm pheromones are similarly non-specific.

Different species of female moths release their pheromones at specific times. Obviously, they are only produced at the proper time during the breeding cycle, and the period or periods of the year when this occurs will vary from species to species. The circadian rhythm is also important. Although there is little known about the variation in the rates of synthesis of pheromones during the course of the day and night, there are many reports on the periods during which the pheromones are active. It is difficult to establish from experiments in which virgin females are caged whether the periods are determined by the times of release by the females, by the periods during which the males are responsive, or by both, but whatever the reason or reasons the periods are often quite short. Rhythmicity of response can be demonstrated however, if sufficient pheromone extract is available for the experimenter to expose the males to a constant concentration at various times during the day and night. The female grape vine moth, *Lobesia batrana*, is known to release her attractant during the evening, the moth *Clysiana ambiguella* between 2 and 6 a.m., and the moth *Sparganothis pilleriana* between about 11 a.m. and 4 p.m. (Beroza and Jacobson, 1963). In experiments with male cabbage looper moths, *Trichoplusia ni*, the peak responsiveness occurred during the second half of a twelve-hour dark period, while that of the tobacco budworm moth, *Heliothis virescens*, is between 4 and 5 a.m. A further complication is the adaptation of the males to the stimulus. The insect often responds to a constant exposure to the pheromone for only a short time. After the stimulus has been removed for some time, higher concentrations at re-exposure are often needed to produce a response.

The responsiveness of the male is also influenced by environmental factors such as temperature. The drone honey-bee, for example, is only attracted by a queen when she is flying at a certain height which varies with the wind velocity. It is possible that more subtle environmental requirements are to be found among other species of insects.

To regard sex attractants as mere attractants is an oversimplification. To that extent at least the name is a misnomer. A prime function, of course, is to bring the two sexually mature sexes together from some distance, but there may be other functions. They may, for example, induce the complete courtship behaviour in the male, this being aroused by the higher concentration in the vicinity of the male. Similarly, the higher concentration would induce the termination of the orientation behaviour.

It was once thought that the male approached the releasing female by following a concentration gradient of the pheromone, but this cannot be accepted. A concentration gradient may well be maintained within a few centimetres of the source (and possibly be of significance over that distance) but the shearing effect of the wind, and turbulence, would destroy any gradient over greater distances. Only insects down-wind of the source orientate towards the source when this is a considerable distance away, and it seems that they are stimulated to fly upwind by the pheromone. If they fly out of the path which contains sufficiently high concentrations of the phero-mone they apparently make turning movements until they re-enter the stream. A similar behaviour would bring back some of those which flew upwind of the source. Some researchers have suggested an alternative mechanism: an orientation towards the mate as an infra-red source, but there appears to be little supporting evidence. This suggestion was probably made to circumvent a difficulty of the anemotaxis hypothesis. Day-flying insects can maintain visual contact with the ground, and thus orientate their flight with respect to the wind current, but it has been suggested that this would be impossible for night-flying insects. Most night-flying insects, however, can probably maintain visual contact with the ground or terrestrial objects, despite the dim light, either by visible light, or by infra-red radiation. Such contact may not even be necessary except in very smooth air flow.

An alternative method of orientation has been suggested by Wright (1958). The disruptive effects of the wind would produce patches of air with comparatively high concentrations of the pheromone, and patches of low concentration. As the source is approached by an insect the distances between the patches of high concentration would become smaller and smaller, and the insect would thus be provided with a method of orientation.

There have been many investigations on the distances travelled by male moths in response to pheromonic stimulation. Marked insects are often used in such studies, the individuals being released at known distances from the source. Unfortunately, when a recipient arrives at the source, there is no way of telling how much of the journey was made under the influence of the pheromone, and how much by chance. It follows that reports of moths being attracted from distances of several kilometres downwind must be treated with reserve. Wright has calculated that if a pheromone is volatilized at the high rate of one-millionth of a gramme a second into an air stream with a velocity of 1·6 km/hr, the concentration of the pheromone in the air 2·5 km downwind would be about 10^{-14} g/litre. If the pheromone concerned has a molecular weight of about 100 this concentration would be equivalent to some 60 molecules/ mm^3. This is considered to be near the lower limit of sensitivity of an insect. Shorey and his colleagues (1968) estimate that most pheromones would be released at a rate very much lower than that used by Wright in his calculations, and thus the effective distances of pheromone communication would be much less than one kilometre. This would not be a grave disadvantage as most species of insects congregate in suitable habitats so that, in general, the males and the females would be reasonably close to one another from the beginning.

Sex attractants that may be of value in control programmes have been detected in a number of other insect groups. A female sex pheromone stimulates the males of the introduced pine sawfly of America, *Diprion similis*, to approach the female. The females of several beetles, including a number of dermestids and elaterids, have been shown to produce male sex attractants, which, in some cases, induce male copulatory behaviour. In contrast, the male of the boll weevil, *Anthonomus grandis*, releases a pheromone which attracts the females. Female house-flies release an olfactory sex pheromone which stimulates the copulatory behaviour of the males, though it does not appear to be essential for successful mating. There is doubt, however, about the production of a pheromone by this insect which attracts the males from a distance. Finally, a number of cockroaches produce sex attractants, but there has been controversy over the chemical structure of the one associated with the American cockroach, *Periplaneta americana*.

Several bark and ambrosia beetles (Scolytidae) produce aggregation pheromones which may prove to be useful in their control. In *Ips confusus* and *I. ponderosae* the pheromone is released by the male: in three species of *Dendroctonus* and in *Scolytus quadrispinosus* and *Trypodendron lineatum*, by the female. Wood and his colleagues (1966) have studied the general mechanism in *I. confusus*. When males find and attack a suitable host tree they secrete the pheromone into the hind gut from whence it is passed out with the faeces. The pheromone released from the faecal pellets orientates flying males and females to the attacked tree; the females enter the galleries containing the established males for mating, while the new males construct new galleries. As the pheromone is released only by males which have established themselves in a suitable host tree, the mechanism serves to concentrate the population in a suitable breeding material. It has been suggested that such pheromones could be used to attract a pest population to a host tree which will not allow breeding. Infested wood from the normal host could be tied to the branches of the substitute species (Gara, Vité, and Cramer, 1965). A further possibility is to attract the insects to trees which serve as lethal traps, by treatment with cacodylic acid (dimethylarsinic acid) (Buffam, 1971).

At present chemical control is the only feasible way of dealing with locust pests (Haskell, 1970). The very size of the locust swarms appears to rule out the use of sterile males, and, in any case, the chemosterilants which are available are too toxic to the insects. Locusts do not, apparently, possess sex attractants, but Haskell suggests that some of the pheromones which have been detected may be of value. It might be possible, for example, to interfere in some way with the action of the maturation hormone so that the swarming is no longer synchronized.

Similarly, if the reported aggregation hormone could be inhibited by some chemical means it should be possible to keep the locusts in the solitary phase. As Haskell says, this would be 'an elegant solution to the problem' but, unfortunately, it remains a possibility only.

Potential uses of attractant pheromones in pest control

Interest in this field has centred on the control of the gypsy moth using the presumed pheromone gyptol, or its synthetic homologue, gyplure. As was said before, gyptol has recently been shown not to be the true pheromone of this moth, and, in fact, the pure form appears to have no sexual activity at all. It is not surprising, therefore, that the trials have proved disappointing, but there are also other probable reasons for failure, some of which are discussed here.

Pheromones may be used in an insect control programme either for survey work, to study the distribution and abundance of the pest, or for direct behavioural control. In surveying they may be employed as an aid to planning the control using other control agents, or to monitor the results of such control measures.

There are two possible methods of direct control using attractants: the sexual behaviour may be stimulated or inhibited. Thus the attractant may be used to bring the pest to some lure or trap, or it may be released in the insect's habitat in such large quantities that the insect cannot orientate to natural sources.

When a trap or lure is baited with an attractant the chemical does not directly damage the pest: some other agent such as a toxicant, a chemosterilant, or a pathogen must be used for this. If a pathogen or a chemosterilant is used then the pheromone has an additional indirect effect upon the members of the population which it does not attract to the trap. It is essential, of course, that the secondary agent, whatever it is, should not repel the insect in any way. The pheromone-baited traps must prove more attractive than the natural sources of the pheromone in the area, and, unfortunately, this does not seem to have been achieved in any trials so far, whether the bait has been virgin females, pheromone extracts, or pheromone homologues (or presumed extracts and homologues). There are many possible reasons for this and these can only be circumvented by more detailed knowledge of the mechanism of attraction. We must, for example, choose a site for a lure which fulfils the other environmental requirements. It should be in an area where the insects naturally congregate before the communication occurs. We must know about the distances over which the communication is effective, so that the lure are dispersed as economically and as effectively as possible. We must also have detailed knowledge of the circadian rhythms of both the male and the female. It may be found, for example, that the males are responsive before the females emerge, or before the females start releasing the pheromone during the night. In such cases it may be possible to use synthetic or extracted pheromones without any competition from the females.

It would seem that the use of large quantities of a pheromone in a trap would be an excellent way of outcompeting with the natural sources, but this may not always be so. It will be remembered that some pheromones at least have multiple effects, which vary with the concentration. When heavily baited traps are used, the concentration at which orientation behaviour ceases and courtship behaviour begins may occur at some distance from the trap. As the courtship behaviour will be abortive in such circumstances the insect will probably eventually become adapted and move away from the area, possibly to mate with a wild female the following night.

Many moths orientate more strongly to a light source than to a pheromone source, and it may be possible to make use of this characteristic in control. The pheromone may attract moths from a greater distance than does a light source, and it could therefore be used to assemble moths in the general area of a light trap. In experiments with the cabbage looper, *T. ni*, Henneberry and Howland (1966) caught more males in blacklight traps situated a short distance from caged virgin females than in blacklight traps located a mile or more away. Even greater catches were obtained when the cage was placed on the light source. Howland (cited in Shorey and Gaston, 1967) speculated that one blacklight trap baited with ninety-six virgin females per 6 acres (2·4 ha) should give reasonable protection to the crop at the centre of the area.

The inhibition method of control would seem to be simpler to use than the technique we have just considered, provided that pheromones, or pheromone synthetic homologues, are available in large enough quantities. The chemical is released in such large quantities that the natural sources become relatively insignificant. The effect is not the confusion of the males (although it has been called the male confusion method from time to time), but the inhibition of the behaviour through the adaptation of the insect to the pheromone. An attempt to control a gypsy moth population on a small island by the broadcasting of gyplure formulations failed. It was suggested at the time that the gyplure effect had been masked by contaminants, but it now seems that the gyplure was, in any case, ineffective, and that the commercial product did not contain enough of the then unknown active material.

Autocidal methods, in general, are promising in that they appear to have few, if any, deleterious side-effects. Insects sterilized by irradiation are not themselves radioactive: those sterilized by chemicals in the laboratory rarely carry any significant residues when they are released. Chemosterilants cannot, of course, be sprayed widely or broadcast in the field, but if used in conjunction with attractants there should be little danger of environmental contamination. The pheromones appear to be without effect on living things other than the target species and, sometimes, its fairly close relatives.

It has been suggested by various authors that if a satisfactory control method is achieved using a pheromone, then there will be no danger of the pest developing a resistance to it, or at least to the pheromonic component. This is a dangerous conclusion. It presumes that there is only one pathway of communication involved in the behaviour concerned. Two kinds of stimuli may initiate the same piece of behaviour; if one of these is removed the behaviour may still be completed through the occurrence of the other stimulus. A control method based on behaviour control, such as attraction by pheromones, could well select for individuals which rely more on some alternative method of communication.

Before concluding this chapter it must be pointed out that there are a number of substances which are effective insect attractants, but which are not pheromones. Since they do not appear to be produced by the insect concerned they cannot be used for autocidal control. Nevertheless, the division between control using pheromones, and control using other attractants, is arbitary. Once again we recognize the inadvisability of trying to divide pest control into mutually exclusive disciplines.

Pheromones, and the attractants of the kind to be described in the next chapter, may well prove to be economically viable products for chemical manufacturers and provide an alternative product to compensate for the probable decline of the insecticide market.

9

Attractants and repellents in
pest control

In the last chapter we discussed the use of sex attractant and aggregation pheromones, but at the conclusion we noted that pheromones are not the only attracting agents which are of use in pest control. Other attractants have potential or proved value, and they include both physical and chemical agents. These will be discussed in this chapter. The attraction may be to a potential mate, as it is with many of the pheromones, or to a feeding or oviposition site.

Chemical attractants

A chemical attractant stimulates the responsive recipient to move towards the source. This behaviour must be distinguished from the stimulation of non-directed movement, which is not orientated towards the source, and from arrestant activity. An arrestant chemical brings about the cessation of movement when the animal encounters the chemical as, for example, when an insect finds by chance, or has been attracted to, a suitable oviposition site or source of food. Cane sugar is, for example, an arrestant for the house-fly. When a fly encounters a supply of sugar it is often stimulated to stop and feed. It is possible that a visual attraction, resulting from the contrasting pattern of dark flies on a white background, stimulates other flies to approach, but the sugar itself is not volatile enough to act as an olfactory attractant.

It is important to be able to distinguish between these responses when chemicals are being screened for attractive properties. It is relatively easy to distinguish between the three effects when an efficient olfactometer is used. A current of air carries the odour towards the insect, and it can be seen if there is significantly more locomotion aroused by such a current compared with a control air stream. If a significantly larger number of insects are found to move upwind than downwind then orientation, and thus attractiveness, is demonstrated. It is unwise to carry out a comparative test by exposing sources of several chemicals to insects in a cage, for the aggregation of insects about one of the sources may be the result of an arrestant activity which is not overcome by any attractant activity which may be present in some of the other chemicals. It should also be noted that a given chemical may show more than one kind of activity, that which is dominant depending upon the concentration, the state of the recipient, and the environmental conditions. Time of day may also influence the effect. At very high concentrations some attractants may even have a repellent effect. It is obvious that the biology of the organism must be well known when screening tests are made.

Other interactions are possible: carbon dioxide is a locomotory stimulant for *Aedes aegypti*, but recent tests show that it may potentiate the activity of true attractants from the body of its 'prey'.

Chemical attractants may be sex attractants (pheromones) which, in general, stimulate males, oviposition lures, which are effective for females, or food lures which may stimulate either sex or both.

The decomposition products of various kinds of organic matter often include oviposition lures. Female house-flies are stimulated to lay their eggs in the vicinity of sources of ammonia, fatty acids, and so forth. Similarly, decomposing fruit, squashed bananas, fermenting molasses, and similar materials are frequently food lures, and are often used in traps. Unfortunately, such chemically undefined materials are usually unspecific and inconsistent in their performance. Often, they only attract a small proportion of the individuals in the area. Nevertheless, they are often used in traps for survey work when nothing better is available, and, indeed, protein hydrolysate/insecticide preparations are employed still for the direct control of some species of fruit fly.

The unreliable performance of such preparations has stimulated chemists and biologists to examine them analytically to find the active components. An initial step was the examination of essential oils for biological activity but although these were often more reliable than the crude preparations, they are still variable in their effects. From these, however, a potent active ingredient may sometimes be isolated. It was this approach which led to the discovery, in Huon oil and citronella oil, of methyleugenol (1) which proved to be an extremely potent attractant of the oriental fruit fly, *Dacus dorsalis*. Its vapours are said to be effective at a distance of at least half a mile (Beroza and Jacobson, 1963).

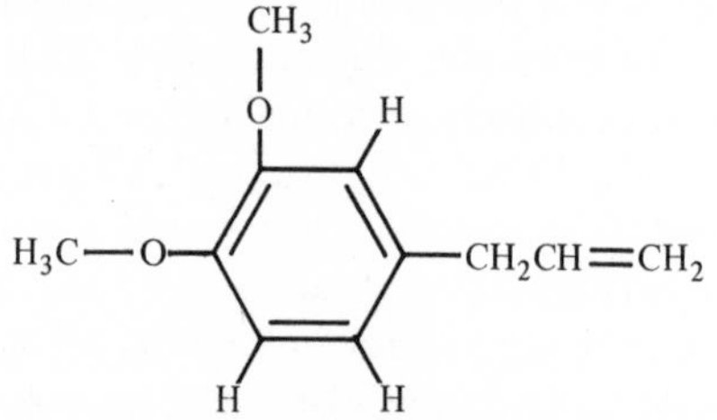

(1) Methyleugenol

A number of synthetic lures for certain economically important fruit flies are now available commercially. Cue-lure (2) is much more attractive to the male melon fly,

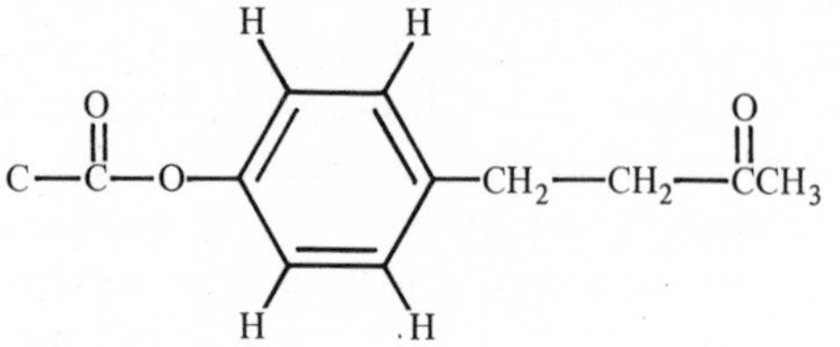

(2) Cue-lure

Dacus curcubitae, than is the related anisylacetone. Siglure (3), medlure (4), and trimedlure (5) have been developed for use against the Mediterranean fruit fly, and survey traps containing trimedlure have been used effectively during eradication campaigns in Florida.

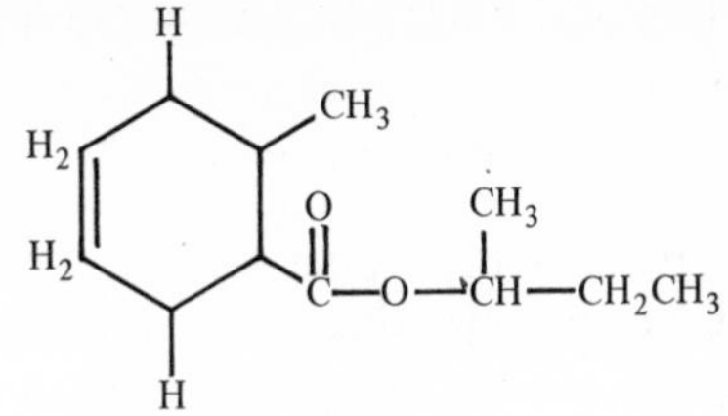

(3) Siglure

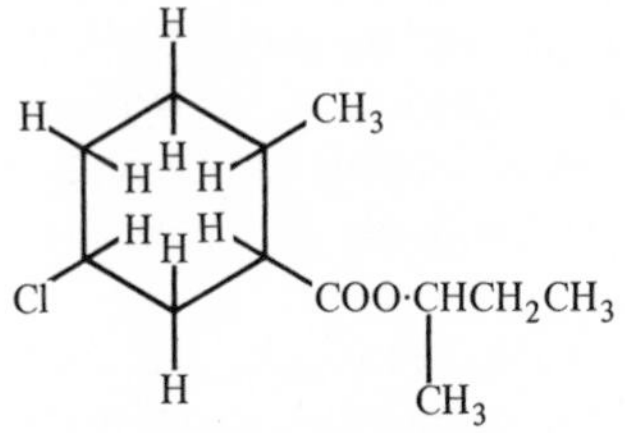

(4) Medlure

Insect attractants

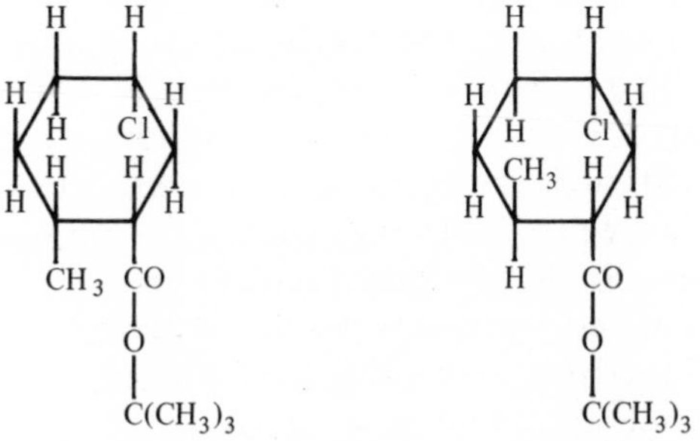

CIS

TRANS

(5) Trimedlure

These compounds are not completely specific. Methyleugenol is also attractive to *D. umbrosus*, anisylacetone to the Queensland fruit fly, *D. tryoni*, and to *D. ochrosiae*, cue-lure to the Queensland fruit fly, siglure to the walnut husk fly, *Rhagoletis completa*, and trimedlure to the Natal fruit fly, *Pterandrus rosae*.

The main value of these attractants has been in survey work in which a variety of specially designed traps have been used. They may also be used in the future for direct control. Absorbent boards, impregnated with methyleuganol and an insecticide, were dispersed on the small island of Rota in the Marianas. The resulting death of the male Oriental fruit flies was followed by the eradication of the pest from the island, at the remarkably low cost of 50 cents per acre (half hectare).

A few attractants have been developed for insects other than fruit flies. Butyl sorbate (6) is an effective lure for the European cockchafer, *Melolontha melolontha* (*Amphimallon majalis*), and methyl linolenate (7) for the bark beetles *Ips typographus*

$$CH_3-CH=CH-CH=CH-CO-OC_4H_9$$

(6) Butyl sorbate

$$CH_3CH_2CH = CHCH_2CH = CHCH_2CH = CH(CH_2)_7COOCH_3$$

(7) Methyl linolenate

and *Hylurgops glabratus*. Presumably, these synthetic attractants are related to natural products which are food or oviposition lures, but the mechanism of attraction is not, as yet, understood. Presumably, too, many more food and oviposition lures await discovery – a discovery that has had to be postponed until the recent instrumental developments of mass spectrophotometry, nuclear resonance, gas chromotography, and similar techniques. A further powerful technique, supplementary to conventional behavioural studies, is the use of electrodes, coupled to amplifiers and oscilloscopes, implanted in the sense organs concerned. Much has been achieved, however, by routine screening of hundreds of chemicals in a search for profitable 'leads'.

Unfortunately, it seems that most attractants will be effective for adult insects only. Immature forms do not seem to be able to detect their food plants from more than a few centimetres distance, and they rely almost completely upon their parent ovipositing on or near their food. The prospects for many public health insects are also discouraging. Parasites such as bed bugs and fleas do not seem to have well-developed olfactory senses and even the house-fly appears to be lacking in this respect, possibly as the result of hundreds of years of domestication. Mosquitoes, however, are attracted by chemical signals from the human body, and this is an active area of research. Blood, sweat, and urine are all attractive to *A. aegypti*, for example, and the attractiveness is enhanced by high contents of moisture and carbon dioxide. Interestingly, this mosquito finds women more attractive than men, and this may be related to high titres of oestrone and oestradiol. A further possibility is the study of the attractiveness of preferred oviposition sites. These factors are being studied for the *Culex pipiens* complex in Rangoon.

There is a further possible application for the study of chemical attractants present in plants. When these have been identified it may be possible to breed varieties which either lack the attractants or which contain them in only small amounts (Beroza and Jacobson, 1963; Jacobson and Beroza, 1964; Jacobson, 1965, 1966; Hocking, 1967; Schoonhoven, 1968).

Physical attractants

Visible and ultra-violet radiation

Light has long been known to be very attractive to many insects, especially nocturnal species. Electric lamps are now usually used as a source in light traps for control or survey purposes, but the various kinds of lamp differ in their spectral outputs. Furthermore, not all parts of the spectrum are equally attractive to insects, and there are also marked differences between the various species.

Electric lamps of various kinds produce wave-lengths from about 1800 Å to 50 000 Å. The region from 1800 to 3800 Å is the ultra-violet (far; 1800 to 2800 Å; middle: 2800 to 3200 Å; near: 3200 to 3800 Å). The visible spectrum is divided into: violet: 3800 to 4300 Å; blue: 4300 to 4900 Å; green: 4900 to 5600 Å; yellow: 5600 to 5900 Å; orange: 5900 to 6300 Å; red: 6300 to 7600 Å. Incandescent lamps, with tungsten filaments, produce a continuous spectrum but with three-quarters or more of the input energy radiated in the infra-red region. Only a small part is radiated in the biologically more useful violet section. A mercury-vapour lamp has a line spectrum (seven lines in the ultra-violet region, and four in the visible) but in fluorescent lamps phosphors are used to produce a more continuous spectrum with a peak at a longer wave-length. The blacklight (BL) fluorescent light has an envelope which converts the 2537 Å line to 3650 Å. The BLB lamp is similar except that it incorporates a filter which removes most of the visible light. Argon glow lamps produce two ranges, 2800 to 5000 Å and 6000 to 7600 Å (National Academy of Sciences, 1969).

The attractiveness of a light source to a receptive insect depends upon several factors; these include the spectral composition of the light, the amount of energy emitted at the various wave-lengths, the brightness, and the size and shape of the light. Environmental conditions are also important, as is the position of the light. In general, photopositive insects are attracted by all parts of the ultra-violet and visible spectrum. For most species, however, the near ultra-violet and middle ultra-violet regions are most effective. These are marked differences between species. The codling moth, *Cydia pomonella*, is attracted by lamps with a high ultra-violet light content in their output, but the region below 3200 Å appears to have little effect on the European corn corer, *Ostrinia nubilalis*, which is particularly stimulated by violet blue light. The pink bollworm, *Pectinophora gossypiella*, is attracted by the near ultra-violet region, from a 2-W argon lamp with a peak at 3650 Å, but the less visible radiation from a BL fluorescent lamp attracted the moths in a rather selective way in a cotton crop.

There are numerous designs of light traps, but a common type has baffles and a funnel below the lamp to direct the insects into the collecting chamber below. When traps are used in survey work a volatile killing agent is placed in the chamber to stun and kill the incoming insects before beetles and other predators damage them beyond recognition. The performance of a trap is often improved by the incorporation of an air-fan. The larger the lamp the better, although the effectiveness does not increase proportionately. Straight-tube fluorescent lamps have been found to attract more insects than circular tubes of an equivalent size and are easier to maintain.

Light traps have been most useful in survey work for pest control. They are also valuable for the detection of the presence of a pest in an area, and in quarantine measures. They have the advantage over chemical traps in that they can be run for longer periods without attention, needing, at the most, a time switch to operate them, and occasional inspection.

They have not been so successful for direct control of pests. Light traps gave some reduction of codling moth damage, but spraying was still needed. Similarly, European corn borer attacks were reduced, but complete control was not achieved. The most successful application has been the control of the tobacco hornworm, *Protoparce* (*Manduca*) *sexta*, and the tomato hornworm *P.* (*M.*) *quinquemaculata*, in tobacco crops in Indiana and North Carolina. One 15-W BL trap per acre gave excellent control of these pests, and the efficiency could be improved by the simultaneous use of sex attractants.

Light traps used for control may be charged with an insecticide, or a specific pathogen. They could also contain a chemosterilant, but as light traps are not highly specific they might be better used as an implement for collecting sufficient pest insects for subsequent sterilization and release. An interesting development, for use indoors or in recreation areas, is the electric-grid light trap which electrocutes the attracted mosquitoes and flies. Unfortunately, the grid may be short-circuited when too many insects are attracted.

Light traps are promising instruments for pest control but little is known about the effects of environmental conditions upon their performance. Such knowledge might be used to improve their efficiency for direct control, but at the moment their main task is to provide the information needed for the application of other control methods. Further details of the use of light traps in insect control will be found in the following publications: LaChance *et al.*, 1967; National Academy of Sciences, 1969. The use of green electroluminescent light for stored-product insects has been studied by Soderstrom (1970).

Infra-red radiation

Some insects are able to perceive infra-red radiation, but little is known of the receptors. One suggestion is that spines on the body may act as dielectric-type antennae.

An infra-red trap which attracts mosquitoes has been developed, and such an instrument might be of use against ectoparasites of warm-blooded animals.

Aedes aegypti adults prefer surfaces which are highly infra-red reflective but blood-fed females and mated insects of both sexes are less attracted than others.

Other visual attractants

Visual signals are undoubtedly very important in the lives of many animals, but they have been little used for behavioural control. House-flies, as was mentioned before, appear to be attracted by dark-coloured objects of approximately the same size as a fly, or by other flies, especially when they are on a contrasting surface. It used to be the custom, when a new fly paper was hung fron a ceiling, to catch two or three flies by hand, and to place them on the paper. Whether the technique has been investigated statistically is doubtful, but the trouble was possibly worth taking. A knowledge of the visual attractants of flies might be useful in deciding upon the best size and colours for granules in fly baits.

Dispersing aphids are known to be attracted towards the blue sky, and experiments have shown that they can be misled by reflective surfaces, such as foil, laid on the ground between crop rows. After dispersal they are attracted by yellowish plants rather than green ones, and may be trapped in brightly coloured yellow pans containing water. Adult cabbage-root flies, *Erioischia brassicae*, are also attracted to bright yellow objects, and it may be possible to divert them on their flight from the headlands and hedgerows, where they mate, to suitable yellow objects treated with chemosterilants or insecticides.

Sounds and vibrations

Various Diptera, Orthoptera, Coleoptera, and Hemiptera (and possibly other insects also) communicate by sound, especially during mating (Hurst, 1967; National Academy of Sciences, 1969). Such sounds are usually of low intensity so they can be reproduced at effective levels quite cheaply. Unfortunately, they must compete with background noise and, since they are usually repellent at high intensities, traps using them are effective only over very short distances. It is considered that the alternative, the jamming of the natural sounds, would be too expensive for use in control. A small-scale trial, in which male mosquitoes were attracted to an electric grid by a recording of the females, was fairly successful.

An interesting experiment is described by Hurst. During the mating period both sexes of the death watch beetle, *Xestobium rufovillosum*, communicate by tapping. It seems that the tapping has an inhibitory effect which prevents other beetles from tapping as long as the signal continues. In a trial in the parish church of Dalham in Suffolk, Hurst played back a recording of tapping on an endless tape over several loudspeakers, with a total power output of half a watt. The fatigue resulting from the endless search for a mate, or possibly from an adaptation to the sound, it was hoped, would reduce the total mating in the population. The results were, apparently, encouraging (Hickin, 1963; Hurst, 1967).

Chemical repellents

The essential characteristic of a repellent is that it stimulates an orientated movement away from a source. It should not be confused with an anti-feedant which prevents feeding, but which does not elicit the moving-away response. Although the final effect is the same, namely the control of damage rather than the killing of the pest, anti-feedants have been considered in chapter 3.

Repellents may be chemical or physical agents, but only the first will be described in this section. In addition to the gustatory and olfactory senses, which are also important in the reception of chemical attractants, the insects also possess a common chemical sense which enables them to respond to high concentrations of irritants such as ammonia, the essential oils, and chlorine. The receptors appear to be scattered over the body, and continue to function even when the olfactory senses are destroyed, but their identity is not certain. In grasshoppers they may be the thick-walled basiconic pegs which are found on many parts of the body and appendages.

Possibly the earliest form of pest control, with the exception of swatting, made use of a chemical repellent – smoke. This was followed by a number of 'folk remedies' which involved herbs and plants of various kinds. A short step led to a number of plant extracts, in particular, oil of citronella, from the grass *Andropogon nardus*, and oil of camphor. Oil of citronella, which contains the active principles geraniol, citronellol, borneol, and various terpenes, repels flying mosquitoes but its residual effect is short.

It was the requirements of tropical warfare that led to an intensive search for repellents for mosquitoes, the mite vectors of scrub typhus, and other carriers of human disease. At one stage during the war it was estimated that the chance of a soldier succumbing to a tropical disease before he encountered the real enemy was about 90 per cent.

Dimethyl phthalate (8) was patented in 1929 as a fly repellent, and it became an important constituent of skin preparations, and of formulations for impregnating clothes. Other compounds used during the war, either alone or in mixtures, were 2-ethyl-1,3-hexanediol (9) and indalone (10).

Possibly the most useful repellent for human use at the present is *N,N*-diethyl-*m*-toluamide (11), which has a fairly wide spectrum of use. A number of other repellents have been marketed but though screening for new compounds still continues, the intensity of search appears to have fallen off. Much of the current research is on methods of improving the cosmetic qualities of the preparations and their

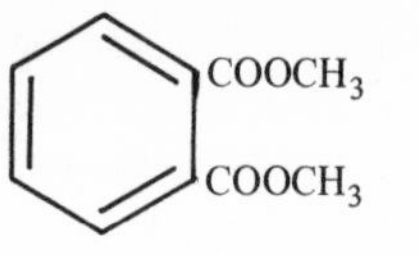

(8) Dimethyl phthalate

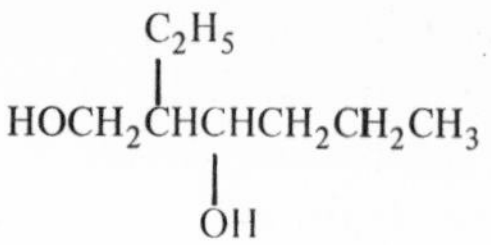

(9) 2-ethyl-1,3-hexanediol

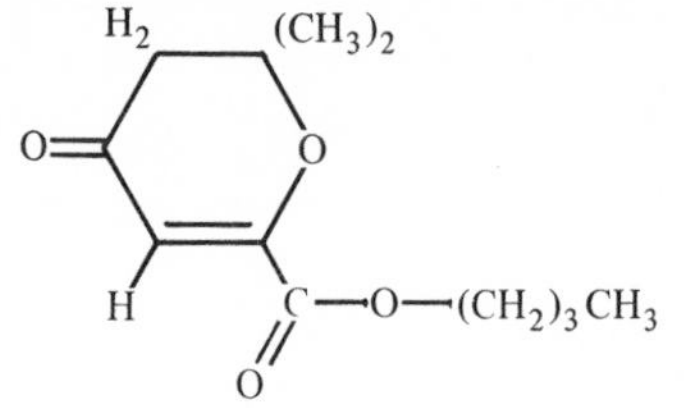

(10) Indalone

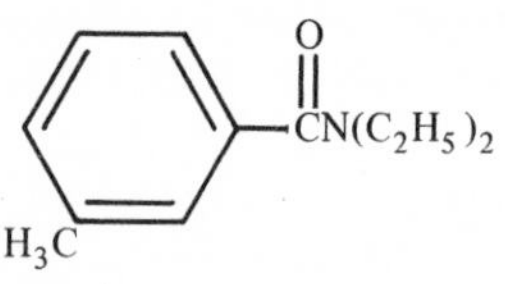

(11) N, N-diethyl-m-toluamide

residual effect. Another goal is the finding of a systemic repellent but it seems that none of the current chemicals which are applied to the skin can be used in this way (Painter, 1967).

Repellents have also been used for the protection of cattle from the attentions of muscid and tabanid flies, but it seems that the preparations have to be applied too frequently for convenience. There has been little success with repellents when they have been used to keep areas free from insect pests, although an aerosol has been marketed in Australia which was said to repel flies and mosquitoes from living rooms. Interestingly, some of the more recent repellents such as 2-hydroxy-n-octyl sulphide are especially active against cockroaches.

Little is known of the mode of action of chemical repellents, or the relationship between chemical structure and activity, Useful tables of repellents with their chemical structures and uses will be found in Metcalf, Flint, and Metcalf, 1962 and Painter, 1967. Other useful references are: Hartley and West, 1969; National Academy of Sciences, 1969.

Chemical repellents are valuable for personal protection, especially out-of-doors where insecticides would be ineffective. They may well play an important part in integrated control as this develops. An interesting finding is that some of them may be toxic to disease organisms as well as repellent to the insect vectors which transmit them. Hocking (1967) reports that thin films of dimethyl phthalate will quickly kill the larval stages of the filaria nematode *Wucheria bancrofti*, and he considers that it would be very unlikely for infection to take place under ordinary conditions through skin that had been treated with this chemical, even several hours previously. Probably all candidate and established vector repellents should be examined for such properties.

In conclusion, naphthalene and paradichlorbenzene (moth balls and similar preparations) are often spoken of as repellents but it seems that their effectiveness is due to insecticidal activity. Certain insecticides with a quick knock-down effect, such as pyrethrum, are also said to have repellent activity, but it is difficult to distinguish between the two effects.

294 Pest Control

Physical repellents

Painter (1967) reports that she was unable to find any reports of repellent properties of light, and she remarks that the repellency reported for yellow light is a misinterpretation of the minimal attractiveness of these wave-lengths. The use of sound is a little more promising, although high-intensity noises, in the 100 to 130 decibel range, have not, so far, been successful. Local residents would, in any case, object to the noise. There have been some interesting experiments on the control of noctuid moths, and of the European corn borer, *Ostrinia nubilalis*, in which crop damage was reduced by the use of ultrasonic sound. This technique makes use of the avoidance action taken by certain moths when hunted by bats.

Bats are surprisingly neglected animals although they are the most widely spread of all terrestrial mammals and, after the Rodentia, their Order is the largest in the Mammalia. The general lack of interest probably stems from their crepuscular and nocturnal habits, and the inaccessibility of their roosting sites. But one feature of bats is common knowledge: their ability to fly in complete darkness, using ultrasonic squeaks for echolocation. This unique aptitude was first discovered by Griffin at Harvard in 1938 (Griffin, 1960; Pye, 1969).

Insectivorous bats use their power of echolocation to find and capture insects flying in the dark or half-light, but many moths avoid capture by falling to the ground or by flying rapidly and erratically when they detect a bat's signals. The receptors involved vary in position from group to group, and since the moths diversified before the bats evolved, Pye suggests this variability indicates that hearing developed as a defence against these mammals.

It has also been discovered that certain moths, particularly the Arctiidae, themselves produce trains of ultrasonic sounds as they detect the approach of a bat. It was first suggested that these moths were jamming the signals of the bats, but it is now thought that these signals advertise the fact that the moths concerned are distasteful.

A few experiments have been carried out in which ultrasonic transducers were disposed around the periphery of a crop with the aim of preventing the moths from entering, and the results have been promising.

Attractants and repellents in the control of vertebrate pests

Behavioural control of vertebrates is considered separately from that of invertebrates as their behaviour is far more complex, and their ability to learn much greater. An initially successful control technique involving signals or scaring devices often becomes ineffective after a time because of the habituation of the animals.

The use of frightening devices is largely based on empirical results, with little analytical background. The oldest device is man himself, but economics and laziness soon led to the substitution of models of men and other organisms, in other words, to scarecrows. These were sometimes made more effective by making them, or parts of them, move in some way, or even, in more recent times, by incorporating loudspeakers. Such improvements, however, usually merely postpone the habituation of the birds or mammals they are supposed to drive away.

Models of hawks, with short necks and long tails, suspended from balloons or poles, have been tried with some success against gallinaceous birds but the method does not seem to have been applied for pest control.

Some animals are scared away by dead specimens of their own kind, and this behaviour may lie behind the gamekeeper's gibbets which were once so common in the British country-side. Dead gulls have been used with some effect on aircraft runways.

Our ignorance of the behavioural mechanisms involved in the scaring of birds and mammals is shown by the unexplained effectiveness of large glass globes on poles against raptorial birds in some areas, and of salted herrings against birds in some European vineyards and orchards! Although birds have poorly developed olfactory senses it is possible that it is the odour of the herrings which the birds find repellent.

Noises (as opposed to communication signals) are now used more commonly than other devices to repel birds from agricultural areas and places such as airfields. An acetylene banger, in which the gas is generated at a chosen rate by metering water on to carbide. is widely used in orchards. Fireworks also scare birds for a time, but the flash and the smoke puffs are believed to play a part. Habituation to noises soon occurs, unless the temporal pattern of the noises is varied, and the kind of firework changed frequently. Unfortunately, fireworks are expensive and usually present a fire risk. It is surprising, therefore, that loudspeakers and tape-recorders have not been used more, for, despite the initial higher costs, they are more economical over long periods. Furthermore, the tapes can be reprogrammed frequently to give the necessary variations in frequency and types of noises.

Ultrasonic sound transducers have been marketed as scaring devices, but it appears that ultrasound is also inaudible to most birds. Ultrasound can elicit autiogenic seizures (which are similar to epileptic seizures in man) in laboratory rodents, and in mice at least these seizures are often fatal. The intensity needed is fairly high, so there will be difficulties in the propagation of the sound, especially over large areas, if this method is developed further. Ultrasound may also be repellent to mammals.

There is growing interest in the use of recordings of communication signals of pest species. Attractive signals include the feeding call of herring gulls, *Larus argentatus*. Gulls which find food attract others to the source and recordings have been used to assemble the birds from distances up to a few miles. It is hardly likely to attract birds which are already feeding, or keep birds in the area if they cannot find food.

More use has been made of communication signals that tend to repel birds, that is distress calls and alarm calls. There is some confusion in the classification of communication signals, but for present purposes a distress call is taken to be the call made by a bird when held in the hands.

Distress calls are not always repellent to birds of the same species which hear them. In French experiments, corvids were found to fly first in groups towards the source, and then to disperse separately. Several European gulls behave in a similar fashion. The oyster catcher, *Haematopus ostralegus*, is attracted towards the sound, and does not fly away again. Flying flocks of starlings, on the other hand, are dispersed and flocks on the ground rise and fly away from the sound. Some birds do not make special movements at all in response to a distress call from one of their fellows. This behaviour, or lack of it, occurs for example in the Laysan albatross, *Diomedea immutabilis* on Midway Island, where it is a menace to aircraft. It is estimated that each year 300 to 400 aircraft collide with albatrosses (*Diomeda immutabilis* and *D. nigriceps*) in these islands, and one-fifth of these collisions results in an abortive take-off. Some 77000 albatrosses nest yearly on the five square kilometres of the atoll so, because of its military implications, the problem of aircraft strikes has been studied more intensively there than anywhere else in the world. Although many methods, including the use of scaring devices, have been tried, the best results have resulted from the levelling of the dunes which interferes with the soaring of the birds (Wright, 1968; National Academy of Sciences, 1970).

Alarm calls are produced by free birds which sight a predator, or which are stimulated by some other object, such as a man carrying a dead bird of the same species. The call may vary within a species with the nature of the alarm stimulus, so that a variety of information can be transmitted. The behaviour of the birds which hear the alarm call varies from species to species. Starlings, again, tend to disperse, while gulls often approach the source before dispersing.

The use of alarm calls and distress calls for bird dispersal is promising, but it is becoming obvious that the operators will have to be well trained, with a sound knowledge of the behaviour of the species with which they are dealing. Environmental factors, such as weather and time of day, will influence the behaviour of the birds. Even more important will be the fidelity of the recordings, and this will call for considerable expertise in the field of acoustical engineering.

In general, the more portable the recording and play-back equipment, the less satisfactory will be the result. This is no great inconvenience when recording distress calls, which can be made in a fully equipped studio, but alarm calls have to be made in the field. One difficulty is that many bird calls include sudden and rapid changes in frequency which are difficult to reproduce. The ultrasonic component, again difficult to reproduce, is of importance although, as said before, most birds are probably not able to hear ultrasound as such. The ultrasonic frequencies in bird calls, however, do alter the aggregate wave form of the sound, and their absence from recordings may be biologically important. It is also essential to avoid the distortions that are introduced by recording at too high a level. The quality of the reproduction apparatus – the tape-deck, the amplifiers, and the loudspeakers – must match that of the recording apparatus.

An allied problem is that of the occurrence of dialects in different populations of some birds. A recording that was effective in Holland, for example, did not stimulate the same species in the United States, so that local recordings had to be made.

There are, therefore, numerous difficulties, a few only having been listed here, in the use of distress and alarm calls, but the techniques are worth elaborating. One advantage is that the responses are largely instinctive so that habituation, if it occurs at all, is slow in developing.

Radar could possibly be of value in controlling aircraft strikes. Even in the early stages of its development radar equipment could detect flocks of birds. Now, it is reported, individual locusts can be detected at distances of several miles. During the war (*The Battle of Barking Creek*, 6 September 1939) a flock of birds brought about the 'scrambling' of a large number of fighter aircraft which searched in vain for, it is tempting to suggest, wild geese, until their fuel was exhausted. Whatever the species was that was involved, this and similar incidents led to the appointment of a number of biologists as radar watchers. Their studies, and those of their successors, have enabled radar operators to recognize a variety of types of bird behaviour on their screens. It is possible to give pilots advance warning of flocks of large birds in their flight path. It should also be feasible to treat flocks of birds, when not too numerous, as 'other aircraft' when giving pilots taking off or landing instructions (Sir Peter Wykeham and Bourne in Murton and Wright, 1968; Gunn and Solman, 1968; Schaefer, 1968).

There have been some reports of a direct effect of radar on flocks of flying birds, and it was suggested that this might be used to regulate their course. Other authors have been unable to detect any response to radar under normal conditions.

The discussion of physical repellents cannot be completed without mentioning the treatment of roosting sites, particularly on public buildings, with slippery materials or substances which are otherwise distasteful to birds. This has been used with some success against starlings and feral pigeons in London. Unfortunately, the birds usually merely change their roosting quarters.

Less attention has been paid to feeding deterrents and repellents. Sublethal doses of insecticides have been applied to seeds to protect them from rodents. The intention is not to kill the animals but to discourage them from feeding on the seed again: this continued existence of conditioned mammals in the area discourages unconditioned individuals from moving in (Rudd, 1964).

Seed dressings that are distasteful to birds are only partially successful, being ineffective when the birds are hungry. This is probably due to their comparatively poorly developed senses of taste and smell. Their vision is extremely well developed however, and the reported effectiveness of colouring seeds a bright green with harmless dyes is worth investigating further.

In conclusion, attractants, repellents, and frightening devices hold great promise in pest control, but significant advances will only come with a greater understanding of animal behaviour. The approach was neglected for many years, apart from the search for repellents for personal application, but they offer the advantage of minimal environmental pollution. The habituation of vertebrates to frightening devices will be a problem, akin to the development of resistance to pesticides, and it would be wise to hold them in reserve so that they are only used at the most opportune time. This must be decided not only on the basis of the danger presented, but also from a consideration of the general biology and ecology of the pest (Frings and Frings, 1967; National Academy of Sciences, 1969; Murton and Wright, 1967).

10

Resistant varieties and miscellaneous methods of pest control

In this chapter, we shall consider a number of pest control techniques which do not fit in conveniently elsewhere. Many of them have been neglected during what has been called the 'DDT Decade', others, such as crop rotation, have been applied because they have become more or less a tradition. Yet others have not been the direct concern of farmers and growers as they have been the business of governments and large organizations. Many of these methods are now being re-examined because of their potential role in integrated control, the subject of the final chapter.

Resistant varieties

If a variety is completely resistant to a pathogen or a pest, there is no immediate need to seek for and to apply other control measures. The grower need take no action, save paying for the development costs in his seed bill, and this is generally much less than the costs of protection for a non-resistant variety.

It may seem surprising, therefore, that greater use has not been made of resistant varieties, but there are several reasons why this is so. Plant breeding can rarely provide a quick solution for a pest problem, for it takes between 10 and 15 years from the formulation of the problem to the distribution of the seed in large quantities. The difficulties are even greater in finding resistant breeds of livestock because of the comparatively slow rate of reproduction, and this difficulty is heightened by the probable confusion between inherited resistance, and immunological resistance. The development of a resistant form of man presents the same problems as any eugenic programme.

The plant or animal breeder is further restricted to the genetic variability present in the populations of the organism, and in closely related organisms which may interbreed. There is the possibility of producing new mutations by X-rays and other agents, but the mutations still occur randomly. In short, it may prove impossible, even after years of careful work, to produce a resistant variety or breed.

The selection for pest or disease resistance is only a part of the breeding process: a new variety must compete successfully with established varieties in such respects as yield in the absence of the pest, uniformity of germination, flowering and maturation, ease of harvesting, and so forth. If the pest concerned is troublesome every year it will be sufficient that the variety outyields the older varieties in the presence of the pest, if not in its absence, but if the pest occurs irregularly this will not be satisfactory.

Plant pathologists are generally more ready than entomologists to search for resistant varieties as an answer to their problems. It is probably true to say that many entomologists only think of using a resistant variety as a last resort, if, indeed, they consider it at all. This is partly because the relationship between a plant and a pathogen is usually much more intimate than that between a plant and an insect pest, but surely it is also partly because of the differing academic training of the two kinds of biologists. The plant pathologist is usually well versed in academic and practical botany, which include the principles of plant breeding, whereas many entomologists have never formally studied such matters.

In the past, possibly too much stress has been laid on the importance of finding or breeding a variety which is completely resistant to a given pest. It is now realized that a partial resistance can be extremely valuable, for it can play a part in an integrated control programme. If, for example, the partial resistance is of the kind which slows down the reproductive rate of an aphid, chemical sprays could be used less frequently or in smaller quantities. Complete resistance, furthermore, is more easily broken down than is partial resistance.

The techniques of plant breeding will be discussed only briefly; an excellent concise introduction will be found in *Plant Breeding* by Lawrence (1968).

Domestic plants and their various varieties can arise in three main ways: by Mendelian segregation and recombination; by interspecific hybridization; by polyploidy.

Mendelian segregation and recombination   Many crop plants have been improved by the simple selection of variants, and some of these improvements have involved resistance. Lawrence lists the following crop plants which have been improved mainly, if not completely, by selection: asparagus, beans, beets, carrots, celery, lettuce, lucerne, lupins, onions, radish, soybeans, and tomatoes.

Interspecific hybridization   This technique has been of great importance in developing resistant varieties, for often close relatives of crop plants show such characteristics. It has also been of value in general crop breeding. Unfortunately, hybridization alone frequently produces sterile hybrids. In some cases this is not of great importance, for the hybrids may be propagated vegetatively (for example, in many ornamental plants, many of which have arisen in this way). Among the cultivated plants in which interspecific hybridization has been important are wheat, apples, grapes, tomatoes, loganberries, maize, and rice.

Polyploidy   Polyploidy, which involves a reduplication of the chromosome number, has been of the utmost importance in the evolution of plants, both wild and cultivated. Autopolyploids, in which identical or almost identical sets of chromosomes are duplicated, often show reduced fertility, but are valued because they often have large fruit and flowers. Allopolyploids result from the doubling of the chromosome number after hybridization of two species. The doubling of the chromosome number restores the fertility of the hybrids which become functional diploids. Such a process – hybridization followed by the doubling of the chromosome number – produces a species which combines the features of the two parent species, often in a favourable way.

Although these three sources of variation have been considered separately, all may have been involved in the development of a species or a variety.

The procedure of selection depends upon whether the plants concerned are outbreeding or inbreeding. Selection in inbreeders will be considered first.

Selection in inbreeders   In the single plant selection technique the breeder starts with a large number of superior plants which he selects from the population available to him, and raises self-progeny from each of these. The net result of continuous self-fertilization is the halving of the heterozygosity in each generation, with increasing homozygosity of recessive and dominant genes. Thus this procedure will develop a large number of pure lines from which the breeder will make an appropriate selection each year, for as many years as necessary. The procedure is slow, and demands a large amount of attention, particularly when the breeder is trying to improve some quantitative character.

In mass selection the crop is rogued out each year, to get rid of the individuals which do not display the desired characteristics. There must be a compromise between insufficient roguing, and too great a reduction of genetic variability in the population. Pedigree breeding, in contrast, starts with the crossing of two desirable varieties, followed by selection in the following generations, until pure lines are obtained. Bulk population breeding relies upon natural selection to do the work. An F_2 generation is obtained by the last procedure, and planted in quantity. Several generations are then planted in bulk, and harvested in bulk, probably in several localities, and natural selection is relied upon to provide plants with the desired characteristics. The method would work well for the production of resistant varieties only if the stock is frequently at risk from the pest.

Backcross breeding has been extremely useful when the desire has been to introduce resistance (or some other valuable characteristic) into an otherwise satisfactory variety, when the desirable characteristic is present in an otherwise inferior 'donor' variety. The donor is continuously backcrossed to the better variety, and their progeny, with selection for the character in each generation. Finally, the last backcross generation is selfed.

Selection in outbreeders The problems of breeding (and hence the procedures) differ for outbreeding plants, as the members of the population share a common gene pool. The aim, therefore, is to select so that a population with a high frequency of desirable gene combinations is obtained. Too much inbreeding often results in a less vigorous population, because of the appearance of a large proportion of homozygotes. It is thus essential to maintain heterozygosity. Single plant selection cannot, therefore, be carried too far, and, indeed, in some species it is impossible because of self-incompatibility. Mass selection, or mass selection followed by line breeding, in which, after several generations, seed is taken from the best plants, sown in isolated plots, and allowed to mate at random, has thus been used more frequently than single plant selection.

In recurrent selection plants are chosen from an heterozygous population, and selfed. The resulting plants are then intercrossed to give progeny for further selection. The cycle is repeated a number of times until a population with the desired qualities is obtained. Backcrossing is also possible with outbreeding plants, but the donor plant must be backcrossed to a large number of plants, rather than to a single plant, as in self-pollinating varieties. For a fuller explanation of these breeding procedures the reader should consult Lawrence's book, of which the above is a partial summary.

Plant resistance to pathogens

The grape vine played a large part in the early development of fungicides: it also furnishes an early example of the deliberate breeding for disease resistance. This was achieved by the hybridization of European and American stocks to confer resistance upon European vines against the downy mildew, *Plasmopara viticola*. This work was particularly noteworthy because it was carried out before the re-discovery of Mendel's work. Another early achievement, which occurred at the beginning of this century, concerned the soil-borne pathogen of potatoes, wart disease (*Synchytrium endobioticum*). This pathogen could not be conveniently controlled by any available method. The disease was first recorded in Britain in Cheshire and Merseyside during 1902, and attempts were made to confine the disease to this area by quarantine. Gough, a Board of Agriculture Inspector, was appointed to survey the incidence of the disease, and during a conversation with a Cheshire farmer, H. S. Daine, he learned that at least one variety of potato, Snowdrop, could be grown successfully in the area. Gough's further inquiries led to the discovery of several other resistant varieties. Trials at various sites confirmed that Golden Wonder, Langworthy, Conquest, Abundance, and Snowdrop displayed complete resistance to the pathogen. Little breeding work, however, was carried out until 1915, although the quarantine measures were maintained (Large, 1940).

It is most unusual to find established varieties which are completely resistant to a disease which has just appeared in a crop. Usually, new varieties have to be produced to provide the necessary resistance, and all too often this painfully acquired resistance disappears after a few seasons when a new strain of the pathogen arises. This has occurred several times among the rusts of cereals, many of which have a sexual stage in the life cycle.

The rust diseases have long been regarded as the most important pathogens of cereals: each spring the Romans sacrificed a red dog, symbolic of the supposedly baleful dog star, to protect their corn from their ravages. For centuries, it was sus-pected that there was some connection between one of these diseases, stem rust of wheat, and the shrub, barberry. It is not known whether this belief was based on the somewhat reddish colour of the shrub (reminiscent of the colour of the fungus), or on some accurate observations made by farmers, but for many years it was thought to be a bad omen if a *red* dog ran through the crop. Whatever the basis for this dislike of the barberry's presence, the city of Rouen ordered its destruction in 1660, and in 1755 the government of Massachusetts decreed that the land-owners should eradicate the shrub by 1760 (Large, 1940).

It was not till the 1860s, however, that Anton de Bary demonstrated the causal relationship between the disease on the wheat, and the cluster-cups on the shrubs, and even he failed to discover the sexual part of the life cycle (Fig. 10.1). It is, of course, this sexuality that endows the fungus, *Puccinia graminis* var. *tritici* with its plasticity which allows new physiological races of the pathogen to develop. At that time, however, this was not of practical importance: what was important was that the barberry, *Berberis vulgaris*, served as a focus of infection in the spring for near-by wheat crops. An important part of the control of stem rust has been, accordingly, the eradication of the shrub wherever possible. This has been more or less completed, for example, in the wheat-growing belts of the Great Plains of America, though the shrub is common enough in more mountainous regions.

The pathogen produces two kinds of spores on wheat. Earlier in the year the urediospores, reddish, bi-nucleate single-celled spores develop, generation after generation, each batch infecting fresh wheat plants. Towards the end of summer and in the autumn, teleutospores appear. These two-celled, dark-coloured spores are, when immature, dikaryotic, that is, each cell contains a pair of nuclei, but as

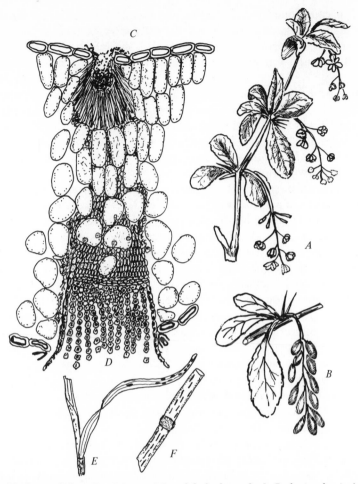

Fig. 10.1. The black rust of wheat *Puccinia graminis*, and the barberry shrub, *Berberis vulgaris*. **A**: flowering shoot; and **B**: fruit of barberry; **C**: stages of fungus on barberry leaf shown in section; **D**: spermagonia on upper surface, aecidia (cluster cups) on lower surface; **E**: summer spores (uredospores) on wheat; **F**: winter spores (teleutospores) on wheat

they mature the nuclei fuse to give diploid cells. This is the overwintering form: after meoitic division to give haploid cells the spores germinate in the spring and release haploid basidiospores which infect the barberry leaves.

On the upper sides of the leaves the spermagonia develop, and haploid pycniospores are produced. The sexual process follows (although the nuclei are not to fuse till much later) when pycniospores of one sex group fuse with hyphal cells of another. From the products of this fusion the aecidiospores, which infect the wheat, develop in the so-called cluster cups on the undersurface of the leaf.

In the absence of barberry the basidiospores develop no further, and wheat in barberry-free areas must be infected from further afield. In Europe, where, fortunately, the disease is rarely epiphytotic one important source, apart from residual barberry bushes, appears to be urediospores in southern Portugal and Spain, which either overwinter there, or which reach the area from somewhere further south in January and February. It seems that overwintering spores alone would not be sufficient to cause epiphytotics in the years that these do occur. In England the disease usually breaks out late in June, shortly after the arrival of the first wind-borne spores from Portugal, and it seems that the infection works its way slowly northwards through Europe in this fashion, reaching Sweden and Norway in late August.

In North America the disease is much more serious. The uredial mycelium and spores do not overwinter in the north, the main focal centres being northern Mexico and southern Texas. In the spring the spores are blown northwards, usually not more than 300 miles at a time, but in cyclonic conditions for much greater distances. In the autumn there is a southward movement of the spores to infest autumn-sown wheat in the overwintering areas.

Modern interest in stem rust originated with the Australian epiphytotic in 1889 which led to a series of international conferences. A few years later a survey by Carleton in the Great Plains and the Mississippi valley showed that the disease was so serious that many districts were abandoning the crop. There is some confusion about the identities of the rusts concerned in some outbreaks, but Carleton was convinced that in North America the most important pathogen was stem rust, and that the others could be safely ignored, for the time being at least. Carleton introduced a number of varieties of durum wheat and Crimean winter wheat from Russia as these, he considered, would be better suited to the semi-arid Great Plains than the pre-dominantly Mediterranean types grown there. The durums in particular proved to be well suited, being hardy and drought resistant. One of them, 'Iumillo' Durum was found to be extremely resistant to rust. Unfortunately, the durums are macaroni wheats, quite unsuited to bread making.

Carleton also turned to the selection of naturally occurring mutations, and en-couraged farmers to take their seed from any plants in their crops which proved resistant to the disease.

The American epiphytotic of 1904 stimulated the United States Department of Agriculture to start a breeding programme for wheat varieties resistant to stem rust. In this hybrid crosses were made between *Triticum vulgare*, bread wheat, in which no source of resistance could be found, and the following species: *T. monococcum*, einkorn; *T. dicoccum*, emmer; *T. turgidum*, rivet or cone wheats; *T. durum*, macaroni wheats. Unfortunately, most crosses proved more or less infertile, but some *durum* × *vulgare* crosses were successful. There were, however, other difficulties, especially in the early stages of the programme. Resistance was often found to be genetically linked to undesirable features such as durum characteristics or suscepti-bility to other diseases. Sometimes the resistance was unstable, breaking down in certain environmental conditions, such as unusually high temperatures.

These were minor difficulties compared with that posed by the plasticity of the pathogen. In 1916 an epiphytotic broke out in North America to which the varieties were not resistant. The new strain of the pathogen had presumably arisen on barberry, and destroyed, it is estimated, about 40 per cent of the potential crop in the USA. This led to the replacement of the popular variety 'Marquis' by 'Ceres' in 1926, but this, in its turn, succumbed to a new strain of the pathogen (race 56) in 1935. This epiphytotic destroyed about half the spring wheat in Minnesota and North and South Dakota, and damaged crops elsewhere. 'Marquis' was replaced by 'Thatcher' and other varieties resistant to race 56 but they failed to resist 15B in 1953 and 1954. The variety 'Selkirk' took their place after these epiphytotics. Fortunately, race 15B had been recognized before the epiphytotics so resistant varieties were soon made available, but such catastrophic losses in a major crop are disastrous.

Vallega (1956, 1959), has suggested that genetic factors for resistance to each of the known races of rust should be introduced into a single variety by backcrossing, and that one of these isogenic lines should be released for general cultivation at a time, to be replaced as soon as signs of infection appeared. Borlaug (1958), however, urged the release of a mixture of such lines so that no single race would cause widespread damage, but it has been pointed out that this would lead to the natural selection of strain pathogens virulent to all the genes present.

The resistance of wheat to a strain of stem rust is apparently given by a single gene which corresponds to a gene for virulence in the pathogen (Flor's gene for gene hypothesis, first formulated for flax rust, Flor (1972)). A single pathogen mutation is sufficient, therefore, to produce a new race which breaks down resistance newly acquired in a wheat variety. It may be possible to avoid the disadvantages of this 'vertical resistance' against specific races by making use of 'horizontal resistance' or general field resistance shown by such varieties as 'Marquis'. This, and similar varieties which were in cultivation before specific breeding for resistance was started, show, as van der Plank (1963) points out, a reasonable level of non-race-specific resistance. Alternatively, wide crossing of wheat with other species, such as members of the genus *Aegilops* – a genus which had an important part in the phylogeny of modern wheat – may produce a resistance which cannot be broken down by a single mutation in the pathogen. Attempts are now being made to introduce resistance to yellow rust, *P. striiformis* (formerly known as *P. glumarum*), and leaf rust, *P. recondita*, using *Aegilops umbellulata* and *A. comosa* as sources.

Other countries have experienced epiphytotics of stem rusts similar to those of North America. In 1963 and 1964, for example, Western Australia suffered heavy losses from the strains 21-Anz-1,2 and 21-Anz-2. It seems that barberry is not functional on the Australian mainland, so the production of new strains by somatic hybridization may have to be accepted. Some new strains may arise in Tasmania where, because of the rough country, the barberry would be difficult to eradicate. Since the discovery of the role of the barberry in wheat rust epidemiology, a second alternative host has been found in the closely related genus *Mahonia*.

Stem rust of wheat has not been as troublesome in North West Europe as the closely related yellow rust, *P. striiformis*. In recent decades, there have been epiphytotics of a series of physiological races which have prevented the growing of a number of otherwise excellent varieties. Thus in successive outbreaks in 1951, 1955, and 1956 different strains attacked the valuable wheat varieties 'Nor Desprez', 'Heine VII', 'Opal', and 'Rothwell Perdix'. There is no known sexual stage in the life cycle of this pathogen, so it is not known exactly how new strains arise. Fortunately, the breeding work has been such that there are a number of varieties which show reasonable field resistance.

There is, naturally, a voluminous literature on cereal rusts and cereal breeding, but the interested reader will find an excellent introduction in the volume edited by Quisenberry and Reitz (1967). The genetics of resistance to *Puccinia* spp. is reviewed by Hooker (1967) while Johnson, Green, and Samborski (1967) discussed the world situation of the cereal rusts in the same year.

The historical and present importance of the cereal rusts is rivalled by that of potato blight, *Phytophthora infestans* and this crop has naturally received much attention from the plant breeders. Here, it is felt, stress should be laid on field or horizontal resistance. 'Champion' was released in the 1880s and soon occupied most of Ireland's potato lands. Unfortunately, it became badly infected with Virus A, but a virus-free stock was released in 1920.

Potato blight can be controlled quite effectively by fungicides so breeding for blight resistance is of less urgency than breeding for resistance to potato root eelworm, *Heterodera rostochiensis* (Nematoda). The only feasible alternative method of combating this pest on a field scale is the unpopular adoption of a very long rotation, as chemical soil fumigation is too expensive. Certain South American clones of the subspecies *andigena* of the potato are resistant by virtue of a single dominant gene which is easily transferable to new varieties. The new varieties resemble the older ones in producing a root diffusate which causes many eggs in the cysts to hatch. The larvae then invade the roots in the normal way. The larvae of the potato root eelworm are unusual in that they can develop into males or females, the sex of the adult depending upon the environment of the larvae. In the roots of resistant varieties most larvae develop into males, so that the populations rapidly decrease because of the paucity of females. The females that do occur either stimulate the root cells to develop into giant cells in the normal way so that the females mature, or they fail to do so, and die. Field trials showed that the developmental behaviour of the eelworms differed from area to area. In most parts of East Anglia the new varieties 'Ulster Glade' and 'Maris Piper' supported few females but in some parts of Lancashire the potatoes were not resistant enough.

The resistant varieties apparently compare favourably with established varieties, even when grown on uninfested land. At Woburn, for example, 'Maris Piper' outyielded three other varieties ('Pentland Dell', 'King Edward', and 'Majestic') when grown on land that had not carried potatoes for ten years. On slightly and on moderately infested land 'Maris Piper' yielded much better than 'Pentland Dell' on both irrigated and unirrigated plots, and on plots that had been fumigated (DD, 400 lb/acre (362 kg/ha)) and on those which had not. Interestingly, fumigation raised the yield of Maris Piper from 16·33 ton/acre to 18·27 ton/acre, presumably by preventing any eelworm attack.

Empson and James (1966) used a computer to forecast population trends in the eelworm, with various rotations which involve resistant varieties, susceptible varieties, or both. They also estimated the yields to be expected under the various systems. It might be expected that the highest mean yield would be obtained by growing resistant varieties only, but it will be remembered that biotypes which can overcome the resistance are already known. The growing of resistant varieties only would select these types fairly quickly, perhaps after a mere five or six years. This could be avoided by alternating susceptible varieties with the resistant ones, which would lessen the selective pressure, but with some loss of yield. Over many years, however, it is expected that the mean yield would be higher.

Field trials in which resistant varieties derived from *andigena* were grown continuously showed that the eelworm population changed from biotypes which did not attack the varieties to two types, *B* and *C* which did. Type *C* was also able to breed successfully on varieties derived from the sub-species *multidissectum* or from both *andigena* and *multidissectum*. Varieties derived from *Solanum vernei*, in which the resistance is polygenic, are resistant to most biotypes, but satisfactory backcrossing with *S. tuberosum* is proving difficult. A new clone, D 40, deriving one gene from *andigena* and two from *multidissectum* is, however, almost as satisfactory with respect to resistance as *S. vernei* (Empson and James, 1966; Jones and Parrott, 1968).

Resistance to viruses may be achieved by breeding for resistance to the pathogen or to its vector. A number of single dominant genetic factors is known for some virus diseases of potatoes, and polygenic factors for potato leaf roll and the two viruses of beet yellows, BYV and BMYV (Lupton, 1967).

The mechanism of plant resistance to pathogens   In the last section we have accepted plant resistance to pathogens as a fact, without discussing the nature of this resistance. This is clearly of great academic and practical importance and it will be discussed in the following paragraphs.

Cruikshank (1966) has pointed out that most plants are susceptible to infection by most fungi, but are resistant to most diseases. The fungi can germinate and penetrate the epidermis, but the resistance mechanism, when it occurs, is a part of the subsequent reactions between the plant and the pathogen. The severity of the disease is, of course, also influenced by other factors such as the parasite's inoculum density, the physiological state of the host, and the environmental conditions.

In a non-resistant host there is generally a reduction in photosynthetic activity, and increases in respiration and in the production of RNA and protein. There is also an accumulation of polyphenols in the tissue near the pathogen. The differences between the reactions of resistant and susceptible hosts appear to be quantitative rather than qualitative. This accumulation of polyphenols is especially marked around the site of infection, and it may be so rapid that it restricts the growth of the pathogen, and may even kill it.

Pathogens also induce their hosts to produce fungitoxic substances through the interaction of the pathogen and hypersensitive cells of the host. There appear to be two distinct groups of such metabolic products: substances which are produced by a wide range of host plants when infected by any one of a wide range of microorganisms (chlorogenic acid, caffeic acid, etc.) and those which are produced by a specific host in response to a wide range of pathogens. The latter have been called phytoalexins, and have been identified in a number of plants. It is too early to generalize about their chemical nature, but they appear to be low molecular weight aromatic compounds. Resistance in the plant appears to be correlated with the rate at which these compounds are produced after infection with the pathogen.

Phytoalexins are wide spectrum fungicides, but they show some selectivity. In trials with pisatin from peas Cruikshank (1966) and his colleagues investigated its toxicity to fifty fungal species. The pathogens were found to be relatively insensitive to the substance, whereas those which were not pathogenic to the pea were very susceptible.

It has already been pointed out that phytoalexins are host specific rather than pathogen specific. Their formation can, in fact, be induced in other ways, as, for instance, by chemical treatment. Heavy metal ions stimulate the formation of pisatin, and the fungicidal activity of some of the products which contain such ions may be partially systemic through the induction of this mechanism. This would be classed as conferred resistance by Grossmann and his colleagues (1968), and further examples of this will be given in the next section.

It may be possible to synthesize phytoalexins and related compounds in the near future and to use them directly as fungicides. Modification of their structure may increase the toxicity and selectivity. It would not be possible to extract them from healthy plants in the way that pyrethrum, for example, is obtained, for these compounds are only synthesized by the plant as a consequence of the interaction of a pathogen and host plant cells. The bioassay or quantitative analysis of phytoalexins will serve as a measure of resistance potential of a variety at the metabolic level, and such techniques will probably take their place in breeding programmes in the future (Cruikshank, 1966; Kuć, 1968; Day, 1968).

The resistance of some crop varieties to nematode attack is similar in many ways to disease resistance. Cells collapse and become brown around the site of attack, effectively isolating the nematode; it is not yet known to what extent nematicidal materials are produced by the injured cells (Dropkin, 1969), but Rhode (1960) reports that one variety at least of asparagus produces an anticholinesterase which possibly inhibits acetylcholinesterase in nematodes. The variety is known to be resistant to nematodes.

Virus resistance can also arise by hypersensitivity reactions, with the isolation of the pathogen, but some resistant varieties can become systemically infested without showing marked symptoms of the disease.

Conferred resistance in the host In the last section we have seen that certain chemical treatments can induce the production of fungitoxic materials by the plant itself, and can thus confer resistance by defence mechanisms to a plant which does not possess them genetically. Other measures can enhance the passive resistance of the plant to the pathogen. Clearly, there are difficulties in drawing a sharp line of distinction between such methods of control and conventional methods of control by chemicals. Copper oxychloride and folpet, for example, are conventional fungicides, but there is evidence that they also strengthen the epidermis and cuticle of poplar and vine leaves, making the invasion by some kinds of fungi more difficult. The number of stomata (which afford a route of entry to many fungi) may be reduced by chemicals such as (2-chloroethyl)-trimethylammonium chloride (CCC). This chemical may also impede the spread of a pathogen within the tissues by increasing the number of cell layers in the parenchyma, and, possibly, by the increased formation of tyloses in infected plants. Other chemicals will close stomata and thus reduce the risk of infection by pathogens such as wheat stem rust. Zelitch (cited in Dimond, 1963) almost completely prevented infection by bean rust by spraying with α-hydroxy-decane sulphonic acid which shows this physiological effect. Growth-regulating herbicides may have a similar effect.

A plant may be made resistant to the effects of a pathogen, rather than to the infection by the pathogen, by treating it with some substance which inactivates the toxins produced by the pathogen. Ferulic acid ($CH_3O . C_6H_3(OH) . CH:CHCO_2H$) counteracts a toxin of *Piricularia oryzae* in rice, and β-indoleacetic acid degrades fusaric acid, a toxin of *Fusarium*, in tomato plants. The difficulty inherent in this approach is that very few pathogens produce only one toxin and, furthermore, most fungi also produce extracellular enzymes which damage the host tissues. It may be possible to introduce anti-enzymatic compounds to counteract these effects. Some amino-acids, for example, appear to suppress enzyme formation, and Grossman (1968) has had some success in treating the effects of wilt in tomatoes with rufianic acid, an inhibitor of pectic enzymes. In some experiments the plants actually recovered from the disease.

The conferring of resistance to plants will become increasingly important as our knowledge of host–pathogen relationships and defence mechanisms grows, but at present the techniques are merely in the experimental stage.

Host-plant resistance to insects

Less practical use has been made of resistance of plants to insects than of resistance to pathogens and nematodes. The method was eclipsed by the coming of the modern synthetic insecticides, but interest is once again growing. It is now realized that a complete resistance is not essential, for even a low level can be of value. There are few, if any, deleterious side-effects when resistant plants are used, and the effect on the pest population, when the resistant variety is grown over a wide area, is cumulative.

The first certain record of a varietal resistance to an arthropod pest concerns the apple 'Winter Majetin' which, in 1831, was found to be resistant to the woolly aphid, *Eriosoma lanigerum*. There were earlier reports of wheat resistance in America to the newly introduced Hessian fly, *Mayietola destructor*. The first major success was, once again, in the vineyards. In 1868 the wine industry of France was threatened by the root-feeding homopteron, *Phylloxera vitifoliae*. Chemical methods were ineffective, but the industry was saved by the introduction of root stocks of American species of *Vitis* on to which the European species, *V. vinifera*, was grafted.

The first extensive search for sources of resistance to an insect pest was among small grains in California. A number of varieties resistant to the Hessian fly were found, but resistant varieties were not released for about twenty-five years. Since 1914, however, breeding for Hessian fly resistance has been carried out continuously in Kansas, so that now more than twenty suitable wheat varieties are available to the farmers. A further outstanding success in wheat breeding has been for the control of the stem sawfly, *Cephus cinctus*. This programme has made possible the growing of wheat on millions of acres of land in Canada and the Great Plains where previously the crop was uneconomical.

There are three components of resistance to arthropod pests, namely non-preference versus preference, antibiosis, and tolerance. Some varieties may escape damage by not being in the susceptible stage when the pest is present, but such escape mechanisms are not usually classed as resistance. Tolerance is, of course, the ability of a variety to grow and produce a good yield when supporting infestations that would destroy other varieties grown under similar conditions. Some biologists would prefer not to class this as true resistance, but to the grower it is just as valuable as non-preference and antibiosis. The distinctions between the components are difficult to define exactly and, of course, the resistance of a single variety may involve more than one of the components.

Non-preference and preference concern the readiness of the pest to utilize the particular variety. There are two extreme forms of non-preference – a non-preference only in the presence of a preferred host, and the absence of utilization even in the absence of the preferred host. The latter is, of course, the more valuable, and occurs in both feeding and oviposition responses.

Phytophagous insects find and utilize plants suitable for feeding or for oviposition by a series of steps, and their responses are modified by their physiological condition. Random searching or certain behavioural responses bring them into the general area of the host, and then visual or chemical stimuli orientate them to the plant. Arrestant stimuli, feeding or oviposition stimuli, swallowing stimuli and the like then initiate the remaining steps in feeding or egg-laying behaviour. Absence of one or more of these stimuli, or the presence of a repellent in the host, would bring about non-preference behaviour in the pest.

A beginning has been made in the study of the biochemical differences between plants which cause their varying attractiveness to different kinds of insects, but most of the work has concerned differences between species of plants; scarcely anything is known of varietal differences. The subject is reviewed by the following authors: Schoonhoven (1968), Beck (1965), Dethier (1970), Wood, Silverstein, and Nakajima (1970).

Antibiosis in a resistant variety is shown when an infesting species does not grow and reproduce as well as it does on a susceptible variety grown in similar conditions. The eggs, or the first instar larvae or nymphs, may die or the adult females may have a lowered fecundity. Sometimes, death may occur later during a period of physiological stress as, for example, just before the adult stage. Lesser manifestations are lower weights of individuals than usual, a shortening of the adult life or a lengthening of the larval life, smaller accumulations of food reserves, and changes in behaviour which are harmful.

Antibiosis may not destroy the infesting insects completely, but as the effects are cumulative from year to year when the variety occupies most of the area, a comparatively small fall in the pest's population increase can lead to the eradication of the pest. The 'Pawnee' variety of wheat, for example, supports an infestation of the Hessian fly which is half of that found on a completely susceptible variety, but this was a sufficiently large effect to lead to the elimination of the fly from large areas of Kansas for fifteen years. Some susceptibility might be an advantage in that the pest is unlikely to break the resistance for many years as the selection pressure for insects completely immune to antibiotic effect would be low.

Little is known of the mechanisms of antibiosis, but a number of possibilities have been put forward. The idea of a direct toxic effect is attractive, especially as a number of potent insecticides, such as pyrethrum, rotenone, and nicotine, are found in plants. Without doubt, plants have developed chemical defences against insects many times, and those insects which infest them are those which have developed a resistance to these substances. The resistance may be biochemical or physiological, or the toxin may be avoided by the mode of feeding. Larvae of the tobacco hornworm moth, *Manduca* (*Protoparce*) *sexta* can excrete nicotine so rapidly that it does not accumulate lethal dosages. The cigarette beetle, *Lasioderma serricorne*, detoxifies the insecticide into less toxic alkaloids. This beetle can also infest pyrethrum flowers in store. The aphid *Myzus persicae*, on the other hand, feeds from the phloem and not from the nicotine-laden xylem vessels and thus avoids poisoning.

In passing, it should be mentioned that the effects of the insecticidal constituents of plants could possibly be enhanced by the application of synergists which need not, in themselves, be toxic. Synergists often appear to act by inhibiting the enzymes which detoxify the insecticide, and should, therefore, control those insects which have overcome the plant resistance in this way. In general, they should be highly selective, unless the predators and parasites of the pest accumulate significant quantities of the plant constituent (Dyte, 1967).

It must be admitted that there are probably many different toxins to be found in the various species of plants, but several compounds are known which potentiate a variety of insecticides. Many of these contain methylenedioxyphenol groups and, interestingly, such compounds are quite common in plants where they have possibly arisen as a secondary chemical defence. In such cases the infesting species of insect have probably developed mechanisms which overcome the combined toxic effect of the toxin and the synergist (Ehrlich and Raven, 1967).

Before leaving the subject of toxins in plants it will be remembered that a number of plants are now known to contain chemicals similar to the juvenile hormones of insects, and these substances may help to defend them from insect attack.

Antibiosis may arise from other chemical substances which, while not lethal, have a harmful effect. Gossypol, in cotton, retards growth of the bollworm, *Heliothis zea*, and a factor in maize, 6-methoxybenzoxazolinone (RFA), has a similar effect on *Ostrinia nubilalis*. This factor has also been identified in other plants.

An infestation may develop slowly, or not at all, because of the complete absence, or deficiency, of some nutritive factor, or as the result of some nutritional imbalance. Fraenkel (1953) opposed this suggestion on the grounds that all phytophagous insects have virtually the same nutritional requirements, so that if a variety would support one species, it would support any other. There is evidence, however, of such a mechanism acting against the pea aphid, *Acyrthosiphon pisum*, on pea varieties which have less free amino-acids than usual. The aphid would grow and reproduce normally, however, when these varieties were perfused with amino-acids, or treated with herbicides which increased the free amino-acid content.

Antibiosis can arise from some morphological characteristics of the resistant variety. Solid-stemmed varieties of wheat resist the attacks of *Cephus cinctus* extremely well, partly because the sawfly's eggs are often damaged when inserted, and partly because the larvae cannot move so freely within the stem as in hollow-stemmed varieties.

Lastly, of course, the plant may be hypersensitive, so that the cells at the site of attack react with the pest in such a way that the damage is isolated, or the pest is incapacitated or killed.

These antibiotic mechanisms may be present in all parts of the plant, or only in those parts subject to attack, and they may be present at all stages of growth, or only at some.

Little is known of the reasons why one variety of a plant can tolerate infestations which would cripple another, but tolerance is undoubtedly important agriculturally. The ability to repair tissue damage, and to put out fresh growth during and after the attack, is, of course, very important. Tolerance is influenced strongly by environmental conditions. It is often found, for instance, towards sucking insects, and here the availability of water plays a large part (Beck, 1966; National Academy of Sciences, 1969).

In summary it can be said that resistance to pests and diseases is extremely common, for most plants are not attacked by a very large number of pathogens and animals. They are, in fact, resistant to most. There has been, during the evolution of both the attacked and the attacker, a long series of new developments, allowing, first, the plant to protect itself, and, second, the pathogen or parasite to attack. The artificial selection of plants since Neolithic times has broken this chain, as man has bred plants for other qualities, but this evolutionary process gives encouragement to plant breeders. Doubtless, the increased interest in the biology of host–pathogen and host–insect relationships will aid in the search.

Most plant breeding for resistance to insects and pathogens has been carried out by government agencies, while many commercial agencies (but not those specializing in particular crops, such as potatoes) have concentrated on more aesthetic projects, such as attractive new varieties of ornamentals. This has been partly because of the length of time usually required to breed a resistant variety, and partly because a variety, once obtained, could be pirated. In recent years, however, it has become possible to patent a variety, so there is no longer this deterrent. It may also encourage commercial breeders to publish the pedigrees of their new varieties, which will help others to identify suitable sources of resistance.

Animal resistance to insect and tick pests
The difficulties inherent in the breeding of varieties of animals resistant to pests and pathogens have already been mentioned, and, not surprisingly, there are only a few cases recorded. Some correlation between the thickness of skin in some breeds of cattle, and their resistance to flies that bite has been reported. The conformation of the wool of Australian Merinos makes them more susceptible to sheep blowfly strike than are most English breeds, and cross-breeding has given some resistance. The Merino pure breeds are, however, otherwise so well suited to the Australian environments that graziers prefer to rely on chemical methods of control, and the simple surgical operation devised by Mules in 1932. As a result of this, the folds of skin around the base of the tail and the anus disappear, so that the skin is drier and less attractive to the flies. This, however, only protects the sheep from crutch strike, and is also expensive.

Cultural and ecological control

There is no sharp distinction between these two approaches. Both involve the manipulation of the environment of the victim so that it is made less favourable to the pest. The manipulation involves only existing factors of the environment, so the application of chemical pesticides or the introduction of resistant varieties, or of natural enemies of the pest, are excluded. So too are most physical methods of control, except those which may be considered to be normal management operations, or extensions of these. A roller which crushes soil insects would be used for cultural control, but electromagnetic energy would not. Within the above definition, cultural control is restricted to the adaptation of generally accepted management procedures, whereas ecological control concerns more radical changes. Clearly, however, some procedures which would be classed as ecological control at present will be regarded simply as cultural control in thirty, forty, or a hundred years from now – hence the division between the two depends largely upon one's attitude. I will therefore refrain from trying to classify rigidly the methods described below.

We can distinguish between methods which are aimed at destroying an existing infestation or source of infection, and those which are directed towards the avoidance of damage. Many of the techniques work best when there is cooperation over a large area, and as this cooperation may be brought about by directions from governments and local authorities, cultural and ecological controls overlap with legislative control.

Without a doubt, many methods of cultural control arose by a process of trial and error, often without any realization that anything was being controlled. It was simply a case of getting better crops when such and such a practice was adopted. Even today, many farmers still continue the practices more from a vague feeling of loyalty to the traditions of husbandry, than as a conscious attempt at pest control. The methods, however, do depend upon the disruption of the pest's normal biology, and rapid advances will come only through a thorough study of this. Finally, cultural and ecological methods of control are basic to the integrated control of pests.

Cultivations and rotations

The main purpose of soil cultivation is the preparation of a seed bed suitable for sowing or planting, but there are a number of secondary functions which include the control of weeds and pests, the incorporation of manures, fertilizers, and crop residues, and the control of temperature, aeration, and water content. The pest control components, may be directly harmful to the pest, or they may promote plant growth so that the effects of the pests are minimized.

Cultivations are integrated with the sequence of cropping, the details of which depend upon the peculiarities of the individual farm – on such factors as soil and climate, the presence or absence of livestock, and the local marketing opportunities. Rotations are partly used for pest control, but there are many other important purposes, such as the increase or maintenance of soil fertility, or the spreading of labour requirements fairly evenly throughout the year. Pest control by cultivations, and by adopting a particular sequence of cropping, must fit in with the other agronomic and economic requirements.

Annual weeds may be destroyed either by burying the plants, or by dragging them out so that they wither and die. Cultivation also encourages the germination of weeds so that they may be conveniently destroyed either by further cultivations or by the application of a weed-killer. Autumn cleaning is effective against such weeds as the speedwells (*Veronica* spp.), the chickweeds (*Stellaria* spp.), wild oats (*Avena fatua*), black grass (*Alopecurus myosuroides*), and poppies (*Papaver* spp.), but, unfortunately, some weeds, such as fat hen (*Chenopodium album*) and knotgrass (*Polygonum* spp.), do not germinate till later and must be controlled chemically or by adopting a bare fallow. The fallow is cultivated to destroy the weeds before the next crop is drilled. The bare fallow was common in the three-field system in Britain (there is also an early Biblical reference: Exodus 23:10–11) and is still common in some dry farming areas in America and Australia, where the function is partly weed control, and partly water conservation. In Australia the rotation used to be cereals followed by fallow, but now pasture with legumes is often included to build up soil fertility. During a fallow period, the nitrogen content of the soil is increased, but there may be some loss of organic matter through oxidation. In areas where the rainfall is normally less than 25 cm during the growing period the fallow is mainly for moisture conservation, but where the rainfall is above about 40 cm the fallowing is mainly for weed elimination and nitrogen build up (Molnar, 1966).

An alternative to a bare fallow is a field crop in which cultivation between the rows is possible. Root crops are excellent examples, and though they make heavy demands on the land, they are generally well dunged and fertilized. Such crops are often planted comparatively late in the spring which gives ample time for cultivations after taking the previous crop. A further feature is their horizontal foliage which, when the plants are well advanced, shades the ground beneath.

Soil cultivation can give efficient control of many insects which have a vulnerable stage in the soil. Quiescent stages, such as pupae, are exposed to dehydration, or the predation of birds, and various stages may be mechanically damaged. Alternatively, the larvae or pupae may be buried so deeply that they cannot emerge.

Autumn ploughing is valuable in parts of the United States in that it results in high mortality in overwintering *Heliothis zea*. Deep ploughing does not destroy all the larvae of the European corn borer, *Ostrinia nubilalis*, but the careful burial of stalks and weed fragments deprives them of refuges above the ground. Ploughing also destroys many grasshopper eggs, and many of the nymphs resulting from the surviving eggs can be starved by the destruction of all green food. Cultivations can also have an indirect effect upon some of the aphid species which are partially dependent upon ants. The stirring of the soil before the planting of maize or cotton, for example, disturbs ant colonies and encourages the good growth of the plants before the establishment of the ants and the corn root aphid, *Anuraphis maidiradicis*.

Certain cropping sequences encourage the increase of certain pests. An unbroken sequence of any single crop will allow the development of large populations of pests associated with them. This is particularly true of nematodes, such as the potato root eelworm, *Heterodera rostochiensis*, and the beet eelworm, *Heterodera schachtii*. The too-frequent cropping with beet or mangolds in Britain (especially in the Fens), Europe, and the United States can result in 'beet' or 'mangold sickness', which is now known to be caused by this pest. An excellent example is Barnfield at Rothamsted Experimental Station where mangolds have been grown continuously (for research reasons) for almost a century.

The nematode attacks and forms cysts on many, if not all, of the crop plants belonging to the beet group and the Crucifera (brassicas, etc.). It is also found attacking a number of weeds including fat hen and its allies, various crucifers, the broad-leaved and the curled docks (*Rumex obtusifolius* and *R. crispus*), persicaria (*Polygonum persicaria*), chickweed (*Stellaria media*), and the hemp nettle (*Galeopsis speciosa*).

The effect of growing a susceptible crop varies with the nature of the crop, and with the initial infestation. In general, cruciferous crops (root or seed) increase the eelworm population much more than do the beet species, although they do not themselves suffer so drastically from the effects of the eelworms. The lower the initial population, the greater is the rate of increase. In small plot trials increases of over 1500 times have been noted when the initial population density was very low, but increases are much lower than this when the initial population is near the critical level for sugar-beet. Typical increases in such circumstances are, for sugar-beet, $\times 2$, for swedes, $\times 11$, for cabbages, $\times 9$. These increases are in comparison to the control (fallow) which is taken to be $\times 1$, although there is, in fact, a decline.

There is little lateral spread of the eelworms, except as a consequence of cultivations or soil movement, so the 'disease' usually has a patchy distribution in a field. Such a pattern, however, indicates that there is probably a widespread infestation in the field, and that the growing of further susceptible crops could lead to large losses.

If the land is fallowed, or if a non-susceptible crop is grown, the infestation drops to about half of its initial value, and this decrease continues year after year for as long as no susceptible crop is grown. In other words, the half-life of a population of beet eelworm, in the absence of susceptible crops, is about one year, unless, of course, there are heavy infestations of susceptible weeds. Chemical methods are, at present, unsatisfactory, so control depends largely upon suitable rotations, and avoiding operations that will spread the eelworm cysts to other areas. As sugar-beet in the

United Kingdom is grown under contract for the British Sugar Corporation, which processes the beet in its factories, rotational clauses are included in the agreement between the Corporation and the farmer. In addition, susceptible crops can only be grown under licence in fields which are known to be infested. In 'infested fields' a susceptible crop may be grown every fourth year only: a susceptible crop may be grown in a 'heavily infested field' only after the passage of five years, and then only if a soil sample shows a sufficient reduction of the pest. The regulations also prohibit the movement of seed potatoes from such fields to other farms.

The rotations described above depend upon the natural disappearance of the eelworm by the hatching of the eggs in the absence of a susceptible crop. In the presence of a susceptible crop the hatching is stimulated by a root diffusate; sometimes, however, a plant may stimulate hatching, yet not support the eelworms. Swinecress (*Coronopus*) and *Beta patellaris* have been tried, but there is difficulty in the establishment of these plants, and they have no other economic value. An alternative method is to grow a susceptible crop (a catch crop) and to destroy it before the cysts mature. The timing is critical; if the crop is destroyed too early only a small proportion of the population will be destroyed, if too late, the population may be increased. The regulations make this method difficult for the farmer, and so far it has only been done experimentally (Jones and Dunning, 1969).

The beet eelworm is particularly troublesome because it has a wide host range which restricts the alternatives open to the farmer when planning his rotations. The related golden or potato root eelworm (*H. rostochiensis*) attacks fewer hosts (the potato, *Solanum tuberosum*, the tomato, *Lycopersicon esculentum*; various nightshades, e.g., *S. dulcamara* and the tomato egg plant, *S. integrifolium*). It is easier, therefore, to lengthen the rotation, and in Yorkshire, for example, good control has been achieved by including peas and beet. In Lancashire, however, there has been difficulty in finding suitable crops, in the absence of beet sugar factories. If potato root and beet eelworms occur in the same field the difficulties are immense. This could occur, for example, in agricultural areas bordering the Wash.

The cereal root eelworm, *Heterodera avenae*, is fairly widely spread in England, being especially troublesome in the light lands of the West Midlands, and the chalks of Hampshire and Wiltshire. Oats are the most affected by the pest, but wheat and barley can also be attacked. Most grasses are comparatively poor hosts, and when the land is grassed down the cyst count falls. On many farms a three-year ley will give good results; on lighter soils a four-year ley may be needed.

The growing of a cereal crop immediately after a long ley or permanent pasture may lead to damage by the so-called ley pests, insects such as larval frit flies, *Oscinella frit*, and rustic moths, *Apamea secalis*, which move from the buried turf to the young cereal plants. The damage can be avoided by ploughing in the grass well before the sowing of the cereal, and ensuring that the turves are well buried.

In the eastern counties of England wheat bulb fly (*Leptohylemyia coarctata*) attacks often follow a bare fallow or a root crop, which leaves the soil surface exposed from July to September. This pest is not, however, a grassland species so populations do not build up while the land is grassed down, and, because of the ground cover, attacks do not occur when cereals follow cereals.

The growing of the same crop year after year on the same land encourages, of course, the build-up of pathogens, although in some cases an equilibrium is established so that a satisfactory harvest can be taken. Rotations will avoid damage from those diseases which are fairly specific, and which cannot exist for long periods (either saprophytically or as some resistant stage) in the absence of the host. Rotations would thus be ineffective against flax wilt, *Fusarium oxysporum* f. *lini*, cabbage yellows, *F. oxysporum* f. *conglutinans*, and related vascular fusarial diseases which, once established, remain in the soil indefinitely. Similarly, it would be difficult to find a sequence of cropping which would eliminate pathogens such as *Rhizoctonia solani* and *Agrobacterium tumefaciens* which have a very wide host range.

Rotations clearly have an important place in pest control, and the method is, in itself, cheap. It cannot, however, be relied upon to control every pest, and it is not always economically feasible; a reasonable proportion of the crops must be cash crops. The actual sequence of cropping, and the associated cultivations, will vary with the pest complex present, and the life-histories of the individual pests in the particular environment. The number and timing of the cultivations will obviously differ in regions where a pest is univoltine and in regions where there are several generations annually.

The spelling of pastures
Very few animal ectoparasites can be controlled by a method akin to rotation, namely keeping the 'crop' away from infested land until the pest is eliminated or greatly reduced in numbers. The technique can be used, however, against those species of ticks which do not have a wild host. The success depends upon the limited powers of dispersal of such ticks. This inability to colonize new areas by their own powers of movement is so great that a field adjacent to a highly infested one can remain free of the pests for years provided that the fences are strong enough to prevent the livestock breaking through. The various species of ticks also have marked environmental requirements when off the host if they are to survive. Thus *Ixodes ricinus* in the United Kingdom is restricted to rough grazing where the characteristics of the basal mat of vegetation are of great importance. The tick is not found on lowland pastures, and thus sheep may be kept in the lowlands till the spring 'rise' of *Ixodes ricinus* in the hills has finished. By the time the autumn rise occurs in the hill pastures, the lambs are well enough grown to withstand the attack.

The numbers of *Boophilus annulatus* on pasture land in the United States have been reduced by pasture rotation. The length of time required for the ticks to die depends upon the climatic conditions. In cool, humid climates the ticks can survive up to ten months, but in arid conditions the time for pasture spelling can be reduced. In the United States this method played a large part in the eradication of the cattle tick, and the redwater disease which it transmits (*Babesia bigemina*).

In Australia the related tick, *B. microplus*, has been reduced in numbers by alternate grazing of stock in adjacent paddocks.

The custom of keeping sheep on the lowlands in winter, and on upland pasture in the summer is widespread, and is of considerable antiquity. Many of the hill farms in Wales, for example, bear the name 'hafod' or summer dwelling, and those in the valleys 'hendre' which implies old township or winter dwelling. The reasons for transhumance, as it is technically known, are partially climatic, but the avoidance of ticks and such diseases as braxy is also important. It did, however, contribute to the spread of other diseases and pests, especially when the animals travelled on the hoof, resting overnight at various places (Barnett, 1961; Walton, 1947).

Sanitation

Any sanitary measures which reduce the population of healthy pests are beneficial. Care must be taken, however, that it is the pest organism that is destroyed, and not its parasite. After the harvesting of cereals, for example, most gout fly adults, *Chlorops taeniopus*, have emerged, and many of the remaining puparia contain parasites. Burning of the crop residues in such situations would be harmful in this respect.

Legislative control is often involved in sanitary measures. In New South Wales it is obligatory, for example, to destroy all fallen fruits in order to reduce the numbers of codling moths and Queensland fruit flies, and to remove and burn all citrus wood that is attacked by the citrus gall wasp, *Eurytoma fellis*. Similarly, in England the Silver Leaf Order of 1923 requires the occupier of any premises on which plums or apples are grown to burn all dead wood on the trees before 15 July of each year, to prevent the spread of the fungus, *Stereum purpureum*.

Some of the more important hygiene methods involving cultivations have already been mentioned. Weed control, for example, includes the destruction of alternative hosts for insect pests and plant pathogens, as does the removal of volunteer plants. Weed destruction is often extended to the headlands and hedgerows, which often provide overwintering sites for pests such as the pea and bean weevils, *Sitona* spp. Unfortunately, such practices often eliminate many of the natural enemies of the pests which find hosts and alternative foods, such as nectar and honey dew, there.

Sometimes, sources of infection or infestation cannot be avoided, but they may be sited where they do the least harm. Mangold clamps provide the peach potato aphid, *Myzus persicae*, with excellent overwintering sites, so they should be placed as far as possible from fields in which it is intended to grow sugar-beet.

Many public health pests can be partially controlled by simple hygienic measures. Cockroaches can be denied access to waste food and to water. House-fly breeding sites may be covered with soil which prevents both fresh oviposition and the emergence of maggots already present. In addition, any fresh material should be added to the centre of a well-made dung heap so that the heat of fermentation destroys any insects present. The numbers of urban mosquitoes such as *Aedes aegypti* and *Culex pipiens fatigans* can be reduced by removing small containers in which they can breed.

Much of the damage caused by stored-product pests can be avoided by a thorough cleaning of silos, railway waggons, and other containers between fillings. Care should be taken in the choice of dunnage material (dunnage is the timber, etc. used to support stacks or to separate material in store and transport), with the rejection of infested wood or substances into which stored-product insects could bore to pupate.

Basically, most sanitary measures reduce the complexity of the habitat, and this often increases ecological instability. Certain environmental manipulations are, however, planned to increase the complexity of the environment in a way which is detrimental to the pest. It may, in fact, appear to the layman that some of the methods of environmental manipulation would be in conflict with good hygiene. Some of these approaches will be discussed later.

The use of clean planting material

As Shakespeare (and, presumably, his Warwickshire farming acquaintances) well knew, if a man sow'd cockle, he reap'd no corn. Various weeds, nematodes (such as ear-cockle, *Anguina tritici*), insects, viruses, and other pathogens can all contaminate planting material. Fortunately, many of these contaminants can be removed or destroyed by mechanical cleaning, hot water treatment, or seed dressings, but some contaminants, such as viruses in vegetative planting material, cannot be removed easily.

Seed potatoes may be raised in an acceptably virus-free condition by growing them in areas in which the aphid vectors of leaf roll and virus Y are scarce, or occur so late in the season that they do little damage, and by the roguing of the odd infested plant. In the United States, arid regions are often chosen for the growing of seed crops free from wet weather diseases which are characteristically soil-borne (anthracnose of beans and curcubits, bacterial blights of legumes, etc.). Irrigation is necessary, but only by furrow, as overhead irrigation helps to spread the pathogens.

A recent development concerns those viruses which are almost invariably present systemically in a cultivar. It involves the culturing of a growing point of the plant removed before it becomes infected. It has been successfully used for carnations in which meristems are taken and cultured in pathogen-free glasshouses. Similarly, vegetative rhizome buds of banana, free from nematodes, viruses, fungi, and bacteria, can be raised successfully in pathogen-free conditions.

The timing of sowing and harvesting
Most plants are only susceptible to the pests which attack them during a limited period of their growth, and many pests are only present for a few days or weeks of the year. It is often possible, therefore, to avoid attack by modifying the sowing date. Wheat bulb flies, for example, do little harm to spring-sown wheat, and wheat sown in October is so well advanced normally at the end of the winter that it can withstand quite severe attacks. Wheat bulb fly is most troublesome in fields situated in areas where there is intensive winter-wheat cultivation, so trials are now in progress in which only spring wheat is grown over several square miles.

The timing of certain cultivations can be important. The eggs of the mangold fly, *Pegomyia hyoscyami* var. *betae* are laid on the leaves of the young plants, the peak usually being about the end of May. If most of the eggs are laid before the plants are singled, then many of them will be destroyed, but if singling is too early the flies immigrating into the field will concentrate on the remaining plants, and the attack will be far more severe.

The time of harvesting can be adjusted in some cases so that pest damage is minimized. Wheat may often be taken before the mines of the wheat stem fly, *Cephus cinctus*, cause severe lodging. Maize can also be harvested before *Ostrinia nubilalis* severely damages the crop. Early harvesting can only be contemplated, however, if there are facilities to deal with grain with a comparatively high moisture content.

The timing of the cutting of alfalfa (lucerne) has been found to influence the insect fauna profoundly in the United States. Early cutting of the first and second crops is effective against the alfalfa weevil (clover leaf weevil), *Hypera postica* (National Academy of Sciences, 1969). Edwards and Heath (1964) also recommend that for hay and silage crops lucerne should first be cut when the plants are still in bud, so that the young larvae of this species and of *Phytonomus posticus* are exposed to drying and starvation.

Variation of planting density
If a pathogen does not spread from plant to plant in a field (as is the case with some of the pathogens present in the soil) losses can be offset by planting more densely than

is usual. This practice is used in the growing of cotton when losses from verticillium wilt are expected. Dense planting, however, increases the humidity within the stand, and this encourages the spread of many diseases.

The spacing of trees in plantations is often important. In the growing of coffee, for example, it is necessary to provide shade as soon as possible, either by appropriate spacing of the coffee trees, or by the growing of other, taller species, in order to reduce attacks by brown eye spot, *Cercospora coffeicola*.

The effects of nutritional factors

It is a widespread belief that vigorous plants are attacked less by pests than are those which are making poorer growth. It is, in fact, one of the basic tenets of some of the compost schools of farming that organic culture is the best insurance against diseases. It is an attractive theory, but there is much evidence against it. Vigorous plants, on the other hand, often withstand the effects of pests and diseases rather better than poorly nourished ones, and recover from the damage more quickly.

Nitrogenous manuring (especially when the readily assimilated nitrates are used) often increases the susceptibility of plants to potato blight, *Phytophthora infestans* or to leaf rust, *Puccinia striiformis*. Presumably the lush growth resulting from the fertilizers is more easily invaded by pathogens. Similarly, succulent, rapidly growing maize is more susceptible to the European corn borer, *Ostrinia nubilalis*.

There is some evidence that manuring with potassium and phosphates reduces the incidence of some pests. Potash applications reduced the percentage of tomato plants attacked by streak disease, and the wire-worm damage to wheat growing in phosphorus-deficient soil. In the United States phosphorus applied alone or with nitrogen at planting time was associated with increased damage by the wheat stem sawfly, but potassium alone or with the other two nutrients reduced the damage in winter-sown wheat.

Fertilizers can help to give a uniform dense stand, and this can discourage pests such as the chinch bug, *Blissus leucopterus*, which congregate in thin patches of the crop.

The undoubted beneficial effect of some organic manures results not merely from nutritional factors but also from antibiotic effects of micro-organisms which it contains. It is, however, difficult to generalize about the effects, good or bad, of nutritional factors, as these effects will vary from site to site, with the soil conditions, the climate, and the crop.

Water management

Most farmers have little control over the amount of water which comes to their land, although they can control that which does by drainage or conservation. In dry areas, however, irrigation is a common practice and the regulation of this can be used for pest control. In the Pacific Slope states of the USA wire-worms are often controlled by either flooding the land for several days or by allowing it to dry out during the summer months, and their incidence in lucerne fields can be reduced by withholding water for some time. Flooding can, however, have detrimental effects on the soil, so the method must be used with care. Flooding of sugar cane fields has been practised in some countries, but is often followed by a resurgence of the pests a year or two later, possibly because many parasites and predators are also destroyed.

In summary, it may be said that the modification of almost any farming operation will have some effect on some pest or disease, and that these modifications cost very little. Unfortunately, few of the methods can be applied by a simple rule of thumb, but most demand a fairly detailed monitoring of the development of the pest or disease concerned. Timing must be precise, as for example in the singling of sugar-beet to reduce mangold fly attacks, and labour may not always be available just when it is needed. It is not surprising, therefore, that many of the formerly widely used practices lapsed when effective pesticides became available. These often gave a complete control of the pest, which cultural control usually cannot do, and although timing is important, substantial control can be achieved even if applications are late.

Ecological control methods

These methods involve altering existing components of the environment in such a way that the pest is harmed. These practices would not, at the moment, be considered to be routine farming or management procedures, although in time many of them will become accepted as such. In many techniques the effects on the pest are indirect, as the procedures encourage natural enemies and pathogens of the pest. This is usually achieved by making the habitat more complex, but sometimes by simplification. Needless to say, these methods may be applied to encourage either parasites, predators, or pathogens which have been introduced into the area, or which are already present. They may be used to benefit both classes of organisms at once. Some ecological control methods may therefore be considered to be a section of general biological control.

The importance of flowering plants growing in the vicinity of a crop as sources of supplementary food for entomophagous insects has been stressed by a number of workers. These plants can be beneficial in several ways. They can serve as a source of nectar or pollen which are essential to many adult forms, or which can tide over predators such as lady-birds in the absence of the prey; they can support populations of aphids which provide honey dew as supplementary food, or they may carry populations of alternative hosts for predators and parasites.

Both nectar and honey dew have been shown to be essential for the maintenance of large populations of *Tiphia popilliavora*, an introduced parasite of the Japanese beetle, in the United States, but it was admitted that the plants could also act as foci of infestation and infection of other pests and diseases. Jepson (1954) has stressed the need for nectar-bearing plants in the biological control of pests of sugar cane, a plant which produces little food for adult parasites. Growers in Mauritius have attempted to fill this gap by taking into the crop suitable food-and-shelter plants.

Pollen appears to be essential for the development of the eggs of entomophagous syrphids, and the various species of syrphids show preferences for different kinds of flowering plants. These flies are strongly flying insects which seldom remain for long in areas which lack suitable plants.

Most of the plants which are used in this way will, of course, be found growing along the edges of fields, roads, and waterways and thus they will be relatively important in areas where the fields are small and irregular in shape. The smaller the fields, the more complex the environment, and the probability of massive pest outbreaks will be lessened. Conversely, the danger of infestations of pests arising from the field margins will be greater, and there is also evidence that the incidence of 'planktonic' insects such as aphids, and of the various virus diseases which some of them transmit, is highest on the leeward side of hedges and windbreaks.

In plantations and orchards nectar-bearing plants can be grown between the trees or bushes, but with field crops this is not often feasible. Russian workers, however, have had some success by interplanting pumpkins with maize to encourage adult corn borer parasites.

The presence of alternative hosts of parasites or prey of predators in the vicinity of the crop offers many advantages, provided, of course, that these alternative species are not more harmful. Most importantly, they lessen the effects of a lack of synchrony between non-specific parasites and predators, and their preferred hosts, the pests. In addition, their presence helps to dampen the extreme oscillations of pest and natural enemy populations that sometimes occur in simplified situations, and they help to maintain a reasonably large natural enemy population when the pest population is low, especially during the winter, or when the attacked crop is not growing. Furthermore, the natural enemies are more widely spread than they would be if alternative hosts or prey did not occur.

Pierce, in 1912, suggested that many plants could support alternative hosts for the parasites of the cotton boll weevil, *Anthonomus grandis*. He also considered that these parasites might, at a suitable time, be encouraged to turn their attentions to the pest on cotton. In some small trials which he carried out he increased the parasitization of the boll weevil by *Eurytoma* spp. by cutting hedges of *Ambrosia trifida* which carried populations of an alternative host, *Lixus tylodermatus*.

An excellent recent example is afforded by the grape leaf-hoppers, *Erythroneura* spp. and their myrmarid egg parasite, *Anagrus epos*, in California. The parasite can maintain economic control of the pests during the growing season, but, unlike the hosts, it does not overwinter in the vineyards. It was eventually discovered that the parasite survived the winter by breeding continuously in the eggs of a non-pest leaf-hopper, *Dikrella cruentata*, which infests wild blackberries. If none of these are near to a vineyard considerable leaf-hopper damage can occur before the parasites move in at the beginning of the season, so many growers are now providing winter refuges of blackberry bushes near to their vineyards whenever these do not occur naturally.

The second requisite of natural enemies which can sometimes be supplemented is cover for nesting or escape from predators and inclement weather.

Paper nest wasps of the genus *Polistes* are efficient predators of various caterpillars in many parts of the world, but their flight range is normally restricted to a few hundred metres. In many agricultural areas there are few suitable nesting places for these wasps so in some countries artificial sites are provided. In North Carolina, for example, the wasps prey on hornworms, *Manduca* spp. When they are absent from the fields growers erect small boxes on poles at the edge of the fields. Predation of the caterpillars has been increased, in some cases, by one-half, and fewer insecticide applications were needed. A similar practice was carried out in the West Indian island of St Vincent, where *P. annularis* preys on the cotton leafworm, *Alabama argillaceae*, and it was said to be so successful that insecticides were not needed for ten years.

Artificial nesting sites are commonly provided for indigenous birds by foresters in various parts of Europe, and the custom is often successful, but it has been suggested that in dry coniferous forests drinking water should also be provided. It must be stressed, however, that birds cannot be expected to eradicate a massive outbreak of pests, but that their presence should be encouraged to prevent the build-up of insect numbers.

An excellent example of the stability of the more complex habitat concerns the coreid bug, *Amblypelta coccophaga*, attacking coconuts in the Solomon Islands. This is attacked by the ant *Oecophylla smaragdina* but the colonies of this predator are eliminated by the widespread species *Pheidole megacephala* when the coconut groves are kept clean. During the war the groves were neglected and it was observed that the scrubby ground cover protected the first ant. O'Connor (1950) suggested that palm fronds should be placed against the trees so that *O. smaragdina* would be able to forage without excessive interference from its enemy. This resulted in the re-establishment of *Oecophylla* in the groves. A second suggestion is that a cover crop with a growth habit similar to that of natural growth should be planted in the groves.

In contrast, the control of mosquito larvae by insectivorous fish can be aided by eliminating aquatic plants which provide thick cover for the larvae, or, when constructing artificial lakes by making the water at the banks so deep that weeds cannot establish themselves.

Stored-product pest entomologists have a greater control over the environment than do their agricultural colleagues. It is now a common practice to 'turn-over' grain periodically by moving it from one silo to another. Grain pests often occur in small groups within the mass of grain, and their metabolic activities cause local increases in temperature, and, consequently, accelerated rates of increase in the numbers of the pests. Turning the grain breaks up these 'hot-spots', especially if the operation is carried out during very cold weather. In such circumstances the pests are often killed directly by the cold, as many species originated in warm climates. Similar results may be obtained by forced aeration of the grain, or by merely opening up the silos or warehouses during long, cold periods.

Cold (by refrigeration) has also been used as a quarantine measure in the United States to prevent the re-introduction of the Mediterranean fruit fly in imported fruit. Cold storage of fruit in general is used to prevent rots of various kinds but it also gives protection against various insect pests.

Humidity management can sometimes be used to control pests as, for example, beneath buildings where termites are likely to be troublesome, or in grain storage. Grain is usually dried until its moisture content is less than about 12 per cent before it is stored, to inhibit the development of insects, mites, and moulds.

Returning once more to the provision of food for natural enemies in order to improve the synchronization of the pest and its attacker, it may be worthwhile considering artificially infesting the crop with the pest at the beginning of the season, before the weather is suitable for a marked natural increase, or after the main infestation has passed. If this is done the natural enemies should be present in numbers sufficient to control the pest when natural infestations begin again. Smith and DeBach (1953) carried out some trials in California with the univoltine black scale on citrus, introducing them during a period when the natural population was critically small. Unfortunately, the parasite concerned, *Metaphycus helvolus*, eliminated the introduced pests too quickly for the method to be successful. Huffaker and Kennett (1953, 1956) carried out similar experiments with the cyclamen mite, *Steneotarsonemus pallidus*, on strawberries, the predators being *Typhlodromus* mites. The results were so encouraging that the authors advocated the adoption of the technique as a routine measure. The main difficulty will be the convincing of the growers about the wisdom of introducing a major pest into a crop. Similar methods are being used to control red spider mites on cucumbers in Britain (see chapter 11).

Modification of the environment has been of the greatest importance in the attempts to control tsetse flies, *Glossina* spp. It will be remembered that each female produces only a few offspring, but these larvae are so advanced at birth that they immediately burrow into the soil to pupate. Chemical control can only be used against the adults, and since these have low population densities, compared with most pestiferous insects, spraying is difficult. Fortunately, tsetse flies are not uniformly spread throughout their respective ranges, but are usually confined to fairly well defined tracts of country known as fly belts. Their ecological and behavioural requirements have been studied intensively for many years, and this research has led to improvements in control. Their pupation sites, for example, must be shaded and have suitable ranges of temperature and humidity. The trees in the area must offer suitable resting sites for the adults, both for the periods between blood meals, and during the night. Suitable food must be within a reasonable flying distance.

It was realized that wild game is an important source of food for the flies, and also that the game often carried the trypanosomes which cause nagana in domestic livestock. Clearing the areas of game would therefore remove the food of the flies, and eliminate the source of infection. Unfortunately, the killing of large game animals was undertaken in many parts of Africa before detailed studies were made of the feeding habits of the flies, and it is now recognized that most flies are selective in their choice of prey. In many cases, therefore, selective killing of the game would have given results as good as those obtained by indiscriminate killing.

Weitz (1964) has described the methods used in the studies of the feeding habits of tsetse flies. The precipitin technique has already been mentioned in the general discussion of predation in chapter 2, but a more precise method, the inhibition test was also employed. This is based on a serological method for the determination of tuberculins. The blood meals of the flies are first screened by the precipitin test to determine the group of mammals attacked, and later the actual species are identified by the second test. In the work summarized by Weitz fifteen tsetse species were examined and, in all, about 22 000 determinations made. It was found that the flies could be divided into five main categories on the basis of the group of mammals attacked. In the first group the flies fed mainly on suids, attacking any available species indiscriminately. In Tanzania, for example, *G. swynnertoni* fed mainly on the warthog, *Phachochoerus aethiopicus*. The second group, which feeds mainly on suids and bovids, contains three savannah sub-species of *G. morsitans*. The prey includes such familiar game animals as kudu, *Strepsiceros* spp. and eland, *Taurotragus oryx*. *G. pallidipes*, *G. longipalpis*, and *G. fusca* are largely restricted to the widespread bushbuck or harnessed antelope, *Tragelaphus scriptus*. This mammal is often a pest of cultivated land and, furthermore, it has been found to carry *Trypanosoma rhodesiense* which causes one of the forms of human sleeping sickness.

Three tsetse flies which are consistently attracted to man, *G. palpalis fuscipes*, *G.p. palpalis*, and *G. tachinoides*, are riverine and lake-side species which transmit Gambian sleeping sickness. They will also attack almost any other mammalian species that comes into the area, and, in addition, crocodiles and aquatic lizards.

The final group contains the two largest tsetse flies, *G. longipennis* and *G. brevipalpis* which, perhaps not unexpectedly, feed mainly on the largest mammals – elephants, rhinoceros, hippopotamus, and buffalo.

Some species of tsetse flies will, of course, attack alternative hosts if their normal prey is absent, but it appears that many mammals such as zebra, wildebeest, water-buck, hartebeest, and impala are rarely, if ever, bitten. Thus these large mammals, and most smaller ones, could be left alone in game eradication programmes designed to starve out tsetse fly populations. The adaptability of the species of tsetse flies concerned is, however, extremely important, and this may vary from population to population. Weitz points out, for example, that the *palpalis* group of tsetse flies might well adapt to new hosts (Weitz, 1964).

Ford (1968) has also discussed the control of tsetse populations by the limitation of their habitat. His paper deals mainly with the factors which prevent tsetse flies from dispersing away from the areas in which they are found, and how a knowledge of these factors might be used to limit the areas in which the flies can survive. The factors include climatic conditions and various requisites for tsetse fly patterns of behaviour.

The climatic factors interact with the effects of game control. In areas near to the upper or lower temperature limits the developing pupae use up most of the food reserves available to them, either because of the increased metabolic rate at high temperatures, or because of the prolongation of the pupal period at low ones. The emerging flies are thus limited in their ability to find blood meals, and game thinning need not be so drastic in such areas as they would be in fly belts where the temperatures are more suitable.

Although tsetse flies can be eradicated by the complete clearing of bush and trees over large areas, less drastic measures may alter the environment enough to interfere with the fly's normal behaviour. In certain areas, for example, it was found that selective felling of *Acacia gerrardii* trees in mixed *A. gerrardii–A. hockii* savannah substantially reduced the populations of *G. morsitans*. It is believed that this species is confined to tree-covered land by visual responses, and that the adults used the first *Acacia* species for resting between meals. In or about 1950, however, the flies apparently changed their behavioural responses in some way so that the selective felling was no longer effective.

The riverine species of *Glossina* are often controlled by felling all the trees along the banks of rivers and lakes in fly belts, but wholesale felling operations in savannah country would be much more difficult. The felling of the trees reduces the shade necessary for the pupae, and also removes some of the requisites of the adults. Glasgow (1963) should be consulted for general information on the distribution and biology of tsetse flies.

Environmental modifications will become more important as integrated control methods develop, although they will often be prevented by the demands of modern mechanical, low-labour farming which stresses the enlargement of fields for easier working. The simplifications of farming in this way brings a simplification of the environment – and it will be difficult to reconcile the two. On the other hand, larger working units will allow suitable methods to be applied over large areas, and this is often extremely important with non-chemical methods of control.

Further examples of environmental modification techniques in agriculture and horticulture will be found in the contribution of van den Bosch and Telford to DeBach's text on Biological Control (van den Bosch and Telford, 1964).

Physical pesticides

Logically, physical pesticides should be considered at the same time as chemical pesticides, for they have the same catastrophic effect on insect populations. Their ecological effects are, however, not so drastic, for there are no residue problems and, in any case, they are, so far, only of value for stored-product pests and similar animals, or for the treatment of planting material.

Electromagnetic radiation

The use of various radiations for sterilizing insects, or as attractants or repellents for pests, has already been discussed, but here we are concerned with their direct lethal effects.

The lowest energy radiation that could be used for insect control is in the radio frequency (RF) range. Extremely low frequency radio waves are not feasible because of the difficulties in their generation, but the high frequency waves are easily obtained as a result of the availability of components used in radio communications, navigational aids, and radar.

The lethal effects of RF are, apparently, the result of the heating of the irradiated material. It has also been suggested that insects could be controlled by some particular frequency which would evoke some resonance phenomenon in the insect, but this has not yet been demonstrated.

The effect of radiation falls off with distance, as the energy decreases in accordance with the inverse square law. It would be extremely uneconomical, even if it were physically possible, to use radiations across wide spaces. In fact, with RF, it appears to be often more satisfactory to introduce the material to be treated to the field between electrodes connected to RF power oscillators, than to subject it to energy radiated from some kind of antenna. It would be possible, however, to radiate the energy when only one side of the material can be approached as, for example, when treating wood panels *in situ*.

As was said above the lethal effects of RF energy arise from heating following the absorption of the energy and it would seem, therefore, that the method would not be better than conventional heating in, for example, kilns. The energy absorbed by the material is, however, dependent upon its electrical properties, and, in particular, its dielectric constant. It is thus possible that when a material containing insects is treated the insects are heated more than is their matrix. When grain was treated so that its temperature reached 38 °C the contained adult rice weevils, *Sitophilus* sp., were killed in a few seconds, although they can normally withstand this temperature for some hours without injury. RF treatment of wood killed *Lyctus brunneus* beetles within it, but the temperature of the wood reached that used in kiln treatment. RF dielectric heating is, however, more uniform and rapid.

This technique is promising for the control of insects and mites in stored products, and it may be of value in the control of boring beetles in cut wood when this is being used in factories, especially if RF is already being used for the curing of glued joints. In general, adult stored-product insect pests are more susceptible than the larval stages, and this may be partly due to the sensitivity of their legs. The larvae may also be shielded to some extent when they are within the kernels of grain.

Radio frequency treatment is considerably more expensive than fumigation, and there are also obvious engineering problems to be solved if it is to be used on a very large scale as, for example, at port terminals. Cornwell (1966) discusses this aspect in relation to ionizing radiation, using the wheat industry of Australia for illustration. Nelson (1967) also summarizes the physical effects of RF, in relation to the electrical properties of the treated material.

Infra-red radiation has also been used to heat grain in order to kill any infesting insects. Tillson (cited in Nelson, 1967) used a tunnel fitted with banks of infra-red lamps down which a shallow layer of grain was passed. He estimated that the grain could be treated at a cost of about 0·5 cents a bushel but other workers concluded that the costs would, in fact, be much higher than this if the control was to be satisfactory.

Experiments have also been carried out on rice using gas-fired ceramic-panel IR sources, as it was thought that these would be more economical than IR lamps. Immature rice weevils were killed at a grain temperature of 68 °C (grain moisture, about 12 per cent), but immature lesser grain borers, *Rhizopertha dominica*, and the Angoumois grain moth, *Sitotroga cerealella*, required higher temperatures.

If the grain moisture content is reaonably low neither RF nor IR treatment at dosages needed to kill insects affect the milling or germination qualities of the grain. RF could also be used to destroy pink bollworm larvae, *Pectinophora gossypiella*, without reducing the germination of the seed. This contrasts with ionizing radiation (see below) which usually reduces germination.

Visible radiation could be used, possibly, to interfere with the development of insects in some detrimental way. Winter diapause is often induced by the shortening of the day-length, and if the daylight can be artificially extended (in, for example, glasshouses) insects may attempt to produce a further generation which fails to complete its development. Short daily exposures to longer wave-lengths interrupts the orderly development of some species, possibly through some effect on the dorsal neurosecretory cells, and this phenomenon may also be of future use in pest control.

The most promising wave-lengths are those which can produce ionization in tissues, namely the X-rays and the gamma rays and both have been used, of course, in sterilization programmes. Although they are particulate, accelerated electrons can also be considered here, for while they are not as penetrating as the electro-magnetic radiations the generating apparatus can be switched on and off as needed. X-ray machines cannot be used for prolonged periods so are unlikely to be used on a commercial scale.

Ionizing radiation can induce sterility, but higher dosages are directly lethal and potentially valuable for the control of insects in stored products, wood, and in various materials passing through quarantine. In stored grain the complete mortality of adults of most species follows a few days after dosages of 200 to 500 krad, or a week or two after dosages of 100 krad. These dosages are considerably less than those needed for bacterial sterilization of food stuffs. The British Working Party on the Irradiation of Food gave the following data for dosages (Ministry of Health, 1964): 4000 to 6000 krad for the sterilization of meat for subsequent storage at room temperature (all micro-organisms, including the extremely dangerous *Clostridium botulinum*); 500 to 1000 krad for the prevention of salmonellae food poisoning in a variety of foods. There is considerable information available therefore about the possible deleterious effects of ionizing radiation on food, at dosages much greater than those used for insect control. There appears to be little effect on the major nutrients, but some of the fat-soluble vitamins may be destroyed. There are sometimes flavour changes, but these are not usually objectionable. Insect-killing dosages do not affect the milling or baking properties of grains, but there is a reduction in the germination and vigour of seed corn.

There appears to be no significant danger of residual radioactivity or toxic products. Care must be taken, however, with packaged foods (a potentially valuable application) as the radiation can produce highly reactive radicals which could interact with the food.

The main difficulty in the use of ionizing radiation will be the construction of a processing unit with a sufficiently high capacity. Cornwell (1966) discusses these aspects and proposes various suitable designs in the concluding chapter of a book devoted to the disinfestation of grain. The International Atomic Energy Agency holds regular symposia on the use of radiation and radioisotopes in agriculture and entomology, and the published proceedings should be consulted (International Atomic Energy Agency, 1962, 1963). Other valuable publications are the report of the Working Committee mentioned before (this contains a convenient glossary of relevant terms (Ministry of Health, 1964) and papers by Nelson (1967), and Nelson and Seubert (1966).

Heat sterilization

Radio frequency and infra-red treatments kill insects by their heating effects, but heat can also be applied in other ways. Even the heat of the sun has been used by primitive societies to destroy insect life in grain. The first application of heating to disinfest large premises in the USA appears to date from 1901 when a miller fitted out his mill for this purpose. Between 1910 and 1913 Dean (1911; see also Cotton, 1956) perfected the technique and found it to be an acceptable alternative to fumigation. In order to achieve a successful kill of the insect population it is necessary to maintain all parts of the mill at a temperature between 120 and 130 °F (49 and 54 °C) for ten to twelve hours. This may be done either by a suitable system of heating pipes, or by unit heaters with forced circulation. The mill is allowed to run empty and all accumulated stock and bagged materials are removed. All ventilators are sealed, the driving belts loosened and treated with neat's foot oil, and steam released on each floor to raise the humidity so that the warping of wood is minimized. The whole operation is best carried out during warm, still weather to lessen losses of heat. After the mill has cooled the machinery should be carefully examined for any damage, and regreased where necessary. In 1956 the costs were comparable with those of fumigation, and in a well-designed mill could be as little as 25 cents for 1000 cubic feet (28 m^3) of space, but the mill is out of operation for one to two days (Cotton, 1956).

Steam sterilization for the treatment of hospital linen will kill both bacteria and vermin, but where insect infestation alone is the problem dry heat treatment is sufficient and much cheaper. Air has a low thermal capacity and is a poor conductor of heat, and it also has a tendency to layer, so modern methods incorporate forced circulation. Heat sterilization was once commonly used for delousing clothing, but chemical treatment has largely replaced it, as insecticides continue to give protection against re-infestation for long periods after treatment.

Wood infested by boring insects can be treated in steam-heated kilns in which both temperature and humidity can be closely regulated. Air that is saturated with moisture is more effective than dry air, but it alters the condition of the wood, and it may damage French polishing.

Dry heat is sometimes used for the control of nematodes and plant pathogens in soil, and a machine has been marketed which scoops up soil as it moves along, heats it to the required degree, then deposits it behind. Dry heat, however, destroys any organic matter in the soil, and alters its structure.

Moist heat – steam or hot water – is, on the other hand, commonly used by horticulturalists to sterilize soil in glasshouse beds. The equipment is cumbersome and the work laborious, but provided that the soil is finely divided and not waterlogged, the method is effective. Furthermore, as the temperature of the soil cannot rise above 100 °C the soil organic matter is not destroyed, although in some cases it may be rendered somewhat toxic to plants for a time. Generally, however, planting can take place almost as soon as the soil has cooled down. Steam is more effective than hot water because of its higher calorific content. Chemicals, such as formalin, may be added to the water or steam to improve the kill.

Planting material can also be subjected to hot-water treatment in order to kill certain contained pathogens and pests. Temperature and timing, which vary with the pest concerned, must be closely controlled (to within half of one degree Centigrade) so that the viability of the plants is not reduced. The material may be given a pretreatment soak in cold water to revive any dehydrated nematodes, and after treatment it may be cooled in a water bath, or allowed to cool slowly in the air.

Aphelenchoides spp. (chrysanthemum and strawberry eelworms) may be destroyed in chrysanthemum stools and strawberry runners by 20 to 30 minutes' treatment at 43·5 °C, but Staniland (1959) suggests that shorter periods at 46 °C damage the plants less. Narcissus bulbs may be treated for 4 hours at 43·5 °C to destroy *Ditylenchus dipsaci* and for 3 hours for iris corms infested with *D. dipsaci*. Fruit rootstocks infected with certain viruses can also be treated with hot water baths.

Endogenic heat, produced by the natural processes of composting and fermentation, has already been mentioned, and it can, when properly managed, destroy many insects, nematodes, and pathogens.

Barriers

The purpose of barriers is simply to keep the pest and the victim from coming into contact. In the discussion of repellents we have already considered one kind of barrier, but these make use of the orientation behaviour of the pest. In this section we shall consider barriers which are more or less passive, or which can be considered as quarantine measures.

Passive barriers

Most of these need no description. Good fences keep out many of the larger mammals, netting can preserve cherished trees and shrubs from birds, mesh fitting to doors and windows can exclude flies and mosquitoes. Well-constructed buildings can be made rodent-proof by grids on ventilators. A less familiar barrier is a metal sheet projecting on all sides in the foundations of buildings to protect the upper parts from subterranean termites. This can often be by-passed, however, by the earthen tubes of the insects. A more expensive method is to construct the building on a concrete raft which must be free from cracks.

Probably the most ambitious use of barriers has been in Australia where thousands of miles of fences have been erected to confine rabbits and dingoes (*Canis antarticus*).

Thomas Austin appears to have been the first grazier in Australia to use wire-netting when he employed it in 1868 to exclude rabbits from his property. His example was followed a few years later by others, but the first large-scale proposal did not come till 1883 when the Canonbar Stock and Pasture directors requested a fence 225 miles long between Dubbo and Bourke in New South Wales. During the remainder of that century, and the first quarter of this, thousands of miles of fencing were erected, usually, however, too late to do much good. The work was often held up for years by border disputes between States, and while the politicians argued the rabbits by-passed the fences already erected. Nor were the fences completely rabbit-proof: in swampy areas they would rot; in arid areas the sand drifted against the fence, bridging the barrier for the rabbits. Emus, wombats, and kangaroos broke down the fences, or blundered their way through, and, consequently, were hunted down themselves. For many years there was a heavy tax on wire-netting which discouraged graziers from erecting fences around their properties.

The dingo fences have been more successful, largely because they have been well maintained by boundary riders who live at intervals along their length. Alone, the fences would not be effective, but the dingoes are hunted by professional doggers, paid on a bounty system, who, over the years have come to understand their cunning quarry. Rolls (1969) relates the history of fencing in Australia and also gives an interesting account of the strategems of various doggers who had to hunt or trap certain dingoes which had charmed lives.

On a very much smaller scale barriers, in the form of insect-proof packaging materials, can prevent the infestation of food products. Certain plastic and metallic foil materials can prevent the entry of most insects and mites, provided that the corners and seams are perfectly sealed. Some insects, such as the cigarette beetle, *Lasioderma serricorne*, and the cadelle, *Tenebroides mauritanicus*, can penetrate any flexible material commonly used in the marketing of foods, and susceptible materials are only completely protected by being enclosed in tin cans or sealed jars. Paper can be made insect-proof for several months by impregnating it with piperonyl butoxide and pyrethrum. The infestation of packaged foods is a serious problem for the manufacturer, even if the insects enter the product long after it has left his care, for the buying public usually holds him to be responsible.

There are a few other types of barriers which can be considered here, although they are hardly passive. Moving currents of air can sometimes prevent insects from entering a building whenever it is essential to keep the doors open for people. The most effective design provides a sheet of moving air across the opening.

The most efficient barriers for dangerous sharks appear to be the regularly inspected fences used off Australian surfing beaches. The fences do not completely enclose the protected area, but lying parallel to the coast they trap many of the sharks as they cruise off shore. Several experiments have been carried out with air curtains released from a submerged perforated tube (National Academy of Science, 1970) but they do not appear to be very effective against sharks, although they might be useful against smaller fish.

Electrical barriers have a limited use. They have been successful in preventing lampreys in the Great Lakes of North America from entering spawning streams. A mechanical barrier of netting would, of course, prevent the flow of water when it becomes clogged with debris. Conventional electrified fences are used to confine livestock and could, presumably, be used to exclude game animals from comparatively small areas. On a much smaller scale electrical barriers can be used on containers in which insects such as cockroaches are cultured.

Quarantine measures

The most effective way of avoiding damage by a particular pest – and the cheapest way – is to be without it, and most countries have set up quarantine measures, and passed the necessary legislation, to try to prevent the entry of new pests.

Quarantine measures are probably the most effective methods of pest control, largely because all the people who are concerned with its execution are usually well trained, and also because when it is successful it is completely successful. Unfortunately, it seems that it usually only postpones the entry of pests. Some still get through, even in countries with extremely efficient services such as the United States of America and Great Britain, and the pressure is becoming continually greater with the introduction of faster and larger jet aircraft. An important part of the quarantine technique is propaganda: one has only to stand on the deck of a liner berthed at Melbourne, say, and watch the pile of confiscated apples, oranges, and other fruit growing at the end of the gangplank, fruit bought in Perth and possibly infested with Mediterranean fruit fly, to realize how unimportant the layman considers these measures to be. Even an Australian Senator once complained in public about an examination that he had been subjected to by quarantine officials, and hinted that the time would have been better spent in the search for illegal drugs. It is not difficult to say which of the two – a single importation of a drug, or of a noxious insect – is the greater threat to a country.

Before the introduction of the United States Plant Quarantine Act of 1912 (extremely late compared with most major countries), approximately ninety-four different kinds of insect pests were introduced into the country, and over one hundred plant diseases had arrived before 1919. There had been earlier laws to prevent the importation of certain plant pests, but these were not comprehensive enough. Since 1912 the rate of introductions has fallen. In 1965 the United States Quarantine Service intercepted the following major pests, the numbers of interceptions being given in parenthesis: European cherry fruit fly, *Rhagoletis cerasi* (104); Mediterranean fruit fly, *Ceratitis capitata* (196); Mexican fruit fly, *Anastrepha ludens* (223); Oriental fruit fly, *Dacus dorsalis* (84); khapra beetle, *Trogoderma granarium* (462); potato root eelworm, *Heterodera rostochiensis* (101); sweet orange scab, *Elsinoë* spp. (299); citrus black spot, *Guignardia citricarpa* (571) (Rainwater and Smith, 1966). In all 32 572 separate plant pest interceptions were made in that one year, at an estimated cost of five cents per head of the United States population.

At ports of entry, material that is likely to be carrying a pest is examined, and if infected or infested, destroyed. Where appropriate, material that could be infested is fumigated. There is considerable cooperation between various countries, so that an exporting country may furnish a certificate to the importing country which clears the product, or the importing country may maintain inspectors who examine the product before it leaves the country of origin. Wherever possible the authorities issue propaganda by various media, often in several languages, to travellers and immigrants, explaining the measures taken, and asking for their cooperation.

Quarantine measures are not restricted to the ports of entry of countries, but are often used within the country to prevent the further spread of a pest already present. There are, for example, strict regulations controlling the movement of cattle within

Australia to prevent the spread of the introduced cattle tick from its strongholds within Queensland. In parts of Africa check points are set up on the peripheries of fly belts where vehicles are examined for tsetse flies which, when present, are frightened away by the flicks of hand flags. Similarly a farmer can take care that soil containing nematodes is not carried from an infested field. This differs from true quarantine only in the self-imposition of the precautions.

It must not be thought that quarantine measures can be successfully applied without a detailed knowledge of the habits and biology of the pests. The kinds of material that are likely to be infested must be known, and also the probability of survival, or the rate of development at the temperatures and humidities to which the material is subjected during transit. Furthermore, the regulations should be such that they inconvenience importers, exporters, and travellers no more than is necessary for the exclusion of the pest or disease. The judgement required for the framing of such regulations, and for their application, can be based only on the same deep knowledge of the biology of the potential invaders.

Mechanical methods of pest control

Many pests are destroyed by mechanical methods – by collecting, crushing, swatting, and so forth. These must surely have been the first kinds of pest control, and they are still used in peasant communities for the destruction of large, easily seen pests such as sphingid caterpillars. When the work is carried out manually it is extremely laborious, and not always very effective, but it is still one of the best methods open to the small gardener.

Various kinds of agricultural machines have been designed to crush or to collect pests, and conventional machines such as heavy rollers are also sometimes used. In the first thirty or so years of this century and in the final fifty years of the last, farmers of the Great Plains region of North America used a horse-drawn, cage-like contraption known as a hopperdozer to clear grasshoppers from their fields. The collected insects were often fed to livestock so that not all of the devoured crop was lost. Similarly, locusts are a common article of food in Africa, and, to judge from an Assyrian bas relief in the British Museum, they were once considered fit for a feast.

Trapping of animal pests is commonly practised throughout the world, and it often makes a significant contribution to their control. The unsightly sticky fly papers once commonly used in houses are not often seen now except, oddly enough, in the insect-breeding rooms of insecticide laboratories and similar institutions where, of course, chemical sprays cannot be used. Sticky traps are sometimes used to protect orchard trees from certain moths whose females are wingless and which must crawl up the trunk in order to lay eggs on the leaves and twigs. Trees are often protected from such pests as codling moth by bands of sacking tied round the trunk in which the pests seek shelter or pupate. Such bands are particularly useful, as natural enemies of the pests also tend to congregate there where, of course, they usually find ample food. A simple trap for earwigs, a plant-pot stuffed with straw and inverted on a stick, makes use of the insect's thigmotactic reaction – its habit of coming to rest with as many of its tactile receptors as possible in contact with an object.

All these traps are designed so that the insects are either killed by natural processes such as dehydration, or mechanically when the traps are inspected. They do not usually contain baits or lures, or insecticides and chemosterilants. A number of modern traps are available which make use of such substances.

Traps are also used in attempts to control pest vertebrates but alone they often do little more than crop the biological surplus, leaving the breeding population intact. Similarly, hunting and shooting usually has little impact on a vertebrate population, unless they are large predators, or gregarious herbivores like the buffalo. In the last case there was a greater incentive than pest control involved in the slaughter: where pest populations are concerned the only incentive for efficient hunting is a bounty paid for each animal destroyed. The bounty system is, of course, open to abuse: the astute hunter will ensure that he takes only the surplus, and by selective killing, leave reproductive females and vigorous males alive (Rudd, 1964; Rolls, 1969).

11

Integrated control, pest management, and the future of pest control

Integrated control

We have now concluded our survey of the methods of pest control, covering briefly both the methods which are already available, and those which may be of value in the future. The methods have so far been considered separately, for this is the way in which they have usually been employed, without any serious attempt at integration.

These *ad hoc* methods have often been very successful in both agricultural and public health applications, but many problems have arisen when pest control has been carried out in this empirical way. In chapter 4, for example, we considered the main drawbacks of chemical control, and it will be convenient to summarize these again. One problem is pest resurgence. The modern pesticides are potent toxicants, usually with a broad spectrum of kill. After their application the pest population, and those of its predators and parasites, and of certain species not directly interacting with it, all fall dramatically. When the residues are no longer sufficient to contain the pest population it increases rapidly in the absence of predation and intra-specific competition. The destruction of organisms which are regulating factors has frequently brought other potential pests from comparative obscurity to major pest status. The more persistent pesticides, such as the organochlorine insecticides and the mercurial fungicides, enter into food chains, or are carried by physical means, and thus spread to places often quite distant from the areas of application.

These are ecological effects and they are, to some extent, reversible. The development of resistance in the target organisms is, however, an evolutionary process, and largely unidirectional. Its occurrence leads first to increased dosages of the chemical concerned, and later to its replacement by chemicals with other modes of action. Indirectly, therefore, the development of resistance has ecological repercussions.

These problems have arisen because pest control was not regarded as applied ecology. The chemicals used since the 1940s have been so efficient that it was often not thought to be worthwhile to consider the biology of the target species, or the functioning of the agro-ecosystem, other than spraying at a suitable stage in the pest's life cycle.

It must not be thought that it is only the modern pesticides that can have far-reaching effects on the ecosystem, or that problems of resistance are restricted to chemicals. Any pest control method, with the exception of quarantine, can affect the ecosystem to which it is applied. The eradication of the screw-worm in the United States is expected to be followed by increases in the numbers of deer and jackrabbits which will compete with livestock. In Britain, the destruction of rabbits by myxomatosis brought great vegetational changes, particularly noticeable near the edges of woodlands where regeneration had previously been prevented. The eradication of a pest, by any means whatever, is a simplification of the environment, and could lead to instability. An insect pest may, for example, support a population of general predators during a certain period, after which they migrate to other crops. Thus the elimination of the first pest by, say, the introduction of a resistant variety over a large area, could increase the damage caused by other pests on other crops in the same district. Furthermore, the eradication of a pest leaves an ecological niche unfilled, and the absent pest may well be replaced by a more noxious species, unless man attempts to introduce a suitable alternative himself. An insect vector of a disease, for example, may be replaced by a more efficient one that had previously been rare.

In general, however, it is the chemical pesticides that have had the greatest disruptive ecological effects, but it is obvious that they are essential. Natural regulatory factors and introduced biological control agents often do not give satisfactory control by themselves, so that chemical measures are necessary regularly or occasionally. It is only commonsense, however, that these additional measures should be no more than is necessary for maintaining the pest population density below the economic threshold. The chemical measures, the natural regulatory factors, and the introduced biological control agents should be integrated in an organized way so that they are compatible.

Integrated control was first concerned with the integration of chemical methods and natural enemies, since these have been the most powerful tools available for pest control, but there is no reason why the control methods should be limited to these two. Ideally, all suitable methods which can be blended together to give a suitable control strategy should be used. A less obvious advantage of this multi-faceted approach is that each method acts upon the pest population less intensely than it would if it were used in isolation, and there is thus less selective pressure leading to a development of resistance.

The aim of integrated control, therefore, is the blending of all suitable techniques to minimize economically the damage caused by pests, with the minimum disturbance of the common environment (Smith and van den Bosch, 1967). But as Geier (1966) points out, this integration must be more than an empirical juxtaposition or super-imposition of techniques. It must, in fact, be based on an understanding of the ecology of the ecosystem involved.

Before the introduction of the modern synthetic pesticides growers used a multiplicity of methods to control pests, but this art was largely forgotten during the 'dark ages' of DDT. Some authors have therefore suggested that just as Molière's *Bourgeois Gentilhomme* had been talking prose for more than forty years without knowing it, farmers had practised integrated control for many years in a similar state of ignorance. If this was integrated control it was so by accident rather than by design: in the modern concept there is a conscious effort to harmonize the pest control measures with each other, and with the environment. Integrated control is, in short, applied ecology.

The term 'integrated control' is thought to have been first used by Bartlett in California, whereas the term 'pest management' originated in Australia. Many authors consider that the two terms have identical or almost identical meanings, although the author of the chapter on integrated systems of pest management in the recent National Academy of Sciences publication (National Academy of Sciences, 1969), considers that pest management would include eradication methods. The eradication of a pest is clearly incompatible with a system in which the preservation of its specific natural enemies is integrated. The reader should also consult the foot-note of page 191 of the text by Clark and his colleagues (1967). In this text the terms will be regarded as synonymous, and will be taken to mean the application of all suitable pest control techniques in an organized way, so that the damage is minimized as economically as possible, with the minimum of harmful effects on the ecology of the environment.

Ideally, integrated control is based on a complete understanding of the ecosystem and of the population dynamics of the pest and all the factors which influence it (that is, of the life system of the pest). In most cases there is a complex of pests to be dealt with, and the life systems of each of these should be understood. In practice, such knowledge has never been collected, and the economics of the situation demand that measures must be taken before comprehensive models are completed. To this extent integrated control is still in the empirical stage. The same failings that are leading to dissatisfaction with unilateral control methods, have, however, served as guides for the integrated approach. The upsurge of formerly negligible pests, for example, has given some understanding of their natural regulation before the introduction of the broad-spectrum pesticides.

The motives for integrated control

It is in California that integrated control has had its greatest support and its greatest successes so far. It was introduced largely because growers were having increasing difficulty in producing certain crops economically while using unilateral methods of pest control, largely because of the resurgences of established pests after spraying, the appearance of new pests, and the development of resistance. A similar situation in the Cañete Valley of Peru caused the abandonment of unilateral control of cotton pests by pesticides.

The British position is different, for satisfactory control can still be achieved, in most cases, with chemical pesticides. Kennedy agrees with Wigglesworth's opinion that the incentive for integrated control in this country stems from the professional biologist's concern with the undesirable side-effects of insecticides and the fear that they will, in time, fail to give satisfactory control (Wigglesworth, 1965; Kennedy, 1968). Kennedy also attributes some of the interest to the strong conservation movements in Britain, though he stresses that modern agriculture and conservation have quite different objectives: conservation is an attempt to maintain diversity for its own sake, whereas the farmer is striving for simplification – the production, as far as possible, of a single kind of organism over a given area, while maintaining the ability of that area to yield productively year after year. It would be wrong to confuse these aims, a mistake to plan integrated control more for the maintenance of diversity for its own sake, than for productive, sustained, and economical cropping.

The interest of the Canadians and the Australians in integrated control stems, in Kennedy's opinion, from the high regard in which the agricultural biologists are held in these countries, and the encouragement they receive for their work. He does not, however, belittle the work and abilities of the British applied entomologists and, in fact, he appears to consider the Australian concept of pest management as rather more primitive than the British or American. Pest management, he considers, is a somewhat static concept, being mainly concerned with agro-ecosystems as they exist now, and paying little heed to the evolutionary changes which will come. This is because resistance to pesticides has not yet become as important in Australia as it has in certain other countries. Significant resistance to one or more chemicals has been reported in about ten arthropod pests of agriculture and horticulture, and in the cattle tick, *Boophilus microplus*. Resistance is also known in the house-fly and other public health pests. Some of the cases are restricted to one or two localities and, so far, alternative chemicals are available for the control of most of the resistant populations. The most serious threat at present is to the chemical control of the red spider mite, *Tetranychus urticae*.

The application of integrated control

The ecological basis   The rational application of integrated control is dependent upon the understanding of the agro-ecosystem concerned, and the population dynamics of the pests and associated organisms within it. An early decision concerns the dimensions of the ecosystem that is to be subjected to the programme. Even when it is delimited it may not be self contained. Both pests and beneficial insects may migrate into it from considerable distances, coccinellids furnishing a particularly good example. If the culture is of a reasonably permanent kind – an apple orchard or a citrus grove – the ecosystem may be small in extent, especially if the main pests are sessile, or do not migrate far. The pests of field crops are, however, characterized by their vagility, and the area must be correspondingly large. The area will thus consist of several fields, often with different crops, and such features as field margins, roadways, ponds, waterways, woodland, scrub, and hedges – and possibly gardens and urban areas. The integrated programme introduced for the control of cotton pests in the Cañete Valley covered several square miles, as did the recent wheat bulb fly trial in eastern England. In the latter, no winter wheat was grown in an entire parish, spring wheat taking its normal place in the rotation.

If attention is being concentrated on pest control in one crop the other crops, and the uncultivated areas, must also be considered as they provide alternative hosts both for some of the crop pests (which are thus able to infest the subject crop early in its development) and quarters and prey for populations of beneficial insects.

In integrated control the whole pest complex is considered at once, whereas in unilateral control this may not be so. The number of arthropod pests that can be found feeding upon a crop at one time or another can be surprisingly high. Massee (1954), for example, lists as 'the more important pests found on apple' in the British Isles, 20 beetles, 52 moths, 3 hymenopterons, 1 earwig, 2 flies, 21 hemipterons and 5 mites – a total of 104 species, and many field entomologists will recall finding species which could not be 'chased down' in the standard reference books. Similarly, in a five-year study of a single orchard a Canadian worker found 92 arthropod species that had been recorded at some time as causing damage to apples. In addition he found over 120 entomophagous insects and 23 insectivorous birds. Furthermore, there are a large number of plant pathogens to be found on apple, the chemical or cultural control of which could affect the integrated control of arthropod pests.

Fortunately, most crops carry only one or two important pests at any one time or place – these are the so-called key pests. On apples in Kent, for example, codling moth and red spider mite are considered to be the key arthropod pests (although codling moth is apparently less important than it was several years ago) and apple scab and mildew the most important pathogens. At the beginning of the Californian integrated control programme on lucerne the key insect pest was the introduced spotted alfalfa aphid, *Therioaphis trifolii*.

The key pests of a crop in a particular region are those organisms which appear, year after year, at such high levels that control measures are essential if economic losses are to be avoided. A pest can be a key species even if it is present in only small numbers, if each individual is potentially very harmful. Vectors of plant virus diseases, or pests which burrow into, or blemish fruit, such as the codling moth, would come under this heading, and would call for control measures at quite low densities.

The key pests form the first of four categories of pests outlined by Chant (1964). Class II contains those occasional pests which in most places and at most times are under adequate biological control or natural regulation. Occasionally, however, they escape these restraints, and the populations rise to economic levels. These, of course, are the pests for which integrated control is particularly suitable. Potential pests (Class III) are those organisms which cause no significant damage in the agro-ecosystem in its current state, but which could become troublesome if this changes. Integrated control must ensure, as far as is possible, that these pests do not change their status.

Class IV is composed of those non-resident and migratory pests which may, from time to time, invade the crop in large numbers. Locusts and army worms are obvious examples, but less familiar ones are the cotton worm, *Alabama argillacea*, and lygus bugs which move into Canadian orchards from other crops. Such pests may cause a disruption of an integrated control system that has been functioning satisfactorily for several years if emergency applications of pesticides are made for their control.

An integrated control system will obviously be built around the key pest species of the agro-ecosystem, a variety of suitable controls being chosen which will give adequate control of these without disturbing too greatly the biological and environmental constraints of the other categories of pests. Ideally, the system will be based on the study of the population dynamics of the key species and of the more important Class II pests. Full use will be made of life-table studies, and of such techniques as Morris's key factor and Watt's systems analysis (chapter 2). Unfortunately, the problems posed by the pests are often too pressing for the ecologist to take the five or ten years necessary for an adequate study before designing the strategy. The system must therefore be based on the knowledge of the pest already gained (and this is often surprisingly meagre), on general ecological principles, and on the lessons learned from past mistakes. Most current integrated control programmes are, at best, compromises between *ad hoc* procedures and applied ecology, but once they are under way the ecological studies should be carried out energetically in order to refine the programme, and to prepare for possible breakdowns. It must be remembered that an agro-ecosystem is a continually evolving system, under pressure from natural changes, and from changes induced by man's techniques and economic demands.

There is one common requirement for the successful prosecution of integrated control, namely an ability to judge the probability of a given pest population developing to economically important levels. This depends, of course, on the nature of the crop – what level of damage is tolerable – as well as upon the biological characteristics of the pest organisms. If key factor studies have been made it will be possible to estimate the risk by an intensive sampling of one life stage, and if this occurs well before the usual period of economic attack, appropriate preparations can be made. It is particularly difficult, however, with organisms that run through several generations in a season, and whose rate of reproduction is greatly influenced by the weather.

The integration of chemical control with the action of natural enemies This is, historically, the most important technique of integrated control, and still is an important part of most strategies. Clearly chemical control is only used when natural enemies – native or introduced – do not prevent economic damage from occurring, and equally clearly, it depends upon a residuum of the pests remaining to act as hosts or prey for the natural enemies. The chemicals are applied or chosen in such a way that the natural enemies are harmed as little as possible.

There are several possible ways of achieving this. Obviously, no chemicals are applied as long as the natural enemies or other factors are preventing the rise of population densities to economic levels, but when applications do become necessary, a highly selective pesticide should be used. Usually, however, no pesticide is available which is sufficiently physiologically or biochemically selective, and a chemical with a wider range of kill has to be used. In such a situation, a non-persistent material is applied so that as few susceptible natural enemies as possible are at risk. This may be done by spraying when the natural enemies are not exposed, or when they are in some comparatively tolerant stage, using as little of the active ingredient as possible, or by applying the chemical to a restricted part of the environment. All these techniques are desirable for reasons other than the conservation of natural enemies – they are less likely to lead to residue problems in the crop, or to pollution of the environment.

Physiologically or biochemically selective compounds are ecologically desirable, but they pose several problems, including the sheer difficulty of discovering them. Unterstenhöfer (1970) discusses the problem from the point of view of the large chemical company. Monotoxic compounds, which kill only one pest species, and which are harmless to all other organisms, are rarely discovered. They are, by their very nature, likely to be rare and, if not rare, many would still be missed because of the small number of kinds of pest organisms used in screening. The rodenticide norbormide and the insecticide 1,3,6,8-tetranitro-carbazole (toxic to the grape berry moths *Clysia ambiguella* and *Polychrosis botrana*) are, probably, the compounds nearest to this ideal which have so far been discovered. The latter replaced arsenic compounds in German vineyards in the 1940s, but was itself displaced by compounds with wider spectra of activity. Interestingly, its short life was caused by the gradual increase of other pests, formerly not of great importance. Reasoning from this example, Unterstenhöfer suggests that we should not accept, without more evidence, the idea that polytoxic compounds almost always encourage the build-up of secondary pests. Most ecologists think, however, that there is ample evidence for the conventional belief.

Compounds which kill a narrow range of insects (oligotoxic compounds) are discovered more frequently than are polytoxic ones, which are those which kill a wide range of insects over a narrow range of doses. For biochemical reasons oligotoxic compounds are likely to be more toxic to closely related organisms, and to be comparatively innocuous to other kinds. Few taxonomic groups are composed of pests alone. A compound which is toxic only to mites, for example, is likely to kill predacious and neutral species as well as pests.

Because of market demands, Unterstenhöfer believes that preference should be given to polytoxic compounds in research and development, although he admits this would conflict with the principles of integrated control. He considers that more attention should be given to using these compounds more selectively.

Manufacturers could approach the problem of using polytoxic compounds in such a way by investigating the effects of the type of formulation on the toxicity of the compound to various kinds of organisms. One possibility that is being investigated (and which shows promise) is enclosing the pesticide in minute capsules which can be eaten by phytophagous insects, but which would be harmless to beneficial and neutral species which merely came into contact with them. An even more sophisticated technique which is being investigated is the encapsulation of useful but highly persistent compounds such as DDT with chemicals which will degrade them. One approach that has been neglected is the development of machinery for application which would give more selective control. Several of the speakers at the 1969 Conference of the Pesticides Group of the (British) Society of Chemical Industry called for a closer cooperation between agricultural machinery designers and manufacturers, and the chemical industry (Society of Chemical Industry, 1970).

Bartlett (1964) lists several ways in which the required selectivity can be achieved:

(i) Preservation of natural enemy reservoirs outside the treated areas. Ideally, the natural enemies should be preserved within the treated area, but this is not always possible, and reliance must be placed on reservoirs outside. The success of this depends upon the powers of dispersal and the searching abilities of the insects; certain *Aphelinus* parasites of scale insects, for example, are believed to disperse no more than a few yards annually.

Reservoirs may be maintained in near-by non-crop environments, or in those parts of the crop where the pest has not reached economic levels. The late Dr A. D. Hanna gives two examples of the latter from his long experience of cotton cultivation in Egypt and the Sudan. In Egypt it was noticed that the infestations by the pink bollworm, *Pectinophora gossypiella*, are highest near villages. The moths emerge from pupation sites within the cotton stalks, which are collected and stored by the villagers for fuel. An appropriate integrated control measure is a chemical application early in the season in the fields surrounding the villages. In the Sudan, where the moth is not so troublesome, the cotton sticks are burned on the spot after pulling, but fairly high infestations were found around ginning factories. In the Sudanese Gezira irrigation area the fields are thus quite bare during the 'dead season' – May and June. Nevertheless, the growing crop suffers from heavy infestations of the cotton jassid, *Empoasca lybica*, and a search was made in order to locate the quarters that they occupied during the dead season. These were found in the large gardens maintained by the inspectors of the scheme – gardens which are scattered throughout the area – and in the vegetation along the river bank. These sites were found to be the only significant sources of infestation, and it should be possible to restrict chemical treatment to these areas at the beginning of the season, thus preserving any beneficial insects which have persisted in the cropping districts (Hanna, 1966).

Several workers have noted that infestations of insects such as aphids which are easily transported by winds are often more serious on the lee side of a field hedge than those in the centre of the field. They also begin earlier. These areas often act as centres of infestation from which the pest spreads to other parts of the field. The sheltering effect of the hedge, which can extend in a leeward direction for a distance of forty times the height of the hedge, if this is dense, but which is appreciable for only about fifteen 'heights' (Caborn, 1971), can influence the microclimate in the crop. This may promote the more rapid development of many pests, and possibly also favour their survival. It is possible that the spraying of these restricted areas of the field, coupled with the effects of natural enemies from the remainder of the field, will give a sufficient control of the pest in the whole crop.

An alternative to the spot treatment of areas with a high population density of the pest is the systematic restricted treatment of the crop in which spray applications are staggered. DeBach and Landi (1959) used this integrated approach successfully when they applied a non-selective oil treatment to alternate pairs of citrus tree rows at six-monthly intervals to control the purple scale, *Lepidosaphes beckii*. Other workers treated only the top leaves of tobacco plants (where the infestation was the highest) to control *Heliothis virescens* and *Protoparce* species. This allowed *Polistes* wasps to continue their attack on the pests at lower levels.

A rather more troublesome method, but perfectly feasible, is to preserve the beneficial insects by culturing them in the laboratory, and releasing them at a suitable time after chemical application.

In the design of an integrated programme for the control of insect and mite pests care must be taken to use fungicides which do not have a markedly toxic effect on beneficial insects and mites. It will be remembered that several fungicides can also be used for the control of red spider mites, and these are likely to be toxic to predatory mites also. A further possibility is that the deposits of fungicides which are not toxic to pest mites may actually stimulate population increases because of their granular nature.

(ii) Selectivity derived from differences between the various stages in their susceptibility to a toxicant: In general, the adult stages of parasitic insects are much more susceptible to insecticides than are the larval and pupal stages, and the applications of chemicals may be arranged to make use of this. The differences in susceptibility are partly due to the concealed and protected positions of the larval stages, but there are probably physiological differences as well. The nymphs of exopterygote insects, on the other hand, are usually more susceptible than the adults. Eggs are generally relatively resistant, even when they come into contact with a pesticide.

(iii) Selectivity derived from the feeding adaptations of natural enemies: Parasitic stages of natural enemies are, as was pointed out in the last section, protected from the action of contact insecticides. Predatory insects are not usually affected by systemic insecticides once these have entered the plants which they are to protect, though, of course, contact action can be harmful. They may, however, be killed if their prey accumulates enough of the toxic material, but Bartlett (1964) does not consider that this is of critical significance.

(iv) Selectivity based on the seasonal life cycles and habitats of the natural enemies: Although the times at which pesticides are applied are dictated by the life history of the pest, these times often do not coincide with times at which parasites are most active. Usually, an attempt is made to kill the pest with chemicals when it has just hatched, or when it is in the adult stage. Unfortunately, predators are usually active at the time of egg hatching, but even then some compromise might be possible. This method of preservation of natural enemies is more likely to work well when there is only one generation of the pest a year, or at least when there is not too much overlapping of generations.

(v) Selectivity based on characteristics of the pesticide, and on its method of application: Even when two chemicals are non-selective in a physiological sense the less persistent one will be preferable. This applies not only to the crop sprayed, but also to any adjacent areas if drift is likely. Natural enemies appear to be very susceptible to accumulations of persistent insecticides which drift into their habitats.

Lowering the dosage also reduces the period during which the crop retains residues harmful to beneficial insects. Although in the past manufacturers have recommended dosages higher than were absolutely necessary, in order to ensure a good kill, and despite the habit of many growers of increasing this dosage still further 'for luck', the drastic reduction of dosages should not be done without care. Parasitic and predatory insects are usually more susceptible to the pesticides than are the pests, and it would be possible to eliminate the natural enemies with a lowered dosage, and leave a greater proportion of the pests alive than there would have been at standard rates of application. In such cases a pest flare-back might occur much more quickly than if standard rates had been used.

Little is known about the effects of formulation of pesticides on their toxicity to beneficial insects. It does seem certain, however, that granulated formulations, and seed dressing where appropriate, are less harmful than most.

In conclusion, it should be noted that Bartlett gives a valuable table showing the toxicity of the chief insecticides and fungicides available in 1964 to the main groups of parasitic and predatory insects and mites. The table also records, where appropriate, the pest upsets caused by certain of these pesticides. Further papers which deal wholly or in part with the integration of chemical control with the actions of beneficial insects are van den Bosch and Stern (1962), Smith and van den Bosch (1967), Chant (1966), National Academy of Sciences (1969), Starý (1970). Important collections of papers presented at symposia, etc., are: FAO (1966), Wigglesworth *et al.* (1965), Empson (1968).

Wider integrated control
Originally, as we have noted above, integrated control was based on two kinds of control or regulating factors, namely chemical pesticides and beneficial insects and mites, but during the 1960s it came to have a wider meaning, as other forms of control were welded into the various programmes. Natural enemies of the pest will, however, almost always play a part, and some chemical measures are usually necessary.

Rotations and various cultivations are used as a means of pest control, and are employed in various programmes. Traditional methods should, however, be examined carefully, since it is sometimes found that they destroy large numbers of beneficial insects. The burning-off of wheat stubble, for example, often kills numerous overwintering pupae of parasitic insects.

In earlier chapters we have often stressed the importance of ecological complexity as a stabilizing factor, and, at the same time, noted that, for economic reasons, farming is becoming simplified. Fields are, in general, larger than they were thirty years ago, with hedges being grubbed out and ditches filled. Traditional rotations are being ignored wherever possible so that the farmer, or manager, specializes in a few crops which he understands well. These trends allow greater mechanization of farming, and reduced labour costs, so that they cannot be lightly abandoned unless there are substantial advantages in doing so. Complexity cannot be reintroduced into the environment unless it can be shown to make a really worthwhile contribution to pest control. Unfortunately, hedgerows and ditch bottoms can be a source not only of beneficial insects, but also of pests and diseases of the near-by crops – in fact, most studies indicate that the gains are just about balanced by the losses (Kennedy, 1967). Nevertheless, use can be made of this kind of complexity if it is properly managed. The floral composition of the field margins and hedgerows could be controlled. To take two obvious examples, spindle and barberry bushes should be grubbed out to reduce the incidence of *Aphis fabae* and wheat stem rust. The mowing of margins should be so timed that the flowering heads which provide food for adult beneficial insects are not removed when these are abundant, especially as this will leave the lower parts of the plants as food for phytophagous general feeders. Another example was mentioned in the last chapter, namely the growing of blackberries near Californian vineyards in order to maintain populations of the parasite of the grape leafhopper. Research will doubtless show that this is not an isolated case.

The importance of complexity of this kind will, however, diminish, as methods are developed which conserve the beneficial organisms within the crop area itself, and, in fact, the Californians are already showing success. The simplification of the environment does, however, bring one great advantage as it allows integrated control to be applied easily over large areas.

There are other ways of introducing complexity which will contribute to the stabilization of pest populations. Integrated control, using several control methods at once, or in succession, in itself introduces such desirable complexity. If one of the methods fails to some extent, the other controls will probably check the rise of the population.

There is also the possibility of introducing some kind of heterogeneity into the crop itself. The Californians now practise strip cropping of lucerne, so that when mowing makes the environment unsuitable for parasites and predators they merely move a comparatively short distance to a more favourable strip. This practice also induces the alfalfa butterfly, *Colias eurytheme*, to concentrate its egg-laying in those strips which are at a favoured stage of development, so that chemical control can be limited to these areas.

The use of resistant varieties has played an important part in integrated control. Formerly, plant breeders attempted to raise varieties which were completely resistant to the given pest, although experience has shown that such varieties often lose their resistance within a few years, because of the development of new biotypes of the attacking organism. Little thought was given to using the varieties which showed only some resistance. In the presence of natural enemies, however, such a partial resistance can be sufficient to tip the balance in favour of satisfactory control. In such situations a completely susceptible variety would suffer economic damage if reliance was placed on natural enemies alone. If the natural enemies are absent even the partially resistant variety suffers economic loss.

In putting forward these ideas, van Emden and Wearing (1965) suggested that a suitable level of resistance could also be induced in the crop plant by some physiological means. In experiments on cabbage they reduced the reproductive rate and fecundity of the aphids *Brevicoryne brassicae* and *Myzus persicae* by reducing the nitrogen supply, and of *Myzus persicae* by inducing drought conditions. These techniques are not consistent with brassica production, but they also depressed both species by providing copious potassium, and irrigating well. They suggested that these more suitable measures may depress the pest population sufficiently to enable natural enemies to bring the damage below economic levels.

The introduction of a new variety is an intricate step which may have unexpected side-effects. It may lead to changes in the method or timing of harvesting which could affect the faunal composition. It may have a habit of growth which differs from that of the earlier varieties, and thus change the nature of the microhabitats available to parasites and predators. Certain new varieties of lucerne grown in California, for example, have some of their leaves growing close to the soil, so that, after mowing, the ground is not left as bare as previously.

A more obvious danger is that the new variety, while resistant to the pest that is being studied, and presumably also resistant to other key pests (this would be routinely checked), could be particularly susceptible to a Class II or Class III pest which would thereby become more important. Certain disease-resistant strawberry varieties, for example, which were introduced into Californian horticulture, proved unsatisfactory because they were found to be particularly susceptible to the cyclamen mite, a pest which was not of great importance on older varieties.

A less direct repercussion would follow the introduction into California of lucerne varieties which are resistant to the pea aphid, *Macrosiphum pisi*. Lucerne, unlike peas, is fairly tolerant of this pest which, in fact, supports a population of coccinellids and other predators which disperse on to other crops associated with lucerne cultivation (Smith and van den Bosch, 1967). The same criticism could, of course, be levelled at any control method which greatly reduced or eradicated a pest population, but in the case of the pea aphid on lucerne the saving gained by the control would be outweighed by the losses induced in the other crops.

The part played by partial resistance when it is integrated with the impact of natural enemies has just been discussed, but it may also be integrated with chemical control. The partial resistance will reduce the number of applications necessary, and the dosage can also be made smaller – two measures which will help to conserve natural enemies. Chemical control could also be integrated with the use of pathogens which are often unaffected by insecticides. Fungicides applied for disease control could, however, lessen the impact of the insect diseases.

Some authors consider that the eradication of a pest and integrated control are incompatible, and, in fact, a few believe that eradication might be undesirable because it reduces the complexity of the environment. Nevertheless, it is useful to consider how a pest might be eradicated more economically by combining several control methods. It is probably safe to say that chemical methods alone would be insufficient to eradicate a widespread, well-established pest, although they have been used, for example, in the elimination of the Mediterranean fruit fly infestations in Florida. Even here, however, chemical control had to be combined with strict quarantine measures which helped to prevent any further spread of the pest. The main difficulty is that chemical control becomes progressively less efficient as the population density falls. The same quantities of material have to be applied over a given area when the pest density is low as when it is high. Furthermore, the percentage kill achieved by a single application does not increase linearly with dosage or even with log dosage, as can be seen from laboratory toxicology trials (chapter 3). Although the percentage kill–log dosage curve is only roughly applicable to field conditions it is clear that over a certain range of kill, the percentage kill rises smoothly as the dosage increases exponentially, but as the kill approaches 100 per cent, the dosage has to be increased even more rapidly. A similar relationship between control achieved and intensity of the pressure by the agent probably applies to other control methods when used in isolation. It has been calculated, for example, that in order to increase control from 30 per cent to 90 per cent by the release of sterile males, the number of insects released would have to be increased, not three times, but 21·42 times (National Academy of Sciences, 1969). Nevertheless, if a constant number of sterile insects are released in each field generation of the pest, the proportion 'killed' will increase as the wild population falls, whereas it will remain roughly constant when fixed amounts of pesticides are applied. Thus, as was suggested in chapter 8, the most efficient procedure would be to reduce the density as much as possible by chemicals before releasing sterile insects. A further refinement would be to use a resistant strain of the pest for the releases.

At the moment it seems that if chemosterilants are to be used in the field, as opposed to their laboratory application as a substitute for ionizing radiation, they will have to be used in conjunction with powerful attractants having reasonable specificity. Such attractants can, of course, be integrated with conventional chemical pesticides as, for example, when baits are used for the control of various tephritid fruit flies.

The interaction between autocidal methods, and natural enemies and pathogens is of importance. Clearly, if the pest population is eliminated or greatly reduced, specific parasites and predators will disappear through starvation or dispersal, while more general feeders will be reduced in numbers. It is less clear, however, how the biological agents will interact during the early stages of the programme.

Shipp and Osborne (1966) have considered the theoretical role of predators in sterile-insect release programmes. In their model they postulated a wild population which normally shows only minor fluctuations in population density. Initially the release should increase the numbers of fertile prey, since the predators will use the released insects as alternative prey. This would be followed by an increase in the numbers of predators. The pest population will fall as a result of the sterile matings, but the predator mortality will lag behind, so that predation pressure will increase, reinforcing the influence of sterile matings.

When the population density of the pest is low the predators will destroy a smaller proportion because of the smaller probability of a predator individual finding a victim. If sterile insects are then released the proportion of the pest population killed will rise, and if the predators feed randomly there will be greater pressure on the wild population.

Shipp and Osborne gave particular attention to the egg predator *Cyrtorhinus mundulus* as this was likely to play an important part in the then proposed autocidal control of the sugar cane leaf-hopper, *Perkinsiella vitiensis*, in Fiji. Although this predator attacks the eggs of other species of leaf-hoppers these are not present in significant numbers in the sugar-growing areas concerned. Such a predator will not, of course, have any effect upon the sterile insects released, but, on the other hand, they do not reduce the longevity of wild individuals. The numbers of such predators will be influenced by the release of sterile females, provided of course that such females produce eggs. Thus, the authors suggest, the presence of predators may offset the disadvantages that would result in some control programmes from the release of sterile insects of both sexes. They also consider that the costs of some programmes may be reduced by integrating the sterile release technique with the pressure of natural enemies, and that their presence will also be of value when sterile releases cease.

It does not seem likely that sterilization procedures will make insects physiologically more susceptible to attack by parasites or predators, although they may affect behaviour or vigour in such a way that the probability of encounter and capture is increased. It is very difficult to predict the effect of the procedures on the susceptibility to pathogens. If the sterile insects are more susceptible than wild ones, the presence of the pathogen in the field population could reduce the effects of the release programme. If, on the other hand, the sterile insects are equally or less susceptible, the two control methods could be used together, or in sequence. Fortunately, we need only be concerned about the susceptibility of the adult insects, and it will be remembered that many of the most promising insect pathogens have little effect on adults.

Jafri has made a series of investigations on the interaction of disease pathogenicity and irradiation. He found an increased susceptibility to infection by *Bacillus thuringiensis* in adult *Tribolium castaneum* and *T. confusum* beetles after they had been subjected to lethal and sublethal doses of ionizing radiation. In contrast to this, the development of two viruses, VND and a nuclear polyhedrosis of *Heliothis*, was inhibited in the larvae of the greater wax moth, *Galleria mellonella*, which had received lethal or sublethal doses of radiation a short time before. Clearly, no generalizations can yet be made. Each pathogen–irradiation interaction will have to be studied separately for each pest species before an integrated control programme can be planned (Jafri, 1965, 1968; Jafri and Khan, 1969).

Integrated control: some successful applications

The Cañete Valley, Peru   Cotton has been grown in this isolated valley for many years, and the incidence of pests has been recorded for more than three decades. Before the war the six species that were considered to be pests were controlled by arsenical- and nicotine-based insecticides, and by hand picking. In the 1950s, however, all insect pest control was carried out by the application of broad-spectrum chlorinated hydrocarbon insecticides. In time, these pesticides lost their potency, so heavier and more frequent applications were needed. There was also a change to organophosphate insecticides such as the parathions. New pests appeared, and some of these also developed resistance. Eventually, in 1956, the valley had its worst crop on record, with almost half of the cotton lost despite very heavy applications of pesticides (National Academy of Sciences, 1969).

This situation developed so rapidly because of the isolated nature of the valley – an irrigated area surrounded, except where it opens on to the Pacific, by arid uplands with indigenous vegetation. Once the natural enemies of the pests had been destroyed in the valley there was little chance of them being re-introduced naturally from outside. Once resistance had developed in the pests there was little to stop their populations from increasing explosively.

Traditionally, the growers in the valley had been very cooperative one with another and, indeed, they had possessed their own research station for some time. They turned to this organization for help in 1956, and certain regulatory measures came into effect within a few months. These were largely directed against the key pest *Heliothis virescens*, but also ensured the control of the older pests, and the new ones which had come into prominence. The use of synthetic organic insecticides was prohibited, except by special dispensation. Even then only systemic compounds were allowed, at rates which were one-quarter to one-half of those recommended by the manufacturers for the control of *Aphis gossypii*. Arsenical and botanical compounds were used in the place of the organic–synthetic compounds for the more routine sprays. Natural enemies of the pests were re-introduced from other Peruvian valleys, or from abroad. There were also certain cultural controls. All techniques that encouraged the speedy establishment, and the rapid growth and maturation of the crop were put to use, and cotton growing was abandoned on the marginal land. Ratooning, the production of additional crops from growth regenerating from cut-back plants, was prohibited in the more newly developed areas. The number of pupae of *H. virescens* killed by cultivation during soil preparation was increased by curtailing irrigation during this period. Irrigation was also reduced during the growing period, as succulent growth attracts female *H. virescens* over wide areas, and the humidity inhibits the chief predator, the anthocorid *Paratriphleps laeviusculus*. Deadlines were set for various operations such as planting and the destruction of crop residues.

The other crops which are grown in the valley – cotton, maize, potatoes, vegetables, and so forth – add to the ecological diversity, and provide alternative hosts for the parasites and predators. Several years before the integrated programme was started Hambleton instigated the introduction of flax as a winter crop to provide hosts during a period when other vegetation was scarce. This crop was followed by maize, beans, sweet potatoes, or weedy fallow. Corn was valuable, as it provided prey for several natural enemies of the bollworm. The flax supported tachinid parasites (*Archytas* spp.) of *Heliothis* which developed on *Agrotis ypsilon* and other cutworms (Bartlett, 1964).

All these measures led to significant increases in cotton production – a mean yield of 789·1 kilogrammes per hectare for 1957 to 1963, compared with a mean of 603·8 kilogrammes per hectare per annum for the organic insecticide period (Smith and van den Bosch, 1967).

Spotted alfalfa aphid in California   The spotted alfalfa aphid, *Therioaphis trifolii*, was first observed in California in 1954, and within two years it had spread explosively to all the important areas where the crop was grown. Clearly, such an invasion called for drastic measures and widespread spraying with broad-spectrum organophosphate insecticides started immediately. Native predators and a fungal disease, *Entomophthora exitialis* began to have an impact upon the pest, but this was lessened by the organophosphate insecticides. Finally, resistance developed and control broke down. This led to a search for a pesticide which would be selective enough to allow the natural enemies of the pest to take their full toll. A suitable compound was found in the systemic insecticide demeton, applied at very low rates. Within a few months satisfactory control was achieved and the damage caused by the pest dropped from nearly ten million dollars in 1957 to less than one-and-three-quarter million in 1958 (Smith and van den Bosch, 1967).

During the emergency period, when parathion and malathion were used in large quantities, a number of minor pests increased in importance. They included a leaf-miner, *Liriomyza* sp. (Agromyzidae), red spider mites, the pea aphid, the beet army-worm, *Spodoptera exigua*, and a leaf roller, *Platynota stultana*, which also invaded cotton. These pests have since declined.

Various cultural methods were also introduced, and one of these has been particularly important, namely strip cropping. Its value in the control of the alfalfa butterfly has already been mentioned, but it also played a part in the control of the aphid. Bartlett (1964) tabulates some data collected by Schlienger and Dietrick which show how this method favours natural enemies of the aphid. Counts of coccinellids (adults and larvae), green lacewing larvae, hymenopterous parasites, big-eyed bugs, and an aphid-eating spider were taken on lucerne under normal cultivation, and on crops subjected to strip farming. All classes of natural enemies were more numerous in the latter. In the first there were 14 parasites and predators on the average in each square foot, and in the strip-farmed lucerne, 56, an increase of 42. The practice also discourages the migration of *Lygus hesperus* to more susceptible crops.

A further measure, and one that has so far reduced even more the need for spraying, has been the introduction of new aphid-resistant varieties.

Apple and pear insects in Nova Scotia This is regarded as the classical example of the application of integrated control methods to pest control. By 1965 it was being applied to over 80 per cent of the apple and pear acreage (Pickett and MacPhee, 1965). Unfortunately, the public demand for blemish-free fruit, and certain economic changes have led to many growers reverting to traditional chemical control (National Academy of Sciences, 1969). The programme was based upon the encouragement of natural enemies of the pests by using only selective chemicals – nicotine sulphate, ryania, and lead arsenate – against insect pests such as codling moth, and captan, dodine, and glyodin against apple scab. Occasionally, other pesticides were used, but at reduced dosages or at periods of minimal natural enemy activity. Red spider mites were not troublesome under these conditions.

In 1965, spray materials (fungicides and insecticides) cost an average of 8 cents for each bushel of fruit produced, compared with 18 cents in the late 1940s.

Pickett's methods have been copied in other apple-growing areas, but not always successfully. This may be partly because the Nova Scotian orchards are relatively isolated ecosystems.

In Australia, codling moth increases under natural conditions to the limits of its food supply (Geier, 1965) so that insecticides are obligatory. An integrated control system, in which ryania is used as a pesticide and overwintering sites for the cocoons are minimized, has been tried and found to work well. The native light-brown apple moth, *Epiphyas postvittana*, tends to become troublesome when broad-spectrum sprays are used. An attempt is being made to conserve its natural enemies but to spray, when it becomes necessary, with *Bacillus thuringiensis*. Other workers in a trial on neglected orchards failed to achieve satisfactory control of codling moth with ryania.

Hoyt (1969) discusses the integrated control used on apples in Washington State where the red spider mite *Tetranychus mcdanieli* is an important pest. It is preyed upon by the mite *Typhlodromus occidentalis*, a predator which survives applications of certain insecticides and fungicides. This mite also feeds upon the apple rust mite, *Aculus schlechtendali* but the regulation is poor. The presence of the rust mite is, however, an advantage as it prevents predator–prey oscillations of *T. mcdanieli* and *T. occidentalis*. *Panonychus ulmi* is controlled by selective sprays at the beginning of the season, and prevented from building up again later by the predator.

Integrated control in Britain
Biological control cannot be said to have been very successful in Britain, but the few species which have been introduced, and the numerous indigenous natural enemies, do have a substantial impact upon pest populations, an impact which can be eliminated by excessive spraying. Integrated control can be very valuable in such circumstances. There is not, however, the same urgency as in California and Peru, for resistance has not, so far, been so troublesome. Most crops can still be protected by conventional spraying programmes, and, as a result, many growers do not see a

need to make any change. Indeed, as Kennedy (1968) points out, some advisory officers have faced scepticism when they have advocated the reduction of dosages: the growers concerned suspected that this was a concession to conservation interests rather than a contribution to agricultural production.

Kennedy has further reservations about the place of integrated control in British agriculture. Although natural enemies do make a contribution to the regulation of pest numbers they are not themselves easily controllable by man, especially under British conditions where their numbers are to some extent dependent upon uncultivated areas of land. In an environment that is greatly simplified by man – for example, Californian irrigated desert valleys, or glasshouses – natural enemies can be manipulated with greater confidence. In this respect Kennedy welcomes the present trend towards simplification in agriculture, with its amalgamation of small units into large ones, often devoted to a small number of products. He also points out that the laboratory studies of population dynamics, so often despised by some field ecologists as 'jam-jar' experiments, will have an important bearing in such simplified agriculture. Diversity may well bring stability, with no population explosions of the pests, but it is sometimes forgotten that the primary aim of agriculture is a population explosion – the population being that of the crop species. In short, conservation, which has as its aim the maintenance of as much ecological diversity as possible, and agriculture, have different goals. To summarize, Kennedy suggests that natural enemies can be integrated with chemical control, but that entomologists should strive for techniques which will give them more direct control over these enemies – techniques such as the artificial rearing and the release of suitable parasites and predators at appropriate times. In other words, he suggests that they should themselves be used in much the same way as insecticides. This is at present possible in glasshouses, but in the future it may also be feasible with crops enclosed in polythene sheeting. As Kennedy himself wrote 'The ecologically simplifying progress of agricultural technology is not something the entomologist need feel embarrassed or hampered by. Quite the contrary. It both demands, and permits, the development and application of his skills as never before.' (Kennedy, 1968.)

At the same conference several advisory entomologists discussed the applicability of integrated control to various crops. White (1968) suggested that studies should be made on the economic significance of cecidomyid midges and aphids in cereal crops, and described the proposed experiment of growing only spring-sown wheat over a large area in an effort to control the wheat bulb fly, *Leptohylemyia coarctata*. Research is being carried out on the feasibility of replacing persistent organochlorine insecticides with less persistent materials for the control of this pest, and of wireworms and tipulids. Cereal pests are unusual in that they are often less troublesome when cereals are grown in succession than when cereals follow, for example, leys. In general, however, little insecticide is used on cereal crops because of the comparatively low return: consequently, there has been little or no trouble with secondary pests assuming greater importance.

There are possibilities of introducing integrated control strategies against the cabbage root fly, *Erioischia brassicae*, as the marketable part of the plant is not attacked by the maggots. Bevan (1968) suggests that chemosterilants or irradiation in the laboratory might be of value, and preliminary work has been started. In some areas organochlorine-resistant strains have developed, but organophosphate substitutes (used as dips before planting) have been phytotoxic; new compounds are needed. Spot treatment with granules near the base of the plant is promising, as many of the natural enemies apparently survive this treatment. The influence of planting density on the number of eggs laid near each plant is also worth investigating. The other key pest on brassicas is the cabbage aphid, *Brevicoryne brassicae*, and the systemic insecticides used for its control are considered to be reasonably selective, though better results still might be achieved by reducing the foliage sprays and substituting soil applications of systemic compounds.

Cogill (1968) does not consider that there will be any great changes in orchard practice in the next few years, though most growers are discarding organochlorine compounds and using organophosphates in their place. Red spider mites are becoming resistant to many of the products that have been used for their control. They are, however, regulated adequately by natural enemies, provided that the spray programmes are modified to conserve these. Codling moth could probably be controlled by two sprays of lead arsenate, followed by one of ryania – the latter being chosen for the final spray before harvest because of its low mammalian toxicity. Winter washes (now somewhat 'out-of-fashion' on apples) should not be used as they kill many natural enemies, as does carbaryl, which has become popular as a fruit-thinning agent.

The interested reader should consult the summary of the papers presented at this meeting for details of other British crops.

Integrated control in British glasshouses

The most successful and promising trials of integrated control in Britain have been on glasshouse crops. There are several reasons for the choice of such crops for trials. The controlled environment present in a glasshouse makes the manipulation of natural enemies simpler than it is in the open air. The crops receive many applications of insecticides and fungicides because of their high value and, consequently, resistance to pesticides soon develops. Despite this intensive cultivation and high value of the crop the glasshouse industry is not a big outlet for the chemical industry, and growers cannot, therefore, expect the chemical companies to spend much time or money in the search for new compounds specifically toxic to pests which are more or less peculiar to their crops. A further incentive for the adoption of integrated methods, with its restriction on the amount of pesticide applied, is the toxic effect of the chemicals on the plants. This is especially troublesome with cucumbers, the yield of which is considerably reduced by the phytotoxic effects of chemical pesticides.

Hussey and his colleagues at the Glasshouse Crops Research Institute in Sussex, England, have carried out many experiments, the results of which are now beginning to be applied in commercial holdings. Two recent papers which describe some of this work, and which include key bibliographies, are Parr and Scopes (1971a, b). The textbook by Hussey, Read, and Hesling (*The Pests of Protected Cultivation*, 1969) also contains much relevant information.

In cucumber houses, the key pest is the glasshouse red spider mite, *Tetranychus urticae*. The next most important animal pest is probably the glasshouse whitefly, *Trialeurodes vaporariorum*. Other pests are aphids, especially the cotton or melon aphid, *Aphis gossypii*, the thrip, *Thrips tabaci*, mildew, *Sphaerotheca fulginea*, and 'French-fly', *Tyrophagus longior*. The last, despite its common name, is really a mite which occurs in the straw placed on the beds.

The red spider mite is controlled by the predatory mite *Phytoseiulus persimilis*.

Shortly after planting the cucumber vines, ten to twenty red spider mites were released on to each plant. Ten to fourteen days later two predators were released on alternate plants. In the earlier experiments the predators were only placed on the plants when the mite damage to the leaves had reached a certain critical stage, but in the larger commercial trials it was decided to 'work by the calendar' to facilitate the production of the predators from the limited resources available. This rule-of-thumb method has the disadvantage that in some nurseries the red spider mites increased so slowly that they were soon eliminated by the released *Phytoseiulus*, and further red spider mites had to be placed on the plants. Another difficulty was caused by the earliness of the season when the experiment started: many of the red spider mites went into diapause instead of breeding. To avoid this difficulty the possibility of using a non-diapausing strain of red spider mite is being tested. In some of the larger trials it was found necessary to make local applications of tetradifon to clear up small patches of severe infestation.

One difficulty in large-scale applications will be, of course, to persuade the growers that it is sensible to actually infest a crop with a pest which they are trying to control. An uncontrolled outbreak of the spider mite generally starts as patches of severe infestation scattered throughout the crop, and it is very difficult to control such advanced infestations biologically. Another difficulty which arises if the control agents have not become established early enough in the season is the congregation of the red spider mites at the tops of the plants, where a comparatively small number can cause severe damage.

The whitefly is also controlled biologically – by the chalcid *Encarsia formosa*. These are introduced into the glasshouses within the blackened nymphs of the pest. In earlier, small-scale trials good control was achieved by introducing four white-flies per plant early in the season, and eight parasites per plant a fortnight later. The control lasted for twelve weeks. In the larger, semi-commercial trials the parasites were not spread as evenly in the house as they were in the smaller trials, so control was not as satisfactory. In still later trials better control was achieved by reducing the numbers of introduced whiteflies, and increasing the numbers of parasites to twenty to each plant. One hundred parasites were released at every fifth plant to ensure a good uniform distribution. Parr and Scopes (1971b) recommend that this control routine should start about six weeks before the whiteflies normally appear.

The other pests were controlled chemically: thrips by a bed drench of 0·02 per cent γ BHC or 0·04 per cent diazinon, at a rate of one gallon per square yard of bed (4·5 litres per 0·8 m^2); French-fly by a bed drench with parathion before planting, but only in those houses where it occurs regularly; aphids by 0·025 per cent pyrimicarb. The mildew was controlled by bed drenches of the selective systemic fungicide dimethirimol or by sprays of dinocap or quinomethionate. The dimethirimol was not entirely satisfactory as it is a protectant rather than a curative fungicide. It was also thought that some strains of mildew may have developed a resistance towards it.

In two houses the average yield was 20–25 per cent higher than in adjacent houses where chemical control only was used. This was partly due to the phytotoxicity caused by the intensive spraying. Preliminary costing shows that integrated control is also far less expensive in labour costs than is traditional chemical control – 40 man-hours per acre plus 25 hours supervision, compared with an average of 250 man-hours per acre for high-volume spraying throughout the season.

The workers at the institute are also investigating the possibilities of integrated control in tomato and chrysanthemum houses. Parr and Scopes (1971b) consider that the larger acreage of tomatoes under glass, and the greater planting density of the plants (1400 per acre) rule out the possibility of artificially infesting the crop with the pest – 'the corner-stone of biological success of biological control on cucumbers'. There is also a greater variation in the methods of growing tomatoes, each technique probably demanding a different approach to the control. The cooler temperatures in tomato houses will, however, allow the successful treatment of patchy naturally occurring infestations. Some trials in commercial houses that were severely infested with red spider mite have been very encouraging.

In chrysanthemums the main pests are the glasshouse red spider mite, the peach-potato aphid, *Myzus persicae*, and the agromyzid leaf-miner *Phytomyza syngenesiae*. When spraying is relaxed, three other aphid species can become serious pests. An integrated programme to control the three key pests must, therefore, cope with the three additional aphids as well.

Green lacewing larvae (*Chrysopa carnea*) are promising for the control of *Myzus persicae* when the pest-to-predator ratio is less than fifty to one, and when all the aphids are within 15 cm of the larvae. *Aphidius matricariae* is being tested for the control of aphids, the parasites being placed in boxes of cuttings so that they will be dispersed in the houses mechanically at planting. The mite *P. persimilis* is, of course, being tested for spider mite control, and two parasites (*Diglyphus isaea*, a larval ectoparasite, and *Rhizarcha* sp., a pupal parasite) for leaf-miner control. Various persistent ovicides are also being tested against the leaf-miner. The secondary aphids are controlled by pyrimicarb.

Glasshouse integrated control is clearly very promising, but much remains to be done to apply it widely. It is one thing to culture sufficient predatory mites for trials, and quite another to provide sufficient numbers for widespread commercial control. These problems are discussed in the papers mentioned above.

Integrated control and vectors of disease

Public health entomologists appear to lay less stress on integrated control than do their agricultural colleagues, largely because their aim is generally eradication, and this is not compatible with integrated control as it is usually understood. In agriculture, the eradication of a pest may well be ecologically unwise, but the gains achieved by eliminating a vector of a serious disease usually outweigh this danger. Even in vector control, however, complete eradication of the vector may not be necessary in order to eliminate the disease. If the patients suffering from the disease do not form a permanent or very persistent reservoir of the disease organism, and especially if they can be treated with curative drugs while the vector control campaign is in progress, it is only necessary to reduce the vector population to a low level until the spread of the disease ceases in the area. In the British Isles, for example, mosquitoes capable of transmitting malaria are still common, but the number of people carrying the parasites is too low for the disease to spread.

There have been some spectacular examples of the emergence of new pests following the insecticidal control of vector mosquitoes. In Sarawak, the persistent household sprays destroyed a hymenopterous parasite of the thatch moth caterpillar, *Herculia* sp., which so increased that many roofs collapsed. In New Guinea the destruction of an *Anopheles* mosquito was followed by an increase in the numbers of biting culicines which bred in the same habitat. Sometimes a pest species may be replaced by a less harmful one, as happened in Sardinia with *Anopheles labranchiae* and *A. hispaniola*.

The resurgence of disease vectors following insecticidal treatment has also been reported. It is probable, for example, that the treatment of streams with persistent insecticides for the control of black-flies, *Simulium* spp., kills many of the predators of the larvae and pupae. In 1944 DDT, after a short period of control, increased the mean emergence on a test section of a Canadian stream seventeen-fold through its effect upon the predators of the pest.

Laird, in 1963, deplored the sketchy knowledge we have of the bionomics and ecology of insect vectors, especially when they are in the larval stages. The breeding sites of mosquitoes, for example, are described in very general terms whereas they are, in fact, highly complex systems with numerous species of plants and animals, many of them which belong to the life system of the mosquito larvae. He stressed the importance of ciliate epibionts such as *Vorticella* spp., and regretted the widespread confusion in the naming of these. Names have multiplied whereas, probably, only a few species are actually involved. These epibionts may help to kill mosquito larvae, especially in polluted water. Laird suggests that the pre-war Malaysian practice of deliberately polluting the water in order to encourage epibionts, and to reduce the oxygen available to the larvae, could be re-introduced, and that ways should be sought to integrate chemical larvicides with these and other natural enemies. Nevertheless, it seems likely that most attention will be given to the use of chemical pesticides for the massive reduction of a population before autocidal methods are begun. But whenever a pest population cannot be eradicated, and must be controlled instead, it would be unwise to rely entirely on one method of control: integrated methods will become as important in vector control as they have in California and Peru.

Integrated control of plant diseases

Plant pathologists do not appear to discuss integrated control as avidly as do entomologists. They have fewer problems to face when using fungicides as these do not destroy the natural enemies of pathogens, or, if they do, this does not appear to have been acknowledged to any great extent. Fungicides are often toxic to predatory mites, and thus the plant pathologist should cooperate with the entomologists when they are planning the integrated control of insect and mite pests. Micro-organisms are, however, extremely adaptable organisms so that resistance to any single control method is possible, if it is applied intensively enough. There appear to have been more difficulties with the breakdown of plant resistance than with the loss of potency of fungicides. A possible integrated control approach might be the use of varieties which show a low level of resistance, based on several loci, to one or more pathogens, together with a reduction in the amounts of fungicides applied. The plant pathologist also makes extensive use of cultural methods of control which could be integrated with chemical methods and the use of resistant varieties.

The wider use of integrated control

The main requisite in integrated control is a trained body of workers. Most growers are far too busy learning and applying their own craft and so have little time to spare to become professional ecologists and entomologists as well, and therefore must rely on the help of others. Integrated control must develop in the richer countries where the government can finance the necessary research and provide the field biologists who will supervise the day-to-day execution of the various control strategies. The poorer countries can afford neither, except, possibly, where some cash crops such as cocoa, coffee, cotton, rubber, and tea are produced. It is true that international organizations such as the FAO can provide the services of visiting experts, but what is really needed is a staff of people who will stay for many years. There is also a distressing tendency for the young students of many of these countries to take up the more remunerative and influential occupations of law and economics, and to despise the profession of agriculture.

There are other difficulties besides the shortage of scientific man-power in the poorer countries, notably the lack of good roads and transport. In fact, in many of the remote areas, farming is at a subsistence level; nothing comes into the farm, and little leaves. The farmer saves some of his crop for seed each year, no artificial fertilizers are bought, and pesticides are very rarely applied. Even when some pesticides are used they have to be the cheapest available, namely the persistent chlorinated hydrocarbons. India, for example, uses more BHC than the United States of America, as it is simple and cheap to manufacture in a crude form. In 1964–5 the amount of active ingredient, in a 13·6 per cent gamma preparation, that was used in plant protection was 6156 tonnes, compared with 600 tonnes of DDT, 111·5 tonnes of parathion and 97 tonnes of malathion – all these figures being weights of active ingredients – although plans were being made to increase the proportions of the organophosphates. Under Indian climatic conditions parathion is particularly dangerous to humans so, although it is one of the cheapest of the organophosphates, emphasis is being placed on the use of malathion (Reddy, 1967).

Clearly, the introduction of integrated control methods into the poorer countries for the protection of food crops will be a slow and difficult task. It is possible, within limits, to introduce non-selective insecticidal control methods after the farmers have been given a little training in their use, but integrated control needs a preliminary ecological study for each crop ecosystem, and the strategies may be quite different in localities 100 km apart.

In the technologically advanced countries the prospects are brighter, since the growers are better educated and richer, and the governments are able to provide the advisory services and to introduce the necessary legislation. Travelling is easier, and communication is further simplified by the use of the press, radio, and television. From an organizational point of view some crops are well suited for the introduction of integrated control methods. Sugar-beet in the United Kingdom, for example, is grown under contract to the British Beet Sugar Corporation which maintains a staff of fieldsmen who are in constant touch with their farmers.

The grower receives advice from several sources. Government advisers are as objective and unbiased as they can be, but often have too much ground to cover, both geographically and professionally. Most commonly, farmers and growers have dealings with general agriculturalists and horticulturalists who cannot be expected to be up-to-date in all the topics that they have to give advice upon. It is true that they can call upon the specialist entomologists and plant pathologists, but these are fewer still in number and, in any case, each one tends to think of pest control in terms of insect pests *or* plant diseases *or* nematodes *or* weeds: there are few, as yet, who can think of control in terms of the whole pest complex as a system. This specialization is a result of their academic training, and will only be remedied by the creation of a new profession of pest management, resembling, in many ways, that of medicine or veterinary science (Beirne, 1967). It is often suggested that a course in agricultural science gives the necessary breadth of outlook, but, in fact, such courses are often too broad and, in any case, the people who undertake such studies frequently specialize after graduation as entomologists or plant pathologists (and generally do extremely well).

Private consultants may also help the grower. Such individuals or teams of individuals will vary in quality, but, presumably, only those which do well will survive. Their advantage is that they can spend more time with the individual grower or cooperative than can the government adviser who must divide his time as fairly as he can among the growers who demand his services. Their advice can be objective if they are not tied financially to some chemical or machinery manufacturer. It seems probable that the profession of 'Ecological Consultant' will soon be listed against 'Soil Analyst' and 'Consultant Chemist' in the directories: such an event has already been reported in Australia.

The grower's most common adviser on pest control matters is, however, the representative of the chemical company or its agent. It would be naïve to expect such people not to advise the use of chemical methods, for their livelihood depends upon them. The grower, in any case, is generally more concerned about protecting his crops than in promoting the general ecological well-being of the countryside and will look for a method which gives quick and reasonably reliable results. Only when chemical methods begin to fail to give such results will he think seriously of trying alternative methods himself, unless prodded into doing so by official pressures or propaganda.

Each chemical company has a limited range of products, and while it is not unknown for a representative to advise the use of a product of another firm when this is clearly superior, it is not a universal practice. Unfortunately, many of the representatives regard themselves as salesmen who happen to sell agricultural chemicals rather than agriculturalists (or biologists) who happen to be salesmen. I clearly remember a discussion with a senior salesman about the relative importance of the research and sales division of a chemical company (an argument doomed to sterility) during which he said that were he not selling agricultural chemicals he would be quite content selling something else. It was presumably a liking for a rural life that made him choose this particular kind of product rather than factory machinery or electronic equipment. This was several years ago, and the larger companies, at least, are now trying to give their staff an understanding of ecological principles. Furthermore, the companies maintain biologically trained staff for research and advisory work but, as the companies must remain solvent, their work is mainly concerned with pesticides. There is, however, a marked tendency for these firms to search for highly selective synthetic chemicals, or to explore the possibilities of such organisms as *Bacillus thuringiensis*. Even so, it would be desirable for more companies to include experienced biologists on their boards of directors, even if it entailed some loss of profits: doubtless, however, their lack of knowledge of management principles and economics would cause some embarrassment to their colleagues.

We have not yet considered who will carry out the research upon which the advice will be based (Beirne (1967) discusses this point at some length). Clearly, this will not be done by the scientific staff of the chemical companies, though if their work provides a succession of materials that can be used selectively their contribution will be invaluable. Their field trials should, if possible, include more studies on the effects of new materials on beneficial organisms than is now the custom. Generally, the vertebrate toxicity and residue aspects are well covered.

The advisory services are, to a large extent, forced to look for quick results: many of their research workers are dealing with a variety of problems and can devote only a limited time to each. Often, however, each one concentrates upon a single crop or a group of closely related crops, and has an unrivalled knowledge of the pest complexes of those crops: an ideal departure point for the development of integrated control methods. Sometimes the cash value of the crop rules out any approach but an integrated one: forest timber and pastures are excellent examples.

The last group of scientists who could possibly carry out the necessary research are employed by universities and agricultural research stations. Unfortunately, this research is, by its nature, long-term: life-tables, to be valuable, must cover several years and, in the meantime, there is little to publish. And, as Beirne points out, a scientist's advancement, both financially and in status, depends upon a steady flow of publications. Ecological work is also uncertain, whereas chemical control trials assure the scientist of some kind of results. Furthermore, ecological work needs a fairly large number of assistants for such duties as sampling and counting, and few universities can provide these. Fortunately, research on unilateral chemical control is less highly regarded than it was several years ago, and the various research institutes are becoming more tolerant of the kind of work needed for the foundations of integrated control (Beirne, 1967).

Much of this research will be carried out by people in well-established professions: entomologists, plant pathologists, geneticists, agronomists, ecologists, statisticians, and so forth. There will, however, be a place for a new kind of professional man, both in the research and in its application. Such a man will be an expert in pest management as such and his status will be comparable with that of the medical and veterinary scientist, for his importance will certainly be no less.

Beirne has given much thought to the training of such specialists. Most people who are now practitioners in the field were originally trained as if they were to take their place in an academic world. Some aspects of pest control, such as applied entomology or plant pathology, would have been covered by their studies, but more stress would have been placed upon such traditional topics as insect structure and classification, or flowering plant morphology. These subjects are, of course, basic to the pest control studies, but they would not be treated with these in mind.

Pest management will be offered as an integrated discipline within special departments of a university in which all groups of pest organisms would be covered. Ecology, and population dynamics in particular, would occupy much of the student's time. Later in the course, control techniques, and the management of pests, would be studied more intensively. After graduation the student would probably devote himself to the pest management of a particular crop or region.

The specialist in pest management will be well grounded in mathematics and biometry, and should be reasonably conversant with computer techniques. In the past, biology has been a refuge for many students who had little aptitude for mathematics, yet wanted to take up some scientific career and, in the past, this has often worked well. There have been many valuable contributions to biological thought which have not depended upon a mathematical presentation: much of Darwin's work can be so described. Now, however, mathematical and biometrical techniques are entering most branches of biological science, even taxonomy, and this will be especially true of pest management in the future. Systems analysis, dynamic programming, and a range of other computer techniques will be used in pest management within a few years time, and it will be necessary for the practitioner to be able to cooperate closely with computer experts and mathematicians.

One difficulty, at least at the beginning, will be to find staff to teach this new discipline. The universities will be able to provide the necessary specialists in certain fields such as insect biochemistry, mycology, the chemistry of pesticides, computer techniques, and so on, but they will have to depend upon help from the advisory services and research institutes for the more applied aspects. Clearly, such people will have little time to spare from their researches, but, in time, they will reap a rich harvest of colleagues whom they have themselves helped to train (Beirne, 1967).

Prospects

There is a story of a well-known entomologist advising his colleagues to collect specimens of all pest species of insects to deposit in museums before they became extinct. Although this advice was probably being given facetiously it shows how dangerous it is to look too far ahead in pest control. Clearly, it will have to become more efficient, more scientifically based, if we are to obtain an ever-increasing proportion of the things we require from our limited environment, and to do so without polluting that environment irrevocably. Pest control will also be influenced by changes in other aspects of agriculture, forestry, and medicine, and it is just as difficult to forecast what these changes will be.

It seems fairly certain, however, that pesticides will be used for the next few decades at least, and probably for longer, and that the pest management and integrated control approach will be more widely applied. Its development in the technologically advanced countries seems to be assured, but there is considerable doubt about the underdeveloped countries. Here it depends upon such things as the spread of literacy, communications, health services, and so forth, which, in turn, depend upon a slowing down of population growth. Otherwise, if pest control is carried out at all, it is likely to be rule-of-thumb application of wide-spectrum pesticides, unless, of course, the government or international agencies can employ a method such as biological control or sterile-male release which does not require the close participation of the growers and agricultural workers.

Basic scientific research will, doubtless, reveal completely new approaches to pest control, and it will certainly bring improvements in techniques which are already known. The current interest in pheromones will allow, within a few years, the use of chemosterilants in the field. Later, perhaps, our increasing knowledge of the biochemistry of inheritance will enable us to produce new strains of organisms 'to order' for improved biological control performance, or for the genetic manipulation of pest populations. There is current speculation about the possibility of creating completely new kinds of organisms for the colonization of other planets (a project which the down-to-earth economic biologist finds disconcerting): the creation of new kinds of natural enemies is a much more modest undertaking.

Pest control in general, and pest management in particular, involve complex systems which are difficult to analyse by traditional methods, but the complexity is of the kind that modern large computers can cope with. Most biologists are unaware of the work of a similar kind that has been carried out in fields other than biology.

Such work could be adapted to pest management and the related discipline of resource management. Watt has been in the forefront of the biomathematicians who have been trying to bring these techniques to the attention of biologists and ecologists, and the following discussion borrows heavily from his writings.

In systems simulation in pest management a mathematical model of the pest's population dynamics is read into the computer, which is then used to explore the results of various strategies, in various environmental conditions in order to see which gives the best control. Alternatively, the model might simulate some resource which is required to give the maximum yield. The resource might be, for example, timber, and the model might include such factors as weather, pest damage and its effects, and various control measures against the pest.

If the model is very simple, and the number of factors small, this exploratory process could be carried out using pencil and paper, and with desk calculators: the computer's only advantage is that it can deal with much more complex models, containing many factors, each at several levels. It can do this because it is able to store basic data (the amount depending on the size of its memory unit); produce intermediate results; and make rapid computations.

Watt (1968) illustrates the technique by considering forest pest control in a tract of woodland 68 000 square kilometres in extent. The model takes into account the dispersal of the pests and of the parasites and predators, likening these to waves of movement radiating from epicentres or foci of infestation. Weather plays a role in the triggering of outbreaks, so data relating to weather conditions over a period of thirty-five years are stored in the memory. Outbreaks can, of course, occur at sites where the weather conditions could not be held responsible for their initiation: these result from the arrival of a wave of the pest from a distant epicentre where at some time previously the weather conditions were suitable. Thus space and time are included as variables in the model.

Certain relationships between pest density, damage, and the host trees are considered: a low level of damage occurring annually for several years might, for example, kill many of the trees and, in fact, be more harmful than a short series of sharp attacks. Density-dependence mechanisms for both the pest species and its natural enemies are incorporated, as are various methods of control. Here again, space and time are influences which must be considered. Insecticides are largely local in action, and the time during which they have a direct mortality effect is limited. Natural enemies, on the other hand, spread from the points of release and can increase in intensity for some years until an asymptote is reached.

The computer is now used to explore the results of various strategies, to determine if one method or combination of methods gives consistently better results than do the others, and also to establish if any strategy is reasonably consistent in its performance even when other factors range over their possible values. Finally, of course, there is always the hope that the investigation will turn up something unexpected, some useful result that might otherwise be missed.

The advantages of computer simulation or gaming of this kind is that when we are faced with a large number of possible strategies, some of which might be very expensive to apply, we have some objective guide for decision making, and, of course, using simulation we can explore a much larger number of strategies than we could test by actual field experimentation. Unfortunately, such simulation studies are 'information hungry' and we have little or no experimental data on such important phenomena as parasite dispersal, and the increase in parasite populations after release. Even crude models are useful, and in addition to indicating the probable consequences of various strategies, the output suggests the kinds of new information needed to improve the simulation.

In this particular simulation study the area is divided into 625 squares, each 4×4 miles in extent. An inner repetitive 'loop' of the program follows the events in each square for each year, taking into account the past events in that square, and in the surrounding squares. A larger loop, or intermediate loop, explores the situation for the total area for a given year, while the outermost loop simulates the events associated with a given set of input parameters for the whole period of the simulated experiment. For simplicity the whole area of 10 000 square miles is considered to carry an evenly spread crop of uniform balsam firs, each with a certain fixed market value.

Fifteen alternative control strategies are employed in the simulation, namely doing nothing, aerial spraying of DDT at three application rates, the ground release of parasites, the aerial spraying of a virus disease of the pest, and nine different combinations of the individual methods. The program is designed to find the particular strategy, and the time at which it should be applied, which will minimize the sum of the control costs and the losses due to inadequate control. The cost of the control should be no more than two dollars per acre per annum.

The actual Fortran program, and the lengthy conclusions cannot be given here (they will be found in chapter 12 of Watt's text), but it is hoped that enough explanation has been given for the appreciation of the potentialities of this technique. Some authors consider that the natural system is far too complex for simulation to be realistic enough to give reliable decision guides, but Watt regards this attitude as somewhat defeatist, and pleads convincingly the need for further exploration and application of the method.

There is a further point to be considered in the use of systems simulation. A model might incorporate six factors (weather conditions, pest density, insecticide, etc.) each of which can exist at ten different levels of intensity. If all possible vectors are considered each year over a sequence of 100 years, the total number of cases would be $10^{6 \times 100}$, or 10^{600} – far beyond the capacity of even the largest computers. If, however, we consider each year in sequence, exploring the consequences of the 10^6 vectors, then ignoring all cases except the optimum, which we use as a starting place for the following year, we have to consider, in all, only $100 \ (10^6)$ cases – a much more reasonable task. This procedure is known as dynamic programming, and was developed by Bellman some years ago. This sequence of decision making has been compared with the actions of a man situated at the top of a mountain who has to make his way to a fixed point below, when most of the intervening country is obscured (Schützenberger, 1954). His best method is to make his way to an easily accessible point within sight further down the mountain and, having reached it, to plan the next stage of his descent.

The appropriateness of this method depends upon the assumption that the situation at year $n+1$ depends only upon that at year n, and that it in no way depends upon the situation at time $n-1$, $n-2$, and so on. In the terms of cybernetics, the process is presumed to be Markovian. This assumption is, in fact, usually unjustified, as Watt illustrates by his analysis of the records for the population density of the moth *Bupalus piniarius* from 1881 to 1940 in the Letzlingen thicket, near Magdeburg in Germany. He found that events from n to $n+1$ were more highly correlated with events at $n-1$ than at n. Morris found a similar relationship for large survival ratios of spruce budworm larvae. Such results are not unexpected when lag phenomena, such as delayed density-dependence mechanisms are considered. Watt used the data in a computer program to investigate the penalty for ignoring this non-Markovian component, and in this case at least two-year 'look-ahead' programming was rarely better than the usual dynamic programming, as the lowest possible infestations two years ahead were only obtained by having very high densities one year ahead – a clearly unsatisfactory state of affairs (Watt, 1963).

A further example of computer simulation was mentioned when the use of eelworm-resistant potatoes was discussed. This simulation showed that, over a number of years, it would be more economical to use both susceptible and resistant varieties in various sequences rather than to use resistant varieties alone, for this would hasten the appearance of eelworm biotypes which would break the resistance.

In conclusion

It is clear that in the advanced countries, at least, we are emerging from a period during which pest control has been an empirical craft to a time when it will be soundly based on scientific knowledge, and, in particular, on ecological knowledge. One encouraging, if exasperating, feature of writing a book such as this is that it rapidly becomes out of date. There is no doubt that efficient pest control is urgently needed, and that it has a great value for the promotion of the well-being of man. It should not be forgotten, however, that it is only one approach to the problem of making the best uses of the resources of sunlight, water, and the land. At the best it is merely postponing the time when the demands upon these limited resources will be far too great, and in some parts of the world this time has already arrived. Pest control receives great moral and financial support from the people of the world: it is a tragedy that its far more important counterpart, population control, is so hindered. The faith that some authorities show in the abilities of the agriculturalists and biologists to continue producing ever more food is gratifying, but such faith cannot make them achieve the impossible.

References

The literature on pest control is vast, and is growing rapidly. This bibliography contains only a small selection of the literature but many of the publications listed, especially those more important ones which are marked with an asterisk, contain good lists of references. Certain journals, some of which are listed below, will help the reader to keep in touch with recent developments.

Advances in Ecological Research; *Advances in Pest Control Research*; *Annual Reviews* (*Entomology, Phytopathology, Microbiology, Pharmacology*, etc.); *Bulletin of Entomological Research*; *Bulletin of the World Health Organization*; *Canadian Entomologist*; *Journal of Applied Ecology*; *Journal of Economic Entomology*; *Journal of Invertebrate Pathology*; *Review of Economic Entomology* (Series A and Series B); *Review of Economic Mycology*.

AGRICULTURAL RESEARCH COUNCIL (1970) *Third Report of the Research Committee on Toxic Chemicals*, HMSO, London.

ALBERT, A. (1968) *Selective Toxicity*, 4th Edn., Methuen, London.

ALEXANDER, J. O'D. 'Mites and Skin Diseases', *Clin. Med.*, April 1972, 14–19.

ALLEE, W. C., O. PARK, T. PARK, A. E. EMERSON, and K. P. SCHMIDT* (1949) *Principles of Animal Ecology*, Saunders, Philadelphia.

ANDREWARTHA, H. G.* (1970) *Introduction to the Study of Animal Populations*, 2nd Edn., Methuen, London.

ANDREWARTHA, H. G.* (1970) *Introduction to the Study of Animal Populations*, 2nd Edn., University of Chicago Press, Chicago.

ANDREWARTHA, H. G. and L. C. BIRCH* (1960) 'Some recent contributions to the study of the distribution and abundance of insects', *Ann. Rev. Entomol.*, 5, 219–42.

ANON* (1957) *Cold Spring Harbour Symp. Quant. Biol.*, 22.

ANON (1960) 'Interfacial phenomena in pesticide application', *Pest Technology*, August 1960, 239–43.

ANON (1969) *Chemical Week*, 104, 36–8 (26 April).

ARTHUR, A. P. (1962) 'Influence of host tree on abundance of *Hoplectis conquisitor* (Say) (Hymenoptera: Ichneumonidae), a polyphagous parasite of the European pine shoot moth *Rhyacionia buoliana* (Schiff) (Lepidoptera: Olethreutidae)', *Can. Ent.*, 94, 337–47.

ARTHUR, D. R. (1969) *Survival*, English Universities Press, London.

ASCHER, K. R. S. and S. NISSIM (1964) 'Organotin compounds and their potential use in insect control', *World Rev. Pest Control*, 3, 188–211.

AUSTRALIAN ACADEMY OF SCIENCE* (1972) *The Use of DDT in Australia, Reports of the Australian Academy of Science No. 14*, Canberra.

AXTELL, R. C. (1963) 'Acarina occurring in domestic animal manure', *Ann. Entomol. Soc. America*, 56, 628–33.

AXTELL, R. C. (1969) 'Machrochelidae (Acarina: Mesostigmata) as biological control agents for synanthropic flies', *Second Internatl Congr. Acarology, Nottingham, 19–25 July 1967 WHO mimeographed document, WHO/VBC/69. 119*.

BAILEY, L. (1971) 'The safety of pest–insect pathogens for beneficial insects'. In Burges and Hussey, 1971a, 491–505.

BAILEY, N. T. J. (1957) *The Mathematical Theory of Epidemics*, Charles Griffin, London.

BAILEY, N. T. J. (1967) *The Mathematical Approach to Biology and Medicine*, John Wiley, London, New York, and Sydney.

BAKER, K. F. and W. C. SNYDER (Eds.) (1965) *Ecology of Soil-borne Pathogens*, University of California Press, Berkeley.

BAKER, R.* (1968) 'Mechanisms of biological control of soil-borne pathogens', *Ann. Rev. Phytopath.*, **6**, 263–94.

BARNETT, S. F. (1961) *The Control of Ticks on Live Stock, FAO Agricultural Studies*, No. 54, Food and Agriculture Organization, Rome.

BARTELS, W. and H. H. CRAMER* (1966) 'Side effects of plant diseases, plant pests and weeds on the health of man and animals and on the quality of harvested products', *Pflanzenschutz-Nachrichten, Bayer*, **19**, 125–86.

BARTLETT, B. R. (1964) 'Integration of chemical and biological control', in *Biological Control of Insect Pests and Weeds*, P. DeBach (Ed.), pp. 489–511, Chapman and Hall, London.

BATES, M. (1949) *The Natural History of Mosquitoes*, Harper and Row, New York.

BAWDEN, F. (1972) Review of '*The Closing Circle*' (Commoner) *New Scientist*, 3 February, 284–5.

BAY, E. C. (1968) 'Mosquito control by predator fish', *International Pest Control*, **10** (1), 23–5.

BECK, S. D.* (1966) 'Resistance of plants to insects', *Ann. Rev. Entomol.*, **10**, 207–32.

BEIRNE, B. P.* (1967) *Pest Management*, Leonard Hill, London.

BENZ, G. (1971) 'Synergism of micro-organisms and chemical insecticides'. In Burges and Hussey, 1971a, 327–55.

BEROZA, M., B. A. BIERL, E. F. KNIPLING, and J. G. R. TARDIF (1971) 'The activity of the gypsy moth sex attractant Disparlure vs. that of the live female moth', *J. Econ. Ent.*, **64**, 1527–9.

BEROZA, M. and M. JACOBSON (1963) 'Chemical insect attractants', *Interntl Rev. Pest Control*, **2** (2), 36–48.

BERRYMAN, A. A. (1967) 'Mathematical description of the sterile male principle', *Can. Ent.*, **99**, 858–65.

BETTS, M. M. (1955) 'The food of titmice in oak woodland', *J. Animal Ecol.*, **24**, 282–323.

BEVAN, W. J. (1968) 'Reports of syndicate: Vegetables and "root" crops'. In D. W. Empson (Ed.), 1968, 502–8.

BIRCH, L. C. (1957) 'The role of weather in determining the distribution and abundance of animals', *Cold Spring Harbour Symp. Quant. Biol.*, **22**, 203–15.

BIRD, F. T. and J. M. BURK (1961) 'Artificially disseminated virus as a factor controlling the European spruce sawfly, *Diprion hercyniae* (Htg.), in the absence of introduced parasites', *Can. Ent.*, **93**, 228–38.

BLISS, C. I. (1970) *Statistics in Biology*, Volume II, McGraw-Hill, New York.

BLUM, M. S.* (1969) 'Alarm pheromones', *Ann. Rev. Entomol.*, **14**, 57–80.

BODENHEIMER, F. S. (1951) *Citrus Entomology in the Middle East*, Dr W. Junk, 'S-Gravenhage.

BOND, R. P. M. and C. B. C. BOYCE (1971) 'The thermostable exotoxin of *Bacillus thuringiensis*'. In Burges and Hussey, 1971a, 275–303.

BONESS, M. (1953) 'Die Fauna der Wiesen und besonderer Berücksichtung der Mahd (ein Beitrag zur Agrarökologie)', *Z. Morph. Oekol. Tiere*, **42**, 225–77.

BORG, K., H. WANNTORP, K. ERNE, and E. HANKS (1966) 'Mercury poisoning in Swedish Wildlife (Summary)', *Pesticides in the Environment and their Effects on Wildlife, J. appl. Ecol.*, **3** (Suppl.), 171–2.

BOŘKOVEC, A. B.* (1966) 'Insect Chemosterilants', *Advances in Pest Control Research*, **7**, 1–140.

BORLAUG, N. E. (1958) *Proc. 1st Int. Wheat Genet. Symp., Winnipeg*, **6**; cited in Lupton, 1967.

BOWERS, W. S. (1968) 'Juvenile hormone: Activity of natural and synthetic synergists', *Science*, **161**, 895–7.

BRENCHLEY, G. H. (1968) 'Aerial photography for the study of plant diseases', *Ann. Rev. Phytopath*, **6**, 1–22.

BRENCHLEY, G. H. (1968) 'Aerial photography in agriculture', *Outlook on Agriculture*, **5**, 258–65.

BRENNAN, R. D., C. T. DE WIT, W. A. WILLIAMS, and E. V. QUATTRIN (1970) 'The utility of a digital simulation language for ecological modeling', *Oecologia (Berl.)*, **4**, 113–32.

BRISTOWE, W. S. (1958) *The World of Spiders*, Collins, London.

BROWN, A. W. A. and R. PAL* (1971) *Insecticide Resistance in Arthropods*, WHO, Geneva.

BUCKNER, C. H.* (1966) 'The role of vertebrate predators in the biological control of forest insects', *Ann. Rev. Entomol.*, **11**, 449–70.

BUFFAM, P. E. (1971) 'Spruce beetle suppression in trap trees treated with cacodylic acid (*Dendroctonus rufipennis*: Col., Scolytidae)', *J. Econ. Ent.*, **64**, 958–60.

BURGERJON, A. and D. MARTOURET (1971) 'Determination of the host spectrum of *Bacillus thuringiensis*'. In Burges and Hussey, 1971a, 305–21.

BURGES, H. D. (1971) 'Possibilities of pest resistance to microbial control agents'. In Burges and Hussey, 1971a, 445–57.

BURGES, H. D. and N. W. HUSSEY (Eds.)* (1971a) *Microbial Control of Insects and Mites*, Academic Press, London and New York.

BURGES, H. D. and N. W. HUSSEY (1971b) 'Past achievements and future prospects'. In Burges and Hussey, 1971a, 687–709.

BURGES, H. D. and E. M. THOMSON (1971) 'Standardization and assay of microbial insecticides'. In Burges and Hussey, 1971a, 591–622.

BUSNEL, R. G. and J. GIBAN (1968) 'Prospective considerations concerning bioacoustics in relation to bird-scaring techniques', in *The Problems of Birds as Pests*, R. K. Murton and E. V. Wright (Ed.), pp. 17–28, Institute of Biology/Academic Press, New York and London.

BUSVINE, J. R. (1951) *Insects and Hygiene*, 1st Edn., Methuen, London.

BUSVINE, J. R.* (1957) *A Critical Review of the Techniques for Testing Insecticides*, Commonwealth Institute of Entomology, London.

BUSVINE, J. R.* (1966) *Insects and Hygiene*, 2nd Edn., Methuen, London.

BUSVINE, J. R.* (1968) 'Impact of insecticide resistance on vector and disease control', *WHO mimeographed document, WHO/VBC/68. 100.*

BUSVINE, J. R.* (1971) 'Cross-resistance in arthropods of public health importance. Part I: The significance of cross-resistance. Part II: A review of data for non-dipterous pests', *WHO mimeographed document, WHO/VBC/71. 307.*

BUTLER, C. G.* (1967) 'Insect pheromones', *Biol. Rev.*, **42**, 42–87.

CABORN, J. M. (1971) 'The agronomic and biological significance of hedgerows', *Outlook on Agriculture*, **6**, 279–84.

CALHOUN, J. B. (1962) 'Population density and social pathology', *Scient. Amer.*, **206**, 139ff.

CALLOW, R. K. (1967) 'The chemical basis of courtship in the insect world', in *Penguin Science Survey 1967. The Biology of Sex*, A. Allison (Ed.), pp. 148–59, Penguin Books, Harmondsworth.

CAMERON, J. W. MACBAIN (1967) 'Suitability of pathogens for biological control', in *Insect Pathology and Microbial Control*, P. A. van der Laan (Ed.), pp. 182–96, North Holland Publishing, Amsterdam.

CARSON, R.* (1962) *Silent Spring*, Houghton Mifflin, Boston (1962); Hamish Hamilton, London (1963).

CASIDA, J. E. (1963) 'Mode of action of carbamates', *Ann. Rev. Entomol.*, **8**, 39–58.

CHANT, D. A. (1964) 'Strategy and tactics of insect control', *Can. Entomol.*, **96**, 182–201.

CHANT, D. A. (1966) 'Integrated control systems', in *Scientific Aspects of Pest Control*, Publication 1402, National Academy of Sciences – National Research Council, Washington, D.C.

CHAPMAN, J. W. and R. W. GLASER (1915) *J. Econ. Ent.*, **8**, 140–50.

CHAPMAN, R. F.* (1969) *The Insects: Structure and Function*, English Universities Press, London.

CHITTY, D. (1960) 'Population processes in the vole and their relevance to general theory', *Can. J. Zool.*, **38**, 277–80.

CHITTY, D. (1965) 'Qualitative changes within fluctuating populations, including genetic variability', *Proc. Int. Congr. Ent.*, **12** (London, 1964), 384–6.

CLARK, L. R. (1947) 'Ecological observations on the small plague grasshopper, *Austroicetes cruciata* (Sauss.) in the Trangie district, Central New South Wales', *Bull. Counc. scient. ind. Res., Melbourne*, No. 228.

CLARK, L. R., P. W. GEIER, R. D. HUGHES, and R. F. MORRIS* (1967) *The Ecology of Insect Populations in Theory and Practice*, Methuen, London.

CLAUSEN, C. P.* (1940) *Entomophagous Insects*, McGraw-Hill, New York and London.

CLAUSEN, C. P. (1951) 'The time factor in biological control', *J. Econ. Ent.*, **44**, 1–9.

COAKER, T. H. (1965) 'The effect of soil insecticides on the natural balance of the cabbage root fly (*Erioischia brassicae* Bouche)', *Proc. XIIth Interntl Congr. Entomol., London, 1964*, p. 590.

COCHRAN, W. G. (1963) *Sampling Techniques*, 2nd Edn., John Wiley, New York.

COGILL, K. J. (1968) 'Reports of syndicates: Top fruit and hops'. In D. W. Empson (Ed.), 1968. 508–10.

COLLESS, D. H. and D. K. MCALPINE (1970) 'Diptera (Flies)', in *Insects of Australia*, CSIRO, Canberra, pp. 656–740, Melbourne University Press, Melbourne.

COLLYER, E. (1952) 'The biology of some predatory insects and mites associated with the fruit tree red spider mite (*Metatetranychus ulmi* Koch) in southeastern England. I: The biology of *Blepharidopterus angulatus* (Fall)', *J. Hort. Sci.*, **27**, 117–29.

COLLYER, E. (1953a) 'The biology of some predatory insects and mites associated with the fruit tree red spider mite (*Metatetranychus ulmi* Koch) in southeastern England. II: Some important predators of the mite', *J. Hort. Sci.*, **28**, 85–97.

COLLYER, E. (1953b) 'The biology of some predatory insects and mites associated with the fruit tree red spider mite (*Metatetranychus ulmi* Koch) in southeastern England. IV: The predator–mite relationship', *J. Hort. Sci.*, **28**, 246–59.

COLYER, C. N. and C. O. HAMMOND (1951) *Flies of the British Isles*, Frederick Warne, London and New York

COOKSEY, K. E. (1971) 'The protein crystal toxin of *Bacillus thuringiensis*: biochemistry and mode of action'. In Burges and Hussey, 1971a, 247–74.

COPE, O. B.* (1971) 'Interactions between pesticides and wildlife', *Ann. Rev. Entomol.*, **16**, 325–64.

CORNWELL, P. B. (Ed.)* (1966) *The Entomology of Radiation Disinfestation of Grain*, Pergamon Press, Oxford.

COTTON, R. T. (1956) *Pests of Stored Grain and Grain Products*, Revised Edn., Burgess, Minneapolis, Minnesota.

CRAGG, J. B. and N. W. PIRIE (Eds.) (1955) *The Numbers of Man and Animals*, Oliver and Boyd, Edinburgh and London.

CRAIG, G. B., Jr. (1963) 'Prospects for vector control through genetic manipulation of populations', *Bull. Wld Hlth Org.*, **29**, Suppl., 89–97.

CRAIG, G. R., Jr. (1967) 'Genetic control of *Aedes aegypti*', *Bull. Wld Hlth Org.*, **36**, 628–32.

CRAIG, G. B., Jr. and W. A. HICKEY (1967) 'Genetics of *Aedes aegypti*', in *Genetics of Insect Vectors of Disease*, J. W. Wright and R. Pal (Eds.). pp. 67–132, Elsevier, Amsterdam.

CRAMER, H. H.* (1967) 'Plant Protection and World Crop Production', *Pflanzenschutz-Nachrichten, Bayer*, **20**, 1–524.

CREIGHTON, C. (1891) *A History of Epidemics in Britain from A.D. 664 to the Extinction of Plague*, Cambridge University Press, Cambridge.

CRISSEY, W. F. and R. W. DARROW (1949) *N.Y. State Conserv. Dept. Div. Fish Game Res.*, Ser. **1**. 28 pp. Cited in Howard, W. E. 1967.

CROOK, J. H. and P. WARD (1968) 'The quelea problem in Africa', in *The Problems of Birds as Pests*, R. K. Murton and E. N. Wright (Eds.), pp. 211–29, Institute of Biology/Academic Press, New York and London.

CROSSE, J. E. (1962) 'Antibiotic sprays for the control of cherry bacteriosis: The bactericidal activity of Streptomycin and Bordeaux mixture on the plant surfaces', in *Antibiotics in Agriculture*, M. Woodbine (Ed.), pp. 86–100, Butterworths, London.

CROWDY, S. H. (1970) 'Crop protection – from within the plant', *Science Journal*, **6** (8), 62–7.

CRUIKSHANK, I. A. M.* (1966) 'Defence mechanisms in plants', *World Rev. Pest Control*, **5** (4), 161–75.

CUELLAR, C. B. (1969) 'A theoretical model of the dynamics of an *Anopheles gambiae* population under challenge with eggs giving rise to sterile males', *Bull. Wld Hlth Org.*, **40**, 205–12.

CURTIS, C. F. (1968) 'A possible genetic method for the control of insect pests, with special reference to tsetse flies (*Glossina* spp.)', *Bull. Ent. Res.*, **57**, 509–23.

CURTIS, C. F. (1969) 'The production of partially sterile mutants in *Glossina austeni*', *Genet. Res.*, **13**, 289–301.

DAVID, W. A. L. and B. O. C. GARDINER (1960) 'A *Pieris brassicae* (Linnaeus) culture resistant to a granulosis'. *J. Insect Pathol.*, **2**, 106–14.

DAVIDSON, G. (1969) 'The potential use of sterile hybrid males for the eradication of member species of the *Anopheles gambiae* complex', *Bull. Wld Hlth Org.*, **40**, 221–8.

DAVIDSON, G., H. E. PATERSON, M. COLUZZI, G. F. MASON, and D. W. MICKS (1967) 'The *Anopheles gambiae* complex', in *Genetics of Insect Vectors of Disease*, J. W. Wright and R. Pal (Eds.), pp. 211–50, Elsevier, Amsterdam.

DAVIDSON, J. and H. G. ANDREWARTHA (1948) 'Annual trends in a natural population of *Thrips imaginis* (Thysanoptera)', *J. Animal Ecol.*, **17**, 193–9.

DAY, P. R. (1968) 'Plant Disease Resistance', *Sci. Prog., Oxf.*, **56**, 357–70.

DEAN, G. A. (1911) 'Heat as a means of controlling mill insects', *J. Econ. Ent.*, **4**, 142–58.

DEBACH, P.* (1964a) 'The scope of biological control', in *Biological Control of Insect Pests and Weeds*, P. DeBach (Ed.), pp. 3–20, Chapman and Hall, London 1964.

DEBACH, P.* (1964b) 'Successes, trends and future possibilities', in *Biological Control of Insect Pests and Weeds*, P. DeBach (Ed.), pp. 673–713, Chapman and Hall, London.

DEBACH, P.* (1966) 'The competitive displacement and coexistence principles', *Ann. Rev. Entomol.*, **11**, 183–212.

DEBACH, P. and K. S. HAGEN* (1964) 'Manipulation of entomophagous species', in *Biological Control of Insect Pests and Weeds*, P. DeBach (Ed.), pp. 429–58, Chapman and Hall, London.

DEBACH, P. and J. LANDI (1959) 'Integrated chemical, biological control by strip treatment', *California Citrograph*, **44**, 324, 345–7, 352.

DEKKER, J.* (1971) 'Agricultural use of antibiotics', *World Rev. Pest Control*, **10**, 9–23.

DEMPSTER, J. P. (1961) 'The analysis of data obtained by regular sampling of an insect population', *J. Anim. Ecol.*, **25**, 1–5.

DEPARTMENT OF EDUCATION AND SCIENCE (1967) *Review of the Present Safety Arrangements for the Use of Toxic Chemicals in Agriculture and Food Storage*, Report by the Advisory Committee on Pesticides and other Toxic Chemicals, HMSO, London.

DEPARTMENT OF EDUCATION AND SCIENCE* (1969) *Further Review of certain persistent Organochlorine Pesticides used in Great Britain*, HMSO, London.

DETHIER, V. G. (1970) 'Some general considerations of insects' responses to chemicals in food plants', in *Control of Insect Behaviour by Natural Products*, D. L. Wood, R. M. Silverstein, and M. Nakajima (Eds.), pp. 21–8, Academic Press, New York and London.

DEXTER, R. (1932) 'The food habits of the imported toad, *Bufo marinus* in the sugar cane sections of Porto Rico', *Interntl Soc. Sugar Cane Technologists, 4th Congr., San Juan, Puerto Rico, Bull.*, *74*, 6 pp.

DIMOND, A. E.* (1963) 'The selective control of plant pathogens', *World Rev. Pest Control*, **2**, (4), 7–17.

DOBZHANSKY, T. (1951) *Genetics and the Origin of Species*, 3rd Edn., Revised, Columbia University Press, New York and London.

DODD, A. P. (1936) 'The control and eradication of prickly pear in Australia', *Bull. Ent. Res.*, **27** (3). 503–17.

DODD, A. P. (1940) *The Biological Campaign Against Prickly Pear*, Prickly Pear Board, Brisbane.

DODD, A. P.* (1959) The biological control of prickly pear in Australia', *Monogr. Biol.*, **8**, 565–77.

DOUTT, R. L.* (1964a) 'The historical development of biological control', in *Biological Control of Insect Pests and Weeds*, P. DeBach (Ed.), pp. 21–42, Chapman and Hall, London.

DOUTT, R. L.* (1964b) 'Biological characteristics of entomophagous adults', in *Biological Control of Insect Pests and Weeds*, P. DeBach (Ed.), pp. 145–67, Chapman and Hall, London.

DOUTT, R. L.* (1967) 'Biological control', in *Pest Control: Biological, Physical and Selected Chemical Methods*, W. G. Kilgore and R. L. Doutt (Eds.), pp. 3–30, Acadmic Press, New York and London.

DOUTT, R. L. and P. DEBACH* (1964) 'Some biological control concepts and questions', in *Biological Control of Insect Pests and Weeds*, P. DeBach (Ed.), pp. 118–42, Chapman and Hall, London.

DOWNES, J. A. (1969) 'The swarming and mating flight of Diptera', *Ann. Rev. Entomol.*, **14**, 271–98.

DROPKIN, V. H. (1969) 'Cellular responses of plants to nematode infections', *Ann. Rev. Phytopathol.*, **7**, 101–22.

DRUMMOND, R. C. (1966) 'Recent developments in the control of commensal rodents', *Chem. and Industry* (1966), 1371–5.

DUDDINGTON, C. L. (1957) *The Friendly Fungi*, Faber and Faber, London.

DUDDINGTON, C. L.* (1962) 'Predacious fungi and the control of eelworms', *Viewpoints in Biology* I, J. D. Carthy (Ed.), pp. 151–201, Butterworths, London.

DUFFUS, J. E.* (1971) 'Role of weeds in the incidence of virus diseases', *Ann. Rev. Phytopath.*, **9**, 319–40.

DULMAGE, H. T. (1967) 'Aspects of the industrial production of microbial insect control agents', in *Insect Pathology and Microbial Control*, P. A. van der Laan (Ed.), pp. 120–4, North-Holland, Amsterdam.

DULMAGE, H. T.* (1971) 'Economics of microbial control'. In Burges and Hussey, 1971a, 581–90.

DULMAGE, H. T. (1971) 'Production of pathogens in artificial media'. In Burges and Hussey, 1971a, 507–40.

DUMAGE, H. T. (1971) 'Production of pathogens in artificial media'. In Burges and Hussey, Burges and Hussey, 1971a, 507–40.

DUNN, P. H. (1970) 'Current progress at the Rome (Italy) Entomology Laboratory of the USDA (United States Department of Agriculture)', *Misc. Publ. Commonw. Inst. Biol. Control, Trinidad (Proc. First Int. Symp. Biol. Cont. Weeds, 1969)*, No. 1, 33–8.

DUTKY, S. R.* (1963) 'The milky diseases', in *Insect Pathology: An Advanced Treatise*, Vol. 2, E. A. Steinhaus (Ed.), pp. 75–116, Academic Press, New York and London.

DUTKY, S. R. and W. S. HOUGH (1955) 'Note on a parasitic nematode from codling moth larvae, *Carpocapsa pomonella* (Lepidoptera, Orethreutidae)', *Proc. Ent. Soc. Washington*, **57**, 244.

DYKSTRA, W. W. (1966) 'The economic importance of commersal rodents', *WHO Seminar on Rodents and Rodent Ectoparasites*, Geneva, 24–28 October, 9–14.

DYTE, C. E. (1967) 'Helping plants protect themselves', *New Scientist*, 16 November 1967, 410–11.

EBLING, J. and K. C. HIGHNAM (1969) *Chemical Communication*, Edward Arnold, London.

EDWARDS, C. A.* (1969) 'Soil pollutants and soil animals', *Scient. Amer.*, **220**, 88–99.

EDWARDS, C. A. and G. W. HEATH (1964) *The Principles of Agricultural Entomology*, Chapman and Hall, London.

EGLER, F. R.* (1964) 'Pesticides – in our ecosystem', *American Scientist*, **52**, 110–36.

EHRLICH, P. R. and A. H. EHRLICH (1970) *Population, Resources, Environment*, Freeman, San Francisco.

EHRLICH, P. R. and P. H. RAVEN (1967) 'Butterflies and plants', *Scient. Amer.*, **216** (6), 104–13,

EHRLICH, P. R. and R. L. HARRIMAN (1971) *How to be a Survivor*. Ballantine Books, New York.

ELTON, C. S. (1942) *Voles, Mice and Lemmings. Problems in Population Dynamics*, Oxford University Press, Oxford.

ELTON, C. S.* (1958) *The Ecology of Invasions by Animals and Plants*, Methuen, London.

ELTON, C. S. (1966) *The Pattern of Animal Communities*, Methuen, London.

EMPSON, D. W. (Ed.)* (1968) 'Integrated Pest Control: Papers presented at a conference of advisory entomologists on 11–12 April 1967', *J. appl. Ecol.*, **5**, 489–519.

EMPSON, D. W. and P. J. JAMES (1966) *N.A.A.S Quarterly Review*, No. 73, 22, Summary in Jones and Parrott, 1968.

ERRINGTON, P. L. (1946) 'Predation and vertebrate populations', *Quart. Rev. Biol.*, **21**, 144–77.

ERRINGTON, P. L. (1963) *Muskrat Populations*, Iowa State University Press, Ames, Iowa.

ERRINGTON, P. L. (1967) *Of Predation and Life*, Iowa State University Press, Ames, Iowa.

ERWIN, D. C. (1969) 'Methods of determination of the systemic and fungitoxic properties of chemicals applied to plants with emphasis on control of verticillium wilt with Thiabendazole and Benlate', *World Rev. Pest Control*, **8**, 6–22.

ESSIG, E. O (1931) *A History of Entomology* (Facsimile Edn., 1965), Hafner Publishing, New York and London.

EVANS, H. E. (1966) 'The behaviour patterns of solitary wasps', *Ann. Rev. Entomol.*, **11**, 123–54.

FALCON, L. A.* (1971) 'Use of bacteria for microbial control'. In Burges and Hussey, 1971a, 67–95.

F.A.O.* (1966) *Proc. F.A.O. Symp. Integrated Pest Control, 1965*, FAO, Rome.

FAY, R. W. and H. B. MORLAN (1959) 'A mechanical device for separating the developmental stages, sexes and species of mosquitoes', *Mosquito News*, **19**, 144–7.

FENNER, F. and F. N. RATCLIFFE* (1965) *Myxomatosis*, Cambridge University Press, London.

FINNEGAN, R. J. (1971) 'An appraisal of indigenous ants as limiting agents of forest pests in Quebec (Canada) (Hymenoptera, Formicidae)', *Can. Ent.*, **103**, 1489–93.

FINNEY, D. J. (1952) *Probit Analysis*, 2nd Edn., Cambridge University Press, Cambridge.

FISHER, R. C.* (1971) 'Aspects of the physiology of endoparasitic hymenoptera', *Biol. Rev.*, **46**, 243–77.

FLANDERS, S. E.* (1971) 'Single factor mortality, the essence of biological control, and its validation, in the field', *Can. Ent.*, **103**, 1351–62.

FLOR, H. H. (1971) 'Current status of the gene-for-gene concept', *Ann. Rev. Phytopath.*, **9**, 275–96.

FORD, J. (1968) 'The control of populations through limitation of habitat distributions as exemplified by tsetse flies', in *Insect Abundance*, T. R. E. Southwood (Ed.), pp. 109–18, Blackwell, Oxford and Edinburgh.

FRAENKEL, G. S. (1953) 'The nutritional value of green plants for insects', *Proc. 9th. Interntl Congr. Ent., Symposia*, 90–100.

FRANZ, J. M.* (1971) 'Influence of environment and modern trends in crop management and control'. In Burges and Hussey, 1971a, 407–44.

FRANZ, J. M. and A. KRIEG (1957) 'Virosen europäischer Forstinsekten', *Zeitschr. Pflanzenkrankh. u. Pflanzenschutz*, **64**, 1–9.

FREEMAN, P. (Ed.)* (1965) *Proc. 12th Interntl Congr. Entomol. London 1964*, Section 9a (part) (Rational Pest Control, 566–603).

FRENCH, D. W. and D. B. SCHROEDER (1969), 'The oak-wilt fungus *Ceratocystis sagacearum* as a selective silvicide', *Forest Science*, **15**, 198–203.

FRINGS, H. and M. FRINGS* (1967) 'Behavioural manipulation (visual, mechanical, and acoustical)', in *Pest Control. Biological, Physical and Selected Chemical Methods*, W. W. Kilgore and R. L. Doutt (Eds.), pp. 387–454, Academic Press, New York and London.

FRONTALI, N. and A. GRASSO (1964) 'Separation of three toxicologically different protein components from the venom of the spider *Lactrodectus tredecimguttatus*', *Arch. Biochem. Biophys.*, **106**, 213–18.

FUKAMI, T., T. MITSUI, K. FUKUNAGA, and T. SHISHIDO (1970) 'The selective Toxicity of Rotenone between Mammal, Fish and Insect', in *Biochemical Toxicology of Insecticides*, R. D. O'Brien and I. Yamamoto (Eds.), pp. 159–78, Academic Press, New York and London.

FYE, R. L. (1968) 'Development of an autosterilization technique for the housefly', *J. Econ. Ent.*, **61**, 1578–81.

GALLEY, R. A. E. (1967) 'Possible new and more selective means of pest control', *Proc. Roy. Soc. Lond.*, B, **167**, 155–63.

GALLEY, R. A. E. (1971) 'The contribution of pesticides used in public health programmes to the pollution of the environment. I: General and DDT', *WHO mimeographed document, WHO/VBC/71. 326.*

GALUN, R. and M. WARBURG (1968) 'Irradiation effects on respiration and blood digestion in the tick *Ornithodoros tholozani* and its importance in the sterile male technique', in *Isotopes and Radiation in Entomology. Symposium of the International Atomic Energy Agency and the FAO, Vienna, 4–8 Dec. 1967*, pp. 249–57, IAEA, Vienna.

GARA, R. J., J. R. VITÉ, and H. H. CRAMER (1965) 'Manipulation of *Dendroctonus frontalis* by use of a population aggregating pheromone', *Contr. Boyce Thompson Inst. Pl. Res.*, **23**, 55–66.

GARFINKEL, D. (1965) 'Computer simulation in biochemistry and ecology', in *Theoretical and Mathematical Biology*, T. H. Waterman and H. J. Morowitz (Eds.), pp. 292–310, Blaisdell Publishing, New York, Toronto, and London.

GEIER, P. W.* (1966) 'Management of insect pests', *Ann. Rev. Entomol.*, **11**, 471–90.

GEORGE, J. L. and D. E. H. FREAR* (1966) 'Pesticides in the Antarctic', *Pesticides in the Environment and their Effects on Wild Life, J. appl. Ecol.*, **3** (Suppl.), 155–68.

GEORGHIOU, G. P.* (1965) 'Genetic studies on insecticide resistance', *Advances in Pest Control Research*, **6**, 171–230.

GERBERICH, J. B. and M. LAIRD* (1968) 'Bibliography of papers relating to the control of mosquitoes by the use of fish. An annotated bibliography for the years 1901–66', *FAO Fish. Tech. Rep., No. 78*, FAO, Rome.

GERBERICH, J. B. (1971) 'Control of mosquitoes by the use of fish. Annotated Bibliography Key Word Index (KWIC-Storage and Retrieval) 1901–68', *WHO mimeographed document, WHO VBC/71. 319.*

GEROLT, P. (1969) 'Mode of entry of contact insecticides', *J. Insect Physiol.*, **15**, 563–80.

GIBB, J. A. and M. M. BETTS (1963) 'Food and food supply of nesting tits (Paridae) in Breckland pine', *J. Animal Ecol.*, **32**, 489–533.

GLASER, R. W. (1932) 'Studies on *Neoaplectana glaseri*, a nematode parasite of the Japanese beetle (*Popillia japonica*)', *New Jersey Dept. Agric. Bur. Plant Industr. Circ.*, **211**, 34 pp.

GLASGOW, J. P. (1963) *The Distribution and Abundance of TseTse*, Pergamon Press, Oxford.

GOEDEN, R. E. (1970) 'Current research on biological weed (Dicotyledones) control in southern California (USA)', *Misc. Publ. Commonw. Inst. Biol. Contr. Trinidad* (*Proc. First Int. Symp. Biol. Contr. Weeds, 1969*), No. 1, 21–4.

GÖNNERT, R. *et al.* (1962) 'Bayluscide *R* a new compound for controlling medically important freshwater snails. (And related topics)', *Pflanzenschutz-Nachrichten Bayer*, **15**, 1–88.

GÖSSWALD, K. (1970) 'Aufruf des Bundes für Waldhygiene', *Waldhygiene*, **8**, 243–7.

GOULD, H. J. (1968) 'Integrated pest control: Glasshouse and ornamental crops', *J. appl. Ecol.*, **5**, 513–16.

GOUGH, H. C., A. WOODS, F. E. MASKELL, and M. J. TOWLER (1961) 'Field experiments on the control of wheat bulb fly, *Leptohylemyia coarctata* (Fall)', *Bull. Ent. Res.*, **52**, 621–34.

GRAHAM, F. (1970) *After Silent Spring*, Hamish Hamilton, London.

GREAVES, J. H. (1970) 'Genetics of resistance to rodenticides', *WHO mimeographed document*, *WHO/VBC/70. 246.*

GRIFFIN, D. R. (1960) *Echoes of Bats and Men*, Heinemann, London, Melbourne, and Toronto.

GRIST, D. H. and R. J. A. LEVER (1969) *Pests of Rice*, Longmans, London.

GROSSMAN, F. (1968) 'Conferred resistance in the host', *World Rev. Pest Control*, **7** (4), 176–83.

GUNN, W. W. H. and V. E. SOLMAN (1968) 'A bird warning system for aircraft in flight', in *The Problems of Birds as Pests*, R. K. Murton and E. N. Wright (Eds.), pp. 88–96, Institute of Biology/Academic Press, London and New York.

GUNTHER, F. A. and L. R. JEPPSON (1960) *Modern Insecticides and World Food Production*, Chapman and Hall, London.

GUSTAFSSON, M.* (1971) 'Microbial control of aphids and mites'. In Burges and Hussey, 1971a, 375–84.

HADLEY, C. H. (1948) *US Dept. Agr. Bur. Entomol. Plant Quarantine EC–4* (cited in Tanada, 1967).

HAGEN, K. S.* (1964) 'Developmental stages of parasites', in *Biological Control of Insect Pests and Weeds*, P. DeBach (Ed.), pp. 168–246, Chapman and Hall, London.

HAGEN, K. S. and R. VAN DER BOSCH* (1968) 'Impact of pathogens, parasites, and predators on aphids', *Ann. Rev. Entomol.*, **13**, 325–84.

HAMON, J. and R. PAL (1968) 'Practical implications of insecticide resistance in arthropods of medical and veterinary importance', *WHO mimeographed document*, *WHO/VBC/68*, 106.

HANNA, A. D. (1966) 'The importance of population ecology studies in approaching insect control problems', *World Rev. Pest Control*, **5**, 16–29.

HARCOURT, D. G.* (1969) 'The development and use of life tables in the study of natural insect populations', *Ann. Rev. Entomol.*, **14**, 175–96.

HARRIS, P. (1969) 'Current research on biocontrol of weeds (Dicotyledones) in Canada', *Misc. Publ. Commonw. Inst. Biol. Contr. Trinidad* (*Proc. First Int. Symp. Biol. Contr. Weeds, 1969*), No. 1, 29–32.

HARRISON MATHEWS, L. (1952) *British Mammals*, Collins, London.

HARTLEY, G. S. and T. F. WEST* (1969) *Chemicals for Pest Control*, Pergamon Press, Oxford.

HARVEY, T. L. and D. E. HOWELL (1965) 'Resistance of the house-fly to *Bacillus thuringiensis* Berliner', *J. Invertebrate Pathol.*, **7**, 92–100.

HASKELL, P. T. (1970) 'The future of locust and grasshopper control', *Outlook on Agriculture*, **6**, 166–74.

HAWKES, F. (1968) 'A virus-like structure in the desert locust', *Die Naturwissenschaften*, 55 Jg., Heft 11, 547.

HAYES, W. J., Jr. (1971) 'The degree of hazard to man of DDT as used against malaria', *WHO mimeographed document*, *WHO/VBC/71. 251*; *WHO/MAL/71. 738.*

HAZEN, W. E. (Ed.)* (1970) *Readings in Population and Community Ecology*, 2nd Edn., W. B. Saunders, Philadelphia, London, and Toronto.

HEADLEY, J. C. (1972) 'Economics of agricultural pest control', *Ann. Rev. Entomol.*, **17**, 273–86.

HEIMPEL, A. M. (1967) 'A critical review of *Bacillus thuringiensis* var. *thuringiensis* Berliner and other crystalliferous bacteria', *Ann. Rev. Entomol*, **12**, 287–322.

HEIMPEL, A. M. and T. A. ANGUS (1959) 'The site of action of crystalliferous bacteria in Lepidoptera larvae', *J. Insect Pathol.*, **1**, 152–70.

HEIMPEL, A. M. and T. A. ANGUS.* 'Diseases caused by certain spore-forming bacteria', in *Insect Pathology: An Advanced Treatise*, Vol. 2, E. A. Steinhaus (Ed.), pp. 21–74, Academic Press, New York and London.

HENNEBERRY, T. J. and A. F. HOWLAND (1966) 'Response of male cabbage loopers to black-light with or without the presence of the female sex pheromone', *J. Econ. Ent.*, **59**, 623–6.

HERFORD, G. V. B. (1961) 'Food lost in store by insect attack', *Span*, **4**, 40–2.

HERSKOWITZ, I. H. (1968) *Basic Principles of Molecular Genetics*, Nelson, London.

HICKEY, J. L. and D. W. ANDERSON (1968) 'Chlorinated hydrocarbons and egg-shell changes in raptorial and fish-eating birds', *Science*, **162**, 271–3.

HICKIN, N. E. (1963) *The Insect Factor in Wood Decay*, The Rentokil Library, Hutchinson, London.

HIGHNAM, K. C. and L. HILL* (1969) *The Comparative Endocrinology of the Invertebrates*, Edward Arnold, London.

HILLEBRANT, P. M. (1960a and b) 'The economic theory of the use of pesticides', Pt. I: *J. Agric. Econ.*, **23**, 464–72; Pt. II: *ibid.*, **24**, 52–61.

HOCKING, B. (1967) 'Use of repellents and attractants in filariasis control', *Bull. Wld Hlth Org.*, **37**, 323–7.

HODEK, I. (Ed.)* (1966) *Ecology of Aphidophagous Insects*, Dr W. Junk, The Hague/Academia, Prague.

HOFFMAN, W. S. (1968) 'Clinical evaluation of the effects of pesticides on man', *Ind. Med. Surg.*, **37**, 289–92.

HOFFMAN, W. S., H. ADLER, W. I. FISHBEIN, and F. C. BAUER (1967) 'Relation of pesticide concentrations in fat to pathological changes in tissues', *Arch. Environm. Hlth*, **15**, 758–65.

HOLAN, G. (1969) 'New halocyclopropane insecticides and mode of action of DDT', *Nature*, **221**, 1025–9.

HOLLING, C. S.* (1961) 'Principles of insect predation', *Ann. Rev. Entomol.*, **6**, 163–82.

HOLLING, C. S. (1965) 'The functional response of predators to prey density and its role in mimicry and population regulation', *Mem. Ent. Soc. Can.*, **45**, 5–60.

HOLLING, C. S.* (1966) 'The functional response of invertebrate predators to prey density', *Can. Entomol. Mem.*, **48**, 1–86.

HOLLING, C. S.* (1968) 'The tactics of a predator', in *Insect Abundance*, T. R. E. Southwood (Ed.), pp. 47–58, Blackwell, Oxford and Edinburgh.

HOLLOWAY, J. K.* (1964) 'Projects in biological control of weeds', in *Biological Control of Insect Pests and Weeds*, P. DeBach (Ed.), pp. 650–70, Chapman and Hall, London.

HOOKER, A. L. (1967) 'The genetics and expression of resistance in plants to rusts of the genus *Puccinia*', *Ann. Rev. Phytopathol.*, **5**, 163–82.

HOWARD, L. O. and W. F. FISKE (1911) 'The importation into the United States of the parasites of the gipsy moth and the brown-tail moth', *U.S. Dept. Agric. Bur. Ent. Bull.*, **91**, 1–312.

HOWARD, W. E.* (1967) 'Biocontrol and chemosterilants', in *Pest Control: Biological, Physical and Selected Chemical Methods*, W. W. Kilgore and R. L. Doutt (Eds.), pp. 343–86, Academic Press, New York and London.

HOYT, S. C. (1969) 'Integrated chemical control of insects and biological control of mites on apple in Washington', *J. Econ. Ent.*, **62**, 74–86.

HUFFAKER, C. B. (1957) 'Fundamentals of biological control of weeds', *Hilgardia*, **27**, 101–57.

HUFFAKER, C. B. and C. E. KENNETT (1953) 'Developments towards biological control of cyclamen mite on strawberries in California', *J. Econ. Ent.*, **46**, 802–12.

HUFFAKER, C. B. and C. E. KENNETT (1956) 'Experimental studies on predation: predation and cyclamen-mite populations on strawberries in California', *Hilgardia*, **26**, 191–222.

HUFFAKER, C. B. and P. S. MESSENGER* (1964a) 'Population ecology – historical development', in *Biological Control of Insect Pests and Weeds*, P. DeBach (Ed.), pp. 45–73, Chapman and Hall, London.

HUFFAKER, C. B. and P. S. MESSENGER* (1964b) 'The concept and significance of natural control', in *Biological Control of Insect Pests and Weeds*, P. DeBach (Ed.), pp. 74–117, Chapman and Hall, London.

HUFFAKER, C. B., M. VAN DE VRIE and J. A. MCMURTRY* (1969) 'The ecology of tetranychid mites and their natural control', *Ann. Rev. Entomol.*, **14**, 125–74.

HUGHES, A. M. (1961) *The Mites of Stored Food*, Ministry of Agriculture, Fisheries and Food Tech. Bull. No. 9, HMSO, London.

HUGHES, R. D. (1962) 'The study of aggregated populations', in *Progress in Soil Zoology*, P. W. Murphy (Ed.), pp. 51–5, Butterworths, London.

HUGHES, T. E. (1959) *Mites or the Acari*, University of London/The Athlone Press.

HUNT, E. G. and A. I. BISCHOFF (1960) 'Inimical effects on wild life of periodic DDT applications to Clear Lake', *Calif. Fish Game*, **46**, 91–106.

HUNTER, C. G., J. ROBINSON, and K. W. JAGER (1967) 'Aldrin and dieldrin – the safety of present exposures of the general populations in the United Kingdom and United States', *Food Cosmet. Toxicol.*, **5**, 781–7.

HURST, H. (1967) 'Sound and vision in the sex life of insects', in *Penguin Science Survey 1967. The Biology of Sex*, A. Allison (Ed.), pp. 160–88, Penguin Books, Harmondsworth.

HUSSEY, N. W. (1965) 'Possibilities for integrated control of some glasshouse pests', *Ann. appl. Biol.*, **56**, 347–50.

HUSSEY, N. W.* (1970) 'Some economic considerations in the future development of biological control', in *Technological Economics of Crop Protection and Pest Control*, Society Chemical Industry Monograph No. 36, pp. 109–18.

HUSSEY, N. W., W. H. READ, and J. J. HESLING (1969) *The Pests of Protected Cultivation*, Edward Arnold, London.

HUXLEY, J. (Ed.) (1940) *The New Systematics*, Oxford University Press, Oxford.

IGNOFFO, C. M. (1967) 'Possibilities of mass-producing insect pathogens', in *Insect Pathology and Microbial Control*, P. A. van der Laan (Ed.), pp. 91–117, North-Holland Publishing, Amsterdam.

IGNOFFO, C. M.* (1968) 'Viruses – living insecticides', *Current Topics Microbiol, Immunol.*, **42**, 129–67.

IGNOFFO, C. M. and W. F. HINK (1971) 'Propagation of arthropod pathogens in living systems'. In Burges and Hussey, 1971a, 541–80.

IMMS, A. D., revised O. W. RICHARDS, and R. G. DAVIES (1957) *A General Textbook of Entomology*, Methuen, London; E. P. Dutton, New York.

INMAN, R. E. (1967) 'Study of phytopathogens as weed control agents: *Final report by Stanford Research Institute for the United States Department of Agriculture*', cited in Wilson, 1969.

INTERNATIONAL ATOMIC ENERGY AGENCY* (1962) *Radioisotopes and Radiation in Entomology. Proc. Symposium sponsored by International Atomic Energy Agency, Bombay, 5–9 Dec. 1960*, IAEA, Vienna.

INTERNATIONAL ATOMIC ENERGY AGENCY* (1963) *Radiation and Radioisotopes Applied to Insects of Agricultural Importance. Proc. Symposium organized by the International Atomic Energy Agency and the F.A.O., Athens, 22–26 April 1963*, IAEA, Vienna.

JACOBSON, M.* (1965) *Insect Sex Attractants*, Wiley, New York.

JACOBSON, M.* (1966) 'Chemical insect attractants and repellents', *Ann. Rev. Entomol.*, **11**, 403–22.

JACOBSON, M. and M. BEROZA (1964) 'Insect attractants', *Scientific American*, August 1964, 20–7.

JACOBSON, M., M. SCHWARZ, and R. M. WATERS* (1970) 'Gypsy moth sex attractants: a re-investigation', *Science*, **63**, 943–5.

JAFRI, R. H. (1965) 'Prospects of integrated radiation and microbial control of harmful insects', *Proc. 12th Interntl Congr. Entomol. Lond., 1964*, 747–8.

JAFRI, R. H. (1968) 'The susceptibility of irradiated larvae of *Galleria mellonella* to VND virus', *J. Invertebrate Path.*, **10**, 355–60.

JAFRI, R. H. and A. A. KHAN (1969) 'A study of the development of a nuclear polyhedrosis virus of *Heliothis* in *Galleria mellonella* larvae exposed to gamma rays', *J. Invertebrate Path.*, **14**, 104.

JARY, S. G. (1965) 'The influence of the development of pesticides on the production of fruit and hops', *World Rev. Pest Control*, **4**, 112–22.

JENKINS, D. W.* (1964) 'Pathogens, parasites and predators of medically important arthropods', *Bull. Wld Hlth Org.*, **30** (Suppl.), 5–150.

JEPSON, W. F. (1954) *A Critical Review of the World Literature on the Lepidopterous Stalk Borers of Tropical Graminaceous Crops*, Commonw. Inst. Entomol., London.

JOHNSON, T., G. J. GREEN, and D. J. SAMBORSKI (1967) 'The world situation of the cereal rusts', *Ann. Rev. Phytopathol.*, **5**, 183–200.

JONES, F. G. W. and R. A. DUNNING (1969) *Sugar Beet Pests*, Min. Agr. Fish. and Food, Bulletin No. 162, HMSO, London.

JONES, F. G. W., R. A. DUNNING, and K. P. HUMPHRIES (1955) 'The effects of defoliation and loss of stand upon yield of sugar beet', *Ann. appl. Biol.*, **43**, 63–70.

JONES, F. G. W. and D. M. PARROTT (1968) 'Potato production using resistant varieties on land infested with potato cyst-eelworm *Heterodera rostochiensis* Woll', *Outlook on Agriculture*, **5** (5), 215–22.

JUDSON, C. L. (1967) 'Alteration of feeding behaviour and fertility in *Aedes aegypti* by the chemosterilant apholate', *Entomologia exp. appl.*, **10**, 387–94.

JUNG, H. F. and H. SCHEINPFLUG (1970) 'Rice growing in Japan with special emphasis on problems of crop protection', *Pflanzenschutz-Nachrichten Bayer*, **23**, 235–63.

JUSTESEN, S. H. and P. M. L. TAMMES (1960) 'Studies of yield losses. I: The self-limiting effect of injurious or competitive organisms on crop yield', *T. Pl. ziekten*, **66**, 281–7.

KARLSON, P. (1960) 'Pheromones', *Ergebn. Biol.*, **22**, 212–25.

KARLSON, P. and A. BUTENANDT* (1959) 'Pheromones (ectohormones) in insects', *Ann. Rev. Entomol.*, **4**, 39–58.

KATZ, B.* (1966) *Nerve, Muscle and Synapse*, McGraw-Hill, New York and London.

KEIDING, J. (1963) 'Possible reversal of resistance', *Bull. Wld Hlth Org.*, **29** (Suppl.), 51–62.

KENNEDY, J. S.* (1968) 'The motivation of integrated control', *J. appl. Ecol.*, **5**, 492–9.

KENNEDY, J. S. and M. J. WAY (1964) 'Intraspecific mechanisms in insect population regulation', *12th Int. Congr. Ent. London*. Not published in proceedings, cited in Klomp, 1966.

KERMACK, W. O. and A. G. MCKENDRICK (1927) 'A contribution to the mathematical theory of epidemics', *Proc. Roy. Soc. London* (*A*), **115**, 700–21.

KLOMP, H.* (1964) 'Intraspecific competition and the regulation of insect numbers', *Ann. Rev. Entomol.*, **9**, 17–40.

KLOMP, H. (1966) 'The dynamics of a field population of the pine looper, *Bupalus piniarius* L.', *Advances Ecol. Res.*, **3**, 207–304.

KNIPLING, E. F. (1955) 'Possibilities of insect control or eradication through the use of sexually sterile males', *J. Econ. Entomol.* **48**, 459–62.

KNIPLING, E. F.* (1960) 'The eradication of the screw-worm fly', *Sci. Amer.*, **203** (4), 54–61.

KNIPLING, E. F.* (1967) 'Sterile technique-principles involved, current application, limitations and future application', in *Genetics of Insect Vectors of Disease*, J. W. Wright and R. Pal (Eds.), pp. 587–616, Elsevier, Amsterdam.

KNIPLING, E. F.* (1968) 'The potential role of sterility for pest control', in *Principles of Insect Chemosterilization*, G. C. LaBrecque and C. N. Smith (Eds.), pp. 7–40, North-Holland Publishing, Amsterdam; Appleton-Century-Crofts, New York.

KNIPLING, E. F., H. LAVEN, G. B. CRAIG, R. PAL, J. B. KITZMILLER, C. N. SMITH, and A. W. A BROWN* (1968) 'Genetic control of insects of public health importance', *Bull. Wld Hlth Org.*, **38**, 421–38.

KRIEG, A. (1971a) 'Possible use of Rickettsiae for microbial control of weeds'. In Burges and Hussey, 1971a, 173–9.

KRIEG, A. (1971b) 'Key publications', *ibid.*, 711–16.

KUĆ, J.* (1968) 'Biochemical control of disease resistance in plants', *World Rev. Pest Control*, **7** (1), 42–96.

LABRECQUE, G. C. and C. N. SMITH (Eds.)* (1968) *Principles of Insect Chemosterilization*, North-Holland Publishing, Amsterdam; Appleton-Century-Crofts, New York.

LACHANCE, L. E., C. H. SCHMIDT, and R. C. BUSHLAND* (1967) 'Radiation – induced sterilization', in *Pest Control: Biological, Physical and selected Chemical Methods*, W. W. Kilgore and R. L. Doutt (Eds.), pp. 148–96, Academic Press, New York and London.

LACHANCE, L. E., D. T. NORTH, and W. KLASSEN* (1968) 'Cytogenetic and cellular basis of chemically induced sterility in insects', in *Principles of Insect Chemosterilization*, G. C.

LaBrecque and C. N. Smith (Eds.), pp. 99–157, North-Holland Publishing, Amsterdam; Appleton-Century-Crofts, New York.

LACK, D. (1954) *The Natural Regulation of Animal Numbers*, Oxford University Press, Oxford.

LAIRD, M. (1963) 'Vector ecology and integrated control procedures', *Bull. Wld Hlth Org.*, **29** (Suppl.), 147–51.

LAIRD, M. (1967) 'Biological and chemical experiments against mosquitoes in the South Pacific', *Int. Pest Control*, **9**, 21–7 and *Chronicle Wld Hlth Org.*, **29** (Suppl.), 147–51.

LAIRD, M. (1971a) 'Microbial control of arthropods of medical importance'. In Burges and Hussey, 1971a, 387–406.

LAIRD, M.* (1971b) 'A Bibliography on diseases and enemies of medically important arthropods 1963–7 with some earlier titles omitted from Jenkins' 1964 list', *ibid.*, 751–89.

LARGE, C.* (1940) *The Advance of the Fungi*, Jonathan Cape, London.

LAVEN, H. (1967a) 'Formal genetics of *Culex pipiens*', in *Genetics of Insect Vectors of Disease*, J. W. Wright and R. Pal (Eds.), pp. 17–66, Elsevier, Amsterdam.

LAVEN, H. (1967b) 'Speciation and evolution in *Culex pipiens*', in *Genetics of Insect Vectors of Disease*, J. W. Wright and R. Pal (Eds.), pp. 251–76, Elsevier, Amsterdam.

LAVEN, H.* (1967c) 'A possible model for speciation by cytoplasmic isolation in the *Culex pipiens* complex', *Bull. Wld Hlth Org.*, **37**, 263–6.

LAVEN, H. (1969) 'Eradicating mosquitoes using translocations', *Nature, Lond.*, **221**, 958–9.

LAWRENCE, W. J. C.* (1968) *Plant Breeding*, Edward Arnold, London.

LAYCOCK, G. (1966) *The Alien Animals*, Natural History Press, New York.

LEROUX, E. J. (1971) 'Biological control attempts on pome fruit (apple and pear) in North America 1860–1970', *Can. Ent.*, **103**, 963–74.

LICHTENSTEIN, E. P. (1966) 'Persistence and degradation of pesticides in the environment'. In National Academy of Sciences 1966, 221–9.

LIPA, J. L.* (1963) 'Infections caused by Protozoa other than Sporozoa', in *Insect Pathology: An Advanced Treastise*, Vol. 2, E. A. Steinhaus (Ed.), pp. 335–62, Academic Press, New York and London.

LORIMER, N., E. HALLIMAN, and K. S. RAI (1972) 'Translocation homozygotes in the yellow fever mosquito, *Aedes aegypti*', *WHO mimeographed document, WHO/VBC/72. 355.*

LUCK, R. F. (1971) 'An appraisal of two methods of analyzing insect life tables', *Can. Ent.*, **103**, 1261–71.

LUPTON, F. G. H.* (1967) 'The use of resistant varieties in crop protection', *World Rev. Pest Control*, **6** (2). 47–58.

LÜTHY, P. and L. ETTLINGER (1967) 'Morphology and chemistry of the parasporal body of *Bacillus fribourgensis*', in *Insect Pathology and Microbial Control*, P. van der Laan (Ed.), pp. 54–8, North-Holland, Amsterdam.

LYSENKO, O. (1963) 'The taxonomy of entomogenous bacteria', in *Insect Pathology: An Advanced Treatise*, Vol. 2, E. A. Steinhaus (Ed.), pp. 1–20, Academic Press, New York and London.

LYSENKO, O. and M. KUČERA* (1971) 'Micro-organisms as sources of new insecticidal chemicals: toxins'. In Burges and Hussey, 1971a, 205–27.

MACARTHUR, R. (1955) 'Fluctuations in animal populations, and a measure of community stability', *Ecology*, **36**, 533–6.

MACARTHUR, R. and J. CONNELL* (1966) *The Biology of Populations*, Wiley, New York, London, and Sydney.

MCCALLAN, S. E. A.* (1967) 'History of fungicides', in *Fungicides: An Advanced Treatise*, Vol. 1, D. C. Torgeson (Ed.), pp. 1–37, Academic Press, New York and London.

MCCLELLAND, G. A. H. (1967) 'Speciation and evolution in *Aedes*', in *Genetics of Insect Vectors of Disease*, J. W. Wright and R. Pal (Eds.), pp. 277–311, Elsevier, Amsterdam.

MCCRAY, E. M., Jr. (1961) 'A mechanical device for the rapid sexing of *Aedes aegypti* pupae', *J. Econ. Ent.*, **54**, 819.

MCDONALD, P. T. and K. S. RAI (1971) 'Population control potential of heterozygous translocations as determined by computer simulations', *Bull. Wld Hlth Org.*, **44**, 829–45.

MCINTOSH, A. H. and D. W. EVELING (1965) 'Tests of aphicides for possible systemic control of potato blight', *Europe Potato J.*, **8**, 98–103.

MCLAUGHLIN, R. E. (1971) 'Use of protozoons for microbial control of insects'. In Burges and Hussey, 1971a, 151–72.

MCLEOD, J. M. (1966) 'The spatial distribution of cocoons of *Neodiprion swainei* Middleton in a jack-pine stand. I: A cartogtaphic analysis of cocoon distribution, with special reference to predation by small mammals', *Canad. Ent..*, **98**, 430–47.

MADEL, W. (1971) 'Ein Denkmal für den Baumwollkapselkäfer', *Z. Angew. Zool.*, **53**, 25–7.

MADELIN, M. F. (1963) 'Diseases caused by Hyphomycetous fungi', in *Insect Pathology: An Advanced Treatise*, Vol. 2, E. A. Steinhaus (Ed.), pp. 233–72, Academic Press, New York and London.

MARGALEF, D. R. (1958) 'Information theory in ecology', originally in Spanish, but reprinted in *General Systems*, **3** (1968), 36–71.

MARTIGNONI, M. E.* (1964) 'Mass production of insect pathogens', in *Biological Control of Insect Pest and Weeds*, P. DeBach (Ed.), pp. 579–609, Chapman and Hall, London

MARTIN, H.* (1964) *The Scientific Principles of Crop Protection*, 5th Edn., Edward Arnold, London.

MARTIN, H.* (1970) 'Crop protection and the pesticides industry: past, present and future', in *Technological Economics of Crop Protection and Pest Control*, Society of Chemical Industry, Monograph No. 36.

MASNER, P., K. SLÁMA, and V. LANDA (1968) 'Sexually spread insect sterility induced by the analogues of juvenile hormone', *Nature, Lond.*, **219**, 395–6.

MASSEE, A. M. (1954) *The Pests of Fruit and Hops*, 3rd Edn., Crosby Lockwood, London.

MATTINGLEY, P. F. (1967) 'The systematics of the *Culex pipiens* complex', *Bull. Wld Hlth Org.*, **37**, 257–61.

MAYR, E. (1942) *Systematics and the Origin of Species*, Columbia University Press, New York.

MEIFERT, D. W. and G. G. LABRECQUE (1971) 'Integrated control for the suppression of a population of houseflies, *Musca domestica* L', (Dipt. Muscidae), *J. Med. Entomol.*, **8**, 43–5.

MEIFERT, D. W., P. B. MORGAN, and G. C. LABRECQUE (1967) 'Infertility induced in male house flies by sterilant-bearing females', *J. Econ. Ent.*, **60**, 1336–8.

MELLANBY, K.* (1967) *Pesticides and Pollution*, Collins, London.

MENDENHALL, W., L. OTT, and R. L. SCHEAFFER (1971) *Elementary Survey Sampling*, Wadsworth, Belmont, California.

METCALF, C. L., W. P. FLINT, and R. L. METCALF (1962) *Destructive and Useful Insects*, 4th Edn., McGraw-Hill, New York and London.

METCALF, R. L. (1965) 'Methods of Estimating Effects', in *Research in Pesticides*, C. O. Chichester (Ed.), Academic Press, New York and London.

MIDDLETON, A. D. (1931) *The Grey Squirrel*, Sidgwick and Jackson, London.

MILANI, R. (1956) *Riv. Parasit.*, **17**, 233–46 (cited in Brown and Pal, 1971).

MILANI, R. (1959) *Genet. Agr.*, **10**, 288–308 (cited in Brown and Pal, 1971).

MILLER, N. C. E. (1956) *The Biology of the Heteroptera*, Leonard Hill, London.

MILNE, A. (1957a) 'The natural control of insect populations', *Can. Ent.*, **89**, 193–213.

MILNE, A. (1957b) 'Theories of natural control of insect populations', *Cold Spring Harb. Symp. quant. Biol.*, **22**, 253–67.

MILNE, A. (1958) 'Perfect and imperfect density dependence in population dynamics', *Nature, Lond.*, **182**, 1251–2.

MILNE, A. (1962) 'On a theory of natural control of insect populations', *J. theor. Biol.*, **3**, 19–50.

MINISTRY OF AGRICULTURE, FISHERIES AND FOOD (1961) *Farm Sprayers and their Use*, Bulletin No. 182, HMSO, London.

MINISTRY OF HEALTH (1964) *Report of the Working Party on Irradiation of Food*, HMSO, London.

MOLNAR, I. (Ed.) (1966) *A Manual of Australian Agriculture*, 2nd Edn., Heinemann, London and Melbourne.

MONRO, H. A. U. (1961) *Manual of Fumigation for Insect Control*, FAO Agricultural Studies No. 56, FAO, Rome.

MONRO, J. (1966) 'Population control in animals by overloading resources with sterile animals', *Science*, **140**, 496–7.

MOODY, D. E. M. and D. P. MOODY (1963) 'Toxic products in groundnuts', *Nature, Lond.*, **198**, 1062–3.

MOOK, L. J. and H. G. W. MARSHALL (1965) 'Digestion of sprucebud worm larvae and pupae in the olive-backed thrush, *Hylochichla ustulata swainsoni* (Tschudi)', *Can. Entomol.*, **97**, 1144–9.

MOORE, N. W. (Ed.)* (1966) *Pesticides in the Environment and their Effects on Wild Life, J. appl. Ecol.*, **3** (Suppl.).

MOORE, N. W.* (1967) 'A synopsis of the pesticide problem', *Adv. Ecolog. Res.*, **4**, 75–130.

MOORE, N. W. (1969) 'The significance of the persistent organochlorine insecticides and the polychlorinated biphenyls', *The Biologist*, **16**, 157–62.

MORETON, B. D. (1969) *Beneficial Insects and Mites*, Min. Agr. Fish. and Food, Bull. No. 20 (6th Edn.), HMSO, London.

MORRIS, R. F.* (1960) 'Sampling insect populations', *Ann. Rev. Entomol.*, **5**, 243–64.

MORRIS, R. F. (Ed.) (1963) 'The dynamics of epidemic spruce budworm populations', *Mem. ent. Soc. Canada*, **31**, 1–332.

MOTT, D. G. (1967) 'The analysis of determination in population systems', in *Systems Analysis in Ecology*, K. E. F. Watt (Ed.), Academic Press, New York.

MULES, J. H. W. (1932) 'Blowfly pest. Surgical treatment of ewes', *Pastoral Rev. Austr.*, **42**, 35–6.

MULLER, H. J. (1954) 'The nature of genetic effects produced by radiation', in *Radiation Biology*, A. Hollaender (Ed.), pp. 351–473, McGraw-Hill, New York.

MULLINS, L. J. (1954) 'Some physical mechanisms in narcosis', *Chem. Rev.*, **54**, 289–323.

MULLINS, L. J. (1955) 'Structure-toxicity in Hexachlorocyclohexane isomers', *Science*, **122**, 118–19.

MURPHY, P. W. (Ed.)* (1962) *Progress in Soil Zoology*, Butterworths, London.

MURTON, R. K.* (1971) *Man and Birds*, Collins, London.

MURTON, R. K. and E. N. WRIGHT (Ed.)* (1968) *The Problems of Birds as Pests*, Institute of Biology/Academic Press, New York and London.

MUSPRATT, J. (1963) 'Destruction of the larvae of *Anopheles gambiae* Giles by a *Coelomyces* fungus', *Bull. Wld Hlth Org.*, **29**, 81–6.

NATIONAL ACADEMY OF SCIENCES* (1966) *Scientific Aspects of Pest Control. Symposium of the National Academy of Sciences National Research Council, Feb. 1–3 1966*, Publication 1402, National Academy of Sciences – National Research Council, Washington, D.C.

NATIONAL ACADEMY OF SCIENCES* (1969) *Principles of Plant and Animal Pest Control. Vol. 1: Plant Disease Development and Control. Vol. 2: Weed Control. Vol. 3: Insect–Pest Management and Control. Vol. 4: Control of Plant–Parasitic Nematodes*, National Academy of Sciences, Washington, D.C.

NATIONAL ACADEMY OF SCIENCES (1970) *Principles of Plant and Animal Pest Control. Vol. 5: Vertebrate Pests: Problems and Control*, National Academy of Sciences, Washington, D.C.

NEILSON, M. M. and R. F. MORRIS (1964) 'The regulation of European spruce sawfly numbers in the Maritime Provinces of Canada from 1937 to 1963', *Can. Entomol.*, **96**, 773–84.

NELSON, S. O. (1967) 'Electromagnetic in Energy' in *Pest Control Biological, Physical and Selected Chemical Methods*, Kilgore, W. E. and R. L. Doutt, (Eds.), pp. 89–145 Academic Press, New York and London.

NELSON, S. O. and J. L. SEUBERT* (1966) 'Electromagnetic and Sonic Energy for Pest Control', in *Scientific Aspects of Pest Control*, Publication 1402, National Academy of Sciences – National Research Council, Washington, D.C.

NEWCOMBE, H. B. (1971) 'The genetic effects of ionizing radiations', *Adv. in Genet.*, **16**, 293–303.

NEWSOM, L. D.* (1967) 'Consequences of insecticide usage on nontarget organisms', *Ann. Rev. Entomol.*, **12**, 257–86.

NICHOLSON, A. J.* (1954) 'An outline of the dynamics of animal populations', *Aust. J. Zool.*, **2**, 9–65.

NICHOLSON, A. J.* (1958) 'Dynamics of insect populations', *Ann. Rev. Entomol.*, **3**, 107–36.

NORRIS, J. R. (1971a) 'The protein crystal of *Bacillus thuringiensis*: biosynthesis and physical structure'. In Burges and Hussey, 1971a, 229–46.

NORRIS, J. R. (1971b) 'Information sources and literature searching in biological control', *ibid.*, 717–22.

NOVÁK, V. J. A. (1967) 'The juvenile hormone and the problem of animal morphogenesis', in *Insects and Physiology*, J. W. L. Beament and J. E. Treherne (Eds.), pp. 119–32, Oliver and Boyd, Edinburgh and London.

O'BRIEN, R. D.* (1967) *Insecticides: Action and Metabolism*, Academic Press, New York and London.

O'BRIEN, R. D. (1970) 'The properties of acetylcholine receptor *in vitro* from torpedo electroplax, housefly head and rat brain', in *Biochemical Toxicology of Insecticides*, R. D. O'Brien and I. Yamamoto (Eds.), pp. 1–12, Academic Press, New York and London.

O'BRIEN, R. D. and I. YAMAMOTO (Eds.)* (1970) *Biochemical Toxicology of Insecticides*, Academic Press, New York and London.

O'CONNOR, B. A. (1950) 'Premature nutfall of coconuts in the British Solomon Islands Protectorate', *Agric. Jour. (Fiji)*, **21** (1–2), 22 pp.

OLBERG, G. (1967) *Das Verhalten der Solitären Wespen Mitteleuropas*, Deutscher Verlag Wissenschaften, Berlin.

ORDISH, G. (1952) *Untaken Harvest*, Constable, London.

ORDISH, G.* (1967) *Biological Methods in Crop Pest Control*, Constable, London.

ORDISH, G and J. F. MITCHELL* (1967) 'World fungicide usage', in *Fungicides: An Advanced Treatise*, D. C. Torgeson (Ed.), pp. 39–62, Academic Press, New York and London.

ORMSBEE, R. A. (1969) 'Rickettsiae (as organisms)', *Ann. Rev. Microbiol.*, **23**, 275–92.

PAINTER, R. R.* (1967) 'Repellents', in *Pest Control: Biological, Physical and Selected Chemical Methods*, W. W. Kilgore and R. L. Doutt (Eds.), pp. 267–86, Academic Press, New York and London.

PARAMONCHIK, V. M. (1968) 'Liver functions engaged in the production of certain organochlorine chemical poisons', *Sovetsk. Med.*, **31**, 62–5. (In Russian, abstract in *WHO mimeographed document, VBC/TOX/69. 3*.)

PARAMONCHIK, V. M. and V. I. PLATANOVA (1968) 'Functional condition of the liver and stomach in persons exposed to effects of organochlorine chemical poisons', *Gig. Truda i Prof. Zabolevaniya*, **12** (3), 27–31. (In Russian, reviewed in *WHO mimeographed documents, VBC/TOX/69. 3 and VBC/TOX/70. 5*.)

PARR, W. J. and N. E. A. SCOPES* (1971a) 'Problems associated with biological control of glasshouse pests, *N.A.A.S Quart. Rev.*. **91** (Spring 1971), 113–21.

PARR, W. J. and N. E. A. SCOPES* (1971b) 'Recent advances in the integrated control of glasshouse pests', *ibid.*, **93** (Winter 1971), 101–8.

PATTERSON, R. S., D. E. WEIDHAAS, H. R. FORD, and C. S. LOFGREN (1970) 'Suppression and elimination of an island population of *Culex pipiens fatigans* with sterile males', *WHO mimeographed document, WHO/VBC/70. 180*.

PAUL, A. H. (1959) 'Dieldrin poisoning – a case report', *New Zealand Med. Jour.*, **58**, 393ff.

PEARSON, O. P. (1955) 'Shrews', in *Twentieth-Century Bestiary*, D. Flanagan *et al.* (Eds.), Simon and Schuster, New York.

PICKETT, A. D. and A. W. MACPHEE (1965) 'Twenty years' experience with integrated control programmes in Nova Scotia apple and pear orchards', *Proc. XIIth Interntl Congr. Entomol. London 1964*, pp. 597–8.

PIERCE, W. D. (1912) 'The insect enemies of the cotton boll weevil', *US Dept. Agric. Bur. Ent. Bull.*, **100**, 99 pp.

PIMENTEL, D. (1961) 'On a genetic feed-back mechanism regulating populations of herbivores, parasites and predators', *American Nat.*, **95**, 65–79.

PIRIE, N. W.* (1969) *Food Resources, Conventional and Novel*, Penguin Books, Harmondsworth.

POINAR, G. O. (1971) 'Use of nematodes for microbial control of insects'. In Burges and Hussey, 1971a, 181–203.

POINAR, G. O. (1972) 'Nematodes as facultative parasites of insects', *Ann. Rev. Entomol.*, **17**, 103–22.

POLISHUK, Z. W., M. WASSERMANN, and D. WASSERMANN (1970) 'Effects of pregnancy on storage of organochlorine insecticides', *Arch. Environnm. Hlth*, **20**, 215–17.

PRICE, P. W. (1971) 'Towards a holistic approach to insect population studies', *Ann. Ent. Soc. America*, **64**, 1399–406.

PROVERBS, M. D.* (1969) 'Induced sterilization and control of insects', *Ann. Rev. Entomol.*, **14**, 81–102.

PYE, D. (1969) 'The diversity of bats', *Science J.*, **5** (4), 47–52.

QUISENBERRY, K. S. and L. P. REITZ (1967) *Wheat and Wheat Improvement*, American Society of Agronomy. Madison. Wisconsin.

RAI, K. S. and SR. M. ASMAN (1969) 'Possible application of a reciprocal translocation for genetic control of the mosquito, *Aedes aegypti*', *Proc. XIIth Interntl Congress Genetics*, I. 164.

RAINWATER, H. I. and C. A. SMITH (1966) 'Quarantine – first-line defense', in *Protecting our*

Food. The Year Book of Agriculture 1966, pp. 216–24, The United States Government Printing Office, Washington, D.C.

RAJAGOPALAN, P. K., M. YASUNO, and G. G. LABRECQUE (1972) 'Dispersal and survival in the field of chemosterilized, irradiated and cytoplasmic incompatible males of *Culex fatigans*', *WHO mimeographed document, WHO/VBC/72. 353.*

REDDY, D. B. (1967) 'The importance of pesticides in Indian food production', *Proc. Roy. Soc. Lond.*, B, **167**, 145–54.

REMINGTON, C. L.* (1968) 'The population genetics of insect introduction', *Ann. Rev. Entomol.*, **13**, 415–26.

RHODE, R. H., J. SIMON, A. PERDOMO, J. GUTIÉRREZ, C. F. DOWLING, Jr., and D. A. LINDQUIST (1971) 'Application of the sterile-insect-release technique in Mediterranean fruit fly suppression (*Ceratitis capitata*: Diptera, Tephritidae)', *J. Econ. Ent.*, **64**, 708–13.

RICHARDS, O. W.* (1961) 'The theoretical and practical study of natural insect populations', *Ann. Rev. Entomol.*, **6**, 147–62.

RICHARDS, O. W. and N. WALOFF (1954) 'Studies on the biology and population dynamics of British grasshoppers', *Anti-Locust Bull.*, **17**, 182 pp.

RICHARDS, O. W. and N. WALOFF (1961) 'A study of a natural population of *Phytodecta olivacea* (Forster) (Coleoptera Chrysomeloidea')', *Phil. Trans.* (B), **244**, 205–57.

RISEBOROUGH, R. W., R. J. HUGGETT, J. J. GRIFFIN, and E. D. GOLDBERG (1968) 'Pesticides: transatlantic movements in northeast trades', *Science*, **159**, 1233–6.

RIVERS, C. F. (1967) 'The natural and artificial dispersion of pathogens', in *Insect Pathology and Microbial Control*, P. A. van der Laan (Ed.), pp. 252–63, North-Holland, Amsterdam.

ROBERTS, D. W. (1967) 'Some effects of *Metarrhizium anisopliae* and its toxins on mosquito larvae', in *Insect Pathology and Microbial Control*, P. A. van der Laan (Ed.), pp. 243–6, North-Holland, Amsterdam.

ROBERTS, D. W. and W. G. YENDOL (1971) 'Use of fungi for microbial control of insects'. In Burges and Hussey, 1971a, 125–49.

ROBINSON, J. (1970) 'Birds and pest control chemicals', *Bird Study*, **17**, 195ff.

ROHDE, R. A. (1960) 'Acetylcholinesterase in plant-parasitic nematodes and an anticholinesterase from asparagus', *Proc. helm. Soc. Wash.*, **27**, 121–3.

ROLLS, E. C.* (1969) *They All Ran Wild: the Story of Pests on the Land in Australia*, Angus and Robertson, Sydney.

ROTH, L. M. and E. R. WILLIS* (1960) 'The biotic associations of cockroaches', *Smithson. misc. Collns.*, **141**, 1–468.

ROTHSCHILD, M. (1967) 'The rabbit flea and hormones', in *Penguin Science Survey 1967. The Biology of Sex*, A. Allison (Ed.), pp. 189–200, Penguin Books, Harmondsworth.

ROTHSCHILD, M. and T. CLAY (1952) *Fleas, Flukes and Cuckoos: A Study of Bird Parasites*, Collins, London.

RUDD, R. L.* (1964) *Pesticides and the Living Landscape*, Faber and Faber, London.

RYDER, W. (1972) 'Pot bugs and the peasant economy', *New Scientist*, 13 April 1972, p. 51.

SALT, G. (1970) *The Cellular Defence Reactions of Insects*, Cambridge University Press Cambridge.

SALT, G. and F. S. J. HOLLICK (1944) 'Studies of wireworm populations. 1. A census of wireworms in pasture', *Ann. appl. Biol.*, **31**, 53–64.

SANDERS, SIR H. (1967) 'The importance of pesticides in British food production', *Proc. Roy. Soc., London* (B), **167**, 141–4.

SARTON, G. (1952) *A History of Science*, John Wiley, New York.

SAVORY, T. H. (1964) *Arachnida*, Academic Press, New York and London.

SCHAEFER, G. (1968) 'Bird recognition by radar: A study in quantitative radar ornithology', in *The Problems of Birds as Pests*, R. K. Murton and E. N. Wright (Ed.), pp. 53–86, Institute of Biology/Academic Press, New York and London.

SCHEINPFLUG, H. and H. F. JUNG (1968) 'Use of organophosphates for the control of fungal diseases of crops', *Pflanzenschutz-Nachrichten Bayer*, **21**, 79–91.

SCHMIALEK, P. (1961) 'Die Identifizierung zweier im *Tenebrio* kot und in der Hefe vorkommender Substanzen mit Juvenilhormonwirkung', *Z. Naturf.*, **16b**, 461–4.

SCHNEIDER, D. (1962) 'Electrophysiological investigation of the olfactory specificity of sexual attracting substances in different species of moths', *J. Insect. Physiol.*, **8**, 15–30.

SCHNEIDER, F.* (1969) 'Bionomics and physiology of aphidophagous syrphidae', *Ann. Rev. Entomol.*, **14**, 103–24.

SCHOONHOVEN, L. M.* (1968) 'Chemosensory bases of host plant selection', *Ann. Rev. Entomol.*, **13**, 115–36.

SCHÜTTMANN, W. (1966) 'Polyneuritis nach befuflichem Kontakt mit DDT', *Z. Ges. Hyg.*, **12**. 307–15.

SCHÜTTMANN, W. (1968) 'Chronic liver diseases after occupational exposure to dichlordiphenyltrichlorethane (DDT) and hexachlorocyclohexane (HCH)' (in German), *Intern. Arch. Gewerbepath. Gewerbehyg.*, **24**, 193–210. Abstract in *WHO mimeographed document, WHO/VBC/TOX/69. 3*.

SCHÜTZENBERGER, M. P. (1954) 'A tentative classification of goal-seeking behaviours', *J. Mental Sci.*, **100**, 97–102. Reprinted in *Systems Thinking*, F. E. Emery (Ed.), Penguin Books, Harmondsworth (1969).

SEECOF, R. (1968) 'The sigma virus infection of *Drosophila melanogaster*', *Current Topics Microbiol. Immunol.*, **42**, 59–93.

SHIPP, E. and A. W. OSBORN (1966) 'The theoretical role of predators in sterile-insect release programs', *Bull. Entomol. Soc. America*, **12**, 115–16.

SHOREY, H. H. and L. K. GASTON* (1967) 'Pheromones', in *Pest Control: Biological, Physical and Selected Chemical Methods*, W. W. Kilgore and R. L. Doutt (Eds.), pp. 241–66, Academic Press, New York and London.

SHOREY, H. H., L. K. GASTON, and R. N. JEFFERSON* (1968) 'Insect sex pheromones', *Adv. Pest Control Res.*, **8**, 57–126.

SHORTEN, M. (1954) *Squirrels*, Collins, London.

SIMMONDS, F. J. (1967) *J. R. Soc. Arts*, **115**, 880ff.

SIMMS, E. (1971) *Woodland Birds*, Collins, London.

SIMON, H. A. (1956) 'Rational choice and the structure of the environment', *Psychological Review*, **63**, 129–38.

SLÁMA, K. (1969) 'Plants as a source of materials with insect hormone activity', *Ent. Exp. and Appl.*, **12**, 721–8.

SLATER, SIR W. (Ed.)* (1967) 'A discussion on pesticides: benefits and dangers', *Proc. Roy. Soc. London (B)*, **167**, 88–163.

SLOBODKIN, L. B. (1962) *Growth and Regulation of Animal Populations*, Vanguard, New York.

SMIRNOFF, W. A. and C. F. MACLEOD (1961) 'Study of the survival of *Bacillus thuringiensis* var. *thuringiensis* Berliner in the digestive tracts and in feces of a small mammal and birds', *J. Insect Pathol.*, **3**, 266–70.

SMITH, C. N. (Ed.)* (1966) *Insect Colonisation and Mass Production*, Academic Press, New York and London.

SMITH, F. E. (1961) 'Density dependence in the Australian thrips', *Ecology*, **42**, 403–7.

SMITH, H. S. and P. DEBACH (1953) 'Artificial infestation of plants with pest insects as an aid in biological control', *Proc. 7th Pacific Sci. Congr.*, **4**, 255–9.

SMITH, K. M.* (1963) 'The cytoplasmic virus diseases', in *Insect Pathology: An Advanced Treatise*, Vol. 1, E. A. Steinhaus (Ed.), pp. 457–98, Academic Press, New York and London.

SMITH, K. M.* (1967) *Insect Virology*, Academic Press, New York and London.

SMITH, K. M. and C. F. RIVERS (1956) 'Some viruses affecting insects of economic importance', *Parasitology*, **46**, 235–42.

SMITH, R. F. and R. VAN DER BOSCH* (1967) 'Integrated control', in *Pest Control: Biological, Physical and Selected Chemical Methods*, W. W. Kilgore and R. L. Doutt (Eds.), pp. 295–340, Academic Press, New York and London.

SOCIETY OF CHEMICAL INDUSTRY* (1970) *Technological Economics of Crop Protection and Pest Control, SCI Monograph No. 36*.

SODERSTROM, E. L. (1970) 'Effectiveness of green electroluminescent lamps for attracting stored-product insects', *J. Econ. Ent.*, **63**, 726–31.

SOLOMON, M. E.* (1957) 'Dynamics of insect populations', *Ann. Rev. Entomol.*, **2**, 121–42.

SOLOMON, M. E. (1964) 'Analysis of processes involved in the natural control of insects', *Adv. in Ecol. Res.*, **2**, 1–58.

SOLOMON, M. E.* (1969) *Population Dynamics*, Edward Arnold, London.

SOUTHWOOD, T. R. E.* (1966) *Ecological Methods with Particular Reference to the Study of Insect Populations*, Methuen, London.

SOUTHWOOD, T. R. E. (Ed.)* (1968) *Insect Abundance*, Blackwell, Oxford and Edinburgh.

SOUTHWOOD, T. R. E. and W. F. JEPSON (1962) 'Studies on the populations of *Oscinella frit* L. (Dipt: Chloropidae) in the oat crop', *J. Anim. Ecol.*, **16**, 139–87.

STABLES, E. R. and N. D. NEW (1968) 'Birds and aircraft: the problems', in *The Problems of Birds as Pests*, R. K. Murton and E. N. Wright (Eds.), pp. 3–16, Institute of Biology/Academic Press, New York and London.

STAIRS, G. R. (1971) 'Use of viruses for microbial control of insects'. In Burges and Hussey, 1971a, 97–124.

STAIRS, G. R. (1972) 'Pathogenic micro-organisms in the regulation of forest insect populations', *Ann. Rev. Entomol.*, **17**, 355–72.

STANILAND, L. N. (1959) 'The principles of the hot-water treatment of plants', in *Plant Nematology*, J. F. Southey (Ed.), *Min. Agr. Fish and Food Tech. Bull. No. 7*, HMSO, London.

STARÝ, P.* (1970) *Biology of aphid parasites with respect to integrated control. Series Entomologia*, Vol. 6, Dr W. Junk NV, The Hague.

STEFFAN, A. W. (1972) 'Möglichkeiten genetischer Bekämpfung von Blattläusen (Homoptera: Aphidina)', *Z. ang. Ent.*, **70**, 267–77.

STEINHAUS, E. A.* (1949) *Principles of Insect Pathology*, McGraw-Hill, New York.

STEINHAUS, E. A. (1956) 'Potentialities for microbial control of insects', *Agri. Food Chem.*, **4**, 676–80.

STEINHAUS, E. A. (1958a) 'Stress as a factor in insect disease', *Proc. 10th Internl Congr. Ent.*, 1956, **4**, 725–30.

STEINHAUS, E. A. (1958b) 'Crowding as a possible stress factor in insect diseases', *Ecology*, **39**, 503–14.

STEINHAUS, E. A.* (1964) 'Microbial diseases of insects', in *Biological Control of Insect Pests and Weeds*, P. DeBach (Ed.), pp. 515–47, Chapman and Hall, London.

STEPHENS, J. M.* (1963) 'Immunity in insects', in *Insect Pathology: An Advanced Treatise*, Vol. 1, E. A. Steinhaus (Ed.), pp. 273–98, Academic Press, New York and London.

STRICKLAND, A. H.* (1961) 'Sampling crop pests and their hosts', *Ann. Rev. Entomol.*, **6**, 201–20.

STRONG, W. M. (1970) 'Pesticides as an agricultural input', in *Technological Economics of Crop Protection and Pest Control. Society of Chemical Industry Monograph No. 36*.

SUDD, J. H. (1967) *An Introduction to the Behaviour of Ants*, Edward Arnold, London.

SWEETMAN, L. L.* (1958) *The Principles of Biological Control*, Wm. C. Brown, Dubuque, Iowa.

TANADA, Y.* (1959) 'Microbial control of insect pests', *Ann. Rev. Entomol.*, **4**, 277–302.

TANADA, Y.* (1963) 'Epizooitology of infectious diseases', in *Insect Pathology: An Advanced Treatise*, Vol. 2, E. A. Steinhaus (Ed.), pp. 423–76, Academic Press, New York and London.

TANADA, Y.* (1964) 'Epizooitology of insect diseases', in *Biological Control of Insect Pests and Weeds*, P. DeBach (Ed.), pp. 548–78, Chapman and Hall, London.

TANADA, Y.* (1967) 'Microbial pesticides', in *Pest Control: Biological, Physical and Selected Chemical Methods*, W. W. Kilgore and R. L. Doutt (Eds.), pp. 31–88, Academic Press, New York and London.

TATTON, J. O'G and J. H. A. RUZICKA (1967) 'Organochlorine pesticides in Antarctica', *Nature*. **215**, 346–8.

TELLE, O. (1970) 'Comments on the use of pesticides from the aspect of physical and physico-chemical properties', *Pflanzenschutz-Nachrichten Bayer*, **23**, 111–23.

THOMPSON, H. V. (1961) 'Myxomatosis', in *Penguin Science Survey 1961. 2*, S. A. Barnett and A. McLaren (Eds.), pp. 102–11, Penguin Books, Harmondsworth.

THOMPSON, W. R.* (1956) 'The fundamental theory of natural and biological control', *Ann. Rev. Entomol.*, **1**, 379–402.

TINKER, J. (1971) 'The PCB story: seagulls aren't funny any more', *New Scientist*, 1 April 1971, 16–18.

TURNBULL, A. L. and D. A. CHANT (1961) 'The practice and theory of biological control of insects in Canada', *Canad. J. Zool.*, **39**, 697–753.

TURNER, J. N. (1959) 'Control of fungus diseases of fruit in storage', *Outlook on Agriculture*, **2**, 229–37.

UNTERSTENHÖFFER, G. (1970) 'Integrated pest control from the aspect of industrial research on crop protection chemicals', *Pflanzenschutz-Nachrichten Bayer*, **23**, 264–72.

VAGO, C. (1968) 'Non-inclusion viruses diseases of invertebrates', *Current Topics Microbiol. Immunol.*, **42**, 24–37.

VALLEGA, J. (1956) *Rep. 3rd Int. Wheat Rust Conf., Mexico*, **32** (cited in Lupton, 1967).

VALLEGA, J. (1959) *Robigo*, **8**, 7 (cited in Lupton, 1967).

VANDENBOSCH, R. and V. M. STERN.* (1962) 'The integration of chemical and biological control of arthropod pests', *Ann. Rev. Entomol.*, **7**, 367–86.

VAN DEN BOSCH, R. and A. F. TELFORD* (1964) 'Environmental modification and biological control', in *Biological Control of Insect Pests and Weeds*, P. DeBach (Ed.), pp. 489–514, Chapman and Hall, London.

VAN DER LAAN, P. A. (Ed.)* (1967) *Insect Pathology and Microbial Control. Proc. Interntl Colloquium on Insect Pathology and Microbial Control, Wageningen – Sept. 5–10 1966*, North-Holland, Amsterdam.

VAN DER PLANK, J. E. (1963) *Plant diseases: epidemics and control*, Academic Press, New York and London.

VAN EMDEN, H. F. and C. H. WEARING (1965) 'The role of the aphid host plant in delaying economic damage levels in crops', *Ann. appl. Biol.*, **56**, 323–4.

VANWIJNGAARDEN, A. (1953) 'Onderzoek van de biologie van de woelrat, *Arvicola terrestris*', *Versl. en Meded. Pl. ziektenk. Dienst.*, No. 120, 227–8.

VANWIJNGAARDEN, A. (1954) *Biologie en Bestrijding van de Woelrat, Arvicola terrestris terrestris (L.) in Nederland*. Doctoral Thesis at the University of Leiden, 7 July 1954. Reprinted in *Collected Reprints* 1954. Plant Protection Service of the Netherlands, Wageningen.

VARLEY, G. C. (1947) 'The natural control of population balance in the knapweed gall-fly (*Urophora jaceana*)', *J. Anim. Ecol.*, **16**, 139–87.

VARLEY, G. C. and G. R. GRADWELL (1965) 'Interpreting winter moth population changes', *Proc. Int. Congr. Ent. 12 London, 1964*, 377–8.

VARLEY, G. C. and G. R. GRADWELL (1968) 'Population models for the winter moth', in *Insect Abundance*, T. R. E. Southwood (Ed.), pp. 132–42, Blackwell, Oxford and Edinburgh.

VARLEY, G. C. and G. R. GRADWELL* (1970) 'Recent advances in insect population dynamics', *Ann. Rev. Entomol.*, **15**, 1–24.

WAGNER, R. H. (1971) *Environment and Man*, Norton, New York.

WALOFF, N. and K. BAKKER (1963) 'The flight activity of Miridae (Heteroptera) living on broom, *Sarothamnus scoparius* (L.) Wimn', *J. Anim. Ecol.*, **32**, 461–80.

WALTON, C. L. (1947) *Farmers' Warfare*, Crosby Lockwood, London.

WASSERMANN, M., D. WASSERMANN, L. ZELLERMAYER, and M. GON (1967) 'Storage of DDT in the people of Israel', *Pesticides Monitoring J.*, **1**, 15–20.

WATERHOUSE, D. F. (1971) 'Insects and Australia', *J. Aust. ent. Soc.*, **10**, 145–60.

WATERHOUSE, D. F. and F. WILSON* (1968) 'Biological control of pests and weeds', *Science Journal*, **4** (12), 31–7.

WATERMAN, T. H. and H. J. MOROWITZ (Eds.)* (1965) *Theoretical and Mathematical Biology*, Blaisdell, New York, Toronto, and London.

WATT, K. E. F.* (1961) 'Mathematical models for use in insect pest control', *Canad. Ent. Suppl.* No. 19, **93**, pp. 195–230.

WATT, K. E. F.* (1963a) 'Mathematical population models for five agricultural crop pests', *Mem. Ent. Soc. Canada*, **32**, 83–91.

WATT, K. E. F.* (1963b) 'Dynamic programming, "look ahead programming", and the strategy of insect pest control', *Can. Ent.*, **95**, 525–36.

WATT, K. E. F.* (1965) 'Community stability and the strategy of biological control', *Can. Ent.*, **97**, 887–95.

WATT, K. E. F. (Ed.)* (1966) *Systems Analysis in Ecology*, Academic Press, New York and London.

WATT, K. E. F.* (1968) *Ecology and Resource Management*, McGraw-Hill, New York.

WEHRHAHN, C. F. and W. KLASSEN* (1971) 'Insect control methods involving the release of of relatively few laboratory-bred insects', *Can. Ent.*, **103**, 1387–96.

WEIDHAAS, D. E.* (1968) 'Field development and evaluation of chemosterilants', in *Principles of Insect Chemosterilization*, G. C. LaBreque and C. N. Smith (Eds.), pp. 275–314, Appleton-Century-Crofts, New York.

WEISER, J.* (1963) 'Sporozoan infections', in *Insect Pathology: An Advanced Treatise*, Vol. 2, E. A. Steinhaus (Ed.), pp. 291–334, Academic Press, New York and London.

WEISER, J.* (1969) *An Atlas of Insect Diseases*, Irish University Press, Shannon.

WEISER, J.* (1970) 'Recent advantages in insect pathology', *Ann. Rev. Entomol.*, 15, 245–56.

WEISER, J. and J. D. BRIGGS (1971) 'Identification of pathogens'. In Burges and Hussey, 1971a, 13–66.

WEISER, J. and J. VEBER (1957) 'Die Mikrosporidie *Thelohania hyphantriae* Weiser des weissen Bärenspinners und anderer Mitglieder seiner Biocönose', *Zeitschr. angew. Ent.*, 40, 55–76.

WEITZ, B. (1964) 'Feeding habits of tsetse flies', *Endeavour*, 13, No. 88, 38–42.

WELCH, H. E.* (1963) 'Nematode infections', in *Insect Pathology: An Advanced Treatise*, Vol. 2, E. A. Steinhaus (Ed.), pp. 363–92, Academic Press, New York and London.

WELCH, H. E.* (1965) 'Entomophilic nematodes', *Ann. Rev. Entomol.*, 10, 275–302.

WEST, I. (1967) 'Lindane and hematologic reactions', *Arch. environm. Hlth*, 15, 97–101.

WHEATLEY, G. A. and T. H. COAKER* (1970) 'Pest control objectives in relation to changing practices in agricultural crop production', in *Technological Economics of Crop Protection and Pest Control, Society of Chemical Industry, Monograph No. 36*.

WHITE, J. H. (1968) 'Reports of syndicates: cereals, grasses and fodder legumes'. In D. W. Empson (Ed.) 1968, 499–502.

WHITE, M. J. D. (1961) *The Chromosomes*, 5th Edn., Methuen, London.

WHITE, G. B. (1967) 'Control of insects by sexual sterilization', in *Penguin Science Survey 1967. The Biology of Sex*, A. Allison (Ed.), pp. 201–21, Penguin Books, Harmondsworth.

WHITE, M. J. D. (1970) 'Cytogenetics', in *The Insects of Australia*, CSIRO, Canberra, pp. 72–82, Melbourne University Press, Melbourne.

WHITTEN, J. L. (1966) *That we may live*, Van Nostrand, New York.

WHITTEN, M. J. (1971) 'Insect control by genetic manipulation of natural populations', *Science*, 171, 682–4.

WHITTEN, M. J. and K. R. NORRIS (1967) 'Booby-trapping as an alternative to sterile males for insect control', *Nature, Lond.*, 216, 1136.

WIGGLESWORTH, SIR V. et al.* (1965) 'Symposium on some approaches towards integrated control of British insect pests', *Ann. appl. Biol.*, 56, 315–50.

WILLIAMS, C. M. (1967) 'Third generation pesticides', *Sci. Amer.*, 217 (1), 13–17.

WILSON, C. L.* (1969) 'Use of plant pathogens in weed control', *Ann. Rev. Phytopath.*, 7, 411–34.

WILSON, E. O.* (1963) 'Pheromones', *Scient. Amer.*, 208, 100–14.

WILSON, F.* (1964) 'The biological control of weeds', *Ann. Rev. Entomol.*, 9, 225–44.

WOOD, D. L., L. E. BROWNE, R. M. SILVERSTEIN, and J. O. RODIN (1966) 'Sex pheromones of bark beetles. I: Mass production, bioassay, source and isolation of the sex pheromone of *Ips confusus* (LeC.)', *J. Insect Physiol.*, 12, 523–36.

WOOD, D. L., R. M. SILVERSTEIN, and M. NAKAJIMA (Eds.)* (1970) *Control of Insect Behaviour by Natural Products*, Academic Press, New York and London.

WOODWELL, G. M. (1967) 'Toxic substances and ecological cycles', *Scient. Amer.*, 216, 24–31.

WHO (1964) *Equipment for Vector Control*, World Health Organization, Geneva.

WHO* (1971) 'The place of DDT in operations against malaria and other vector-borne diseases' *WHO mimeographed document, WHO/VBC/TOX./71.7*.

WHO EXPERT COMMITTEE ON INSECTICIDES (1957) *Wld Hlth Org. Techn. Rep. Ser.*, 125.

WORMALD, H. (1955) *Diseases of Fruits and Hops*, 3rd Edn., Crosby Lockwood, London.

WRIGHT, D. P., Jr. (1963) 'Antifeeding compounds for insect control', in *New Approaches to Pest Control and Eradication. Advances in Chemistry*, Series 41, S. A. Hall (Ed.), pp. 56–63, American Chemical Society, Washington, D.C.

WRIGHT, D. P., Jr.* (1967) 'Antifeedants', in *Pest Control: Biological, Physical and Selected Chemical Methods*, W. W. Kilgore and R. L. Doutt (Eds.), pp. 287–94, Academic Press, New York and London.

WRIGHT, D. W., R. D. HUGHES, and J. WORRAIL (1960) 'The effect of certain predators on the numbers of cabbage root fly *E. brassicae* Bouche and the subsequent damage caused by the pest', *Ann. appl. Biol.*, 48, 756–63.

WRIGHT, E. N. (1968) 'Modification of the habitat as a means of bird control', in *The Problems of Birds as Pests*, R. K. Murton and E. N. Wright (Eds.), pp. 97–105, Institute of Biology/ Academic Press, New York and London.

WRIGHT, J. W., R. F. FRITZ, and J. HAWORTH* (1972) 'Changing concepts of vector control in malaria eradication', *Ann. Rev. Entomol.*, **17**, 75–102.

WRIGHT, R. H. (1958) 'The olfactory guidance of flying insects', *Can. Ent.*, **90**, 81–9.

WRIGHT, J. W. and R. PAL* (1967) *Genetics of Insect Vectors of Disease*, Elsevier, Amsterdam.

WRIGHT, S. (1941) 'On the probability of fixation of reciprocal translocations', *Am. Nat.*, **75**, 513–22.

WURSTER, C. F. (1968) 'DDT reduces photosynthesis by marine plankton', *Science*, **159**, 1474–5.

WYNNE-EDWARDS, V. C.* (1962) *Animal Dispersion in Relation to Social Behaviour*, Oliver and Boyd, Edinburgh and London.

YAMAMOTO, I. (1970) 'Problems in the mode of action of pyrethroids', in *Biochemical Toxicology of Insecticides*, R. D. O'Brien and I. Yamamoto (Eds.), pp. 115–30, Academic Press, New York and London.

YASUMATSU, K. and T. TORII* (1968) 'Impact of parasites, predators and diseases on rice pests', *Ann. Rev. Entomol.*, **13**, 295–324.

ZWÖLFER, H. (1963) 'The structure of the parasite complexes of some Lepidoptera', *Z. angew. Ent.*, **97**, 887–95.

ZWÖLFER, H. (1970) 'Current investigations on phytophagous insects associated with thistles and knapweeds (Dicotyledones)', *Misc. Publ. Commonw. Inst. Biol. Contr. Trinidad* (*Proc. 1st Int. Symp. Biol. Contr. Weeds, 1969*), No. 1, 63–7.

ZWÖLFER, H. and P. HARRIS (1971) 'Host-specificity determination of insects for biological control of weeds', *Ann. Rev. Entomol.*, **16**, 159–78.

Supplementary references

ASCHER, K. R. S.* (1964) 'A review of chemosterilants and oviposition inhibitors in insects', *World Rev. Pest Control*, **3** (1), 7–27.

BEIRNE, B. P.* (1967) 'Biological control and its potential', *World Rev. Pest Control*, **6** (1), 7–20.

BERRIE, A. D. (1970) 'Snail problems in African schistosomiasis', *Advances in Parasitol*, **8**, 43–96.

BIRCH, L. C. (1963) 'Population ecology and the control of pests', *Bull. Wld Hlth Org.*, **29** (Suppl.), 141–6.

BOŘKOVEC, A. B. (1969) 'Aokylating agents as insects chemosterilants', *Ann. N.Y. Acad. Sciences*, **163** (Art. 2), 860–8.

CAMERON, J. W. MACBAIN* (1963) 'Factors affecting the use of mocrobial pathogens in biological control', *Ann. Rev. Entomol.*, **8**, 265–86.

DEWILDE, J. and L. M. SCHOONHOVEN (Ed.)* (1969) *Proc. 2nd Interntl Symposium*, 'Insect and the Host Plant', *Ent. exp. and appl.*, **12** (5), 471–810.

DOBROTWORSKY, N. V. (1967) 'Hybridization in the *Culex pipiens* complex', *Bull. Wld Hlth Org.*, **37**, 267–70.

FRAENKEL, O. H. *et al.** (1965) 'New perspectives in insect control', *Australian J. of Science*, **28**, 217–47.

HAYES, W. J., Jr. (1964) 'The toxicology of chemosterilants', *Bull. Wld Hlth Org.*, **31**, 721–36.

HEIMPEL. A. M.* (1965) 'Microbial control of insects', *World Rev. Pest Control*, **4**, 150–61.

HUFFAKER, C. B. (1970) 'Life against life – Nature's pest control scheme', *Environmental Research*, **3**, 162–75.

INTERNATIONAL ATOMIC ENERGY AGENCY* (1968) *Control of Livestock Insect Pests by the Sterile-Male Technique*, International Atomic Energy Agency, Vienna.

JENSON, A. G. (1965) 'Proofing of buildings against rats and mice', Ministry of Agriculture, Fisheries and Food, *Tech. Bull. No. 12*, HMSO, London.

KENNEDY, J. S. (1965) 'Mechanisms of host plant selection', *Ann. appl. Biol.*, **56**, 317–22.

LEE, D. L. (1965) *The Physiology of Nematodes*, Oliver and Boyd, Edinburgh.

MADELIN, M. F.* (1966) 'Fungal parasites of insects', *Ann. Rev. Entomol.*, **11**, 423–48.

MILANI, R. (1963) 'Genetical aspects of insecticide resistance', *Bull. Wld Hlth Org.*, **29** (Suppl.), 77–87.

OPPENOORTH, F. J. (1965) 'Biochemical genetics of insecticide resistance', *Ann. Rev. Entomol.*, **10**, 185–206.

SMITH, C. N. (1963) 'Prospects for vector control through sterilization procedures', *Bull. Wld Hlth Org.*, **29** (Suppl.), 99–106.

SMITH, C. N. (1967) 'Possible use of the sterile-male technique for control of *Aedes aegypti*', *Bull. Wld Hlth Org.*, **36**, 633–5.

WATT, K. E. F.* (1973) *Principles of Environmental Science*, McGraw-Hill Book Co., New York, London, and Sydney.

VAN EMDEN, H. F. (1966) 'Plant insect relationship and pest control', *World Rev. Pest Control*, **5** (2), 115–23.

Subject index 1

O,O-Diethyl *S*-(2,5-dichlorophenylthiomethyl) phosphorothiolothionate (*see* Phenkapton)

Diethyl *S*-[2-(ethylthio) ethyl] phosphorothiolothionate (*see* Disulfoton)

Diethyl-2-isopropyl-6-methyl-4-pyrimidinyl phosphorothionate (*see* Diazinon)

Diethyl-4-nitrophenyl phosphorothionate (*see* Parathion, ethyl)

Diethyl-*O*-2-pyrazinyl phosphorothionate (*see* Thionazin)

N,N-diethyl-*m*-toluamide, 293, 294

2-3-Dihydro-5-carboxanilido-6-methyl-1,4-oxathiin, (*see* Vitavax®)

9,10-Dihydro-8a,10a-diazoniaphenanthrene cation (*see* Diquat)

S-(3,4-Dihydro-4-oxobenzo[*d*] (1,2,3)-triazin-3-yl) dimethyl phosphorothiolothionate (*see* Azinphos methyl)

Dimethirimol, 350

Dimethoate, 76

Dimethrin, 84

Dimethylarsinic acid, 284

4-Dimethylamino-3,5-xylyl-*N*-methyl carbamate (*see* Zectran®)

1,1′-Dimethyl-4,4′-bipyridylium ion (*see* Paraquat)

Dimethyl S-(*N*-methylcarbamoylmethyl) phosphorothiolothionate (*see* Dimethoate)

Dimethyl phthalate, 293, 294

Dimethyl 2,4,5-trichlorophenyl phosphorothionate (*see* Fenchlorphos)

4′(Dimethyltriazino) acetanilide (*see* DTA)

Dinex, 104

Dinitro-o-cresol (*see* DNOC)

Dinitro-*O*-cyclohexylphenol (*see* Dinex)

Dinocap, 87–8

Dinocton-O, 88

Dinoseb, 114

Diphacinone, 107

2-Diphenylacetyl indane-1,3-dione (*see* Diphacinone)

Diquat, 113–14

Disease vectors, integrated control, 351–2

Diseases:
  human, transmission, 4–5

immunity in insects:
  cellular, 229–31
  humoral, 229–31
  latent, 225
  outbreak prediction, 240

Disparlure, 281

Distress calls, 296–7

Disul, 112

Disulfoton, 76

Dithiocarbamate fungicides, 96

DMF, 91, 256

DNA synthesis, 254

DNOC, 80, 114

cis-7-Dodecenyl acetate, 281

Dodecylguanadine acetate (*see* Dodine)

Dodine, 97

Dominican Republic, pesticide poisoning, 134

Dowco 199®, 98

2,4-DP, 112

DTA, 81

Dusts, 65

Earthworms, effects of soil pesticides, 142

Ecdysones, 90–2

EC 50, 254

Ecological control, 312, 320–4

Economic injury level, 13

Economic threshold, 1

Economic threshold level, 13

Economics of control:
  biological control, 18–19
  chemical control, 12–20
  microeconomic models, 15–18
  social costs, 19

Ecosystem stability, 10

Ectohormones (*see* Pheromones)

Education for integrated control practitioners, 355–6

Egg shells, thinning by pesticides, 147

Egypt:
  bilharzia, 11
  biological control, 192

Electromagnetic radiation, 325–7

Endrin, 72
  human poisoning, 134

Engineering works – ecological effects, 11–12

Environment:
  meaning of the concept, 22, 34
  importance in population dynamics, 34–5
  pesticides, impact, 140–52

Environmental modification:
  to control tsetse flies, 323–4
  to harm pests, 320–4

Enzootic disease, definition, 225

Epidemiology (*see also* Epizootiology):
  mathematical models, 233–4

Epizootic, definition, 225

Epizootic disease, 225

Epizootic strain, 226

Epizootiology, 225–34
  influence of biotic factors, 232
  influence of humidity, 231–2
  influence of sunlight, 231–7
  influence of temperature, 231–7
  mathematical models, 233–4
  role of the host, 228–31
  spatial and temporal development, 232–6

cis-7,8-Epoxy-2-methyl-octadecane (*see* Disparlure)

Ergotism, 5–6

Ethylene chlorobromide, 92

Ethylene dibromide, 92, 103

Ethylene oxide, 93

2-Ethyl-1,3-hexanediol, 293, 294

Ethyl-phenyl ethylphosphonodithioate (*see* Fonofos)

S-[2-(Ethylsulphinyl) ethyl] dimethyl phosphorothiolate (*see* Oxydemeton methyl)

2-(Ethylthio)-ethyl dimethyl phosphorothiolate (*see* Demeton-O-methyl)

Eulan CN®, 82

Europe:
  cereal rusts, 303–5
  ecological control, 321
  rotations, 313

Farm insects (Curtis), 12

Farnesol, 90

Fecundity, 25

Feedback, negative, in population dynamics, 28

Feeding calls, 296

Feeding repellents:
  birds, 298
  insects, 293–4
  mammals, 298

Fences: rabbit, dingo and shark, 328–9

Fenchlorphos, 77

Fenson, 86

Fentin acetate, 81

Fertility, 25

Fertilizers, effects on susceptibility, 319

Ferulic acid, 308

N-trityl morpholine (*see* Frescon®)
Tsutsugamushi disease, economic importance, 9
Typhus (*see also* Naples), 151
  economic importance, 9
Typhus, scrub, economic importance, 9

Uganda, biological control, 207
Ultrasonic transducers, 295, 296
Ultrasound, bats and moths, 295
Ultra-violet radiation as attractant, 290–2
'Undercrowding', 25
United Kingdom:
  apple pests, 336–7
  biological control, 182, 197, 218, 244–5, 246
  hops, 20
  insect predation by birds, 201–2
  integrated control, 335, 336, 347–51
  nematode control, 314–16
  population replacement, 277
  quarantine, 330
  resistant varieties:
    potato root eelworm, 306
    wart disease of potatoes, 302
  rotations and nematodes, 314–15
  sterile-male technique, 263
  transhumation, 316
United Nations Fund for Drug Abuse Control, 171
'Untaken acre', 7
USA (*see also* various states and dependencies):

biological control, 183, 187, 190, 197, 209, 210, 215, 218, 240, 241, 243, 246
cereal rusts, 304–5
cultivations, 314, 318
ecological control, 321
genetic control, 274, 276
integrated control, 347
introduced pests, 11
light traps, 291
lucerne (alfalfa) cutting, 318
loss of fruit in storage and transit, 8
pasture spelling, 316
quarantine, 330
resistant varieties:
  cereal rusts, 302–5
  hessian fly, 309, 310
  stem sawfly, 309, 310
sterile-male technique, 249, 261–2, 263, 264, 265
water management, 319
US Plant Quarantine Act, 330
USSR, biological control, 183, 209, 241, 246

Venoms, 92
Vertebrates:
  biological control, 204–6
  chemical control, 105–8
Vibration as an attractant, 292
Victim, definition, 4
Virion, 212
Virulence, 226
Viruses:
  commercial production, 215
  effects on insect behaviour, 213, 228
  granuloses, 213–14
  plant, transmission, 4
  polyhedroses, 212–13
Visual attractants, 292
  (*see also* Light, Ultra-violet radiation)

Vitamin K, 107
Vitavax®, 100–1
Wales, endrin poisoning, 134
WARF antiresistant, 130
Warfarin, 107
  resistance, 130–1
Water pollution for mosquito control, 352
Weed Control Handbook, 109
Weeds (*see also* Prickly pear):
  alternative hosts for pests, 5
  biological control, 170–1, 246–7
    dangers of host-plant transfer, 168–9
    vertebrate agents, 170–1
  growth inhibition of competitors, 5
Wepsyn®, 98
West Indies and Bahamas, biological control, 184, 188, 205, 219
Whitten effect in mice, 280
Windbreaks, influence on pests, 339
Wisconsin Alumni Research Foundation, 107
Wood borers:
  heat control, 327
  physical pesticides, 325–6
  sinking ship, 4
Wool-proofing, 82
World Health Organization, 131, 134, 205
World Review of Pest Control, 73, 115

Xenic culture, 237

Yellow fever, 4, 137

Zambia, biological control, 216
Zectran®, 79
Zinc phosphide, 106
Zineb, 96
Ziram, 96

Scientific index 2

The scientific names of those species which have well-known vernacular names are given first. This list is followed by the names of species listed under genus, and of higher taxa. Taxa higher than superfamily are printed in capitals.

Many of the more important economic species of insect, such as the codling moth, have a variety of scientific names in common use though only one is correct according to the International Rules on Zoological Nomenclature. I have made no attempt to ensure that all the names given are completely up to date; I have, instead, tried to use the name which is probably the best known to economic biologists. In some cases, however, two alternative names may be listed, e.g., *Cydia pomonella* and *Carpocapsa pomonella*.

Viruses are listed in the second part of the index under 'virus'.

Some taxa are also listed in the subject index. These listings refer to the more important sections on the organisms concerned (e.g., prickly pears).

Gulls, sea, *Laridae*
Guppy, *Lebistes reticulatus*
  (*Poecilia reticulata*)

Hare:
  blue, *Lepus timidus*
  brown, *L. europaeus*
  snowshoe, *L. americanus*
Hawk, sparrow, *Accipiter nisus*
Heather, *Calluna vulgaris*
Hemp, *Cannabis sativa*
Hemp nettle, *Galeopis speciosa*
Hyacinth, water, *Eichornia crassipes*

Isle of White disease, *Acarapis woodii*

Jassid, cotton, *Empoasca lybica*

Kestrel, *Falco tinnunculus*
Killifish, salt-marsh, *Fundulus confluentus*
Knotgrass, *Polygonum aviculare*
Krill, *Ephausia* spp.
Kudu, *Strepsiceros* spp.

Lacewings:
  brown, Hemerobiidae
  green, Chrysopidae
  powdery, Coniopterygidae
Lamprey, *Petromyzon marinus*
Leaf-curl, peach, *Taphrina deformans*
Leaf-hopper, cotton, *Psallus seriatus*
Leaf-hopper, sugar cane, *Perkinsiella saccharicida*
Leaf-hoppers, grape, *Erythroneura* spp.
Lice, PHTHIRAPTERA
Lizard, mangrove monitor, *Varanus indicus*
Locusts, Acrididae; (*see also* *Austroicetes cruciata* *Locusta migratoria* *Locustana pardalina* *Nomadacris septemfasciata* *Schizocerca gregaria*)
Louse, human body, *Pediculus humanus humanus*
Lucerne flea, *Sminthurus viridis*

Manatee, *Trichechus manatus*
Melanose, citrus, *Diaporthe citri*
Midges, biting, Ceratopogin-idae

Mildew:
  American gooseberry mildew, *Sphaerotheca mors-uvae*
  apple powdery mildew, *Podosphaera leucotricha*
  cereal mildew, *Erisyphe graminis*
  grape downy mildew, *Plasmopara viticola*
  grape powdery mildew, *Oidium tuckeri* (*Uncinula necator*)
  hop powdery mildew, *Sphaerotheca macularis*
  powdery mildews, Erisyphaceae
Milky diseases of Japanese beetle:
  Type A, *Bacillus popilliae*
  Type B, *B. lentimorbus*
Millipedes, DIPLOPODA
Miners, leaf, Agromyzidae
Mink, *Mustela vison*
Mite:
  apple rust mite, *Aculus schlechtendali*
  citrus red spider mite, *Panonychus citri*
  cyclamen mite, *Steneotarsonemus pallidus*
  depluming mite of poultry, *Chemidocoptes laevis* var. *gallinae*
  follicle mite, *Demodex folliculorum*
  french fly mite, *Tyrophagus longior*
  fruit-tree red spider mite, *Panonychus ulmi*
  glasshouse red spider mite, *Tetranychus urticae*
  red-legged earth mite, *Halotydeus destructor*
  scabies mite, *Sarcoptes scabiei*
  scarlet velvet mite, *Allothrombidium* spp.
Mites, ACARI
Mites, red spider, Tetranychidae
Mole, European, *Talpa europea*
Mole-cricket, African, *Gryllotalpa africana*
Mongoose, *Mungos birmanicus*
Mormon cricket, *Anabrus simplex*
Mosquito, yellow-fever, *Aedes aegypti*
Mosquito fish, *Gambusia affinis*
Mosquitoes, Culicidae

Moth:
  African army-worm, *Spodoptera exempta*
  Angoumois grain moth, *Sitotroga cerealella*
  beet army-worm, *Spodoptera exigua*
  bollworm, *Heliothis zea*
  brown-tail moth, *Nygmia phaeorrhoea*
  cabbage looper, *Trichoplusia ni*
  coconut moth, *Levuana iridescens*
  codling moth, *Cydia pomonella*
  corn ear-worm, *Heliothis zea*
  cotton leaf-worm (Old World), *Spodoptera* (*Prodenia*) *litura*
  cotton leaf-worm (USA), *Alabama argillacea*
  dark cutworm, *Agrotis ypsilon*
  dark sword-grass moth, *Agrotis ypsilon*
  diamondback moth, *Plutella maculipennis*
  European corn-borer, *Ostrinia nubilalis*
  eye-spotted bud moth, *Spilonota ocellana*
  fall webworm, *Hyphantria cunea*
  fir bud-worm, *Choristoneura murinana*
  forest tent caterpillar, *Malacosoma disstria*
  fruit-tree leaf-roller, *Archips argyrospilla*
  grape-berry moth, *Clysia ambiguella* and *Polychrosis botrana*
  grape vine moth, *Lobesia botrana*
  gypsy moth, *Porthetria dispar*
  light-brown apple moth, *Epiphyas* (*Tortrix*) *postvittana*
  Mediterranean flour moth, *Anagasta kühniella*
  nun moth, *Lymantria monacha*
  Oriental fruit moth, *Grapholita* (*Cydia*) *molesta*
  pea moth, *Laspeyresia nigricana*
  pine looper, *Bupalus piniarius*
  pine processionary moth, *Thaumaetopoea pityocampa*

Spider, Sydney funnel-web, *Atrax robustus*
Spiders, ARANEAE
Spindle, *Euonymus europaeus*
Springtails, COLLEMBOLA
Squill, red, *Urginea (Squilla) maritima*
Starling, European, *Sturnus vulgaris*
Stoat, European, *Mustela erminea*
Sucker, apple, *Psylla mali*
Sundews, *Drosera* spp.
Swinecress, *Coronopus* spp.

Thrips, THYSANOPTERA
Thrush, olive-backed, *Hylochichla ustulata swainsoni*
Tick:
    cattle (USA), *Boophilus annulatus*
    cattle (Australia), *B. microplus*
    sheep (UK), *Ixodes ricinus*
Ticks, ACARI

Titmice, Paridae
Toads, *Bufo* spp.

Vole, water, *Arvicola terrestris terrestris*

Wart disease of potatoes, *Synchytrium endobioticum*
Warthog, *Phacochoerus aethiopicus*
Wasp:
    citrus gall wasp (Australia), *Eurytoma fellis*
    European wasps, *Vespa* spp.
    wood wasps, *Sirex* spp.
    'Yellow Jack', *Polistes* sp.
Water lily, *Nymphaea odorata*
Water shield, *Brasenia schreberi*
Weasel, European, *Mustela nivealis*
Weasel, Japanese, *M. sibricica itatsi*
Weevil:
    apple-blossom weevil, *Anthonomus pomorum*

apple twig cutter, *Rhynchites coeruleus*
banana root weevil, *Cosmopolites sordidus*
boll weevil, *Anthonomus grandis*
clover leaf weevil, *Hypera postica*
New Guinea sugar cane weevil, *Rhabdoscelus obscurus*
rice weevil, *Sitophilus oryzae*
Whiteflies, Aleyrodidae
Whitefly, glasshouse, *Trialeurodes vaporariorum*
Wire-worms, Elateridae
Wilt of flax, *Fusarium oxysporum* f. *lini*
Wilt of oak, *Ceratocystis fagacearum*
Woodlice, ISOPODA

Yellows, cabbage, *Fusarium oxysporum* f. *conglutinans*

Index 3

Aspergillus flavus, 6
A. parasiticus, 6
Aspergillus spp., 219
Aspidiotiphagus citrinus, 182
Aspidiotus destructor, 182, 192
Athalia rosae, 12
Athene noctua, 200
Atrax robustus, Plate 11
Austroicetes cruciata, 31–2
Automeris spp., 282
Avena fatua, 313
AVES, 200–2, 321–2
Azolla spp., 247

BA 068, 212
Babesia bigemina, 316
Bacillus cereus, 210, 229
B. fribourgensis, 210
B. lentimorbus, 210, 227, 240–1
B. l. var. australis, 210
B. popilliae, 210, 227, 235, 240–1
Bacillus spp., 210, 238
B. subtilis, 246
B. thuringiensis, 210–12, 226, 230, 232, 236–9, 241, 345, 347, 354
B. t. var. dendromilus (var. sotto), 227
BACTERIA, 209–12, 226, 228, 246
Bacterium Strain A 180, 246
Bacterium amylovorus, 11, 99, 246
Balantidium spp., 223
Barbus spp., 207
Barrouxia spp., 221
Beauveria bassiana, 209, 218–19, 227, 236, 240
Bembidion lampros, 190
Berberis vulgaris, 5, 302–5, 303 (Fig. 10.1). 342
Bessa remota, 188, 193
Beta patellaris, 315
Biscirus lapidarius, 199
Blaesoxipha filipjevi, 189
B. lineata, 189
Blarina brevicauda, 203
Blastothrix sericea, 183
Blatella germanica, 128, 253
BLATTARIA, 72, 80, 129, 185, 206, 223, 279, 280, 294, 317
Blepharidoptrus angulatus, 132, 195
Blissus leucopterus, 209, 219, 240, 319
Bombylidae, 189
Bombyx mori, 90, 209, 212, 222, 230, 281, Plate 27
Boophilus annulatus, 316
B. microplus, 316, 336
Boophilus spp., 125
Bothrops atrox, 205

Botrytis cinerea, 96
Botrytis spp., 97, 99
Braconidae, 180–1
Bradynema spp., 224
Brahmaea spp., 282
Brasenia schreberi, 247
Brasenia spp., 247
Brevicoryne brassicae, 6, 342, 349
Brugia patei, 264
Bufo bufo, 206
B. marinus, 205, 206
Bupalus piniarius, 33–4, 358–9

Cacoecia argyrospila, 36
Cactoblastis cactorum, 162–3, 168, 179, 222, 246
Calosoma sycophanta, 190
Callimeris arcufer, 193
Calliphoridae, 189
Calluna vulgaris, 33
Campylomma verbasci, 195
Canis antarcticus, 328–9
Canis latrans, 204
Cannabis sativa, 171
Cantheconidea furcellata, 195
Carabidae, 190
Carpocapsa pomonella (see Cydia pomonella)
Catharacta skua maccormicki, 149
Cecidomyidae, 189
Cephalosporium diospyri, 219, 247
Cephus cinctus, 6, 309, 311, 318, 319
C. pygmaeus, 6
Ceratitis capitata, 118, 253, 262, 277, 288, 322, 330, 343
Ceratocystis fagacearum, 247
Ceratopoginidae, 144
Cercospora coffeicola, 319
C. eupatorii, 247
Ceromasia sphenophori, 161
Ceroplastes destructor, 81, 118
C. rubens, 81, 118
Chagasella spp., 221
Chalcidoidea, 181–4
Chaoborus spp., 144
Chemidocoptes laevis var. gallinae, 88
Chenopodium album, 313, 314
Cheyletiella parasitivorax, 199
Cheyletus eruditus, 198
Chilo spp., 196
Chilo suppressalis, 183–4
Chironomidae, 207, 223–4
CHIROPTERA, 295
Chlorops taeniopus, 317
Choristoneura muriana, 165, 201
C. fumiferana, 45, 222
Chortophila brassicae (see Erioischia brassicae)

Chryomphalus aonidum, 118
Chrysolina quadrigemina, 170
Chrysanthemum cinerariaefolium, 83
Chrysopa californica, 193–4
C. carnea, 193, 351
Chrysopa spp., 194 (Fig. 6.7), 213
Chrysopidae, 193–4
Cicadidae, 218
CILIATA, 223
Cimex lectularius, 69, 125, 129, 195–6, 290
Circulifer tenellus, 163
Claviceps purpurea, 5
Cleridae, 193
Cloaca cloaca var. acridiorum, 212
Clostridium botulinum, 245, 326
Clunio sp., 272
Clysia ambiguella, 338
Clysiana ambiguella, 282
COCCIDIA, 221
Coccinella septempunctata, 191
Coccinellidae, 183, 190–3, 279, Plate 20
Coccoidea, 73, 81, 117, 172, 181, 183, 184, 185–7, 189, 191, 192, 194, 196, 219, 322, 339
Coccophagus gurneyi, 182
Coccus hesperidum, 118
Cochliomyia hominovorax, 249–53, 261–2
Coelomyces indicus, 216
Coelomyces spp., 216
Coleophora inaequalis, 192
Coleophora malivorella, 36
COLEOPTERA, 73, 74, 77, 79, 177, 187, 189–93, 195, 212, 214, 216, 221, 224, 260, 280, 292
Colias eurytheme, 173, 215, 239, 240, 342
COLLEMBOLA, 142
Colletotrichum lagenarium, 246
C. linicola, 96
C. xanthii, 247
Columba livia, 298
C. oenas, 143
C. palumbus, 108, 143
Compositae, 187
Compsilura concinnata, 187
Coniopterygidae, 193–4
Contwentzia pineticola, 194
C. psociformis, 194
Cordyceps spp., 219
Coreidae, 162, 196
Coronopus spp., 315
Corvus frugilegus, 143
Cosmopolites sordidus, 193
Cratichneumon gangis, 218
Creophilus erythrocephalus, 190
Cryptochaetidae, 188–9

MICROSPORIDIA, 222,
Plate 21
Microtus pennsylvanicus
terraenovae, 203
Miridae, 194–5
Misocyclops marchali, 184
MOLLUSCA, 80, 82, 103–5,
148, 193
Monilia fructicola, 98
Mormoniella spp., 272
Mugli curema, 140
Mungos birmanicus, 204–5
Mus musculus, 105–8, 130–1,
203, 212, 243, 267, 275
280
Musca autumnalis, 84, 225
M. domestica, 11, 52, 75, 79,
80, 84, 124, 127–8, 129,
198–9, 216–17, 222, 253,
254, 263–4, 270, 278,
283–4, 290, 336
Muscidae, 193, 222
Musgraveia sulciventris, 118
Mustela erminea, 205
M. nivealis, 205
M. sibrica itatsi, 205
M. vison, 204
Myrmecia gulosa, 92, 184,
Plate 14
Myrmeleontidae, 185
Myzus persicae, 75, 181, 310,
342, 351

Nabidae, 196
Necrobia rufipes, 193, 223–5
NEMATOCERA, 41
NEMATODA, 4, 26, 102–3,
198, 220, 226–7, 308
NEMERTINEA, 148
Neoaplectana carpocapsae, 26,
224
N. glaseri, 224, 226
Neoaplectana spp., 224
Neodiprion sertifer, 215, 227,
240
Neodiprion spp., 212
N. swainei, 203
NEOGREGARINIDA, 221
NEUROPTERA, 193–4, 212,
259, 280
Nezara viridula, 184
Nicotiana robustum, 85
N. rustica, 85
Nicotiana sp., 275
Noctuidae, 193, 337
Nomadacris septemfasciata,
200
Nosema bombycis, 222, 228, 230
N. cactoblastis, 222
N. carpocapsae, 222
N. melolonthae, 222
Nosema spp., 222, 228
N. stegomyiae, 222
Nothobranchius spp., 207

Nyctotherus spp., 223
Nygmia phaeorrhoea, 173, 187,
218, 227, 241
Nymphaea odorata, 247
Nymphaea spp., 247
Nymphalis io, 10, 146

Octosporea muscaedomesticae,
222
ODONATA, 196
Odynerus spp., 184
Oechalia consocialis, 195
Oecophylla smaragdina, 322
Oidium tuckeri, 94
Oligota flavicornis, 190
Oncopeltus spp., 89
Ondatra zibethica, 203
Oophagomyia spp., 189
Operophthera brumata, 45, 80,
201, Plate 2
OPHIDIA, 205
Ophion spp., 179
Opius longicaudatus, 167
O. oophilis, 167
O. vandenboschi, 167
Opuntia aurantiaca, 162–3
O. imbricata, 162–3
O. inermis, 162–3
O. monacantha, 161–3
Opuntia spp., 161–3, 199, 246
O. stricta, 162–3
Orgyia anartoides, Plates 22
and 23
Orius spp., 196
Ornithodoros tholozani, 251
ORTHOPTERA, 185, 187, 199,
218, 292
Orussus spp., 178
Oryctolagus cuniculus, 131,
242, 244–5, 328–9
Oscinella frit, 224, 315
Ostrinia nubilalis, 36, 82, 183,
198, 219, 222, 224, 239,
253, 276, 291, 295, 311, 314,
318, 319
Oxynychus marmottani, 192
Oxyuroidea, 223

Paederus fuscipes, 190
Pandion haliaetus, 147
Panonychus citri, 214
P. ulmi, 80, 115–17, 190, 194,
195, 197, 215, 347, Plate 4
Papaver somniferum, 171
Papaver spp., 313
Parasitylenchus spp., 224–5
Paratriphleps laeviusculus, 346
Paregle spp., 189
Paridae, 201–2
Parlatoria oleae, 163
Passer domesticus, 202, 204
Pasteurella spp., 242
PAUROPODA, 142
Pectinophora gossypiella, 120,

253, 266, 281, 291, 326,
339
Pediculus humanus humanus,
92, 151, 215
Pegomyiahyoscami var.
betae, 2, 76, 318
Penicillium digitatum, 8, 130
P. griseofulvum, 99
P. puberulum, 6
Pentatomidae, 195
Perilloides biomaculatus, 195
Periplaneta americana, 63,
185, 284
P. australasiae, 185
Perkinsiella saccharicida, 161,
168, 195, 344
Perrisia pyri, 184
Petromyzon marinus, 206
Phacochoerus aethiopicus, 323
Phalaenoides sp., 195
Phaonia miribalis, 189
Phasianus colchicus, 143
Pheidole megacephala, 322
Pheidole spp., 224
Philonthus aeneus, 190
Phlebotominae, 216
Phlebotomus spp., 215
Phormia regina, 53
Phorodon humuli, 20
PHTHIRAPTERA, 74, 260,
Plate 9
Phyllactinia corylea, 94
Phyllophaga spp., 201, 205,
206
Phylloxera spp., 194
P. vitifoliae, 94, 309
Physorhynchus linnaei, 195
Phytomyza syngenesiae, 351
Phytonomus posticus, 318
Phytophthora infestans, 13, 95,
96, 98, 305–6, 319
P. palmivora, 95
P. parasitica, 119
Phytoseiulus persimilis, 197,
199, 349–51
Phytoseius spoofi, 197
Picromerus bidens, 195
Pieris brassicae, 166, 213, 227–9,
239
P. oleracea, 277
P. rapae, 215, 222, 224, 227–8,
239, 277
Pieris spp., 181 (Fig. 6.3)
Piophila casei, 193
Piricularia oryzae, 97–101, 308
PISCES, 140–1, 145, 206–8
Plaesius javanus, 193
Planococcus citri, 192
Plasmodium cynomolgi
bastianelli, 264
P. gallinaceum, 264
Plasmodium spp., 221
Plasmopara viticola, 94, 302
Platygaster hiemalis, 178